Lecture Notes in Computer Science 10167

Commenced Publication in 1973
Founding and Former Series Editors:
Gerhard Goos, Juris Hartmanis, and Jan van Leeuwen

Sheung-Hung Poon · Md. Saidur Rahman
Hsu-Chun Yen (Eds.)

WALCOM: Algorithms and Computation

11th International Conference and Workshops, WALCOM 2017
Hsinchu, Taiwan, March 29–31, 2017
Proceedings

 Springer

Editors
Sheung-Hung Poon
Universiti Teknologi Brunei
Bandar Seri Begawan
Brunei Darussalam

Hsu-Chun Yen
National Taiwan University
Taipei
Taiwan

Md. Saidur Rahman
Bangladesh University of Engineering
  and Technology
Dhaka
Bangladesh

ISSN 0302-9743   ISSN 1611-3349   (electronic)
Lecture Notes in Computer Science
ISBN 978-3-319-53924-9   ISBN 978-3-319-53925-6   (eBook)
DOI 10.1007/978-3-319-53925-6

Library of Congress Control Number: 2017932439

LNCS Sublibrary: SL1 – Theoretical Computer Science and General Issues

Printed on acid-free paper

This Springer imprint is published by Springer Nature
The registered company is Springer International Publishing AG
The registered company address is: Gewerbestrasse 11, 6330 Cham, Switzerland

Preface

The 11th International Conference and Workshops on Algorithms and Computation (WALCOM 2017) was held in Hsinchu, Taiwan, during March 29–31, 2017. The conference covered diverse areas of algorithms and computation, namely, approximation algorithms, computational complexity, computational geometry, combinatorial optimization, graph algorithms, graph drawing, and space-efficient algorithms. The conference was organized by National Chiao Tung University, Taiwan.

This volume of *Lecture Notes in Computer Science* contains 35 contributed papers that were presented at WALCOM 2017. There were 83 submissions from 30 countries. Each submission was reviewed by at least three Program Committee members, with the assistance of external referees. The volume also includes the abstracts and extended abstracts of three keynote lectures presented by Francis Chin, Peter Eades, and Etsuji Tomita.

We wish to thank all who made this meeting possible: the authors for submitting papers, the Program Committee members and external referees (listed in the proceedings) for their excellent work, and our three invited speakers. We acknowledge the Steering Committee members for their continuous encouragement. We also wish to express our sincere appreciation to the sponsors, local organizers, Proceedings Committee, and the editors of the *Lecture Notes in Computer Science* series and Springer for their help in publishing this volume. We especially thank Chun-Cheng Lin and his team for their tireless efforts in organizing this conference. Finally, we thank the EasyChair conference management system, which was very effective in handling the entire reviewing process.

March 2017

Sheung-Hung Poon
Md. Saidur Rahman
Hsu-Chun Yen

WALCOM 2017 Organization

Steering Committee

Kyung-Yong Chwa	KAIST, Korea
Costas S. Iliopoulos	KCL, UK
M. Kaykobad	BUET, Bangladesh
Petra Mutzel	TU Dortmund, Germany
Shin-ichi Nakano	Gunma University, Japan
Subhas Chandra Nandy	Indian Statistical Institute, Kolkata, India
Takao Nishizeki	Tohoku University, Japan
C. Pandu Rangan	IIT, Madras, India
Md. Saidur Rahman	BUET, Bangladesh

Program Committee

Sang Won Bae	Kyonggi University, Korea
Tiziana Calamoneri	Sapienza University of Rome, Italy
Subhas Chandra Nandy	ISI, India
Kun-Mao Chao	National Taiwan University, Taiwan, ROC
Jianer Chen	Texas A&M University, USA
Yijia Chen	Fudan University, China
Mordecai Golin	Hong Kong University of Science and Technology, Hong Kong, SAR China
Seok-Hee Hong	University of Sydney, Australia
Md. Rezaul Karim	University of Dhaka, Bangladesh
Naoki Katoh	Kwansei Gakuin University, Japan
Mohammad Kaykobad	Bangladesh University of Engineering and Technology, Bangladesh
Marc van Kreveld	Utrecht University, The Netherlands
Van Bang Le	University of Rostock, Germany
Minming Li	City University of Hong Kong, Hong Kong, SAR China
Chun-Cheng Lin	National Chiao Tung University, Taiwan, ROC
Giuseppe Liotta	University of Perugia, Italy
Jerome Monnot	Paris Dauphine University, France
David Mount	University of Maryland, USA
Petra Mutzel	Technical University of Dortmund, Germany
Shin-ichi Nakano	Gunma University, Japan
Rolf Niedermeier	Technical University of Berlin, Germany
Martin Nollenburg	Technical University of Vienna, Austria
Nadia Pisanti	University of Pisa, Italy

Sheung-Hung Poon (Co-chair)	Universiti Teknologi Brunei, Brunei Darussalam
Md. Saidur Rahman (Co-chair)	Bangladesh University of Engineering and Technology, Bangladesh
C. Pandu Rangan	IIT Madras, India
Kunihiko Sadakane	The University of Tokyo, Japan
Takeshi Tokuyama	Tohoku University, Japan
Antoine Vigneron	Ulsan National Institute of Science and Technology, Korea
Sue Whitesides	University of Victoria, Canada
Prudence Wong	University of Liverpool, UK
Chee K. Yap	New York University, USA
Hsu-Chun Yen (Co-chair)	National Taiwan University, Taiwan, ROC
Siu-Ming Yiu	The University of Hong Kong, Hong Kong, SAR China
Guochuan Zhang	Zhejiang University, China

Organizing Committee

Chun-Cheng Lin (Co-chair)	National Chiao Tung University, Taiwan, ROC
Sheung-Hung Poon (Co-chair)	Universiti Teknologi Brunei, Brunei Darulssalam
Hung-Lung Wang	National Taipei University of Business, Taiwan, ROC

Proceedings Committee

Sheung-Hung Poon	Universiti Teknologi Brunei, Brunei Darulssalam
Md. Saidur Rahman	Bangladesh University of Engineering and Technology, Bangladesh
Hsu-Chun Yen	National Taiwan University, Taiwan, ROC

Additional Reviewers

Muhammad Abdullah Adnan	Basile Couëtoux	Laurent Gourves
Hee-Kap Ahn	Gautam K. Das	Hasnain Heickal
Greg Aloupis	Minati De	Petr Hlineny
Anna Bernasconi	Matteo Dell'Orefice	Xuangui Huang
Bhaswar Bhattacharya	Thomas C. Van Dijk	Frank Kammer
Edouard Bonnet	Xavier Défago	Ken-Ichi Kawarabayashi
Robert Bredereck	Till Fluschnik	Matias Korman
Henning Bruhn	Robert Ganian	Denis Kurz
Yi-Jun Chang	Mattia Gastaldello	Kazuhisa Makino
Xujin Chen	Nicolas Gastineau	David Manlove
	Petr Golovach	Andrea Marino

Debajyoti Mondal
Christopher Morris
Amer Mouawad
Krishnendu
  Mukhopadhyaya
Haiko Muller
Toshio Nakata
Nhan-Tam Nguyen
André Nichterlein
Benjamin Niedermann
Yoshio Okamoto

Sebastian Ordyniak
Yota Otachi
Daniel Paulusma
Rossella Petreschi
Timo Poranen
Giuseppe Prencipe
Günther Raidl
Dieter Rautenbach
Bernard Ries
Sasanka Roy
Ignaz Rutter

Toshiki Saitoh
Newaz S. Shah
Tetsuo Shibuya
Akira Suzuki
Rossano Venturini
Haitao Wang
Mathias Weller
Bernd Zey
Chihao Zhang
Yong Zhang

Sponsoring Institutions

Ministry of Science and Technology, Taiwan, ROC
National Chiao Tung University, Taiwan, ROC
National Taiwan University, Taiwan, ROC
The Institute of Electronics, Information and Communication Engineers (IEICE), Japan

Invited Talks

A Few Steps Beyond Planarity

Peter Eades

University of Sydney, Camperdown, Australia
peter.eades@sydney.edu.au

We discuss algorithms for graphs that are not planar, but not far from planar. We present recent combinatorial characterisations of straight-line drawings of some such classes of graphs. In some cases, these characterisations lead to efficient algorithms; in other cases, they lead to NP-completeness results.

In particular, we investigate three classes of topological graphs: "1-skew" graphs, "1-plane" graphs, and "RAC" graphs.

1-Skew: A topological graph $G = (V, E)$ is *1-skew* if it has an edge (s, t) such that $G' = (V, E - \{(s, t)\})$ is planar. Some 1-skew topological graphs are illustrated in Fig. 1.

Suppose that G is a 1-skew topological graph with a straight-line drawing D; suppose that the edge (s, t) crosses edges $e_1, e_2, \ldots, e_k$. It is simple to observe that for each i, one endpoint of e_i is left of (s, t) and the other is right of (s, t). Note that the vertex x in Fig. 1(b) is both left and right of (s, t); thus this topological graph has no straight-line drawing.

This simple observation leads to an elegant theorem: a 1-skew topological graph on the sphere that a straight-line drawing in the plane if and only if no vertex is both left and right [3]. The proof of this theorem is a linear time algorithm. It involves an inelegant characterisation of maximal 1-skew topological graphs in the plane that have a straight-line drawing in the plane.

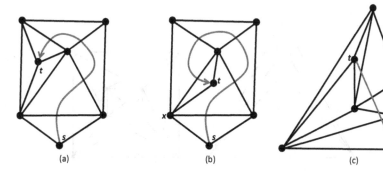

Fig. 1. Some 1-skew topological graphs. Graph (b) has no straight-line drawing. Graph (c) is a straight-line drawing of graph (a).

1-Plane: A *1-plane graph* is a topological graph in which each edge has at most one crossing. Thomassen [6] characterised those 1-plane graphs that admit a straight-line drawing, in terms of the two forbidden subgraphs illustrated in Fig. 2.

bulgari   gucci

Fig. 2. A 1-plane topological graph has a straight-line drawing if and only if it does not contain either a bulgari or a gucci subgraph.

A linear-time algorithm which tests for these forbidden subgraphs, and constructs a straight-line drawing if the forbidden subgraphs are absent, is described in [5].

RAC: A *RAC (right-angle crossing) drawing* of a topological graph is a straight-line drawing in which each edge crossing forms a right angle. Figure 3(a) illustrates a RAC drawing of the complete graph with 5 vertices; Fig. 3(b) shows a graph that has no RAC drawing. Human experiments that have shown that right-angled edge crossings do not inhibit human understanding of diagrams. These experiments have motivated wide-ranging research on RAC drawings. In particular, it has been shown that a RAC graph with n vertices has at most $4n - 10$ edges [2]; further, if it has exactly $4n - 10$ edges then it is 1-planar [4].

An interesting open question is to characterise those 1-planar topological graphs that have a RAC drawing. It is easy to observe that none of the topological graphs in Fig. 4 has a straight-line drawing. Dehkordi [1] conjectures that the six forbidden subgraphs in Figs. 2 and 4 characterise those 1-plane graphs that have RAC drawings.

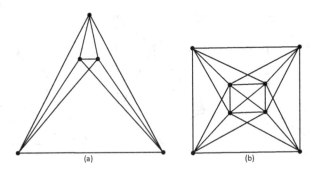

(a)   (b)

Fig. 3. (a) a RAC drawing of K_5. (b) a graph that does not have a (straight-line) RAC drawing.

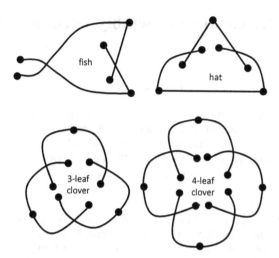

Fig. 4. Four forbidden subgraphs for RAC drawings. It is conjectured that a 1-plane graph has a (straight-line) RAC drawing if and only if it does not contain any subgraph isomorphic to either bulgari, gucci (Fig. 2), fish, hat, 3-leaf-clover, or 4-leaf-clover graph.

References

1. Dehkordi, H.R.: On algorithmic right angle crossing drawing. Master's thesis, University of Sydney (2012)
2. Didimo, W., Eades, P., Liotta, G.: Drawing graphs with right angle crossings. Theor. Comput. Sci. **412**(39), 5156–5166 (2011). doi:10.1016/j.tcs.2011.05.025
3. Eades, P., Hong, S., Liotta, G., Katoh, N., Poon, S.: Straight-line drawability of a planar graph plus an edge. In: Dehne, F., Sack, J.-R., Stege, U. (eds.) WADS 2015. LNCS, vol. 9214, pp. 301–313. Springer, Switzerland (2015)
4. Eades, P., Liotta, G.: Right angle crossing graphs and 1-planarity. Discrete Appl. Math. **161** (7–8), 961–969 (2013). doi:10.1016/j.dam.2012.11.019
5. Hong, S., Eades, P., Liotta, G., Poon, S.: Fáry's theorem for 1-planar graphs. In: Gudmundsson, J., Mestre, J., Viglas, T. (eds.) COCOON 2012. LNCS, vol. 7434, pp. 335–346. Springer, Heidelberg (2012)
6. Thomassen, C.: Rectilinear drawings of graphs. J. Graph Theory **12**(3), 335–341 (1988). doi:10.1002/jgt.3190120306

Why Genome Assembly Is Difficult?

Francis Chin[1,2]

[1] Department of Computing, Hang Seng Management College,
Siu Lek Yuen, Shatin, N.T., Hong Kong
`francischin@hsmc.edu.hk`
[2] The University of Hong Kong, Pok Fu Lam, Hong Kong
`chin@cs.hku.hk`

Abstract. It has been about 60 years since Watson and Crick first discovered the double-helix structure of DNA. Each genome (about 3 billion long) define every human uniquely (e.g. hair colour, eye colour, etc.) as well as ones genetic diseases. Consequently, there is a need to find the genome of each individual for assessing the genetic risk of potential diseases. At the same time, research groups are sequencing the DNA of all kinds of organisms, e.g., the rice genome in search of higher production yields, the bacteria genome in search of a more effective cure, and the orchid genome in search of more varieties and higher financial returns. To sequence a genome, Next-Generation Sequencing (NGS) technology is commonly used to output billions of overlapping DNA fragments (known as reads) from the genome, but without information on how these reads link together to form the genome. Then, effective sequencing software tools are used to combine these reads to form the genome. This process is called "genome assembly".

Theoretically, genome assembly is an easy task as the chance of mismatching two reads is extremely low if they overlap 30–40 positions (because $4^{\wedge}30 \ggg 3 \times 10^{\wedge}9$). In this talk, we shall review past developments and difficulties of genome assembly and explain why some of the straight-forward approaches fail. The most successful approach borrows the idea of **De Bruijn graph** problem which transforms an NP-hard problem to an efficient polynomial Eulerian graph problem. This approach is counter-intuitive by breaking the reads into smaller parts (length-k substrings called k-mers) before assembly.

This De Bruijn graph approach has other advantages: handling **errors** in reads (error rate is about 1–2%) and **repeated patterns** in genome, which is not random. however the choice of k, i.e., how short about the substrings, is very important. As the rule of thumb, smaller k (shorter substrings) can tolerate more errors while larger k (longer substrings) can resolve repeated patterns (**branches**) in the graph. Thus, there is always a tradeoff in the selection of k in balancing these two conflicting needs. The traditional approach is to try a range of k and to choose the k which demonstrates the best performance. If k is too large, there will be many "gaps" or "discontinuities" in the graph. If k is too small, there will be many branches in the graph. So, it is crucial for the traditional assemblers to find a specific value of k. Our **IDBA** (iterative De Bruijn Graph Assembler) does not use only one specific k but a range of k values to build the

De Bruijn Graph iteratively. It can keep all the information in the graphs with different k values and can capture the advantages of all ranges of small and large k values.

In practice, the Next-Generation Sequencing (NGS) technology might not produce the same number of reads (depth) at different positions of a genome evenly, some parts of the genome might be sequenced more and some less. **IDBA-UD** is an extension of IDBA algorithm with two additional techniques which can handle highly **uneven** sequencing **depths** of NGS genomic data. The first technique is similar to IDBA to filter out erroneous reads, IDBA-UD also iterates from small k to a large k. In each iteration, short and low-depth contigs (partially assembled contiguous genomic segments) are removed iteratively with cutoff threshold from low to high to reduce the errors in low-depth and high-depth regions. The second technique is to make use of paired-end reads (two generated reads separated by a range of insert sizes) which are aligned to contigs and assembled locally to generate some missing k-mers in low-depth regions. With these two techniques, IDBA-UD can iterate k values of de Bruijn graph to a very large value with less gaps and less branches so as to form long contigs in both low-depth and high-depth regions. IDBA-UD has 500 citations and thousands downloads and has been used widely by many medical and biological researchers since its launching about five years ago.

Recent work to develop more effective algorithms and software tools for genome assembly will also be discussed.

Contents

Space-Efficient Algorithms

Computational Complexity

Approximation Algorithms

Graph Algorithms II

Invited Talk

Efficient Algorithms for Finding Maximum and Maximal Cliques and Their Applications

Etsuji Tomita[✉]

The Advanced Algorithms Research Laboratory,
The University of Electro-Communications,
Chofugaoka 1-5-1, Chofu, Tokyo 182-8585, Japan
tomita@ice.uec.ac.jp

Abstract. The problem of finding a maximum clique or enumerating all maximal cliques is very important and has been explored in several excellent survey papers. Here, we focus our attention on the step-by-step examination of a series of branch-and-bound depth-first search algorithms: Basics, MCQ, MCR, MCS, and MCT. Subsequently, as with the depth-first search as above, we present our algorithm, CLIQUES, for enumerating all maximal cliques. Finally, we describe some of the applications of the algorithms and their variants in bioinformatics, data mining, and other fields.

1 Introduction

Given an undirected graph G, a clique is a subgraph in which all pairs of vertices are mutually adjacent in G. The so-called maximum clique problem is one of the original 21 problems shown to be NP-complete by Karp [18]. The problem of finding a maximum clique or enumerating all maximal cliques in G is very important and significant work has been done on it, both theoretically and experimentally [7,10,17,34,49,54].

An excellent review on recent various algorithms for the maximum clique problems can be found in [54] by Wu and Hao. Herein, we focus our attention on a series of branch-and-bound depth-first search algorithms — Basics [13,43], MCQ [45], MCR [47], MCS [48], and MCT [52] — for finding a maximum clique, such that their progress can be understood easily.

Subsequently, similarly to the depth-first searches above, we present our $O(3^{n/3})$-time algorithm, CLIQUES [46], for enumerating all maximal cliques, that is optimal with respect to the number of vertices n.

Finally, we outline some of the applications of the previous algorithms or their variants to fields including bioinformatics, data mining, and image processing.

2 Preliminaries

(1) Throughout this paper, we are concerned with a simple undirected graph $G = (V, E)$ with a finite set V of vertices and a finite set E of *unordered* pairs

© Springer International Publishing AG 2017
S.-H. Poon et al. (Eds.): WALCOM 2017, LNCS 10167, pp. 3–15, 2017.
DOI: 10.1007/978-3-319-53925-6_1

(v, w) $(= (w, v), v \neq w)$ of distinct vertices called edges. V is considered to be *ordered*, and the i-th element in V is denoted by $V[i]$. A pair of vertices v and w are said to be adjacent if $(v, w) \in E$.

(2) For a vertex $v \in V$, let $\Gamma(v)$ be the set of all vertices adjacent to v in $G = (V, E)$, *i.e.*, $\Gamma(v) = \{w \in V | (v, w) \in E\}$. We call $|\Gamma(v)|$ the degree of v. In general, for a set S, the number of elements in S is denoted by $|S|$.

(3) For a subset $R \subseteq V$ of vertices, $G(R) = (R, E \cap (R \times R))$ is an *induced subgraph*. An induced subgraph $G(Q)$ is said to be a *clique* if $(v, w) \in E$ for all $v, w \in Q \subseteq V$ with $v \neq w$. In this case, we may simply state that Q is a clique. In particular, a clique that is not properly contained in any other clique is called *maximal*. A maximal clique with the maximum size is called a *maximum clique*. The number of vertices of a maximum clique in an induced subgraph $G(R)$ is denoted by $\omega(R)$.

3 Efficient Algorithms for Finding a Maximum Clique

3.1 Basic Algorithms

3.1.1 A Basic Branch-and-Bound Algorithm

One standard approach for finding a maximum clique is based on the branch-and-bound depth-first search method. Our algorithm begins with a small clique and continues finding larger and larger cliques until one is found that can be verified to have the maximum size. More precisely, we maintain global variables Q and Q_{max}, where $Q = \{p_1, p_2, ..., p_d\}$ consists of vertices of a current clique and Q_{max} consists of vertices of the largest clique found so far. Let $R = V \cap \Gamma(p_1) \cap \Gamma(p_2) \cap \cdots \cap \Gamma(p_d) \subseteq V$ consist of *candidate* vertices that can be added to Q. We begin the algorithm by letting $Q := \emptyset$, $Q_{max} := \emptyset$, and $R := V$ (the set of all vertices). We select a certain vertex p from R and add p to Q ($Q := Q \cup \{p\}$). Then, we compute $R_p := R \cap \Gamma(p)$ as the new set of candidate vertices. This procedure, EXPAND(), is applied recursively while $R_p \neq \emptyset$.

Here, **if** $|Q| + |R| \leq |Q_{max}|$ **then** $Q \cup R$ can contain only a clique that is smaller than or equal to $|Q_{max}|$, hence searching for R can be pruned in this case. This is a *basic bounding condition*.

When $R_p = \emptyset$ is reached, Q constitutes a *maximal* clique. If Q is maximal and $|Q| > |Q_{max}|$ holds, Q_{max} is replaced by Q. We then backtrack by removing p from Q and R. We select a new vertex p from the resulting R and continue the same procedure until $R = \emptyset$.

This is a well-known basic algorithm for finding a maximum clique and is shown in Fig. 1. We call it **Algorithm #0** [13] and it serves as a reference algorithm. This process can be represented by a *search tree* with root V; whenever $R_p := R \cap \Gamma(p)$ is applied, then R_p is a child of R.

3.1.2 Ordering of Vertices

If the vertices are sorted in an ascending order with respect to their degrees prior to the application of Algorithm #0 and the vertices are expanded in this order,

procedure Algorithm #0 $(G = (V, E))$
begin
    global $Q := \emptyset$;   *global* $Q_{max} := \emptyset$;
    EXPAND(V);
    output Q_{max}
end {of Algorithm #0}

procedure EXPAND(R)
begin
    while $R \neq \emptyset$ **do**
        $p :=$ a vertex in R;   { a vertex for expansion }
        if $|Q| + |R| > |Q_{max}|$ **then**
            $Q := Q \cup \{p\}$;
            $R_p := R \cap \Gamma(p)$;
            if $R_p \neq \emptyset$ **then** EXPAND(R_p)
            else {*i.e.,* $R_p = \emptyset$} **if** $|Q| > |Q_{max}|$ **then** $Q_{max} := Q$ **fi**
            fi
            $Q := Q - \{p\}$
            else return
        fi
        $R := R - \{p\}$
    od
end {of EXPAND}

Fig. 1. Algorithm #0

then the above Algorithm #0 is more efficient. This fact was experimentally confirmed in [13]. Algorithm #0 combined with this vertex-ordering preprocessing is named **Algorithm #1** [13].

Carraghan and Pardalos [11] also employed a similar technique successfully.

3.1.3 Pruning by Approximate Coloring: *Numbering*

One of the most important points for improving the efficiency of the basic Algorithm #0 is to strengthen the *bounding condition* to prune unnecessary searches.

For a set R of vertices, let $\chi(R)$ be the chromatic number of R, *i.e.*, the *minimum* number of colors such that all pairs of adjacent vertices are colored by different colors, and $\chi'(R)$ be an *approximate* chromatic number of R, *i.e.*, a number of colors such that all pairs of adjacent vertices are colored by different colors. Then we have $\omega(R) \leq \chi(R) \leq \chi'(R) \leq |R|$. An appropriate chromatic number $\chi'(R)$ could be a better upper bound on $\omega(R)$ than $|R|$, and might be obtained with low overhead. Here, we employ a very simple *greedy* or *sequential* approximate coloring to the vertices of R, as introduced in [41]. Let positive integral numbers 1, 2, 3, ... stand for different colors. *Coloring* is also called *Numbering*. For each vertex $q \in R$, *sequentially* from the first to the last, we assign a positive integral *Number* $No[q]$ which is as small as possible. That is, for the vertices in $R = \{q_1, q_2, \ldots, q_m\}$, first let $No[q_1] = 1$, and subsequently, let $No[q_2] = 2$ if $q_2 \in \Gamma(q_1)$ else $No[q_1] = 1, \ldots$, and so on.

We select p (at the 4th line in **procedure** EXPAND(R) in Fig. 1) so that $No[p] = \mathrm{Max}\{No[q] \mid q \in R\}$, where $No[p]$ is an approximate chromatic number of R. Thus, we modify the basic bounding condition:

$$\text{if } |Q| + |R| > |Q_{max}| \text{ then}$$

in Fig. 1 Algorithm #0, to the following *new bounding condition*:

$$\text{if } |Q| + No[p] > |Q_{max}| \text{ then.}$$

In addition, to make the above bounding condition more effective, we sort the vertices in R in descending order with respect to their degrees prior to *Numbering*.

We have now an improved algorithm, named **Algorithm #2** [13], as follows: First, sort the vertices in $R = V$ as above. Second, give *Numbers* to the sorted vertices in $R = V$. Subsequently, apply the modified Algorithm #0 as described above. Note that the sorting and *Numbering* are applied only once prior to the first application of EXPAND() at depth 0 of the search tree and that the *Numbers* are inherited in the following EXPAND(). In general, Algorithm #2 is more efficient than Algorithm #1 [13].

We can reduce the search space more effectively by applying the sorting and *Numbering* of vertices prior to every application of EXPAND(), but with the potential for more overhead and thus, more overall computing time. We confirmed that *adaptive control* of the application of sorting and/or *Numbering* is effective in reducing the overall computing time [21, 29, 37]. By restricting the application of vertex sorting, as in Algorithm #2, but applying the *Numbering* to vertices prior to every EXPAND(), we obtain another efficient algorithm, **MCLIQ** [43].

3.2 Algorithms MCQ, MCR, MCS, and MCT

3.2.1 Algorithm MCQ

The algorithm MCQ [45] is directly improved from MCLIQ. At the beginning of MCQ, vertices are sorted in descending order with respect to their degrees. Subsequently, we apply *Numbering* and sorting prior to each EXPAND() operation, where vertices are sorted in an ascending order with respect to their *Numbers*. Then, the last vertex with the *maximum Number* is expanded step-by-step. This sorting can be carried out with little overhead. Hence, MCQ is very simple and efficient.

3.2.2 Algorithm MCR

Algorithm MCR [47] is an improved version of MCQ, where the improvements mainly address the *initial* vertex sorting. First, we alter the order of the vertices in $V = \{V[1], V[2], \ldots, V[n]\}$ so that in a subgraph of $G = (V, E)$ induced by a set of vertices $V' = \{V[1], V[2], \ldots, V[i]\}$, it holds that $V[i]$ always has the minimum degree in $\{V[1], V[2], \ldots, V[i]\}$ for $1 \leq i \leq |V|$ as in [11]. Here, the degrees of adjacent vertices are also considered. In addition, vertices are assigned initial *Numbers*. This improvement is described precisely in the steps from {SORT} to just above EXPAND(V, No) in Fig. 4 (Algorithm MCR) in [47], called *EXTENDED INITIAL SORT-NUMBER* to V.

3.2.3   Algorithm MCS

Algorithm MCS [39, 48, 50] is a further improved version of MCR that introduces the following techniques:

(1) **Re-NUMBER.** Because of the *bounding condition* mentioned above, if $No[r] = \text{Max}\{No[q] \mid q \in R\} \leq |Q_{max}| - |Q|$ then it is not necessary to search from vertex r. Let $No_{th} := |Q_{max}| - |Q|$. When we encounter a vertex p with $No[p] > No_{th}$, we attempt to change its *Number* as follows: Try to find a vertex q in $\Gamma(p)$ such that $No[q] = k_1 \leq No_{th} - 1$, with $|C_{k_1}| = 1$. If such q is found, then try to

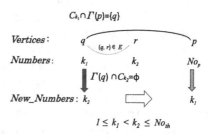

Fig. 2. Re-NUMBER

find *Number* k_2 such that no vertex in $\Gamma(q)$ has *Number* k_2. If such *Number* k_2 is found, then exchange the *Numbers* of q and p so that $No[q] = k_2$ and $No[p] = k_1$. When this is possible, *it is no longer necessary to search from p.* See Fig. 2 for an illustration.

The above procedure is named Re-NUMBER to p and is very effective.

(2) **Adjunct ordered set of vertices for approximate coloring.** The ordering of vertices plays an important role in the algorithms as described in Sects. 3.1.2 and 3.1.3. In particular, the procedure *Numbering* strongly depends on the order of vertices, since it is a *sequential* coloring. In our new algorithm, we sort the vertices in the same way as in the first stage of MCR [47]. However, the vertices are *disordered* in succeeding stages, owing to the application of Re-NUMBER. To avoid this difficulty, we employ another *adjunct ordered set V_a of vertices for approximate coloring* that preserves the order of vertices appropriately sorted in the first stage.

We apply *Numbering* to vertices from the first (leftmost) to the last (rightmost) in the order maintained in V_a, while we select a vertex p for expansion in R in which vertices are sorted in ascending order with respect to their *Numbers* as in MCQ and MCR, for *searching* from the last vertex with the *maximum Number*. Finally, we reconstruct the adjacency matrix in MCR just after the EXTENDED INITIAL SORT-NUMBER to establish a more effective use of the cache memory.

The individual contributions of the above techniques in MCS can be found in Tables 2–4 in [50].

3.2.4 Algorithm MCT

An improved algorithm MCT [15,52] is obtained by modifying MCS in the following ways:

(1) **An approximate solution as an initial lower bound.** We turn back to our original MCS [39] that initially employs an approximation algorithm,

Table 1. CPU time [sec] for benchmark graphs

Graph				KLS [19]		MCR [47]	MCS [48]	MCT [52]	MCX [36]	MaxC [23]	I&M [25]	BG14 [6]
Name	n	$d.$	ω	sol	t.							
brock400_1	400	0.75	27	25	0.1	729	288	**116**	150	205	188	302
brock800_1	800	0.65	23	21	0.2	7,582	4,122	**1,950**	2,690	4,560	4,000	4,220
brock800_4	800	0.65	26	20	0.2	3,248	1,768	**819**	1,100	1,850	1,680	1,870
C250.9	250	0.90	44	44	0.1	15,386	1,171	404	713	**268**		
gen400_p0.9_65	400	0.90	65	65	0.3	$> 6 \times 10^6$	57,385	**0.74**	66,100	36,700	2,130	19
gen400_p0.9_75	400	0.90	75	75	0.3	$> 3 \times 10^6$	108,298	**0.33**	47,200	9,980	84	7.8
MANN_a45	1035	0.99	345	344	22	653.3	53.4	75.5	32.0	22.7	**17.3**	55.1
p_hat700-3	700	0.75	62	62	0.5	25,631	900	**216**	680	879	552	767
p_hat1000-3	1000	0.74	68	68	1.0	$> 2 \times 10^6$	305,146	**38,800**				
p_hat1500-1	1500	0.25	12	11	0.0	2.22	1.82	**1.40**	1.95	10.00	478	422
p_hat1500-2	1500	0.51	65	65	0.7	268,951	6,299	**1,560**	3,850	8,030	5,350	5,430
san1000	1000	0.50	15	10	0.1	2.16	1.02	**0.21**	0.68	0.72	449	158
sanr400_0.7	400	0.70	21	21	0.1	158.7	77.3	**40.7**	44.5	81.2	86.2	81.4
DSJC500.5	500	0.50	13	13	0.0	1.9	1.5	1.2	**0.8**	2.8		
DSJC1000.5	1000	0.50	15	15	0.1	182	141	**93**	102	265		
keller5	776	0.75	27	27	0.3	45,236	82,421	10,000	30,300	**4,980**	5,780	82,500
r200.8	200	0.80	24–27	24–27	0.0	4.56	1.66	**0.78**	0.95	1.08		
r200.95	200	0.95	58–66	58–66	0.1	218.2	21.1	10.3	30.2	**2.5**		
r300.8	300	0.80	28–29	28–29	0.1	528	161	**61**	89	87		
r400.7	400	0.70	21–22	20–22	0.1	150.1	73.9	**34.9**				
r500.6	500	0.60	17–18	16–17	0.1	27.1	18.0	11.4	**10.1**	22.1		
r500.7	500	0.70	22–23	21–22	0.1	1,533	723	**340**	423	564		
r1000.5	1000	0.50	15–16	14–15	0.1	177	134	**92**	103	231		
r2000.4	2000	0.40	13–14	12	0.2	548	460	**366**				
r3000.2	3000	0.20	9	7–8	0.1	3.94	3.67	**3.42**	4.34	34.40		
r3000.3	3000	0.30	11	10–11	0.2	138	121	**107**				
r3000.4	3000	0.40	14	12–13	0.5	7,834	6,392	**5,152**				
r5000.2	5000	0.20	9–10	7–8	0.2	46.8	44.6	**39.0**	69	578		
r5000.3	5000	0.30	12	10–11	0.5	2,636	2,284	**1,875**				
r10000.1	10000	0.10	7	5–6	0.6	15	**14**	**14**	20	684		
r10000.2	10000	0.20	10	8–9	0.9	1,475	1,303	**1,139**				
r15000.1	15000	0.10	8	6	1.3	80	**62**	**62**	115	2,749		
r20000.1	20000	0.10	8	6–7	2.3	307	**234**	**234**				

init-lb, for the maximum clique problem, to obtain an initial lower bound on the size of the maximum clique. When a sufficiently large near-maximum clique Q'_{max} is found, we let $Q_{max} := Q'_{max}$ at the beginning of MCS [39]. Then $No_{th} := |Q_{max}| - |Q|$ becomes large and the bounding condition becomes more effective. Our init-lb is a local search algorithm based on our previous work [44]. Here, we choose another approximation algorithm, called *k-opt local search* (KLS) [19], by Katayama et al. Recently, Batsyn et al. [6] and Maslov et al. [25] also demonstrated the effectiveness of an approximate solution, independently.

(2) **Adaptive application of the sorting and/or *Numbering* of vertices.** The effectiveness of this approach was already confirmed, as described at the end of Sect. 3.1.3 [21,29,37]. We modify MCS so that we sort the set of vertices by the EXTENDED INITIAL SORT-NUMBER at the first stage near and including the root of the search tree. Konc and Janežič [22] were also successful in improving MCQ in a similar way as in [21], independently.

In contrast, mainly near the leaves of the search tree, to lighten the overhead of preprocessing before expansion of vertices, we only inherit the order of vertices from that in their parent depth, and we merely inherit the *Numbers* from those assigned in their parent depth if their *Numbers* are less than or equal to No_{th}. For vertices whose inherited *Numbers* are greater than No_{th}, we give them new *Numbers* by sequential *Numbering* combined with $Re - Numbering$.

Table 1 shows the progression of the running times required to solve some benchmark graphs using the above algorithms within these ten years [52]. Here, *d.* indicates the *density* of the graph, and *sol* and *t.* show the solution and the computing time of KLS in MCT. In the last half of the table, *rn.p* stands for a random graph, with the number of vertices $= n$ and the edge probability $= p$. The results of the state-of-the-art algorithm BBMCX (MCX *for short*) [36] by Segundo et al. and of other algorithms [6,23,25] are also included for reference [52]. Note that MaxCLQ (MaxC *for short*) [23] by Li and Quan is fast for dense graphs. ILS&MCS (I&M *for short*) [25] and BG14 [6] require more time than MCT for most of the instances tested. One reason for this difference comes from the fact that our approximation algorithm, KLS, takes only small portion of the whole algorithm's computing time, whereas their approximation algorithm, ILS, [2] in I&M and BG14 consumes a considerable part of the whole computing time.

4   Efficient Algorithm for Enumerating All Maximal Cliques

In addition to finding one maximum clique, enumerating all maximal cliques is also important and has diverse applications. We present a depth-first search algorithm, **CLIQUES** [42,46], for enumerating all maximal cliques of an undirected graph $G = (V, E)$. All maximal cliques enumerated are output in a tree-like form. The basic framework of CLIQUES is almost the same as that of Algorithm #0 *without the basic bounding condition*. We maintain a global variable $Q = \{p_1, p_2, ..., p_d\}$ that consists of the vertices of a current clique, and let

$SUBG = V \cap \Gamma(p_1) \cap \Gamma(p_2) \cap \cdots \cap \Gamma(p_d)$. We begin the algorithm by letting $Q := \emptyset$ and $SUBG := V$ (the set of all vertices). We select a certain vertex p from $SUBG$ and add p to Q ($Q := Q \cup \{p\}$). Then, we compute $SUBG_p := SUBG \cap \Gamma(p)$ as the new set of *candidate* vertices. In particular, the initially selected vertex $u \in SUBG$ is called a *pivot*. This EXPAND() procedure is applied recursively while $SUBG_p \neq \emptyset$.

We describe two methods to prune unnecessary parts of the search tree, which happen to be the same as in the Bron-Kerbosch algorithm [8]. We regard the set $SUBG$ ($= V$ at the beginning) as an *ordered* set of vertices, and we continue to enumerate maximal cliques from vertices in $SUBG$ step-by-step in this order.

First, let $FINI$ be a subset of vertices of $SUBG$ that have already been processed by the algorithm ($FINI$ is short for *finished*). Then we denote by $CAND$ the set of remaining candidates for expansion: $CAND = SUBG - FINI$. Initially, $FINI := \emptyset$ and $CAND := SUBG$. In the subgraph $G(SUBG_q)$ with $SUBG_q := SUBG \cap \Gamma(q)$, let

$$FINI_q := SUBG_q \cap FINI,$$
$$CAND_q := SUBG_q - FINI_q.$$

Then only the vertices in $CAND_q$ can be candidates for expanding the clique $Q \cup \{q\}$ to find *new* larger cliques.

Second, for the initially selected pivot u in $SUBG$, any maximal clique Q' in $G(SUBG \cap \Gamma(u))$ is not maximal in $G(SUBG)$, since $Q' \cup \{u\}$ is a larger clique in $G(SUBG)$. Therefore, searching for maximal cliques from $SUBG \cap \Gamma(u)$ should be excluded.

Taking the previously described pruning method into consideration, the only search subtrees to be expanded are from vertices in $(SUBG - SUBG \cap \Gamma(u)) - FINI = CAND - \Gamma(u)$. Here, to minimize $|CAND - \Gamma(u)|$, we choose the pivot $u \in SUBG$ that *maximizes* $|CAND \cap \Gamma(u)|$, which is *crucial* to establish the *optimality* of the worst-case time-complexity of the algorithm. This kind of pivoting strategy was first proposed by Tomita et al. [42].

The algorithm CLIQUES [42, 46] is shown in Fig. 3, which enumerates all maximal cliques based upon the above approach, where all maximal cliques enumerated are presented in a tree-like form. Here, if Q is a *maximal* clique that is found at statement 2, then the algorithm only prints out the string of characters "*clique*," instead of Q itself at statement 3. Otherwise, it is impossible to achieve the optimal worst-case running time. Instead, in addition to printing "*clique*" at statement 3, we print out q followed by a *comma* at statement 7 every time q is picked out as a new element of a larger clique, and we print out the string of characters "*back*," at statement 12 after q is moved from $CAND$ to $FINI$ at statement 11. We can easily obtain a tree representation of all the maximal cliques from the sequence printed by statements 3, 7, and 12. The tree-like output format is also important *practically*, since it saves space in the output file.

procedure CLIQUES(G)
begin
1 : EXPAND(V,V)
end {of CLIQUES}

    procedure EXPAND($SUBG$, $CAND$)
    begin
2 :   **if** $SUBG = \emptyset$
3 :   **then print** (*"clique,"*)
4 :   **else** $u :=$ a vertex u in $SUBG$ which maximizes $\mid CAND \cap \Gamma(u) \mid$;{*pivot*}
5 :   **while** $CAND - \Gamma(u) \neq \emptyset$
6 :   **do** $q :=$ a vertex in $(CAND - \Gamma(u))$;
7 :   **print** (q, *","*);
8 : $SUBG_q := SUBG \cap \Gamma(q)$;
9 : $CAND_q := CAND \cap \Gamma(q)$;
10:    EXPAND($SUBG_q, CAND_q$);
11:    $CAND := CAND - \{q\}$;
12:    **print** (*"back,"*)
    od
  fi
end {of EXPAND}

Fig. 3. Algorithm CLIQUES

We have proved that the worst-case time-complexity of CLIQUES is $O(3^{n/3})$ for an n-vertex graph [42, 46]. This is *optimal* as a function of n, since there exist up to $3^{n/3}$ cliques in an n-vertex graph [27]. The algorithm was also demonstrated to run fast in practice through computational experiments [46]. An example run of CLIQUES can be found in [51] together with those of [53] by Tsukiyama et al. and [24] by Makino-Uno applied to the same graph. By combining a bounding rule with CLIQUES, we obtained a simple $O(2^{n/2.863})$-time algorithm, **MAXCLIQUE** [38], for finding a maximum clique. It was experimentally shown in [38] that MAXCLIQUE runs faster than Tarjan and Trojanowsky [40]'s $O(2^{n/3})$-time algorithm.

In this approach, Eppstein et al. [12] proposed an algorithm for enumerating all maximal cliques that runs in time $O(dn3^{d/3})$ for an n-vertex graph G, where d is the *degeneracy* of G which is defined to be the smallest number such that every subgraph of G contains a vertex of degree at most d. If the graph G is *sparse*, d can be much smaller than n; hence $O(dn3^{d/3})$ can be much smaller than $O(3^{n/3})$.

Exact cliques are often too restrictive for practical applications as has been pointed in [35]. A useful algorithm for enumerating pseudo-cliques in large scale networks (graphs) has recently been proposed [57].

5   Applications

Many applications of maximum and maximal cliques can be found in [7,10,34,49, 54], and others. Thefore, we refer only to some of the literature in the following fields:

(a) Boinformatics
    (a-1) Analysis of protein structures [1,3–5,9]
    (a-2) Analysis of glycan structures [14,28]
(b) Data mining
    (b-1) Basic algorithms
        • Structural change pattern mining [31,32]
        • Pseudo clique enumeration [33,56,57]
    (b-2) Practical applications
        • Data mining for related genes [26]
        • Structural analysis of enterprise relationship [55]
(c) Image processing
    • Face detection [16]
(d) Design of quantum circuits [30]
(e) Design of DNA and RNA sequences for bio-molecular computation [20]

Acknowledgments. The author would like to express his sincere gratitude to H. Ito, T. Akutsu, M. Haraguchi, Y. Okubo, T. Nishino, H. Takahashi and many others for their fruitful joint work and kind help. This work was supported by JSPS KAKENHI Grant Numbers JP16300001, JP19300040, JP19500010, JP21300047, JP22500009, JP25330009, Kayamori Foundation of Informational Science Advancement, Funai Foundation for Information Technologies, and others.

References

1. Akutsu, T., Hayashida, M., Bahadur, D.K.C., Tomita, E., Suzuki, J., Horimoto, K.: Dynamic programming and clique based approaches for protein threading with profiles and constraints. IEICE Trans. Fundam. Electron. Commun. Comput. Sci. **E89–A**, 1215–1222 (2006)
2. Andrade, D.V., Resende, M.G.C., Werneck, R.F.: Fast local search for the maximum independent set problem. J. Heuristics **18**, 525–547 (2012)
3. Bahadur, D.K.C., Akutsu, T., Tomita, E., Seki, T., Fujiyama, A.: Point matching under non-uniform distortions and protein side chain packing based on an efficient maximum clique algorithm. Genome Inf. **13**, 143–152 (2002)
4. Bahadur, D.K.C., Tomita, E., Suzuki, J., Akutsu, T.: Protein side-chain packing problem: a maximum edge-weight clique algorithmic approach. J. Bioinform. Comput. Biol. **3**, 103–126 (2005)
5. Bahadur, D.K.C., Tomita, E., Suzuki, J., Horimoto, K., Akutsu, T.: Protein threading with profiles and distance constraints using clique based algorithms. J. Bioinform. Comput. Biol. **4**, 19–42 (2006)
6. Batsyn, M., Goldengorin, B., Maslov, E., Pardalos, P.M.: Improvements to MCS algorithm for the maximum clique problem. J. Comb. Optim. **27**, 397–416 (2014)

7. Bomze, I.M., Budinich, M., Pardalos, P.M., Pelillo, M.: The maximum clique problem. In: Du, D.-Z., Pardalos, P.M. (eds.) Handbook of Combinatorial Optimization, Supplement vol. A, pp. 1–74. Kluwer Academic Publishers (1999)
8. Bron, C., Kerbosch, J.: Algorithm 457, finding all cliques of an undirected graph. Commun. ACM **16**, 575–577 (1973)
9. Brown, J.B., Bahadur, D.K.C., Tomita, E., Akutsu, T.: Multiple methods for protein side chain packing using maximum weight cliques. Genome Inf. **17**, 3–12 (2006)
10. Butenko, S., Wilhelm, W.E.: Clique-detection models in computational biochemistry and genomics - invited review-. Eur. J. Oper. Res. **173**, 1–17 (2006)
11. Carraghan, R., Pardalos, P.M.: An exact algorithm for the maximum clique problem. Oper. Res. Lett. **9**, 375–382 (1990)
12. Eppstein, D., Löffler, M., Strash, D.: Listing all maximal cliques in large sparse real-world graphs. J. Exp. Algorithmics **18**, 3.1:1–3.1:21 (2013)
13. Fujii, T., Tomita, E.: On efficient algorithms for finding a maximum clique. Technical report IECE, AL81-113, pp. 25–34 (1982)
14. Fukagawa, D., Tamura, T., Takasu, A., Tomita, E., Akutsu, T.: A clique-based method for the edit distance between unordered trees and its application to analysis of glycan structure. BMC Bioinform. **12**(S–1), S:13 (2011)
15. Hatta, T., Tomita, E., Ito, H., Wakatsuki, M.: An improved branch-and-bound algorithm for finding a maximum clique. In: Proceedings of the Summer LA Symposium, no. 9, pp. 1–8 (2015)
16. Hotta, K., Tomita, E., Takahashi, H.: A view-invariant human face detection method based on maximum cliques. Trans. IPSJ **44**(SIG14(TOM9)), 57–70 (2003)
17. Johnson, D.S., Trick, M.A. (eds.): Cliques, Coloring, and Satisfiability. DIMACS Series in Discrete Mathematics and Theoretical Computer Science, vol. 26. American Mathematical Society (1996)
18. Karp, R.: Reducibility among combinatorial problems. In: Miller, R.E., Thatcher, J.W. (eds.) Comlexity of Computer Computations, pp. 85–103. Plenum Press, New York (1972)
19. Katayama, K., Hamamoto, A., Narihisa, H.: An effective local search for the maximum clique problem. Inf. Process. Lett. **95**, 503–511 (2005)
20. Kobayashi, S., Kondo, T., Okuda, K., Tomita, E.: Extracting globally structure free sequences by local structure freeness. In: Chen, J., Reif, J. (eds.) Proceedings of Ninth International Meeting on DNA Based Computers, p. 206 (2003)
21. Kohata, Y., Nishijima, T., Tomita, E., Fujihashi, C., Takahashi, H.: Efficient algorithms for finding a maximum clique. Technical report IEICE, COM89-113, pp. 1–8 (1990)
22. Konc, J., Janežič, D.: An improved branch and bound algorithm for the maximum clique problem. MATCH Commun. Math. Comput. Chem. **58**, 569–590 (2007)
23. Li, C.M., Quan, Z.: Combining graph structure exploitation and propositional reasoning for the maximum clique problem. In: Proceedings of IEEE ICTAI, pp. 344–351 (2010)
24. Makino, K., Uno, T.: New algorithms for enumerating all maximal cliques. In: Hagerup, T., Katajainen, J. (eds.) SWAT 2004. LNCS, vol. 3111, pp. 260–272. Springer, Heidelberg (2004). doi:10.1007/978-3-540-27810-8_23
25. Maslov, E., Batsyn, M., Pardalos, P.M.: Speeding up branch and bound algorithms for solving the maximum clique problem. J. Glob. Optim. **59**, 1–21 (2014)
26. Matsunaga, T., Yonemori, C., Tomita, E., Muramatsu, M.: Clique-based data mining for related genes in a biomedical database. BMC Bioinform. **10**, 205 (2009)
27. Moon, J.W., Moser, L.: On cliques in graphs. Israel J. Math. **3**, 23–28 (1965)

28. Mori, T., Tamura, T., Fukagawa, D., Takasu, A., Tomita, E., Akutsu, T.: A clique-based method using dynamic programming for computing edit distance between unordered trees. J. Comput. Biol. **19**, 1089–1104 (2012)

29. Nagai, M., Tabuchi, T., Tomita, E., Takahashi, H.: An experimental evaluation of some algorithms for finding a maximum clique. In: Conference Records of the National Convention of IEICE 1988, p. D-348 (1988)

30. Nakui, Y., Nishino, T., Tomita, E., Nakamura, T.: On the minimization of the quantum circuit depth based on a maximum clique with maximum vertex weight. Technical report RIMS, 1325, Kyoto University, pp. 45–50 (2003)

31. Okubo, Y., Haraguchi, M., Tomita, E.: Structural change pattern mining based on constrained maximal k-plex search. In: Ganascia, J.-G., Lenca, P., Petit, J.-M. (eds.) DS 2012. LNCS (LNAI), vol. 7569, pp. 284–298. Springer, Heidelberg (2012). doi:10.1007/978-3-642-33492-4_23

32. Okubo, Y., Haraguchi, M., Tomita, E.: Relational change pattern mining based on modularity difference. In: Ramanna, S., Lingras, P., Sombattheera, C., Krishna, A. (eds.) MIWAI 2013. LNCS (LNAI), vol. 8271, pp. 187–198. Springer, Heidelberg (2013). doi:10.1007/978-3-642-44949-9_18

33. Okubo, Y., Haraguchi, M., Tomita, E.: Enumerating maximal isolated cliques based on vertex-dependent connection lower bound. In: Perner, P. (ed.) MLDM 2016. LNCS (LNAI), vol. 9729, pp. 569–583. Springer, Heidelberg (2016)

34. Pardalos, P.M., Xue, J.: The maximum clique problem. J. Glob. Optim. **4**, 301–328 (1994)

35. Pattillo, J., Youssef, N., Butenko, S.: Clique relaxation models in social network analysis. In: Thai, M.T., Pardalos, P.M. (eds.) Handbook of Optimization in Complex Networks: Communication and Social Networks. Springer Optimization and Its Applications, vol. 58, pp. 143–162. Springer, Heidelberg (2012)

36. Segundo, P.S., Nikolaev, A., Batsyn, M.: Infra-chromatic bound for exact maximum clique search. Comput. Oper. Res. **64**, 293–303 (2015)

37. Shindo, M., Tomita, E., Maruyama, Y.: An efficient algorithm for finding a maximum clique. Technical report IECE, CAS86-5, pp. 33–40 (1986)

38. Shindo, M., Tomita, E.: A simple algorithm for finding a maximum clique and its worst-case time complexity. Syst. Comput. Jpn. **21**, 1–13 (1990). Wiley

39. Sutani, Y., Higashi, T., Tomita, E., Takahashi, S., Nakatani, H.: A faster branch-and-bound algorithm for finding a maximum clique. Technical report IPSJ, 2006-AL-108, pp. 79–86 (2006)

40. Tarjan, R.E., Trojanowski, A.E.: Finding a maximum independent set. SIAM J. Comput. **6**(3), 537–546 (1977)

41. Tomita, E., Yamada, M.: An algorithm for finding a maximum complete subgraph. In: Conference Records of the National Convention of IECE 1978, p. 8 (1978)

42. Tomita, E., Tanaka, A., Takahashi, H.: The worst-case time complexity for finding all the cliques. Technical report, University of Electro-Communications, UEC-TR-C5(2) (1988). (Reference [238] in [34], Reference [308] in [7]). http://id.nii.ac.jp/1438/00001898/

43. Tomita, E., Kohata, Y., Takahashi, H.: A simple algorithm for finding a maximum clique. Technical report, University of Electro-Communications, UEC-TR-C5(1) (1988). (Reference [239] in [34], Reference [309] in [7]). http://id.nii.ac.jp/1438/00001899/

44. Tomita, E., Mitsuma, S., Takahashi, H.: Two algorithms for finding a near-maximum clique. Technical report, University of Electro-Communications, UEC-TR-C1 (1988). (Reference [240] in [34], Reference [310] in [7]). http://id.nii.ac.jp/1438/00001900/

45. Tomita, E., Seki, T.: An efficient branch-and-bound algorithm for finding a maximum clique. In: Calude, C.S., Dinneen, M.J., Vajnovszki, V. (eds.) DMTCS 2003. LNCS, vol. 2731, pp. 278–289. Springer, Heidelberg (2003)

46. Tomita, E., Tanaka, A., Takahashi, H.: The worst-case time complexity for generating all maximal cliques and computational experiments. Theoret. Comput. Sci. **363**, 28–42 (2006). (Special Issue on COCOON 2004)

47. Tomita, E., Kameda, T.: An efficient branch-and-bound algorithm for finding a maximum clique with computational experiments. J. Glob. Optim. **37**, 95–111 (2007). J. Glob. Optim. **44**, 311 (2009)

48. Tomita, E., Sutani, Y., Higashi, T., Takahashi, S., Wakatsuki, M.: A simple and faster branch-and-bound algorithm for finding a maximum clique. In: Rahman, M.S., Fujita, S. (eds.) WALCOM 2010. LNCS, vol. 5942, pp. 191–203. Springer, Heidelberg (2010). doi:10.1007/978-3-642-11440-3_18

49. Tomita, E., Akutsu, T., Matsunaga, T.: Efficient algorithms for finding maximum and maximal cliques: Effective tools for bioinformatics. In: Laskovski, A.N. (ed.) Biomedical Engineering, Trends in Electronics, Communications and Software, pp. 625–640. InTech, Rijeka (2011). http://cdn.intechopen.com/pdfs-wm/12929.pdf

50. Tomita, E., Sutani, Y., Higashi, T., Wakatsuki, M.: A simple and faster branch-and-bound algorithm for finding a maximum clique with computational experiments. IEICE Trans. Inf. Syst. **E96–D**, 1286–1298 (2013). http://id.nii.ac.jp/1438/00000287/

51. Tomita, E.: Clique enumeration. In: Kao, M.-Y. (ed.) Encyclopedia of Algorithms, 2nd edn, pp. 313–317. Springer, Heidelberg (2016)

52. Tomita, E., Yoshida, K., Hatta, T., Nagao, A., Ito, H., Wakatsuki, M.: A much faster branch-and-bound algorithm for finding a maximum clique. In: Zhu, D., Bereg, S. (eds.) FAW 2016. LNCS, vol. 9711, pp. 215–226. Springer, Heidelberg (2016)

53. Tsukiyama, S., Ide, M., Ariyoshi, H., Shirakawa, I.: A new algorithm for generating all the maximal independent sets. SIAM J. Comput. **6**, 505–517 (1977)

54. Wu, Q., Hao, J.K.: A review on algorithms for maximum clique problems - invited review-. Eur. J. Oper. Res. **242**, 693–709 (2015)

55. Yonemori, C., Matsunaga, T., Sekine, J., Tomita, E.: A structural analysis of enterprise relationship using cliques. DBSJ J. **7**, 55–60 (2009)

56. Zhai, H., Haraguchi, M., Okubo, Y., Tomita, E.: Enumerating maximal clique sets with pseudo-clique constraint. In: Japkowicz, N., Matwin, S. (eds.) DS 2015. LNCS (LNAI), vol. 9356, pp. 324–339. Springer, Heidelberg (2015). doi:10.1007/978-3-319-24282-8_28

57. Zhai, H., Haraguchi, M., Okubo, Y., Tomita, E.: A fast and complete algorithm for enumerating pseudo-cliques in large graphs. Int. J. Data Sci. Anal. **2**, 145–158 (2016). Springer

Computational Geometry I

Efficient Enumeration of Flat-Foldable Single Vertex Crease Patterns

Koji Ouchi[✉] and Ryuhei Uehara

School of Information Science, Japan Advanced Institute of Science
and Technology (JAIST), Nomi, Japan
{k-ouchi,uehara}@jaist.ac.jp

Abstract. We investigate enumeration of distinct flat-foldable crease patterns with natural assumptions. Precisely, for a given positive integer n, potential set of n crease lines are incident to the center of a sheet of disk paper at regular angles. That is, every angle between adjacent lines is equal to $2\pi/n$. Then each line is assigned one of "mountain," "valley," and "flat (or consequently unfolded)." That is, we enumerate all flat-foldable crease patterns with up to n crease lines of unit angle $2\pi/n$. We note that two crease patterns are equivalent if they are equal up to rotation and reflection. In computational origami, there are two well-known theorems for flat-foldability: the Kawasaki Theorem and the Maekawa Theorem. The first one is a necessary and sufficient condition of crease layout, however, it does not give us valid mountain/valley assignments. The second one is a necessary condition between the number of "mountain" and that of "valley." However, sufficient condition(s) is(are) not known. Therefore, we have to enumerate and check flat-foldability one by one using other algorithm. In this research, we develop the first algorithm for the above stated problem by combining these results in a nontrivial way, and show its analysis of efficiency. We also give experimental results, which give us a new series of integer sequence.

1 Introduction

Recent origami is a kind of art, and origamists around the world struggle with their problems; what is the best way to fold an origami model. One of the problems is that a unit of angle that appears in the origami model. Some origamists restrict themselves to use only multiples of 22.5°, 15° or some other specific angle which divides 360°. A nontrivial example, which was designed by the first author, is shown in Fig. 1. It is based on a unit angle of 15°. Once origamists fix the unit angle as $(360/n)°$ for suitable positive integer n, their designs are restricted to one between quite real shapes and abstract shapes, which is the next matter in art.

When we are given a positive integer n, we face a computational origami problem which is interesting from the viewpoints of mathematics and algorithms. We consider the simplest origami model; all crease lines are incident to the single vertex at the center of origami, and each angle between two creases is a multiple of $(360/n)°$. We are interested in only flat-foldable crease patterns.

© Springer International Publishing AG 2017
S.-H. Poon et al. (Eds.): WALCOM 2017, LNCS 10167, pp. 19–29, 2017.
DOI: 10.1007/978-3-319-53925-6_2

Fig. 1. "Maple leaf" designed and folded by the first author. Its crease pattern is based on 15° unit angle.

When mountain or valley folding is assigned to every crease pattern, the flat-foldability can be computed in linear time (see [2,5]). However, its rigorous proof is not so simple, which is the main topic of Chap. 12 in [5]. Roughly speaking, the algorithm repeatedly folds and glues the locally smallest angle in each step. In other words, we have no mathematical characterization for this problem, and we have to check one by one.

The problem of computing a folding for a crease pattern that does not contain a specification of whether folds are mountains or valleys is very different. Hull investigated this problem [8] from the viewpoint of counting. Precisely, he considered the number of flat-foldable assignments of mountain and valley to a given crease pattern of n lines which were incident to the single vertex. He gave tight lower and upper bounds. These bounds are given in two extreme situations; one is given in the case that all n angles are different, and the other is given in the case that all n angles are equal to each other. From the viewpoint of origami design, we are interested in the case between these two extreme situations. To deal with reasonable situations between extreme ones, we slightly modify the input of the problem. The input of our problem is a positive integer n, and we restrict ourselves to the single vertex folding of unit angle $(360/n)°$. In order to investigate our problem, we assign one of three labels—"mountain," "valley," and "flat"—to each of n creases. When a crease line is labeled "flat," this crease line is not folded in the final folded state. In this way, we can deal with the single vertex crease patterns of unit angle equal to $(360/n)°$, which is more realistic situation from the viewpoint of origami design.

Our aim is to enumerate all distinct flat-foldable assignments of the three labels to n creases. In other words, our algorithm eventually enumerates all flat-foldable crease patterns with labels of "mountain" and "valley" of unit angle $(360/n)°$. We consider the sheet of paper is a disk, the vertex is at the center of the disk, and two crease patterns are considered to be equivalent if they can be equal up to rotation and reflection (i.e., including turning over and exchanging all mountains and valleys). Our algorithm enumerates all distinct crease patterns under this assumption.

For flat-foldability of a given crease pattern, there are two well-known theorems in the area of computational origami, which are called the Kawasaki Theorem and the Maekawa Theorem (see [5, Chap. 12] for further details):

Theorem 1 (The Kawasaki Theorem). *Let θ_i be an angle between the ith and the (i + 1)th crease lines. A single-vertex crease pattern defined by angles $\theta_1 + \theta_2 + \cdots + \theta_{n'} = 360°$ is flat-foldable if and only if n' is even and the sum of the odd angles θ_{2i+1} is equal to the sum of the even angles θ_{2i}, or equivalently, either sum is equal to $180°$: $\theta_1 + \theta_3 + \cdots + \theta_{n'-1} = \theta_2 + \theta_4 + \cdots + \theta_{n'} = 180°$.*

We note that the Kawasaki Theorem gives a necessary and sufficient condition for flat-foldability, but mountain-valley assignments are not given. That is, we have to compute foldable assignments for foldable crease pattern satisfying the Kawasaki Theorem. In order to compute a flat-foldable assignment, we can use the Maekawa Theorem:

Theorem 2 (The Maekawa Theorem). *In a flat-foldable single-vertex mountain-valley pattern defined by angles $\theta_1 + \theta_2 + \cdots + \theta_{n'} = 360°$, the number of mountains and the number of valleys differ by ± 2.*

We again note that the Maekawa Theorem is a necessary but not sufficient condition.

In the last decades, enumeration algorithms have been well investigated, and many efficient enumeration algorithms have been given, e.g., [1,10,11], and so on. Using techniques that follow above properties of origami, we construct an enumeration algorithm for flat-foldable crease patterns for given n, where n is the maximum number of crease lines of unit angle $(360/n)°$. As far as the authors know, this is the first algorithm for the realistic computational origami problem. As a result, we succeeded to enumerate flat-foldable crease patterns up to two $n = 32$ in a reasonable time.

2   Preliminaries and Outline of Algorithm

Based on the Kawasaki Theorem and the Maekawa Theorem, for given n, we can design the outline of our enumeration algorithm as follows:

(1) Assign "crease" or "flat" to each of n crease lines incident to a single vertex so that the Kawasaki Theorem is satisfied.
(2) For each "crease", assign "mountain" or "valley" so that the Maekawa Theorem is satisfied.
(3) Output the pattern if this crease pattern is flat-foldable.

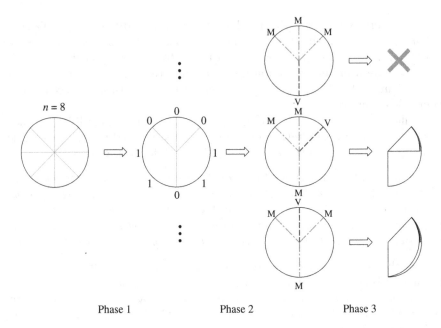

Fig. 2. Simple example for $n = 8$.

Essentially, the outline consists of two different kinds of enumeration problems in phases 1 and 2, and flat-foldability checking in phase 3.

A simple example is given in Fig. 2. For $n = 8$, we first generate all possible crease lines in phase 1 which is described in binary string (in the figure, we only show one, but there are exponentially many). Here "0" and "1" denote "crease" and "flat" respectively. Therefore, for a string 00011011, we have four crease lines in the shape in Fig. 2. In phase 2, we assign mountain (M) or valley (V) to each crease line. In phase 3, we check whether each crease pattern with M/V assignments is flat-foldable or not, and output the pattern if it is flat-foldable.

We have different issue for each phase. Especially in phases 1 and 2, we have to consider two different problems about symmetry (to reduce redundant output), and enumeration.

3   Description of Algorithm

Now we describe more details in each phase.

3.1   Phase 1: Assignment of "crease"/"flat"

In phase 1, we are given n crease lines, and we have to assign "crease" or "flat" to them so that the assignment satisfies the Kawasaki Theorem. Since the crease pattern cannot be flat-folded for odd number n, without loss of generality, we assume that n is even hereafter.

We first describe "crease" by 0 and "flat" by 1, and consider binary string. Then it is easy to see that, before checking the Kawasaki Theorem, we have to generate all binary strings over $\Sigma = \{0,1\}$ efficiently reducing equivalent rotations and reflections. To consider this problem, we introduce the bracelet problem, which is a classic and basic problem in combinatorics. A *bracelet* is an equivalence class of strings, taking all rotations and reversals as equivalent. This is a special case of *necklace* whose equivalence is rotation only. In this paper, let the word *bracelet* also denote the lexicographically smallest string of the equivalence class and so does *necklace*. It is easy to observe that our problem is now enumeration of binary bracelet of length n. For bracelets, we have an optimal enumeration algorithm [9]:

Theorem 3 (Sawada 2001). *Bracelets of length n can be enumerated in constant amortized time.*

That is, the algorithm in [9] runs in a time proportional to the number of bracelets of length n.

We note that the values of the function $B(n)$ are listed in the OEIS (The On-line Encyclopedia of Integer Sequences; http://oeis.org/) as A000029, and it is given as

$$B(n) = \sum_{d \text{ divides } n} \frac{2^{n/d}\phi(d)}{2n} + 2^{n/2-1} + 2^{n/2-2} \tag{1}$$

for an even number n, where ϕ is Euler's totient function.

3.2 Phase 1: Satisfying the Kawasaki Theorem

After assigning "crease" or "flat" to each crease, we have to check whether these crease lines satisfy the Kawasaki Theorem or not. The Kawasaki Theorem states that the alternating sum of angles should be equal to 0. This notion corresponds to a kind of necklace in a nontrivial way as follows. We first observe that each angle θ_i is $k \times \frac{360°}{n}$ for given even n. That is, θ_i consists of k unit angles. Now we regard θ_i as the integer k, and we consider $\theta_1, \theta_3, \ldots$ as "white," and $\theta_2, \theta_4, \ldots$ as "black." Then, each sequence of angles corresponds to a necklace with n beads such that the number of white beads is equal to the number of black beads. That is, each sequence of n' creases satisfying the Kawasaki Theorem corresponds to a necklace with n beads such that (1) the necklace consists of $n/2$ white beads and $n/2$ black beads, and (2) the number of runs of white beads (and hence black beads) is n'. This notion is investigated as "balanced twills on n harnesses" in [7], and listed in OEIS as A006840. For $k = n/2$, the number is given as follows:

Theorem 4 (Hoskins and Street 1982). *The number of distinct balanced twills on* $n = 2k$ *harnesses is*

$$B'(2k) = \frac{1}{8k} \left\{ \sum_{\substack{d \text{ divides } n \\ d = 2e}} \phi\left(\frac{k}{e}\right)\binom{2e}{e} + \sum_{d \text{ divides } k} \phi\left(\frac{2k}{d}\right) 2^d + 2k\binom{2\lfloor k/2\rfloor}{\lfloor k/2\rfloor} + k2^k \right\}.$$

(2)

We note that Eq. (2) just gives us the numbers for each n and no concrete sets of creases. Therefore, we have to enumerate them by ourselves. A straightforward approach is to insert a test of the Kawasaki Theorem into Sawada's algorithm [9]. The test computes $\sum_{i=1}^{n'}(-1)^i\theta_i$ and checks whether the value is 0 or not. Note that n' is the number of "crease" and θ_i is the angle between the ith and $(i+1)$th creases as defined in Theorem 1.

Now we have the following theorem:

Theorem 5. *For a given even number n, phase 1 can be done in $O(nB(n))$ time, where $B(n)$ is the number of bracelets of length n.*

3.3   Phase 2: Assignment of "mountain"/"valley"

In this phase, we inherit a binary string of length n from the phase 1, which describes "crease" ($=0$) or "flat" ($=1$). We note that the binary string is the lexicographically smallest one among rotations and reversals. Then we translate it to a set of other strings that represent the assignments of "mountain" and "valley" and the angles between adjacent creases. The first step can be described as follows:

(2a) For each adjacent pair of 0s, replace 1s between them by the number of 1s plus 1. For example, the string 00011011 in Fig. 2 is replaced by 0<u>1</u>0<u>1</u>0<u>3</u>0<u>3</u>, where the positive (underlined) numbers describe the number of unit angles there.

Then we assign mountain ($=M$) and valley ($=V$) to each 0, but here we only consider the assignments that satisfies the Maekawa Theorem. The Maekawa Theorem says that the number of Ms and the number of Vs should differ by 2. To avoid symmetry case, we can assume that (the number of Ms)$-$(the number of Vs) $= 2$. Thus next step is described as follows:

(2b) For the resulting string over $\{0, 1, 2, \ldots, n-1\}$, assign all possible Ms and Vs to each 0 such that the number of Ms is 2 larger than the number of Vs. For example, for the string 01010303, we obtain the set of strings $\{V1M1M3M3, M1V1M3M3, M1M1V3M3, M1M1M3V3\}$.

For a string s generated by step 2a, we can have equivalent assigned crease patterns. Precisely, if some rotation(s) or reversal(s) of s is (are) equal to s, the result of step 2b may contain equivalent assigned crease patterns. For example, in the set of strings $\{V1M1M3M3, M1V1M3M3, M1M1V3M3, M1M1M3V3\}$, we can observe that $V1M1M3M3$ is a crease pattern which is the mirror image of a crease pattern $M1M1V3M3$, hence we consider they are equivalent. (In Fig. 2, after phase 2, the crease pattern at the center has its mirror image, and it should be omitted.) To avoid such equivalent patterns, we perform the following:

(2c) For the resulting string s' over $\{M, V, 1, 2, \ldots, n-1\}$ after step 2b, generate the lexicographically smallest string among rotations and reversals of s', which we call s'_{small}, and store all s'_{small}. s' is discarded if s'_{small} has been already obtained. Note that $M < V < 1 < 2 < \ldots$.

In this process, we take a caching strategy to detect duplications; For every s', we generate and store a representative of the bracelet equivalence class to which s' belongs, and we refer to the representatives generated so far to check whether we have obtained an equivalent of s' or not. The string s'_{small} can be one of such representatives because the lexicographically smallest string is easy to be generated and unique among rotations and reversals. Because of the exponential number of strings to be cached, we use a trie (see [4,6], a.k.a. prefix tree) that is a space-efficient data structure for storing many strings.

To generate s'_{small}, we use Booth's least circular string algorithm [3]. It is a linear time algorithm to find the smallest string among rotations of given string. Note that the algorithm doesn't care about reversals. Precisely, Booth's algorithm finds the *right index* of the lexicographically smallest string for a given circular string of length n in linear time. *Right index* is the start index of a circular string that may be larger than (or on the "right" side of) the original start index 0, which is a conventional description in the field of string algorithms. To deal with both rotation and reversal, the step 2c can be implemented as follows:

(2c-1) For the resulting string s' over $\{M, V, 1, 2, \ldots, n-1\}$ after step 2b, let s'^R is the reverse string of s'. Prepare an empty trie.

(2c-2) Using Booth's algorithm, find the right index i of circular string s' such that the string starting from the index i is the lexicographically smallest string among all rotations of s'.

(2c-3) Similarly, find the right index j of the lexicographically smallest string among all rotations of s'^R. The index j gives the smallest string among the equivalents of reversals.

(2c-4) Select the smallest string as s'_{small} from the result of (2c-2) and (2c-3): rotation of s' starting from i and rotation of s'^R starting from j. If s'_{small} is already in the trie, discard s'. Otherwise append s'_{small} to the trie and s' goes to phase 3 to be processed.

This test takes $O(n)$ time because the steps don't contain loops and recursions but it runs linear time sub routines just constant times, which are Booth's algorithm, string comparison, and operations on a trie. Summarizing, we have the following theorem:

Theorem 6. *For a given crease pattern from phase 1 based on n unit angles, we can generate all distinct assignments of mountain and valley that satisfies the Maekawa Theorem in $O(nC(n))$ time with space linear in the product of n and the number of such assignments, where $C(n)$ is $\binom{n}{n/2-1}$.*

Proof. The number of lines in the crease pattern is at most n, and the number of Ms is 2 larger than the number of Vs. Thus, the number of strings s' over $\{M, V, 1, 2, \ldots\}$ of length at most n with the constraint for the number of Ms and Vs is at most $\binom{n}{n/2-1}$. The other management can be done in linear time, which implies the time complexity in the theorem. The space complexity is linear in the maximum number of nodes in the trie used in the algorithm, which can be suppressed by the product of n and the number of desired assignments.   □

3.4   Phase 3: Test of Flat-Foldability

In this phase, we check if the resulting string s' over $\{M, V, 1, 2, \ldots\}$ is flat-foldable or not. For this problem, Demaine and O'Rourke give a linear time algorithm [5, Chap. 12]. Therefore we can finish this phase in linear time. Roughly, the algorithm is simple; it finds a local minimal angle, folds two creases on the boundary of the small fan-shape, glues it, and repeats until all creases are folded. However, showing the correctness of this algorithm is not easy; as mentioned at the footnote in [5, p. 204], the rigorous proof is first done by Demaine and O'Rourke in [5, Chap. 12].

We obtain the following obvious upper bound of the number of the outputs in this phase by integration of the observations in Sects. 3.2 and 3.3:

Theorem 7. *For a given even number n, the number of distinct flat-foldable mountain and valley assignments with unit angle $(360/n)°$ is $O\left(B'(n)\binom{n}{n/2-1}\right)$ where $B'(n)$ is the number of distinct balanced twills on $n = 2k$ harnesses (see Eq. 2).*

3.5   Analysis of Algorithm

The correctness of our algorithm relies on the algorithms used in each phase as described above. Here we consider its time complexity and space complexity for computing all outputs. Our main theorem is the following:

Theorem 8. *For a given even number n, all distinct flat-foldable mountain and valley assignments with unit angle $(360/n)°$ can be done in $O\left(n^2 B(n)\binom{n}{n/2-1}\right)$ time with $O\left(n\binom{n}{n/2-1}\right)$ space, where $B(n)$ is the number of bracelets of length n (see Eq. 1).*

We note that the order of space complexity may be far from strict one because the actual required space for the computation depends on the behavior of the trie used in phase 2.

Table 1. The number of enumerated patterns. The number of lines in a pattern is even number from 2 to n.

n	Phase 1	Phase 2	Phase 3
4	2	2	2
6	3	7	6
8	7	27	20
10	13	143	87
12	35	837	420
14	85	5529	2254
16	257	38305	12676
18	765	276441	73819
20	2518	2042990	438795
22	8359	15396071	2649555
24	28968	117761000	16217883
26	101340	912100793	99888892
28	361270	7139581543	621428188
30	1297879	56400579759	3893646748
32	4707969	449129924559	24548337096
34	17179435	-	-
36	63068876	-	-
38	232615771	-	-
40	861725794	-	-
42	3204236779	-	-

Table 2. Distribution of the patterns obtained at phase 1.

n	# line of each pattern										Sum
	2	4	6	8	10	12	14	16	18	20	
4	1	1									2
6	1	1	1								3
8	1	3	2	1							7
10	1	3	6	2	1						13
12	1	6	13	11	3	1					35
14	1	6	26	30	18	3	1				85
16	1	10	46	93	74	28	4	1			257
18	1	10	79	210	275	145	40	4	1		765
20	1	15	124	479	841	716	280	56	5	1	2518

4   Experimental Results

As shown in Theorem 7, the upper bound of the number of distinct flat-foldable mountain and valley assignments is exponential if $(360/n)°$ unit angle is introduced. Exact values for each n are difficult to estimate theoretically. Therefore, we here show experimental results. The program is written in C++ using its default STL library.

4.1   The Number of Crease Patterns

Table 1 and Fig. 3 show the exact number of distinct patterns obtained at each phase. Table 2 explains more details of each data in Table 1. As mentioned in Sect. 3.2, the result of phase 1, which enumerates "crease"/"flat" assignments satisfying the Kawasaki theorem, coincide with the sequence listed in OEIS as A006840. The counting results at the other phases are different from any existing sequences in OEIS, that is, we find totally new sequences in this study (Table 1).

4.2   Solution Space

We measure the rate of the number of solutions against that of possible patterns at each phase (see Table 3 and Fig. 4), which suggests how difficult the problems are. We can see that the solution spaces are very sparse for all phases. There

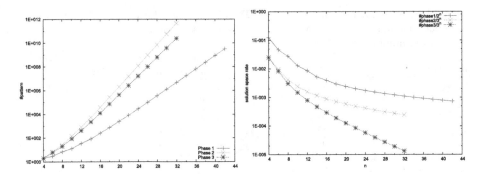

Fig. 3. The number of enumerated patterns. The number of lines in a pattern is even number from 2 to n.

Fig. 4. The rate of solutions against possible patterns at each phase.

Table 3. #solution/#possible at each phase.

n	$\#Phase1/2^n$	$\#Phase2/3^n$	$\#Phase3/3^n$
4	0.125	0.024691358	0.024691358
6	0.046875	0.009602195	0.008230453
8	0.02734375	0.004115226	0.003048316
10	0.012695313	0.002421718	0.001473353
12	0.008544922	0.001574963	0.000790304
14	0.005187988	0.001155977	0.000471255
16	0.003921509	0.000889847	0.000294471
18	0.002918243	0.000713543	0.000190540
20	0.002401352	0.000585924	0.000125845
22	0.001992941	0.000490617	8.44317E-05
24	0.001726627	0.000416957	5.74228E-05
26	0.001510084	0.000358831	3.92975E-05
28	0.001345836	0.000312088	2.71641E-05
30	0.001208744	0.000273934	1.89112E-05
32	0.001096159	0.000242377	1.32477E-05
34	0.000999975	-	-
36	0.000917773	-	-
38	0.000846251	-	-
40	0.000783735	-	-
42	0.000728559	-	-

are 2^n possible "crease"/"flat" assignments at phase 1. Only about 4.7% is the solution for phase 1 if $n = 6$. It decreases significantly and gets less than 1% for $n \geq 12$. The rates at phase 2 and phase 3 are against 3^n since we consider "mountain"/"valley"/"flat" assignments at that phases. The two rates tend to decrease similarly to that of phase 1, and are much smaller, e.g., 2.5% for phase 2 when $n = 6$. Such rate at every phase seems to be exponential to n according to Fig. 4.

5 Concluding Remarks

We develop the first algorithm for enumerating distinct flat-foldable single vertex crease patterns. We also experimentally show how many such patterns there are, which is done the first time as well. Improving the algorithm and investigating further for the counting problems are the future works. For example, rather than Sawada's algorithm in Theorem 3, enumeration of the sequences stated in Theorem 4 directly could improve the running time of our algorithm drastically.

We also examine the rates in each phase; experimentally, they seem to decrease exponentially. Nevertheless, we conjecture that there are exponentially many flat-foldable crease patterns. Showing theoretical lower and upper bounds also remains open.

Acknowledgement. We would like to thank Yota Otachi for his fruitful discussions and comments. This work is partially supported by MEXT/JSPS Kakenhi Grant Number 26330009 and 24106004.

References

1. Avis, D., Fukuda, K.: Reverse search for enumeration. Discrete Appl. Math. **65**(1), 21–46 (1996)
2. Bern, M., Hayes, B.: The complexity of flat origami. In: SODA, vol. 96, pp. 175–183 (1996)
3. Booth, K.S.: Lexicographically least circular substrings. Inf. Process. Lett. **10**(4–5), 240–242 (1980)
4. De La Briandais, R.: File searching using variable length keys. In: Papers Presented at the Western Joint Computer Conference, 3–5 March 1959, pp. 295–298. ACM (1959)
5. Demaine, E.D., O'Rourke, J.: Geometric Folding Algorithms: Linkages, Origami, Polyhedra. Cambridge University Press, Cambridge (2007)
6. Fredkin, E.: Trie memory. Commun. ACM **3**(9), 490–499 (1960)
7. Hoskins, W., Street, A.P.: Twills on a given number of harnesses. J. Aust. Math. Soc. (Ser. A) **33**(01), 1–15 (1982)
8. Hull, T.: Counting mountain-valley assignments for flat folds. Ars Comb. **67**, 175–187 (2003)
9. Sawada, J.: Generating bracelets in constant amortized time. SIAM J. Comput. **31**(1), 259–268 (2001)
10. Uno, T., Asai, T., Uchida, Y., Arimura, H.: An efficient algorithm for enumerating closed patterns in transaction databases. In: Suzuki, E., Arikawa, S. (eds.) DS 2004. LNCS (LNAI), vol. 3245, pp. 16–31. Springer, Heidelberg (2004). doi:10. 1007/978-3-540-30214-8_2
11. Zaki, M.J.: Efficiently mining frequent trees in a forest. In: Proceedings of the Eighth ACM SIGKDD International Conference on Knowledge Discovery and Data Mining, pp. 71–80. ACM (2002)

Dynamic Sum-Radii Clustering

Nicolas K. Blanchard[1,2(✉)] and Nicolas Schabanel[3,4]

[1] U. Paris Diderot, Paris, France
Nicolas.K.Blanchard@gmail.com
http://www.irif.univ-paris-diderot.fr/users/nkblanchard
[2] ENS Paris, Paris, France
[3] CNRS, U. Paris Diderot, Paris, France
http://www.irif.univ-paris-diderot.fr/users/nschaba
[4] IXXI, U. Lyon, Paris, France

Abstract. Real networks have in common that they evolve over time and their dynamics have a huge impact on their structure. Clustering is an efficient tool to reduce the complexity to allow representation of the data. In 2014, Eisenstat *et al.* introduced a dynamic version of this classic problem where the distances evolve with time and where coherence over time is enforced by introducing a cost for clients to change their assigned facility. They designed a $\Theta(\ln n)$-approximation. An $O(1)$-approximation for the *metric* case was proposed later on by An *et al.* (2015). Both articles aimed at minimizing the sum of all client-facility distances; however, other metrics may be more relevant. In this article we aim to minimize the sum of the *radii of the clusters* instead. We obtain an asymptotically optimal $\Theta(\ln n)$-approximation algorithm where n is the number of clients and show that existing algorithms from An *et al.* (2015) do not achieve a constant approximation in the metric variant of this setting.

Keywords: Facility location · Approximation algorithms · Clustering · Dynamic graphs

1 Introduction

Context. During the past decade, a massive amount of data has been collected on diverse networks such as the web (pages and links), social networks (e.g., Facebook, Twitter, and LinkedIn), social encounters in hospitals, schools, companies, conferences as well as in the wild [1–3]. These networks evolve over time, and their dynamics have a considerable impact on their structure and effectiveness [4]. Understanding the dynamics of evolving networks is a central question in many applied areas such as epidemiology, vaccination planning, anti-virus design, management of human resources, and viral marketing. A relevant clustering of the data is often needed to design informative representations of massive data sets. Algorithmic approaches have already yielded useful insights on real

This work was supported by Grants ANR-12-BS02-005 RDAM and IXXI-Molecal.

S.-H. Poon et al. (Eds.): WALCOM 2017, LNCS 10167, pp. 30–41, 2017.
DOI: 10.1007/978-3-319-53925-6_3

networks such as the social interaction networks of zebras [5]. In most experiments, data is recorded first and analyzed next, see [2]. The complete evolution of the network is thus known from the beginning, as opposed to the online setting where one must continuously adapt a partial solution to new incoming data [6].

Previous Work. Given a set of facilities, a set of clients, and a measure of distances between them, the facility location problem consists in opening a subset of facilities and assigning the clients to open facilities so as to minimize a trade-off between the cost of opening the facilities and the cost corresponding to the distance between the clients and their assigned facilities. This problem and its many variants have been extensively studied since the 1960s, using tools such as LP-rounding [7], primal-dual methods [8] or greedy improvements [9]. The uncapacited version, where any number of clients can connect to a facility, is considered here as it is known to be a successful approach to clustering when the number of clusters is not known a priori.

In 2014, [10] introduced a dynamic version of this classic problem to handle situations where the distances evolve with time and where one looks for an assignment consistent with the evolution of the distances. To achieve a balance between the stability of the solution and its adaptability, they introduced a cost to be paid every time a client is assigned to a new facility. As shown in [10], in many natural scenarios the output solutions follow the observed dynamic better than independent optimizations of consecutive snapshots of the evolving distances. This has been further refined in [11], yielding an $O(1)$-approximation algorithm when the distances are metric (i.e., follow the triangular inequalities).

Our Approach: Dynamic Sum-Radii Clustering. In both articles [10,11], the distance cost in the objective consisted of the sum of all distances between every client and its assigned facility over all time steps. Whereas this distance cost makes perfect sense in the case where clients need to physically connect to a facility, other metrics are preferred in the context of clustering. The present article introduces a dynamic version of the problem studied in [8]. We aim at minimizing the *radii of the clusters,* i.e. the sum over all open facilities of their distances to their farthest assigned client at each time step. This objective focuses on the closeness of the clients to their assigned facility regardless of the number of clients assigned to each open facility. It is thus better suited to situations with clusters of very different sizes which are typically observed in nature where groups tend to follow power laws in size [1]. Optimal solutions for this objective cost have been explored in [12], where it was shown that even in the 1-dimensional euclidean space, optimal solutions can have surprisingly complex structures.

In the general setting, we introduce a primal LP-rounding algorithm that achieves a logarithmic approximation, which is shown to be asymptotically optimal unless $P = NP$. We then turn to metric distances and show that existing algorithms from [11] do not achieve a constant approximation in this setting, as the lack of cooperation between the clients is not being absorbed by the sum-of-radii objective anymore. The next section presents a formal definition of the problem and states our main results, proved in the following sections. The ommited proofs can be found at https://hal.archives-ouvertes.fr/hal-01424769.

2   Definition and Main Results

2.1   Definitions

Dynamic Sum-Radii Clustering (DSRC). Given a set F of m facilities, a set C of n clients, their respective distances $(d_{ijt})_{i \in F, j \in C, t \in [T]}$ for each time step $t \in [T] = \{1, \ldots, T\}$, an opening cost $f_{it} \geqslant 0$ for each facility i at time t, and a changing cost $g \geqslant 0$, the goal is to open at each time step t a subset $O_t \subseteq F$ of facilities and to assign each client $j \in C$ to an open facility $\varphi_{jt} \in O_t$ so as to minimize the sum of:

Opening cost: $\sum_{t \in [T]} \sum_{i \in O_t} (f_{it} + r_{it})$, where for each facility $i \in O_t$, r_{it} denotes its open radius: $r_{it} = \max\{d_{ijt} : j \in C$ s.t. j is assigned to i at time $t\}$.

Changing cost: $\sum_{t \in [T-1]} \sum_{j \in C} g \cdot \mathbb{1}_{\{\varphi_{jt} \neq \varphi_{j(t+1)}\}}$.

Precisely, this problem is strictly equivalent to the linear program (1), inspired by [8,10], when its variables $x_{ijt}, y_{irt}, z_{ijt}$ are restricted to integral values in $\{0, 1\}$. Their integral values are interpreted as follows: $x_{ijt} = 1$ iff Client j is assigned to Facility i at time t (Constraint (1.a)); $y_{irt} = 1$ iff Facility i is open with radius r at time t (Constraint (1.b)); $z_{ijt} = 1$ iff Client j is assigned to Facility i at time $t + 1$ and was not assigned to i at t (Constraint (1.c)). Note that one can restrict the total number of y_{irt} variables to mnT as one shall only consider the radii r equal to some distance d_{ijt} for some $j \in C$, for each facility i and time t.

$$
\left.
\begin{array}{lll}
\text{Minimize} & \displaystyle\sum_{i \in F, r \geqslant 0, t \in [T]} y_{irt} \cdot (f_{it} + r) \;+\; g \cdot \displaystyle\sum_{i \in F, j \in C, t \in [T-1]} z_{ijt} \\[2ex]
\text{such that (1.a)} & \displaystyle\sum_{i \in F} x_{ijt} \geqslant 1 & (\forall j \in C, t \in [T]) \\[2ex]
\text{(1.b)} & \displaystyle\sum_{r : r \geqslant d_{ijt}} y_{irt} \geqslant x_{ijt} & (\forall i \in F, j \in C, t \in [T]) \\[2ex]
\text{(1.c)} & z_{ijt} \geqslant x_{ij(t+1)} - x_{ijt} & (\forall i \in F, j \in C, t \in [T-1]) \\[2ex]
\text{and } x_{ijt}, y_{ijrt}, z_{ijt} \geqslant 0.
\end{array}
\right\}
\quad (1)
$$

We denote by LP the optimum (fractional) value of (1), and for each time period $U \subseteq [T]$, by $\mathrm{openCost}_U(x, y, z) = \sum_{i \in F, r \geqslant 0, t \in U} y_{irt} \cdot (f_{it} + r)$ the fractional opening cost of solution (x, y, z) during the time period U, and by $\mathrm{changeCost}_U(x, y, z) = g \cdot \sum_{i \in F, j \in C, t \in U \setminus \{\max U\}} z_{ijt}$ the fractional changing cost of (x, y, z) during U. The index U is omitted when $U = [T]$.

2.2   Preprocessing

As in [10], our algorithm first preprocesses an optimal solution to this LP in order to obtain some useful properties. This preprocessing, ommited here, uses a rounding scheme for the z_{ijt} to determine at which discrete time the clients must change their assigned facility. This is achieved by the following Lemma.

Lemma 1 (Direct adaptation from [10]). *Given an optimal solution to LP* (1), *one can compute a feasible solution* (x, y, z) *together with a collection of time intervals* $I_{1,1}, \ldots, I_{1,\ell_1}, \ldots, I_{n,1}, \ldots, I_{n,\ell_n}$ *such that:*

- *for all* $j \in C$: $I_{j,1}, \ldots, I_{j,\ell_j}$ *form a partition of* $[T]$; *and*
- *for all* $i \in F$, $j \in C$ *and* $k \in [\ell_j]$: x_{ijt} *is constant during each time interval* I_{jk}; *and*
- *for all* $j \in C$: $\ell_j - 1 \leq 2 \sum_{i \in F, t \in T} z_{ijt}$; *and*
- *the new solution costs at most twice as much as the original.*

Moreover, one can assume that for all i, j *and* t: $x_{ijt} \leqslant 1$ *and* $\sum_r y_{irt} \leqslant 1$.

2.3   Our Main Results

Let us first recall that thanks to a standard reduction from the Set Cover problem (folklore) to the (static) Facility Location problem, the Dynamic Sum-Radii Clustering problem has no $(1 - o(1)) \ln n$-approximation unless $P = NP$.

We then present three algorithms for the DSRC problem. Algorithms 1 and 2 (Sects. 3.1 and 3.2) allow us to obtain a randomized approximation with optimal approximation ratio $\Theta(\ln n)$ for the general (non-metric) case:

Theorem 1 (Algorithm). *With probability at least* $1/4$, *Algorithm 2 (page 7) outputs in polynomial expected time a valid solution to the DSRC problem, with cost at most* $8 \ln(4n) \cdot \text{OPT}$.

Note that the success probability and the approximation ratio can be improved by independent executions of the algorithm. The techniques in Sect. 3.2 also apply to the algorithm in [10] in the non-metric setting, improving its approximation ratio from $\Theta(\log nT)$ to $\Theta(\log n)$. We then turn to the metric case and propose a candidate approximation algorithm based on the work [11], but show, by exhibiting a hard metric instance family, that its approximation ratio is no better than $\Omega(\ln \ln n)$ for the sum-of-radii objective.

Theorem 2 (Hard metric instance). *There is a metric instance family for which the Sum-of-radii ANS algorithm (Algorithm 3, page 8) outputs solutions with cost* $\Omega(\log \log n) \, \text{OPT}$ *w.h.p.*

3   Tight Approximation Algorithm for the General Case

3.1 $O(\log(nT))$-Approximation

As in [10], the first step consists in preprocessing an optimal solution to the LP in order to determine when clients should change the facility they're assigned to. Lemma 1 allows us to focus only on the opening cost within each time interval I_{jk} independently for each client j. Indeed, if one can assign a unique facility φ_{jk} to client j during each interval I_{jk}, then the changing cost for j is at most the number of intervals minus one times g. As Lemma 1 ensures that for all $j \in C$: $\ell_j - 1 \leq 2 \sum_{i \in F, t \in T} z_{ijt}$, the resulting changing cost is at most twice the amount paid by the optimal solution in the original LP. It is worth noting, though, that the intervals are not the same for each client and are not synchronized. The dynamic dimension of the problem is hence simplified but not eliminated.

From now on, we can assume that the clients don't change facilities inside each of their intervals (which is verified by our algorithms). Hence, we shall focus on deciding which facilities to open, when, and with which radius, and how to assign each client to one of them during each of their time intervals. Algorithm 1 does that by combining $\log nT$ partial solutions, each of expected cost LP and obtained by opening a set of random facilities according the y_{irt}.

Algorithm 1. $O(\log nT)$-approximation

Preprocess an optimal solution to LP (1) to obtain a feasible solution (x, y, z) as in Lemma 1.

Let $Z = \ell_1 + \cdots + \ell_n$ be the total number of time intervals I_{jk} associated to (x, y, z) by Lemma 1.

Set $r_{it} := -\infty$ for all $i \in F$ and $t \in [T]$.

repeat $\ln(2Z)$ *times*

    for *each facility i* **do**

        Draw a random variable Y_i uniformly and independently in $[0, 1]$.

        for *every time t* **do**

            Let $\rho_{it} := \max\{\rho : \sum_{r \geqslant \rho} y_{irt} \geqslant Y_i\}$ ($\rho_{it} = -\infty$ if the set is empty)

            Set $r_{it} := \max(r_{it}, \rho_{it})$ and open Facility i with radius r_{it} at time t if $r_{it} \geqslant 0$.

    for *each client j and time interval I_{jk} during which j is not yet covered* **do**

        Connect j to any open facility i (if there is one) that covers j during the whole time interval I_{jk} (i.e., s.t. $d_{ijt} \leqslant r_{it}$ for all $t \in I_{jk}$).

We first analyse the cost of the algorithm, then prove that the solution is indeed correct.

Lemma 2. *The expected increase in total opening cost at each iteration of the repeat loop is at most $\sum_{irt} y_{irt}(f_{it} + r)$.*

Proof. The probability that Facility i is open with radius r at each iteration of the repeat loop is: $\Pr\{\rho_{it} = r\} = \Pr\{Y_i \leqslant \sum_{\rho \geqslant r} y_{i\rho t} \text{ and } Y_i > \sum_{\rho > r} y_{i\rho t}\} = \Pr\{Y_i \in (\gamma, \gamma + y_{irt}]\} = y_{irt}$ where $\gamma = \sum_{\rho > r} y_{i\rho t}$ and recalling that $\gamma + y_{irt} \leqslant 1$ by Lemma 1. It follows that the expected opening cost for Facility i at time t is precisely $\sum_r y_{irt}(f_{it} + r)$. As the radius of each facility i increases by at most ρ_{it} at each iteration of the repeat loop, the expected total added opening cost of each loop is thus at most: $\sum_{it} \sum_r y_{irt}(f_{it} + r)$.

Lemma 3. *For each client j and each time interval I_{jk}, at the end of each iteration of the repeat loop, the probability that j is not covered during I_{jk} is at most $1/e$.*

Proof. Fix a client j and a time t. Client j is covered if there is an open facility i with radius at least d_{ijt}, i.e. s.t. $Y_i \leqslant \sum_{r \geqslant d_{ijt}} y_{irt}$. As $x_{ijt} \leqslant \sum_{r \geqslant d_{ijt}} y_{irt}$ by constraint (1.b), j is thus covered by i as soon as $Y_i \leqslant x_{ijt}$ which happens with probability x_{ijt}. As the Y_is are independent, j is not covered by any facility at time t with probability at most $\prod_i (1 - x_{ijt}) \leqslant \left(1 - \sum_i x_{ijt}/m\right)^m \leqslant (1 - 1/m)^m \leqslant 1/e$ by concavity of the logarithm and constraint (1.a). Since the x_{ijt}s are constant for $t \in I_{jk}$, this also bounds from above the probability that j is not covered during the whole time interval I_{jk}.

Theorem 3. *With probability $1/4$, Algorithm 1 outputs a valid assignment of clients to open facilities with cost at most:*

$$8\ln(2Z) \cdot \mathrm{LP} \leqslant 8\ln(2Z) \cdot \mathrm{OPT} \leqslant 8\ln(2nT) \cdot \mathrm{OPT}.$$

Proof. As the iterations of the repeat loops are independent, each client j has a probability at most $1/e^{\ln(2Z)} = 1/2Z$ of not being covered during each interval I_{jk}. The union bound taken over all intervals I_{jk} ensures that the probability that some client is not covered at some time t by an open facility is at most $Z/2Z = 1/2$ at the end of the algorithm. Let A be the event that all clients are covered at all time steps by the assignment φ computed by Algorithm 1, and $\bar{A}$ its complementary event. Then, the $\mathbb{E}[\mathrm{cost}(\varphi)|A] = (\mathbb{E}[\mathrm{cost}(\varphi)] - \mathbb{E}[\mathrm{cost}(\varphi)|\bar{A}] \Pr\bar{A})/\Pr A \leqslant \mathbb{E}[\mathrm{cost}(\varphi)]/\Pr A \leqslant 2 \cdot \ln(2Z) \cdot 2\,\mathrm{LP}$ by the previous lemmas. By Markov's inequality, we conclude that with probability at least $1/4$, Algorithm 1 produces a valid assignment of the clients to open facilities with total cost at most $2 \cdot 4\ln(2Z)\,\mathrm{LP} \leqslant 8\ln(2nT)\,\mathrm{OPT}$, since $Z \leqslant nT$ obviously.

3.2 $O(\log n)$-Approximation

Concatenating two partial assignments around time t does not change the opening cost of each partial assignment and increases the changing cost by at most $g \cdot n$. We can greedily split the instance into several time periods, making sure that at least n and no more than $2n$ intervals I_{jk} end in each time period (except for the last). Doing so, the cost of stitching together two consecutive partial assignments is at most $n \times g$, hence no higher than the changing cost already paid within each part. By running Algorithm 1 on each partial solution corresponding to a time period and stitching the different solutions, we at most double the changing cost, increasing the bound to $4\,\mathrm{changeCost}(x, y, z)$. On each time period with T' intervals, the opening cost is at most $8\ln(2T')\,\mathrm{openCost}(x, y, z)$, with $T' \le 2n$. This implies that the overall approximation ratio is $8\ln(4n)$ for Algorithm 2 on the facing page, proving Theorem 1.

Note that this technique also applies to the algorithm in [10], improving the approximation ratio in the non-metric hourly sum-of-distances setting from $O(\log nT)$ to $O(\log n)$.

Algorithm 2. Batch $O(\log n)$-approximation

Preprocess an optimal solution to LP (1) to obtain a feasible solution (x, y, z) as in Algorithm 1.

if $Z \le 2n$ **then**

  Run Algorithm 1

else

  Partition time greedily into Q periods $U_q = [t_q, t_{q+1})$ where Q and $(t_q)_{q \in [Q+1]}$ are defined as follows: $t_1 = 1$, and t_q is defined inductively as the largest $t \le T$ such that at most n intervals I_{jk} end between t_{q-1} and $t-1$. Set $t_{Q+1} = T+1$.

  for $q = 1..Q$ **do**

    Run several times Algorithm 1 with (x, y, z) on the instance restricted to time period U_q until it outputs a valid solution with opening cost at most $8\ln(4n)\,\mathrm{openCost}_{U_q}(x, y, z)$.

  Output the concatenation of the computed assignments in each time period U_q.

Proof (Proof of Theorem 1). Assume $Z > 2n$. As the instance restricted to interval U_q contains $Z_q \le 2n$ overlapping intervals I_{jk}, Algorithm 1 outputs a solution for this restriction with opening cost at most $8\ln(4n)\,\mathrm{openCost}_{U_q}(x, y, z)$ with probability at least $1/4$. It follows that Algorithm 1 is run at most four times on expectation for each q, hence the polynomial expected time. The changing cost paid for the concatenating of the solutions is then at most:

$$g \cdot (Z_1 + \cdots + Z_Q + n(Q-1)) \le g(Z + n \cdot \tfrac{Z}{n}) \le 3g(Z-n) \le 6\,\mathrm{changeCost}(x, y, z)$$

It follows that the solution output by Algorithm 2 costs at most:

$$8\ln(4n)\,\mathrm{openCost}_{U_q}(x, y, z) + 6\,\mathrm{changeCost}(x, y, z) \le \max(6, 8\ln(4n))\,\mathrm{LP}$$
$$\le 8\ln(4n)\,\mathrm{OPT}.$$

4   Lower Bounds for the Metric Case

In this section, we focus on the *metric* case, i.e. where the distances d_{xyt} (with $x, y \in F \cup C$) verify the triangle inequalities at all times. Exploiting this additional property, [11] proposed an $O(1)$-approximation (referred to here as the

ANS algorithm) for the Metric Dynamic Facility Location problem with the *sum-of-distances* objective. For the *sum-of-radii* objective studied here, it is unclear whether an $O(1)$-approximation exists when the distances are metric. Indeed, we were not able to obtain such an $O(1)$-approximation algorithm for metric DSRC. However, we show in this section that the natural adaptation of the ANS algorithm to the sum-of-radii setting cannot achieve any approximation ratio better than $\Omega(\ln \ln n)$ by exhibiting a hard metric instance family. This example demonstrates that the main issue is that clients have to collaborate to make the right choices in order to avoid rare errors that would be absorbed by the sum-of-distances objective but not by the sum-of-radii objective.

Adapting the ANS Algorithm. The original ANS algorithm preprocesses the solution of the LP further so that every variable in the LP only takes one positive value besides 0. This is obtained by duplicating each facility at most nT times, so that only one client x_{ijt}-variable contributes to each of the copies of the y_{irt}-variables and for one radius r only.

Lemma 4 [11]. *Given an optimal solution (x^*, y^*, z^*) to LP (1), one can compute an equivalent instance together with a feasible solution (x', y', z') to the corresponding LP s.t.:*

- *each facility i is replaced in the new instance by a set of (at most nT) virtual facilities located at the same position as i at all times and with opening cost f_{it}; and*
- *(x', y', z') verifies the properties in Lemma 1; and*
- *for each virtual facility i', there is a constant $c_{i'}$ and a client j such that for all time steps t, $x'_{i'jt} \in \{0, c_{i'}\}$, $y'_{i',d_{ijt},t} \in \{0, c_{i'}\}$ and $y'_{i'rt} = 0$ for all $r \neq d_{ijt}$; and*
- *the solution to the original LP is obtained for each facility by summing up the fractional solutions over its virtual copies.*

Algorithm 3 on page 8 presents the transcription of the ANS algorithm to the sum-of-radii objective. The only difference lies in using LP (1) instead of the linear program with the sum-of-distance objective in [11].

4.1   A Hard Instance Family

The key to the performance of the ANS algorithm for the sum-of-distances objective in [11] is that the Y_is and X_js drop exponentially when one follows the directed path originating from a client, which ensures that just enough facilities are open, and that all the clients are a constant factor away from their ideal facility on expectation. Deviations from the expectation are absorbed by the summation in the objective. In the following, we will exhibit a metric instance showing that the adaptation to the sum-of-radii objective (Algorithm 3) cannot obtain an approximation ratio better than $\Omega(\ln \ln n)$.

Algorithm 3. Sum-of-radii ANS algorithm (from [11])

Preprocess an optimal solution to LP (1) according to Lemma 4.
For each virtual facility $i' \in F'$: draw a random variable $Y_{i'}$ according to the
exponential distribution of parameter $c_{i'}$ independently.
For each client j: draw a uniform random variable X_j from $[0, 1]$ independently.
for *each time step* $t \in [T]$ **do**

> Starting from an empty bipartite Clients-Facilities graph G_t:
> - add an arc from each client j to the facility i' with minimal $Y_{i'}$ among those
>   with $x_{i'jt} > 0$;
> - add an arc from each facility i' to the client j with smallest X_j among those
>   with $x_{i'jt} > 0$.
>
> Open at time t every facility whose virtual copy belongs to a circuit in G_t with
> the corresponding radius, and assign each client j to the open facility at the
> end of the directed path originating from j enlarging its radius accordlingly.

The Static Arborescent Instance T_h. Consider for now the *static* (one time-step)
instance T_h where the metric distance is defined by the L_∞ norm over $\mathbb{R}^h$, where
the facilities are positioned at $(\pm 2^0, \pm 2^{-1}, \dots, \pm 2^{-k+1}, 0, \dots, 0)$ for $0 \leqslant k < h$
and where the clients are positioned at $(\pm 2^0, \pm 2^{-1}, \dots, \pm 2^{-h+1})$. Facilities with
coordinates in $(\pm 2^0, \pm 2^{-1}, \dots, \pm 2^{-k+1}, 0, \dots, 0)$ are said to be of *level* k; there
are 2^k of them. We denote by λ_i the level of Facility i. We organize the instance
as a tree by declaring that the client or facility located at $u = (u_1, \dots, u_h)$ is
a *descendant* of all the facilities located at $(u_1, \dots, u_k, 0, \dots, 0)$ for $0 \leqslant k < h$.
The instance T_h consists thus of $n = 2^h$ clients and $m = 2^h - 1$ facilities. The
distance between any two locations u and v in the tree is equal to 2^{-k+1} where
k is the level of their lowest common ancestor. All facilities have zero opening
cost. Figure 1 shows a flat representation of T_5.

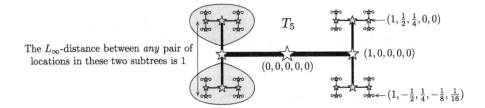

The L_∞-distance between *any* pair of locations in these two subtrees is 1

T_5 $(1, \frac{1}{2}, \frac{1}{4}, 0, 0)$

$(0, 0, 0, 0, 0)$ $(1, 0, 0, 0, 0)$

$(1, -\frac{1}{2}, \frac{1}{4}, -\frac{1}{8}, \frac{1}{16})$

Fig. 1. A flat representation of the instance T_5 where each level of the tree lies in
a different dimension. The clients and facilities are represented by circles and stars
respectively. The levels of the facilities are represented by stars of decreasing size and
edges of decreasing thickness.

Lemma 5 (Proof omitted). *The optimal solutions to LP (1) for the static
instance T_h have value 1 and the uniform solution, which opens a fraction $1/h$ of
every facility i with radius $2^{-\lambda_i}$ and assigns each client to each of the h facilities
covering it with fraction $1/h$, is optimal.*

We will first show that the adaptation of the ANS algorithm outputs a solution with cost $\Omega(\log \log n) \gg 1$ w.h.p. when presented with the uniform solution to LP (1) for T_h, and then show how to design a *dynamic* instance that forces LP (1) to output this uniform solution.

Running the Adapted ANS Algorithm on T_h with the Uniform Solution. The preprocessing leaves the uniform solution unchanged and the random variables Y_i are i.i.d. according to an exponential law of parameter $1/h$. In order to improve readability, let us introduce $U_i = 1 - \exp(-Y_i/h)$ so that the U_is are uniformly distributed and ordered as the Y_is. The arcs in the graph G built by the algorithm at time 1 then consist of an arc for each client j, pointing to its ancestor facility i with the smallest U_i, and of an arc for each facility i, pointing to its descendant client j with the smallest X_j.

Lemma 6 (Proof omitted). *The directed paths starting from a client in G have length at most 4 as illustrated by Fig. 2.*

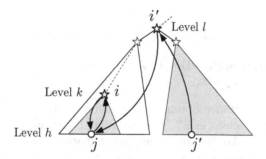

Fig. 2. Paths in the graph G built by ANS algorithm from the uniform solution for T_h.

To prove the $\Omega(\ln \ln n)$ lower bound (conditioned to the production of the uniform solution when solving LP (1)), we first need a combinatorial lemma. Let's consider a complete rooted binary tree A_q of height q where each node is labelled by a uniform random real chosen from $[0, 1]$ independently.

Lemma 7 (Proof omitted). *The probability $p_q(x)$ that there is a branch in A_q where all the nodes have label $> x$ verifies:*

- *if $x < \frac{1}{2}$, then $2 - \frac{1}{1-x} < p_q(x) < 2 - \frac{1}{1-x} + \frac{(2x)^{q+2}}{4(1-x)}$ and $p_q(x) \searrow 2 - \frac{1}{1-x}$.*
- *if $x > \frac{1}{2}$, then $0 < p_q(x) < (1-x)(2(1-x))^q$ and $p_q(x) \searrow 0$.*

W can prove the following:

Lemma 8 (Proof omitted). *The expected opening cost of a facility i of level k is at least $2^{-k}(\ln k - \beta)/8h$ for a universal constant β.*

Which allows us to conclude that:

Lemma 9. *The expected opening cost of the solution output by the sum-of-radii ANS Algorithm 3 from the uniform solution to LP (1) for T_h is $\Omega(\ln \ln n)$.*

Proof. By linearity of expectation and the lemmas above, $\mathbb{E}[\text{openCost}] \geqslant \sum_{k=10}^{h} 2^k \cdot 2^{-k} (\ln k - \beta)/8h = \Theta(\sum_{k=1}^{h} (\ln k)/h) = \Theta(\ln h) = \Theta(\ln \ln n)$ since $n = 2^h$.

4.2   Forcing the Uniform Fractional LP Solution

Our lower bound for the T_h instance relies on running the algorithm on the uniform solution. Unfortunately, this solution is not a vertex of LP (1) and will not be output by any linear solver. We thus extend the instance T_h to a *dynamic* instance D_h whose optimal solution is unique and uniform, concluding the proof of Theorem 2. This D_h instance consists in several initial time steps with very low cost where the clients and facilities are mixed together (enforcing then the need for uniformity in the optimal solution), and a final time step equivalent to T_h.

Lemma 10 (Proof omitted).   *All optimal solutions to the instance D_h are uniform on the last time step.*

We can now conclude the proof of Theorem 2 through two corollaries.

Corollary 1 (Proof omitted).   *Algorithm 3 produces the same output for the last time step of D_h as for T_h.*

Let $D_h^{n^2}$ be the instance obtained by making n^2 independent copies of D_h located at distant locations in $\mathbb{R}^h$. The Hoeffding bound allows us to strengthen the result above by showing that the approximation ratio sum-of-radii ANS Algorithm 3 on this new instance is at least $\Omega(\ln \ln n)$ with high probability, when run from the uniform solution to LP (1):

Corollary 2.   *The opening cost of the solution output by the sum-of-radii ANS Algorithm 3 from the uniform solution to LP (1) for $D_h^{n^2}$ is $\Omega(\ln \ln n)$ with probability $1 - 2^{-n}$.*

Proof. We directly apply the Hoeffding bound, observing that the cost of the solution output by sum-of-radii ANS Algorithm 3 on D_h is at most twice the cost on T_h, hence at most $O(\log n)$.

5   Conclusion and Open Problems

We have obtained an asymptotically optimal $O(\log n)$-approximation algorithm for DSRC in the general case, with a technique that translates to the sum-of-distances case. We have also shown that the approximation ratio for the algorithm in [11] is no better than $\Omega(\ln \ln n)$ for metric instances. This leaves open the question of whether an $O(1)$-approximation algorithm exists in the metric case. Further experimental work has to be conducted to evaluate how these algorithms can help improve the representation of real dynamic graphs such as the ones in [2]. One final remark is that our algorithms all rely on the primal formulation of LP (1) while the algorithms in [8] for the static setting rely on the dual. Unfortunately, the dual variables seem to act evasively with respect to time in the dynamic setting. Understanding these dual variables is a promising direction towards an $O(1)$-approximation, if it exists.

References

1. Newman, M.E.J.: The structure and function of complex networks. SIAM Rev. **45**(2), 167–256 (2003)
2. Stehlé, J., Voirin, N., Barrat, A., Cattuto, C., Isella, L., Pinton, J., Quaggiotto, M., den Broeck, W.V., Régis, C., Lina, B., Vanhems, P.: High-resolution measurements of face-to-face contact patterns in a primary school. PLoS ONE **6**(8), e23176 (2011)
3. Sundaresan, S.R., Fischhoff, I.R., Dushoff, J., Rubenstein, D.I.: Network metrics reveal differences in social organization between two fission-fusion species, grevy's zebra and onager. Oecologia **151**(1), 140–149 (2007)
4. Pastor-Satorras, R., Vespignani, A.: Epidemic spreading in scale-free networks. Phys. Rev. Lett. **86**, 3200–3203 (2001)
5. Tantipathananandh, C., Berger-Wolf, T.Y., Kempe, D.: A framework for community identification in dynamic social networks. In: SIGKDD, pp. 717–726 (2007)
6. Fotakis, D., Koutris, P.: Online sum-radii clustering. In: Rovan, B., Sassone, V., Widmayer, P. (eds.) MFCS 2012. LNCS, vol. 7464, pp. 395–406. Springer, Heidelberg (2012). doi:10.1007/978-3-642-32589-2_36
7. Li, S.: A 1.488 approximation algorithm for the uncapacitated facility location problem. Inf. Comput. **222**, 45–58 (2013)
8. Charikar, M., Panigrahy, R.: Clustering to minimize the sum of cluster diameters. J. Comput. Syst. Sci. **68**(2), 417–441 (2004)
9. Guha, S., Khuller, S.: Greedy strikes back: improved facility location algorithms. J. Algorithms **31**(1), 228–248 (1999)
10. Eisenstat, D., Mathieu, C., Schabanel, N.: Facility location in evolving metrics. In: Esparza, J., Fraigniaud, P., Husfeldt, T., Koutsoupias, E. (eds.) ICALP 2014. LNCS, vol. 8573, pp. 459–470. Springer, Heidelberg (2014). doi:10.1007/978-3-662-43951-7_39
11. An, H., Norouzi-Fard, A., Svensson, O.: Dynamic facility location via exponential clocks. In: SODA, pp. 708–721 (2015)
12. Fernandes, C.G., Oshiro, M.T., Schabanel, N.: Dynamic clustering of evolving networks: some results on the line. In: AlgoTel, pp. 1–4, May 2013
13. Mahdian, M., Ye, Y., Zhang, J.: Approximation algorithms for metric facility location problems. SIAM J. Comput. **36**(2), 411–432 (2006)
14. Behsaz, B., Salavatipour, M.R.: On minimum sum of radii and diameters clustering. Algorithmica **73**(1), 143–165 (2015)
15. Hochbaum, D.S.: Heuristics for the fixed cost median problem. Math. Program. **22**(1), 148–162 (1982)
16. Dinur, I., Steurer, D.: Analytical approach to parallel repetition. In: STOC, pp. 624–633 (2014)
17. Lee, Y.T., Sidford, A.: Path finding methods for linear programming: solving linear programs in Õ(vrank) iterations and faster algorithms for maximum flow. In: Proceedings of the 2014 IEEE 55th Annual Symposium on Foundations of Computer Science, FOCS 2014, pp. 424–433. IEEE Computer Society, Washington (2014)

How to Extend Visibility Polygons by Mirrors to Cover Invisible Segments

Arash Vaezi[1](✉) and Mohammad Ghodsi[1,2]

[1] Department of Computer Engineering, Sharif University of Technology,
Tehran, Iran
avaezi@ce.sharif.edu, ghodsi@sharif.edu
[2] Institute for Research in Fundamental Sciences (IPM), Tehran, Iran

Abstract. Given a simple polygon $\mathcal{P}$ with n vertices, the visibility polygon (VP) of a point q $(VP(q))$, or a segment $\overline{pq}$ $(VP(\overline{pq}))$ inside $\mathcal{P}$ can be computed in linear time. We propose a linear time algorithm to extend VP of a viewer (point or segment), by converting some edges of $\mathcal{P}$ into mirrors, such that a given non-visible segment $\overline{uw}$ can also be seen from the viewer. Various definitions for the visibility of a segment, such as weak, strong, or complete visibility are considered. Our algorithm finds every edge such that, when converted to a mirror, makes $\overline{uw}$ visible to our viewer. We find out exactly which interval of $\overline{uw}$ becomes visible, by every edge middling as mirror, all in linear time.

1 Introduction

Many variations of visibility polygons have been studied so far. In general, we have a simple polygon $\mathcal{P}$ with n vertices, and a viewer which is a point (q), or a segment $(\overline{pq})$ inside $\mathcal{P}$. The goal is to find the maximal sub-polygon of $\mathcal{P}$ visible to the viewer $(VP(q)$ or $VP(\overline{pq}))$. There are linear time algorithms to compute $VP(q)$ [7] or when the viewer is a segment [5].

It was shown in 2010 that VP of a given point or segment can be computed in presence of *one* mirror-edge in $O(n)$ [6]. Also, it was shown in the same paper that the union of two visibility polygons can be computed in $O(n)$.

We consider different problems of finding every edge e such that when converted to a mirror (and thus called *mirror-edge*) can make at least a part of a specific invisible segment visible (also called *e-mirror-visible*) to a given point or segment. We propose linear time algorithms for these problems. Considering a segment as a viewer, we deal with all different definitions of visibility, namely, weak, complete and strong visibility, which was introduced by [3]. Also, we can easily find mirror-visibile intervals of the invisible segment $(\overline{uw})$ considering all edges as mirrors in linear time corresponding to the complexity of $\mathcal{P}$.

This paper is organized as follows: In Sect. 2, notations are described. Next in Sect. 3, we present a linear time algorithm to recognize every mirror-edge e

M. Ghodsi—This author's research was partially supported by the IPM under grant No: CS1392-2-01.

S.-H. Poon et al. (Eds.): WALCOM 2017, LNCS 10167, pp. 42–53, 2017.
DOI: 10.1007/978-3-319-53925-6_4

of $\mathcal{P}$ that makes a given segment $\overline{uw}$ e-mirror-visible to q. In Sect. 4 we will show that e-mirror-visible interval of $\overline{uw}$ to q can be computed in constant time. In Sect. 5, we deal with a given segment instead of a point. And finally, Sect. 6 contains some discussions and future works.

2   Notations and Assumptions

Suppose $\mathcal{P}$ is a simple polygon and $int(\mathcal{P})$ denotes its interior. Two points x and y are visible to each other, if and only if the open line segment $\overline{xy}$ lies completely in $int(\mathcal{P})$. The visibility polygon of a point q in $\mathcal{P}$, denoted as $VP(q)$, consists of all points of $\mathcal{P}$ visible to q. Edges of $VP(q)$ that are not edges of $\mathcal{P}$ are called *windows*. The weak visibility polygon of a segment $\overline{pq}$, denoted as $WVP(\overline{pq})$, is the maximal sub-polygon of $\mathcal{P}$ visible to at least one point (not the endpoints) of $\overline{pq}$. The visibility of an edge $e = (v_i, v_{i+1})$ of $\mathcal{P}$ can be viewed in different ways [3]: $\mathcal{P}$ is said to be *completely visible* from e if for every point $z \in \mathcal{P}$ and for any point $w \in e$, w and z are visible (denoted as CVP short from completely visible polygon). Also, $\mathcal{P}$ is said to be *strongly visible* from e if there exists a point $w \in e$ such that for every point $z \in \mathcal{P}$, w and z are visible (SVP). These different visibilities can be computed in linear time (see [5] for WVP and [3] for CVP and SVP).

Suppose an edge e of $\mathcal{P}$ is a mirror. Two points x and y are *e-mirror-visible*, if and only if they are directly visible with one specular reflection through a mirror-edge e. Specular reflection is the mirror-like reflection of light from a surface, in which light from a single incoming direction is reflected into a single outgoing direction. The direction in which light is reflected is defined by the law-of-reflection, which states that the incident, surface-normal and reflected directions are coplanar [2].

Since only an interval of a mirror-edge is useful, we can consider the whole edge as a mirror, and there is no need to split an edge.

We assume that n vertices of $\mathcal{P}$ are ordered in clockwise order (CWO).

3   Expanding Point Visibility Polygon

We intend to find every mirror-edge e of $\mathcal{P}$ that causes a given point q see any interval of a given segment $\overline{uw}$ inside $\mathcal{P}$. We will find the exact interval of $\overline{uw}$ which is e-mirror-visible to q for every mirror-edge e of $\mathcal{P}$ in the next section.

3.1   Overview of the Algorithm

Obviously, any potential mirror-edge e that makes $\overline{uw}$ visible to q should lie on $VP(q) \cap WVP(\overline{uw})$ which can be computed in linear time. If the goal is to check e-mirror-visibility of the whole $\overline{uw}$, we should instead compute the complete visibility polygon of $\overline{uw}$ (i.e. $VP(q) \cap CVP(\overline{uw})$).

Suppose that e is intersected by $VP(q) \cap WVP(\overline{uw})$ from $v_1(e)$ to $v_2(e)$ in CWO. We use this part of e as mirror. We will find out whether any part of $\overline{uw}$

is e-mirror-visible. Let $L_1(e)$ and $L_2(e)$ be two half-lines from the ray-reflection of q at $v_1(e)$ and $v_2(e)$ respectively. Also let $q'(e_i)$ be the image of q considering e_i from $v_1(e_i)$ to $v_2(e_i)$ (e_i is the ith potential mirror-edge in CWO).

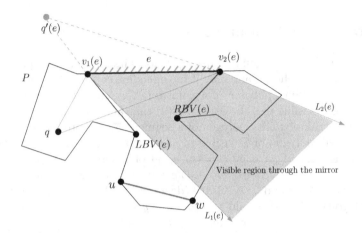

Fig. 1. The region between $L_1(e)$ and $L_2(e)$ is the visible area by q through e being a mirror from $v_1(e)$ to $v_2(e)$.

If $\overline{uw}$ intersects the region between $L_1(e)$ and $L_2(e)$ and no part of $\mathcal{P}$ obstructs $\overline{uw}$, then $\overline{uw}$ is e-mirror-visible (see Fig. 1). Since $\mathcal{P}$ is simple, e-mirror-visibility can only be obstructed by reflex vertices.

For each mirror-edge e, we define $LBV(e)$ (for Left Blocking Vertex of e) and $RBV(e)$ (for Right Blocking Vertex of e) as below. In Subsect. 3.2, we will prove that no other reflex vertex can block e-mirror-visibility area except for these two reflex vertices.

3.2 *LBV*s and *RBV*s

Definition 1. *Assume that $p_1, p_2, \ldots, p_k$ are the reflex vertices we meet when tracing $WVP(\overline{uw})$ starting from u in CWO before we reach a mirror-edge e. We define $LBV(e)$ to be that vertex p_j such that if $\overline{p_j q'(e)}$ (i.e. from p_j to $q'(e)$) holds all other p_i ($i \neq j$ $1 \leq i \leq k$) reflex vertices on its left side. In another word, if we move from p_j to $q'(e)$ all other p_i reflex vertices are on our left side (see Fig. 2(a)). If more than one vertex has this property, we choose the one with the lowest index. If no such vertex exits, we set $v_1(e)$ as $LBV(e)$. $RBV(e)$ is defined similarly when we trace $WVP(\overline{uw})$ from w in counter-clockwise order (CCWO).*

Different mirror-edges may have the same LBVs or RBVs. And, obviously through Definition 1, for each mirror-edge e, $LBV(e)$ and $RBV(e)$ is unique.

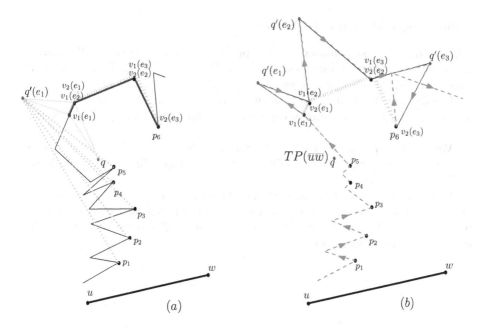

Fig. 2. (a) From Definition 1, vertex p_5 is $LBV(e_1)$. (b) Constructing $TP(\overline{uw})$, which is useful to distinguish LBV vertices for all mirror-edges. p_1, p_2, ... , p_6 are the reflex vertices of $\mathcal{P}$.

Algorithm 1 (to check whether q can see any interval of $\overline{uw}$ through mirror-edge e). Assuming that e, from $v_1(e)$ to $v_2(e)$ is in $VP(q) \cap WVP(\overline{uw})$ and $L_1(e)$ and $L_2(e)$ are as defined above, the following cases are considered:

1. If $L_1(e)$ and $L_2(e)$ both lie in one side of $\overline{uw}$, then $\overline{uw}$ is not in the e-mirror-visible area. That is, q cannot see $\overline{uw}$ through e.
2. Otherwise, if $\overline{uw}$ is between $L_1(e)$ and $L_2(e)$. I.e., it is in the middle of the mirror-visible area, q can see $\overline{uw}$ through the mirror-edge e. Because e is visible to $\overline{uw}$, and the visibility area from $L_1(e)$ to $L_2(e)$ is a continuous region.
3. Otherwise, $L_1(e)$ or $L_2(e)$ crosses $\overline{uw}$. In this case, we check whether any part of $\mathcal{P}$, obstructs *the whole* visible area through e (In case of $CVP(\overline{uw})$, it is sufficient to check $L_1(e)$ and $L_2(e)$ not to cross $\overline{uw}$, except in its endpoints.) For this, it is checked whether $\mathcal{P}$ blocks the rays from the right or left side of e. If $LBV(e)$ lies on the left side of $L_2(e)$, and $RBV(e)$ lies on the right side of $L_1(e)$, then q can see $\overline{uw}$ through e.
   Otherwise, q and $\overline{uw}$ are not e-mirror-visible.

Obviously, collision checking of a constant number of points and lines can be done in $O(1)$ for any mirror-edge.

Computing LBV and RBV Vertices

Algorithm 2. First consider the computation of LBV vertices. We already know that the potential mirror-edges lie in $VP(q) \cap WVP(\overline{uw})$. To make an easy understanding, these edges are numbered in CWO as $e_1, e_2, \ldots$

Considering $WVP(\overline{uw})$ we construct a new polygon by adding $\overline{q'(e_i)v_1(e_i)}$ and $\overline{q'(e_i)v_2(e_i)}$ to each mirror-edge e_i, and eliminating $\overline{v_1(e_i)v_2(e_i)}$ interval from e_i.

We call this polygon $TP(\overline{uw})$ (Tracing Polygon). Obviously, $TP(\overline{uw})$ may not be a simple polygon, and has $O(n)$ vertices corresponding to the complexity of $\mathcal{P}$ (see Fig. 2(b)).

Starting from u (the left endpoint of $\overline{uw}$) we trace $TP(\overline{uw})$ in clockwise order. While doing so, we construct a convex shape on the reflex vertices of $\mathcal{P}$ we visit, using an algorithm similar to Graham's scan [4] in $\mathcal{P}$'s order of vertices. We consider u as one reflex vertex.

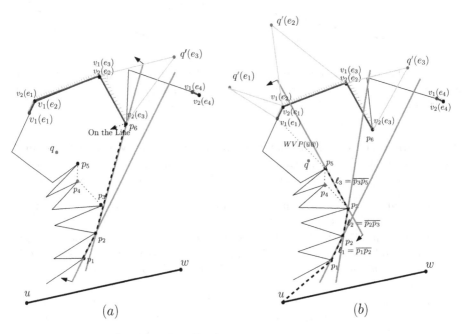

Fig. 3. (a) Updating the convex shape while tracing $TP(\overline{uw})$ and facing with new reflex vertices. p_5 is chosen as $LBV(e_1)$, p_3 and p_2 as $LBV(e_2)$ and $LBV(e_3)$, respectively. If we consider u for the fourth mirror-edge, first we select p_1. But, later we should change $LBV(e_4)$ to be $v_1(e_4)$, because p_1 cannot block the e_4-mirror-visibility. (b) Constructing the convex hull to distinguish LBV vertices for all mirror-edges. p_1, p_2, p_3 and p_5 are the reflex vertices that are used in the convex hull construction. Four mirror-edges e_1 to e_4 are shown. In this figure, p_5 is $LBV(e_1)$.

As we meet a new reflex vertex, we push the line containing the new constructed edge of the convex shape into a stack named S and update the stack as we move forward. When $q'(e_i)$ of a mirror-edge e_i is reached in our trace, $q'(e_i)$ is compared with the line on the top of the stack called ℓ. If $q'(e_i)$ lies on the right side of ℓ, ℓ is popped from S. Otherwise, if $q'(e_i)$ lies on, or on the left side of ℓ, then we assign ℓ as the *chosen line* for e_i, denoted as $cl(i)$. ℓ (Top(S)) is then check with $q'(e_{i+1})$, $q'(e_{i+2})$, ..., to become their possible chosen lines, or popped up. If the stack is empty when we visit $q'(e_i)$, we assign $LBV(e_i) = v_1(e_i)$.

Obviously, when a new reflex vertex is met, the convex shape and the stack is updated accordingly (see Fig. 3(a)) and the algorithm continues.

See Fig. 3(b), for an example. Here, the stack contains 3 lines (ℓ_1, ℓ_2, ℓ_3, the last on top) when we reach $v_1(e_1)$. We check $q'(e_1)$ with ℓ_3, which is on Top(S), to see if it has $q'(e_1)$ on its left. If $q'(e_1)$ lies on the right, then $q'(e_1)$ is checked with ℓ_2. Here, $cl(1) = \ell_3$.

At the end, for each mirror-edge e_i, we consider the two reflex vertices of $cl(i)$, say re_1 and re_2 (in $CCWO$). If $q'(e_i)$ lies on the left of $cl(i)$, then $LBV(e_i) = re_2$. Otherwise, it lies on $cl(i)$, then $LBV(e_i) = re_1$. If there are more than two reflex vertices consider the last on $cl(i)$ (re_1).

The RBV vertices are computed similarly by tracing $TP(\overline{uw})$ in counterclockwise direction starting from w.

At the end, since there may be some false vertices chosen as LBV or RBV vertices, we will trace $WVP(\overline{uw})$ in both directions to correct these cases. First each $LBV(e_i)$ chosen by previous algorithm is compared with the segment $d = \overline{v_2(e_i)u}$. If $LBV(e_i)$ lies on the left side of d, or if $LBV(e_i) = u$, then $LBV(e_i)$ was falsely chosen since it is not obstructing the mirror-visibility area. The correction is made in this case by setting $LBV(e_i) = v_1(e_i)$. We proceed similarly for RBV vertices in the other direction.

Obviously, all these operations can be performed in $O(n)$ time. For more justification, do not consider the stack and see what happens to the lines (see [1] for more details).

Proof of Correctness and Analysis of the Algorithm. In this subsection we present the proof and the analysis of the algorithm.

Theorem 1. *Suppose $\mathcal{P}$ is a simple polygon with n vertices, q is a given point inside $\mathcal{P}$, and $\overline{uw}$ is a given segment which is not directly visible by q. Every edge e that makes $\overline{uw}$ e-mirror-visible to q can be found in $O(n)$ time.*

Remark 1. We will prove this theorem assuming that $\overline{uw}$ is a diagonal of $\mathcal{P}$. Since the assertion that $\overline{uw}$ is actually a diagonal is not used in the proof, the stated proof holds for any segment inside $\mathcal{P}$. To start tracing $TP(\overline{uw})$, instead of the endpoints of the diagonal, we can use one endpoint of the closest edge of $\mathcal{P}$ to the given segment. Let at least one endpoint of this edge be upon the given segment inside the polygon.

Remark 2. Note that the algorithm covers some situations where $\overline{uw}$ does not have their endpoints on the boundary of $\mathcal{P}$. In these cases there might be some mirror-edge e which can see $\overline{uw}$ from its behind. In another word, e may see a part of the invisible target segment from w to u, and w is on the left side of the e-mirror-visible interval when we are standing on $\overline{uw}$ and facing to e (see Fig. 4(a)). And, we need to swap the position of u and w and run the above-mentioned algorithms one more time to see if these kind of mirror-edges exist that may make an interval of the target mirror-visible to q. So, we need to use $\overline{wu}$ instead of $\overline{uw}$. And, we need to run all above-mentioned algorithms one more time using $\overline{wu}$, which takes an additional $O(n)$ time complexity. Note that these two runs do not have any conflict with each other, and they find absolutely independent mirror-edges. This is because, a mirror-edge e which sees $\overline{uw}$ from behind will be eliminated in the first run. And this is because, in the first run, using $\overline{uw}$, w is placed on the left side of $L_1(e)$, and e will be eliminated through case 1 of Algorithm 1. Without lost of generality, for simplicity we assume that no mirror-edge can see $\overline{uw}$ from behind.

Proof

1. The algorithm correctly computes all LBV's and RBV's in $O(n)$. This is clear from Definition 1 and Algorithm 2. This algorithm constructs two convex hulls.
2. Algorithm 1 correctly checks whether each mirror-edge e can make at least a part of the given segment $\overline{uw}$ e-mirror-visible to q. For this, we only need to prove that the algorithm is correct if case 3 occurs. Other cases are obvious. That is, if $L_1(e)$ or $L_2(e)$ or both cross $\overline{uw}$, and if $LBV(e) = p_j$ does not cross $L_2(e)$ where we decide that $\overline{uw}$ is e-mirror-visible from q, then no other reflex vertices can completely obstruct the e-mirror-visible area. Suppose on the contrary, that another vertex p_l completely obstructs the visible area while p_j does not. In this case, $\overline{q'(e)p_l}$ is on the right side of $L_2(e)$ and thus is on the right side of $\overline{q'(e)p_j}$ which contradicts p_j being $LBV(e)$. Similar arguments hold for RBV. We can also prove that no other reflex vertices (other than the left and right chains that appear when we trace the $WVP(\overline{uw})$) can obstruct the visibility.

4   Specifying the Visible Part of $\overline{uw}$

In this section we present an algorithm to determine the visible interval of the given segment ($\overline{uw}$) which is e-mirror-visible by middling of a given mirror-edge (e).

Lemma 1. *We have a simple polygon $\mathcal{P}$, a point q as a viewer, and a segment $\overline{uw}$, inside $\mathcal{P}$. In linear time corresponding to the complexity of $\mathcal{P}$, for every mirror-edge e, we can compute the exact interval of $\overline{uw}$ that is e-mirror-visible.*

Proof. We will show for a specified mirror-edge e, while we have $LBV(e)$, we can find e-mirror-visible part of $\overline{uw}$ in constant time. Therefore, it takes $O(n)$ time to distinguish the visible intervals of $\overline{uw}$, for every mirror-edge.

Consider a mirror-edge e, without loss of generality suppose we know $\overline{uw}$ is e-mirror-visible. We can find the visible part of $\overline{uw}$ using the following algorithm:

Algorithm 3 (to find the visible part of $\overline{uw}$ through mirror-edge e). Let $u'(e)$ and $w'(e)$ corresponding to u and w, be the endpoints of the visible interval of $\overline{uw}$, respectively.

Note that Algorithm 2 provides all LBV and RBV vertices of all mirror-edges.

1. If $LBV(e) = v_1(e)$: Then the intersection of $L_1(e)$ and $\overline{uw}$ determines $u'(e)$. Clearly, if $L_1(e)$ places in the left side of $\overline{uw}$ then u itself is $u'(e)$.
2. If $LBV(e) \neq v_1(e)$: If $LBV(e)$ does not lie on the right side of $L_1(e)$, then again the intersection of $L_1(e)$ and $\overline{uw}$ determines $u'(e)$. Otherwise, we compute the intersection of the protraction of $\overline{q'(e)LBV(e)}$ and $\overline{uw}$. The intersection point is $u'(e)$.

Acting the same way we can find w'.

Correctness and Analysis of Algorithm 3

First step is obvious because there is nothing to obstruct the mirror-visibility area, and it takes constant time. About the second step, if $LBV(e)$ lies–on or–on the left side of $L_1(e)$, the intersection point of $L_1(e)$ and $\overline{uw}$ is $u'(e)$. Note that we know $L_1(e)$ is not in the right side of w because we knew $\overline{uw}$ is e-mirror-visible to q. If $LBV(e)$ lies on the right side of L_1, then from Definition 1 we know $LBV(e)$ is e-mirror-visible. We only need to prove that the protraction of $\overline{q'(e)LBV(e)}$ determines $u'(e)$. There may be several reflex vertices on the right side of $L_1(e)$. Suppose on the contrary, $u''(e)$, the intersection of $\overline{uw}$ and $\overline{q'p_j}$ ($p_j \neq LBV(e)$ is a reflex vertex on the right side of L_1), is closer to u. Then, the line $\overline{q'p_j u''(e)}$ must be on the right side of $LBV(e)$, which contradicts Definition 1 (see Fig. 4(b)).

Since no direction for $L_1(e)$, or property of q being in the left side of e was used, the same proof holds for $RBV(e)$.

5 Extending a Segment Visibility Polygon

In this section, we deal with different cases of the problem of making two invisible segments mirror-visible to each other.

Lemma 2. *We are given a simple polygon $\mathcal{P}$ and two segments, say $\overline{xy}$ and $\overline{uw}$, inside $\mathcal{P}$. Assume that $\overline{uw}$ is not visible to $\overline{xy}$. For every mirror-edge e, we can find out if $\overline{uw}$ is weakly, completely, or strongly mirror-visible to $\overline{xy}$, in linear time corresponding to the complexity of $\mathcal{P}$.*

Proof. To prove Lemma 2 we simply use Algorithm 1 in Sect. 3. Here, as we deal with a segment as a viewer, we encounter more difficulties than the previous sections. For instance, we need to consider different vertices in place of $v_1(e)$, or

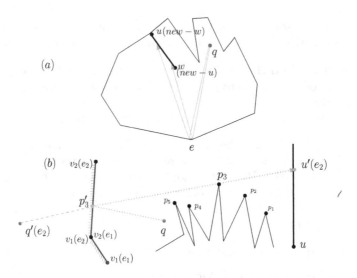

Fig. 4. (a) Mirror-edge e sees $\overline{uw}$ from its behind. And, we need to replace w with u and run all algorithms one more time in order to find these kinds of mirror-edges. (b) $p_3 = LBV(e_2)$, and the intersection of the protraction of $\overline{q'(e_2)p_3}$ and $\overline{uw}$ is $u'(e_2)$.

$v_2(e)$ in Algorithm 1. And, to find these verices the intersection of different visibility polygons maybe required. Also, different half-lines may be as replacement for $L_1(e)$ and $L_2(e)$.

We have the following cases:

1. The whole $\overline{xy}$ can see the whole $\overline{uw}$.
2. The whole $\overline{xy}$ can see at least one point of $\overline{uw}$.
3. $\overline{xy}$ can see the whole $\overline{uw}$ in a weak visible way.
4. At least one point of $\overline{xy}$ can see at least one point of $\overline{uw}$.

We deal with these cases in the following subsections. Without loss of generality, consider a mirror-edge e on $\mathcal{P}$. In each subsection, we find appropriate substitutes for $v_1(e)$, $v_2(e)$, $L_1(e)$, and $L_2(e)$.

5.1   The Whole $\overline{xy}$ Can See the Whole $\overline{uw}$

First, we compute the intersection visibility polygon of the endpoints of $\overline{xy}$ (x and y). Then, while tracing the completely visibility polygon of $\overline{uw}$ ($CVP(\overline{uw})$), we select the common part of each edge with the intersection visibility polygon of the endpoints. As a result, we have $v_1(e)$ and $v_2(e)$ for every mirror-edge e. Obviously, this step only takes $O(n)$ time complexity.

Consider x as a viewer, let the reflective ray from $v_1(e)$ be $L_{1,x(e)}$, and the reflective ray from $v_2(e)$ be $L_{2,x(e)}$. Similarly, we define $L_{1,y(e)}$ and $L_{2,y(e)}$.

We should use $L_{1,x(e)}$ as $L_1(e)$, and $L_{2,y(e)}$ as $L_2(e)$ in Algorithm 1. Since we know any potential mirror-edge from $v_1(e)$ to $v_2(e)$ is completely visible for $\overline{xy}$, it is sufficient to check $L_{1,x(e)}$ to lie in the left side of u, and $L_{2,y(e)}$ to lie in the right side of w.

5.2 The Whole $\overline{xy}$ Can See at Least One Point of $\overline{uw}$

In this subsection, we want to find out if there is any point on $\overline{uw}$ which is e-mirror-visible to the whole $\overline{xy}$.

We can use a method similar to the previous subsection, only now the strongly visibility polygon of $\overline{uw}$ ($SVP(\overline{uw})$) is required. We use $L_{1,x(e)}$ as $L_1(e)$, and $L_{2,y(e)}$ as $L_2(e)$.

Considering $SVP(\overline{uw})$, there is an interval or at least a point on $\overline{uw}$ which holds the property of being strongly visible.

For the last step, we need to find out if this point or segment has intersection with the interval from $u'(e)$ to $w'(e)$.

5.3 $\overline{xy}$ Can See the Whole $\overline{uw}$ in a Weak Visible Way

There may be no point on $\overline{xy}$ to see the whole $\overline{uw}$ by itself. Here, we want to find out if $\overline{uw}$ is completely e-mirror-visible considering all the points on $\overline{xy}$.

We use the intersection of $WVP(\overline{xy})$ and $CVP(\overline{uw})$, to find all the potential mirror-edges (v_1 and v_2 vertices).

Since we deal with the weak visibility polygon, we may face some mirror-edges which are visible to none of the endpoints of $\overline{xy}$, but to an interval of $\overline{xy}$ in the middle. We need to find this interval for each mirror-edge. In fact different mirror-edges may have different points on $\overline{xy}$, to make their L_1 and L_2 half-lines. It is sufficient to check these half-lines with the endpoints of $\overline{uw}$ to make sure that the mirror-visibility region covers $\overline{uw}$ completely.

For a specific mirror-edge e_i, let $x(e_i)$ and $y(e_i)$ be the points on $\overline{xy}$ corresponding to x and y respectively. We can use the ray reflection of $x(e_i)$ on e_i as $L_1(e_i)$, and the ray reflection of $y(e_i)$ as $L_2(e_i)$ in Algorithm 1. In $O(n)$ time we can find these points on $\overline{xy}$ for all mirror-edges through the following way:

Definition 2. *Consider a potential mirror-edge e (from $v_1(e)$ to $v_2(e)$) such that there are two reflex vertices that block the visibility of a portion of $\overline{xy}$ before $v_1(e)$ and after $v_2(e)$ in $\mathcal{P}$'vertex order. Define $r_1(e)$ and $r_2(e)$ to be these reflex vertices, respectively.*

Obviously, if there is no $r_1(e)$ or $r_2(e)$ then there is no obstruction, and we can use corresponding $v_1(e)$ and $v_2(e)$, to find $L_1(e)$ and $L_2(e)$.

See Fig. 5(a), in this figure we have $r_1(e)$ and $r_2(e)$ vertices. The blue sub-segment of $\overline{xy}$ can see e completely, but all the points–from $x(e)$ to the blue sub-segment, and from the blue sub-segment to $y(e)$–cannot see at least some part of e. For the points on the other side of these yellow points, e is not visible. The reflected rays from e, which is between the green half-lines, is the area which segment $\overline{xy}$ can see, in a weak visible way, through e. We call these half-lines $L_{1,y(e)}$ and $L_{2,x(e)}$.

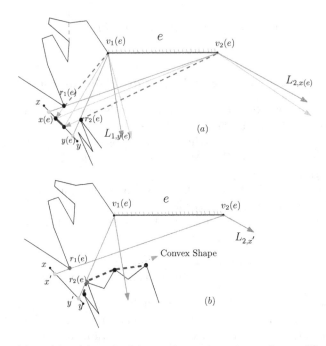

Fig. 5. (a) $r_1(e)$, $r_2(e)$, $x(e)$ and $y(e)$ are shown for mirror-edge e. (b) Constructing convex shape similar to Algorithm 2 (Color figure online).

In order to find $x(e)$ and $y(e)$, we only need $r_1(e)$ and $r_2(e)$, because we can protract $\overline{v_2(e)r_1(e)}$ and $\overline{v_1(e)r_2(e)}$ to find their intersection with $\overline{xy}$. The intersection points are $x(e)$ and $y(e)$.

Suppose there are m potential mirror-edges, we should find $r_1(e_j)$ and $r_2(e_j)$ $1 \le j \le m$. The idea is similar to Algorithm 2.

Computing $r_1(e)$ and $r_2(e)$ Reflex Vertices for All Mirror-Edges:
To compute these reflex vertices we use two convex shapes over the reflex vertices in two directions. For a particular mirror-edge e, $\overline{r_1(e)v_2}$ should hold all left-side reflex vertices on its left, and of course $\overline{r_2(e)v_1}$ should hold all the right-side reflex vertices on its right. Note that it is not important if there were more than one reflex vertex on either $\overline{r_1(e)v_2}$ or $\overline{r_2(e)v_1}$ (see Fig. 5(b)).

In this subsection, we use $L_{1,y(e)}$ and $L_{2,x(e)}$ instead of $L_1(e)$ and $L_2(e)$ respectively. Also, while using Algorithm 2, we need $CVP(\overline{uw})$ in place of $WVP(\overline{uw})$ to construct $TP(\overline{uw})$.

5.4 At Least One Point of $\overline{xy}$ Can See at Least One Point of $\overline{uw}$

Here we can behave similar to the previous subsection except that we need $WVP(\overline{xy}) \cap WVP(\overline{uw})$ to find potential mirror-edges. And, considering a mirror-edge e, we use $L_{1,x(e)}$ and $L_{2,y(e)}$ half-lines to be used in Algorithm 1.

Also, we need $WVP(\overline{uw})$ in the construction of $TP(\overline{uw})$ because it is sufficient to make e-mirror-visible any point on $\overline{xy}$ to any point on $\overline{uw}$.

6 Discussion

We dealt with the problem of extending the visibility polygon of a given point or a segment in a simple polygon, so that another segment becomes visible to the viewer.

We tried to achieve this purpose by converting some edges of the polygon to mirrors. The goal is to find all such kind of edges, and the mirror-visible part of the target segment by each of these edges individually. Using the algorithm we proposed, this can be done in linear time corresponding to the complexity of the simple polygon.

We covered all the possible types of visibility when we dealt with a given segment as a viewer, and we wanted to extend its visibility to see another given segment. We proved all the possible cases need just $O(n)$ time.

We only discussed finding the edges to be mirrors, but it is shown that having two mirrors, the resulting visibility polygon, may not be a simple polygon [7]. Also, having h mirrors, the number of vertices of the resulting visibility polygon, can be $O(n + h^2)$, and for h mirrors, each projection, and its relative visibility polygon can be computed in $O(n)$ time, which leads to overall time complexity of $O(hn)$.

The problem can be extended as; put mirrors inside the polygon, a point with a limited visibility area, find some edges which can give the point a specific vision or different visions and so on.

References

1. Vaezi, A., Ghodsi, M.: Extending visibility polygons by mirrors to cover specific targets. In: EuroCG 2013, pp. 13–16 (2013)
2. Aronov, B., Davis, A., Day, T., Pal, S.P., Prasad, D.: Visibility with one reflection. Discrete Comput. Geom. **19**, 553–574 (1998)
3. Avis, D., Toussaint, G.T.: An optional algorithm for determining the visibility of a polygon from an edge. IEEE Trans. Comput. **C–30**, 910–1014 (1981)
4. de Berg, M., van Kreveld, M., Overmars, M., Schwarzkopf, O.: Computational Geometry Algorithms and Applications, 3rd edn., pp. 13–14. Department of Computer Science Utrecht University (2008)
5. Guibas, L.J., Hershberger, J., Leven, D., Sharir, M., Tarjan, R.E.: Linear-time algorithms for visibility and shortest path problems inside triangulated simple polygons. Algorithmica **2**, 209–233 (1987)
6. Kouhestani, B., Asgaripour, M., Mahdavi, S.S., Nouri, A., Mohades, A.: Visibility polygons in the presence of a mirror edge. In: Proceedings of the 26th European Workshop on Computational Geometry, vol. 26, pp. 209–212 (2010)
7. Lee, D.T.: Visibility of a simple polygon. Comput. Vis. Graph. Image Process. **22**, 207–221 (1983)

On Guarding Orthogonal Polygons
with Sliding Cameras

Therese Biedl[1], Timothy M. Chan[1], Stephanie Lee[1], Saeed Mehrabi[1(✉)],
Fabrizio Montecchiani[2], and Hamideh Vosoughpour[1]

[1] Cheriton School of Computer Science, University of Waterloo, Waterloo, Canada
{biedl,tmchan}@cs.uwaterloo.ca, {s363lee,smehrabi,hvosough}@uwaterloo.ca
[2] Department of Engineering, University of Perugia, Perugia, Italy
fabrizio.montecchiani@unipg.it

Abstract. A sliding camera inside an orthogonal polygon P is a point
guard that travels back and forth along an orthogonal line segment γ
in P. The sliding camera g can see a point p in P if the perpendicular
from p onto γ is inside P. In this paper, we give the *first* constant-
factor approximation algorithm for the problem of guarding P with the
minimum number of sliding cameras. Next, we show that the sliding
guards problem is linear-time solvable if the (suitably defined) dual graph
of the polygon has bounded treewidth. On the other hand, we show that
the problem is NP-hard on orthogonal polygons with holes even if only
horizontal cameras are allowed. Finally, we study art gallery theorems
for sliding cameras, thus, give upper and lower bounds in terms of the
number of sliding cameras needed relative to the number of vertices n.

1 Introduction

Let P be a (not necessarily orthogonal) polygon with n vertices. The art gallery
problem, posed by Victor Klee in 1973 [25], asks for the minimum number of
point guards required to guard P, where a point guard g sees a point $p \in P$
if the line segment connecting g to p lies inside P. Chvátal [7] was the first to
answer the question by giving the tight bound $\lfloor n/3 \rfloor$ on the number of point
guards that are needed to guard a simple polygon with n vertices. For polygons
with holes, Hoffmann et al. [15] proved that $\lfloor (n+h)/3 \rfloor$ point guards are always
sufficient and occasionally necessary, where h is the number of holes. For *orthogonal polygons*, it was proved multiple times [16,23,25] that $\lfloor n/4 \rfloor$ point guards
are always sufficient and sometimes necessary to guard the interior of a simple
orthogonal polygon with n vertices.

Finding the minimum number of guards is NP-hard on simple polygons [22],
even on simple orthogonal polygons [28] or monotone polygons [21]. A number
of results concerning approximation algorithms are also known [13,20,21].

T. Biedl and T.M. Chan—Research of the authors is supported by NSERC.

F. Montecchiani—Research of the author supported in part by the MIUR project
AMANDA, prot. 2012C4E3KT_001. Research was done while the author was visiting
the University of Waterloo.

S.-H. Poon et al. (Eds.): WALCOM 2017, LNCS 10167, pp. 54–65, 2017.
DOI: 10.1007/978-3-319-53925-6_5

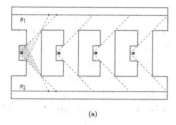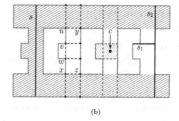

(a)   (b)

Fig. 1. (a) An orthogonal polygon P that can be guarded with two orthogonal mobile guards, but requires $\Theta(n)$ sliding cameras to be guarded since no two crosses can be seen by one sliding camera. (b) Sliding camera s sees the rising-shaded subpolygon of P. We also show parts of the pixelation induced by rays from reflex vertices $\{u, v, w, x, y, z\}$, and the cross c whose supporting horizontal slices is downward shaded. Segments s_1 and s_2 are guard-segments.

Mobile Guards and Sliding Cameras. A *mobile guard* is a point guard that travels along a line segment γ inside P. Guard γ can see a point p in P if and only if there exists a point $g \in \gamma$ such that the line segment pg lies entirely inside P. If the line segment γ must be orthogonal, then we call it an *orthogonal mobile guard*. Moreover, if the line segment pg is required to be perpendicular to γ, then we call γ a *sliding camera*. Note that an orthogonal mobile guard travelling along γ may see a larger area of P than a sliding camera travelling along γ, see also Fig. 1(a). The notion of mobile guards was introduced by Avis and Toussaint [2]. O'Rourke [26] proved that $\lfloor n/4 \rfloor$ (not necessarily orthogonal) mobile guards are sufficient for guarding arbitrary polygons with n vertices. For orthogonal polygons with n vertices, $\lfloor (3n + 4)/16 \rfloor$ mobile guards are always sufficient and sometimes necessary [1].

In this paper we study the *Minimum Sliding Cameras (MSC)* problem, i.e., we want to guard an orthogonal polygon P with the minimum number of sliding cameras. We also consider the variant *Minimum Horizontal Sliding Cameras (MHSC)* where only horizontal cameras are allowed. These problems were introduced by Katz and Morgenstern [18], who proved that MHSC can be solved in polynomial time in the special case where the polygon is simple (has no holes). It was shown later that MSC is NP-hard in polygons with holes [12,24]; NP-hardness in simple polygons is open. Durocher et al. [11] claimed a (3.5)-approximation algorithm for MSC problem on simple orthogonal polygons, but this was later discovered by the authors to be incorrect (private communication). For the special case of monotone orthogonal polygons, Katz and Morgenstern [18] gave a 2-approximation algorithm, which was later improved by de Berg et al. [4] to a linear-time exact algorithm.

Our Results. In this paper, we give hardness results and algorithms for both MSC and MHSC. Specifically, we give two (conceptually very different) algorithms. The first works by constructing a small ε-net for the hitting set problem that naturally arises from MSC. This gives then an $O(1)$-approximation algorithm for the MSC problem on orthogonal polygons. Note that no constant-factor approximation algorithm was known previously, and, as opposed to previous attempts at such approximation-algorithms [11], our algorithm works even on orthogonal polygons with holes. The second algorithm uses a tree-decomposition approach. We show that if the dual graph of the so-called pixelation of the polygon has bounded treewidth, then MSC can be solved in polynomial time. In particular this holds in so-called thin polygons that have no holes.

Both the above approaches also work (and become even simpler) for MHSC where only horizontal cameras are allowed. We also establish NP-hardness of MHSC for polygons with holes. The same proof also works for MSC and is different, and perhaps simpler, than the previous NP-hardness proof for MSC [12].

Finally, we consider art gallery theorems for sliding cameras, i.e., theorems that bound the number of guards relative to the number of vertices. We present the following results for an orthogonal polygon P with n vertices: (i) $\lfloor (3n + 4)/16 \rfloor$ sliding cameras are always sufficient and sometimes necessary to guard P entirely, (ii) $\lfloor n/4 \rfloor$ only-horizontal sliding cameras are always sufficient and sometimes necessary to guard P, and (iii) if sliding cameras are not allowed to intersect each other, then $\lfloor (n + 1)/5 \rfloor$ cameras are always sufficient to guard P.

Due to space constraints, some proofs will be given in the full version of this paper.

2 Preliminaries

Throughout the paper, P denotes an orthogonal polygon with n vertices. The *horizontal* (respectively *vertical*) *segmentation* of P consists of extending a horizontal (vertical) ray inward from any reflex vertex of P until it hits another vertex or edge. The rectangles in the resulting partition of P are called the horizontal (vertical) *slices* of P. Each slice can be represented by the horizontal (vertical) line segment that halves the slice; we call these the *slice-segments* and denote them by Σ.

The *pixelation* of P is obtained by doing both the horizontal and the vertical segmentation of P. The resulting rectangles are called *pixels*. The pixelation may well have $\Theta(n^2)$ pixels. Notice that the pixels are in 1-1-correspondence with pairs of slices that cross. We can hence identify each pixel with a *cross* c, which is the point where the two slice-segments σ_H and σ_V of these two slices cross. We say that σ_H and σ_V *support* c. Denote the set of crosses by X.

A *sliding camera* γ is a horizontal or vertical line segment that is inside P. (We will frequently omit "sliding", as we study no other type of camera.) The region visible from γ is the set of all points p such that the perpendicular from p to γ is inside P. Note that doing a *parallel shift* (i.e., translating a horizontal camera vertically or a vertical camera horizontally) does not change its visibility region for as long as we stay inside P. We may hence assume that any camera runs along pixel-edges. We may also restrict our attention to cameras that are maximal line segments within P (all others would see a subset). Let Γ be the set of *guard-segments* which are maximal horizontal and vertical line segments within P that run along pixel edges. See Fig. 1(b) for an illustration.

The following lemma (whose proof is given in the full version of the paper) is a straightforward re-formulation of what guarding means, but casts the problem into a discrete framework that will be crucial later.

Lemma 1. *A set S of k sliding cameras guards polygon P if and only if there exists a set of k guard-segments $S' \subseteq \Gamma$ such that for every cross $c \in X$, at least one of the slice-segments that support c is intersected by some $\gamma \in S'$.*

We say that a guard-segment γ *hits* a cross c if and only if γ intersects one of the slice-segments supporting c. Lemma 1 can then be re-stated as that S' hits all crosses. In fact, the algorithms we design later will allow further restrictions: we can specify exactly which crosses should be hit and which cameras may be used as guards. So assume we are given some $X' \subseteq X$ and some $\Gamma' \subseteq \Gamma$. The (X', Γ')-*sliding cameras problem* consists of finding a minimum subset of cameras in Γ' that hit all crosses in X'. Note that with a suitable choice of Γ' this encompasses both MSC and MHSC.

3   Approximation Algorithms via ε-Nets

In this section, we give approximation algorithms for MSC and MHSC that are based on phrasing the problem as a hitting set problem and then using ε-nets. We do this first for MHSC, and then later re-use those ε-nets for MSC.

Hitting Sets. A *set system* is a pair $\mathcal{R} = (\mathcal{U}, \mathcal{S})$, where $\mathcal{U}$ is a universe set of objects and $\mathcal{S}$ is a collection of subsets of $\mathcal{U}$. A *hitting set* for the set system $(\mathcal{U}, \mathcal{S})$ is a subset of $\mathcal{U}$ that intersects every set in $\mathcal{S}$.

For the (X', Γ')-sliding camera problem, we construct a set system as follows. Let $\mathcal{U} = \Gamma'$ be all potential sliding cameras. For each cross $c \in X'$ that needs to be hit, define S_c to be all the cameras in $\mathcal{U}$ that hit c, and let $\mathcal{S}$ be the collection of these sets. From the definitions, finding a hitting set for this set system is the same as solving the (X', Γ')-sliding-camera problem.

An ε-*net* for a set system $\mathcal{R} = (\mathcal{U}, \mathcal{S})$ is a subset N of $\mathcal{U}$ such that every set S in $\mathcal{S}$ with size at least $\varepsilon \cdot |\mathcal{U}|$ has a non-empty intersection with N. Brönnimann and Goodrich [6] showed that ε-nets can be used to derive approximation algorithms as follows. Define a *net finder* to be a (poly-time) algorithm that, for a given set system $\mathcal{R} = (\mathcal{U}, \mathcal{S})$ and any given $r > 0$, computes an $(1/r)$-net of $\mathcal{R}$ whose size is at most $s(r)$ for some function s. Also, a *verifier* is a poly-time algorithm that, given a subset $H \subset \mathcal{U}$, states (correctly) that H is a hitting set, or returns a non-empty set $R \in \mathcal{S}$ such H does not hit S.

Lemma 2 *[6]. Let $\mathcal{R}$ be a set system that admits both a poly-time net finder and a poly-time verifier. Then there is a poly-time algorithm that computes a hitting set of size at most $s(4 \cdot OPT)$, where OPT stands for the size of an optimal hitting set, and $s(r)$ is the size of the $(1/r)$-net.*

Thus, the lemma gives an $O(1)$-approximation algorithm for as long as we can find an ε-net whose size is $O(1/\varepsilon)$. (Clearly the hitting set problems defined by MHSC and MSC both have a polynomial-time verifier.)

An ε-net for the MHSC Problem. We now show the existence of such a small ε-net for MHSC. For this, we need (yet another) reformulation that simplifies the problem.

Lemma 3. *A set S of horizontal guard-segments hits all crosses in a set U' if and only if S intersects all the vertical slice-segments that support crosses in U'.*

Proof. If camera γ hits cross c, then it intersects either its horizontal supporting slice-segment σ_H or its vertical supporting slice-segment σ_V. But if γ intersects σ_H, then since both are horizontal and γ is maximal we have $\sigma_H \subseteq \gamma$, in case of which γ also contains point c and therefore intersects σ_V. So either way γ intersects σ_V.   □

For MHSC, it hence suffices to represent every cross by its vertical slice-segment and so reduce the problem to the following: Given a set of horizontal line segments $\mathcal{H}$ and a set of vertical line segments $\mathcal{V}$, find a minimum set $S \subseteq \mathcal{H}$ such that every line segment in $\mathcal{V}$ is intersected by S. This problem is also known as the *Orthogonal Segment Covering* problem and was shown to be NP-complete [17]. We hence have:

Corollary 1. *MHSC reduces to the Orthogonal Segment Covering problem.*

The following lemma shows that the Orthogonal Segment Covering problem has a small ε-net; by the above this immediately implies a small ε-net for the hitting set problem for MHSC.

Lemma 4. *The Orthogonal Segment Covering problem has a $(1/r)$-finder with size-function $s(r) \in O(r)$.*

Proof. Here, we sketch the proof; the full proof appears in the full version of the paper. Observe that a horizontal line segment $[x, x'] \times y$ intersects a vertical line segment $a \times [b, b']$ if and only if the point (x, y, x') lies in the range $(-\infty, a] \times [b, b'] \times [a, \infty)$. The union of these ranges forms a geometric object that (as one can argue) has complexity $O(n)$. Clarkson and Varadarajan [8] showed that ε-nets of small size can be found for hitting set problems in such a geometric object, using random sampling.   □

Combining the above results gives:

Theorem 1. *There exists a poly-time $O(1)$-approximation algorithm for the Orthogonal Segment Covering problem and the MHSC problem.*

An ε-net for the MSC Problem. Using the ε-net for MHSC, we can easily find one for MSC and hence have an approximation algorithm for this as well.

Theorem 2. *There exists a poly-time $O(1)$-approximation algorithm for the MSC problem.*

Proof. Fix a polygon P and consider the (X', Γ')-sliding camera problem for P. It suffices to show that for any $r > 0$ there exists a $1/r$-net T of size $O(r)$ for the corresponding hitting set problem $\mathcal{R}$. Let T_H be a $1/2r$-net for the hitting set of MHSC for P, X' and the horizontal cameras in Γ'. Let T_V be a $1/2r$-net for the hitting set of MVSC (i.e., when we want to guard the polygon using only vertical sliding guards) for P, X' and the vertical cameras in Γ'. Set $T := T_H \cup T_V$. We claim that T is a $1/r$-net for $\mathcal{R}$.

So assume some set S_c in the hitting-set problem satisfies $|S_c| \geq |\mathcal{U}|/r$. Translating back, this means that some cross $c \in X'$ is hit by at least $|\Gamma'|/r$ guard-segments. Assume w.l.o.g. that at least half of these hitting guard-segments are horizontal. Then the vertical slice-segment σ_V that supports c intersects at least $|\Gamma'|/2r$ horizontal guard-segments in Γ'. By definition of a $(1/2r)$-net, therefore there is a line segment $\gamma \in T_H$ that intersects σ_V. Therefore $\gamma \in T$ hits c as required. $\qquad\square$

4   Polygons with Bounded-Treewidth Pixelation

Recall that the *pixelation* of a polygon is obtained by cutting the polygon horizontally and vertically at all reflex vertices. The *dual graph D of the pixelation* is obtained by interpreting the pixelation as a planar graph and taking its weak dual (i.e., dual graph but omit the outer face). Thus, D has a vertex for every pixel of P, and two pixels are adjacent in D if and only if they share a side. We now show how to solve MSC and MHSC under the assumption that D has small treewidth. (Our approach was inspired by a similar result for a different guarding problem [5], but the construction here is simpler.)

2-Dominating Set in an Auxiliary Graph. By Lemma 1, the (X', Γ')-sliding camera problem is equivalent to finding a set of guard-segments that hits at least one supporting slice-segment of each cross. This naturally gives rise to an auxiliary graph H as follows: The vertices of H are $X' \cup \Sigma \cup \Gamma'$. For any $c \in X'$, add an edge from c to each of its two supporting slice-segments. For any guard-segment γ, add an edge to any slice-segment that it intersects. From Lemma 1, and since there are no edges from X' to Γ', one immediately sees the following:

Lemma 5. *The minimum guard set for the (X', Γ')-sliding-cameras problem corresponds to a subset $S \subseteq \Gamma'$ of vertices in H such that all vertices in X' are within distance 2 from S.*

The above lemma means that the sliding-camera problem reduces to a graph-theoretic problem that is quite similar to the 2-dominating set (the problem of finding a set S such that all other vertices have distance at most 2 from S); the only change is that we restrict which vertices may be used for S and which vertices must be within distance 2 from S. 2-dominating set is an NP-hard problem in general, but is easily shown to be polynomial in graphs that have bounded treewidth, which we define next.

Treewidth. A *tree decomposition* $\mathcal{T} = (I, \mathcal{X})$ of a graph $G = (V, E)$ consists of a tree I and an assignment $\mathcal{X} : V(I) \to 2^V$ of *bags* of vertices of G to nodes of I such that the following holds: (a) for every vertex $v \in V$, the set of bags containing v forms a non-empty connected subtree of I, (b) for every edge $e \in E$, at least one bag contains both ends of e. The *width* of a tree decomposition is the maximum bag-size minus 1, and the *treewidth* $tw(G)$ of a graph G is the smallest possible width over all tree decompositions of G. In particular, a tree has treewidth 1. We prove the following lemma in the full version of the paper.

Lemma 6. *Let P be a polygon whose dual graph D of the pixelation has treewidth at most k. Then for any choice of $X' \subseteq X$ and $\Gamma' \subseteq \Gamma$, the auxiliary graph H has treewidth at most $7k + 6$.*

To apply this treewidth-result, we must show that the problem can be expressed as a suitable logic-formula. (See e.g. [10, Chap. 7.4] for more details.) In particular, the following formula will do: A set S of guard-segments that guards X' satisfies $S \subseteq \Gamma' \quad \wedge \quad \forall u \in X' (\exists \sigma \in \Sigma \quad \mathrm{adj}(u, \sigma) \wedge \exists \gamma \in S \quad \mathrm{adj}(\sigma, \gamma))$ (where adj is a logic-formula to encode that the two parameters are adjacent in H). Since H has bounded treewidth, we can find the smallest set S that satisfies this (or report that no such S exists if Γ' was too small) in linear time using Courcelle's theorem [9]. Putting everything together, we hence have:

Theorem 3. *If P is a polygon whose dual graph has bounded treewidth, then the (X', Γ')-sliding-cameras problem can be solved in linear time.*

We give one application of this result. A *thin* polygon is a polygon for which no pixel-corner is in the interior. MSC and MHSC are NP-hard even for thin polygons with holes (as we will see in the next subsection). However, for thin polygons without holes, the dual graph of the pixelation is clearly a tree, hence has bounded treewidth, and both MSC and MHSC can be solved in linear time.

Corollary 2. *If P is a thin polygon without holes, then MSC and MHSC can be solved in linear time.*

This result is not directly comparable to existing results [4, 19]: it is stronger than these since it works for MSC and does not require monotonicity, but it is weaker than these since it requires a thin polygon. A natural question is whether this result for bounded treewidth could be used to generate a PTAS, by splitting the polygon (hence the planar graph) suitably and applying the "shifting technique" (see [3] or [10, Chap. 7.7.3]). We have not been able to develop such a PTAS, principally because the cameras are not "local" in the sense that they can guard pixels that have arbitrarily large distance in D. Creating a PTAS (or proving APX-hardness) hence remains an open problem.

5   NP-Hardness of MHSC

Recall that the MHSC problem is polynomial-time solvable on simple orthogonal polygons [18]. We show in this section that the problem becomes NP-hard on

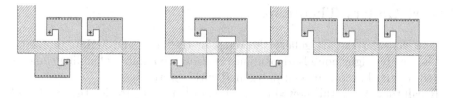

Fig. 2. The NP-hardness construction. Vertex-bars are red (dotted). Edge-strips are green (falling pattern). At each vertex-bar we attach an "elephant-gadget" (gray) that requires a sliding camera of its own (dashed) to guard the point inside the "trunk". (Color figure online)

orthogonal polygons with holes. We note that the hardness proof of Durocher and Mehrabi [12] does not apply to the MHSC problem because they require both horizontal and vertical sliding cameras.

The reader may recall that we showed that MHSC reduces to the Orthogonal Segment Covering problem, which is known to be NP-hard [17]. However, this does not prove NP-hardness of MHSC, because not every instance of Orthogonal Segment Covering can be expressed as MHSC. Instead we give a different reduction from Minimum Vertex Cover on max-deg-3 planar graphs. This problem (which is NP-hard [14]) consists of, given a planar graph $G = (V, E)$ with at most 3 incident edges at each vertex, find a minimum set $C \subseteq V$ such that for every edge at least one endpoint is in C.

Given a max-deg-3 planar graph G, we first compute a *bar visibility representation* of G, that is, we assign to each vertex a horizontal line segment (called *bar*) and to each edge (v, w) a vertical *strip* of positive width that joins the corresponding bars and that is disjoint from all other strips. It is well-known that every planar graph has such a representation (see e.g. Tamassia and Tollis [29]), and it can be found in linear time. By making strips sufficiently thin, we can ensure that no two strips of edges occupy the same x-range. From this visibility representation, we can construct in polynomial time an orthogonal polygon P such that the following holds (see Fig. 2; the proof of the following lemma will appear in the full version of the paper):

Lemma 7. *The following are equivalent: (i) G has a vertex cover of size k; (ii) P can be guarded with $k + 3N$ horizontal sliding cameras; (iii) P can be guarded with $k + 3N$ sliding cameras.*

The constructed polygon is thin. NP-hardness of guarding problems in thin polygons (albeit with other models of guards and visibility) have been studied before [5,30]. The NP-hardness holds for both MSC and MSHC; NP-hardness of MSC was known before [12], but the constructed polygon was not thin. We summarize:

Theorem 4. *MSC and MHSC is NP-hard on thin orthogonal polygons with holes.*

6   Art Gallery Theorems

We now consider the art gallery theorems for the MSC and MHSC problems; that is, we give tight bounds, depending on n, on the number of sliding cameras needed to guard an orthogonal polygon P with n vertices.

Recall that Aggarwal showed a tight bound $\lfloor (3n + 4)/16 \rfloor$ for the number of mobile guards necessary and sufficient to guard P [1]. Closer inspection reveals that the lower bound construction (see Fig. 3) actually works for sliding cameras, since no two of the $(3n+4)/16$ pixels marked with a cross can be guarded by one camera. The upper bound, indeed, also works for sliding guards. We very briefly review the approach taken in [1]. The idea is to guard first a small portion of P using one or two mobile guards, cutting a guarded region out of P, and then guarding the rest of P by an induction hypothesis. There are numerous cases, but in all of them one can establish that indeed a sliding camera would have achieved the same as the mobile guard used. So we have the following result.

Theorem 5 (Based on [1]**).** *Given a simple orthogonal polygon P with n vertices, $\lfloor (3n+4)/16 \rfloor$ sliding cameras are always sufficient and sometimes necessary to guard P entirely.*

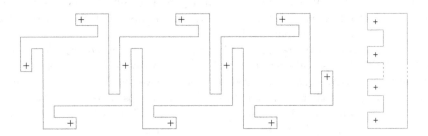

Fig. 3. (Left) A polygon that requires $(3n + 4)/16$ cameras. (Right) A polygon that requires $n/4$ horizontal cameras.

For the MHSC problem, Fig. 3 shows a polygon that requires $\lfloor n/4 \rfloor$ horizontal sliding cameras. We show in the full version of the paper that this is tight.

Theorem 6. *Given an orthogonal polygon P with n vertices, $\lfloor n/4 \rfloor$ horizontal sliding cameras are always sufficient and sometimes necessary to guard P entirely.*

Non-crossing Sliding Cameras. There are cases in the upper bound approach of Aggarwal (Theorem 5 and [1]) in which the trajectories of mobile guards intersect. We show using a different approach that $\lfloor (n + 1)/5 \rfloor$ non-crossing sliding cameras are always sufficient to guard a simple orthogonal polygon P with n vertices that is in *general position* in the sense that no two vertical edges of P have the same x-coordinate.

Recall that earlier we considered the dual graph of the pixelation of a polygon P. In this section, we again consider a dual graph, but this time of one of the segmentations of P. Thus, consider (say) the vertical segmentation obtained after extending vertical rays from all reflex vertices. Interpret the segmentation as a planar graph, and let G be its weak dual graph obtained by defining a vertex for every vertical rectangle and connecting two rectangles if and only if they share (part of) a side. If P is simple, then this dual graph is a tree T; we know that $|T| = n/2 - 1$ since P is in general position [27]. In the following, we show that $\lfloor 2/5 \cdot |T| + 3/5 \rfloor$ non-crossing sliding cameras are sufficient to guard P entirely, which therefore gives the desired bound.

We partition T into a set of disjoint subtrees as follows. Root T at a leaf. Let u be the lowest node in T that has degree two (i.e., u has only one child) and u is not the parent of a leaf. Let $T(u)$ be the subtree rooted at u, and partition $T - T(u)$ recursively. Let T_0 be the tree remaining in the base case (when no u exists). T_0 may have just a single node; this will be treated separately. Any other subtree has the form $T(u)$ for some node u and at least 3 nodes, and we will argue now that we can guard it with at most $2/5 \cdot |T(u)|$ cameras.

To guard one such $T(u)$, we consider it to consist of the following components (see also Fig. 4): (i) Vertex u is the root of $T(u)$, (ii) Let Y be all those leaves of $T(u)$ whose parent have only one child, and set $y = |Y|$, (iii) Let $X = T(u) - \{u\} - Y$ and set $x = |X|$. Since u had only one child and Y consists of leaves, X forms a tree. By choice of u and Y, no inte-

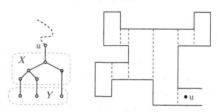

Fig. 4. An example of tree $T(u)$.

rior node of X has degree 2. One can show that X can have at most one vertex of degree 4 since it corresponds to an orthogonal polygon (we formally prove this in the full version). Hence, X forms a tree that is a rooted binary tree except that one node may have three children. Thus X has at most $x/2 + 1$ leaves, and $y \leq x/2 + 1$. The following is shown in the full version:

Lemma 8. *Let c be the node of $X \cup \{u\}$ whose corresponding rectangle $R(c)$ has the maximum height. Let s be a maximal vertical line segment inside $R(c)$. Then s guards all rectangles corresponding to $X \cup \{u\}$.*

Since every leaf in Y can be covered using a single sliding camera, the sub-polygon corresponding to $T(u)$ can hence be guarded with $y + 1$ sliding cameras. In fact, one can show that y sliding cameras suffice if y is close to the maximum, i.e., $y = x/2 + 1$ or $y = (x+1)/2$ (we formally prove this in the full version of the paper). Elementary calculations show that with this we use at most $2/5 \cdot |T(u)|$ cameras for $T(u)$ if $x \geq 6$. For $x < 6$, the only cases where the number of cameras is too large is $(x, y) = (1, 1)$ or $(4, 2)$ which can be dealt with by analyzing their structure directly. Finally tree T_0 may have an empty X, it then can always be guarded with $1 \leq 2/5 \cdot |T_0| + 3/5$ cameras. We hence have:

Theorem 7. *Given a simple orthogonal polygon P in general position with n vertices, $\lfloor (n + 1)/5 \rfloor$ sliding cameras are always sufficient to guard P such that no two sliding cameras intersect each other.*

7   Conclusion

In this paper, we studied the problem of guarding an orthogonal polygon with the minimum number of sliding cameras. We gave the first constant-factor approximation algorithm for this problem, which works even if the polygon has holes. We also showed how to solve the problem optimally if the polygon is thin and has no holes, and we gave art-gallery-type results bounding the number of sliding cameras that are always sufficient and sometimes required. The most interesting remaining question is whether guarding an orthogonal polygon with sliding guards is polynomial if the polygon has no holes. Also, the factor in our $O(1)$-approximation algorithm (which we did not compute since it is hidden in the machinery of [6,8]) is likely large. Can it be improved? Even better, could we find a PTAS or is the problem APX-hard?

References

1. Aggarwal, A.: The art gallery theorem: its variations, applications and algorithmic aspects. Ph.D. thesis, Johns Hopkins University, A summary can be found in [27], Chap. 3 (1984)
2. Avis, D., Toussaint, G.T.: An optimal algorithm for determining the visibility of a polygon from an edge. IEEE Trans. Comput. **30**(12), 910–914 (1981)
3. Baker, B.: Approximation algorithms for NP-complete problems on planar graphs. J. ACM **41**(1), 153–180 (1994)
4. de Berg, M., Durocher, S., Mehrabi, S.: Guarding monotone art galleries with sliding cameras in linear time. In: Zhang, Z., Wu, L., Xu, W., Du, D.-Z. (eds.) COCOA 2014. LNCS, vol. 8881, pp. 113–125. Springer, Heidelberg (2014). doi:10.1007/978-3-319-12691-3_10
5. Biedl, T., Mehrabi, S.: On r-guarding thin orthogonal polygons. In: 27th International Symposium on Algorithms and Computation (ISAAC 2016), LIPIcs, vol. 64, pp. 17:1–17:13 (2016)
6. Brönnimann, H., Goodrich, M.T.: Almost optimal set covers in finite VC-dimension. Discret. Comput. Geom. **14**(4), 463–479 (1995)
7. Chvátal, V.: A combinatorial theorem in plane geometry. J. Comb. Theory Ser. B **18**, 39–41 (1975)
8. Clarkson, K.L., Varadarajan, K.R.: Improved approximation algorithms for geometric set cover. Discret. Comput. Geom. **37**(1), 43–58 (2007)
9. Courcelle, B.: The monadic second-order logic of graphs. I. Recognizable sets of finite graphs. Inf. Comput. **85**(1), 12–75 (1990)
10. Cygan, M., Fomin, F.V., Kowalik, L., Lokshtanov, D., Mark, D., Pilipczuk, M., Pilipczuk, M., Saurabh, S.: Parameterized Algorithms. Springer, Heidelberg (2015)
11. Durocher, S., Filtser, O., Fraser, R., Mehrabi, A.D., Mehrabi, S.: A (7/2)-approximation algorithm for guarding orthogonal art galleries with sliding cameras. In: Pardo, A., Viola, A. (eds.) LATIN 2014. LNCS, vol. 8392, pp. 294–305. Springer, Heidelberg (2014). doi:10.1007/978-3-642-54423-1_26

12. Durocher, S., Mehrabi, S.: Guarding orthogonal art galleries using sliding cameras: algorithmic and hardness results. In: Chatterjee, K., Sgall, J. (eds.) MFCS 2013. LNCS, vol. 8087, pp. 314–324. Springer, Heidelberg (2013). doi:10.1007/978-3-642-40313-2_29

13. Eidenbenz, S., Stamm, C., Widmayer, P.: Inapproximability results for guarding polygons and terrains. Algorithmica **31**(1), 79–113 (2001)

14. Garey, M.R., Johnson, D.S.: The rectilinear Steiner tree problem in NP-complete. SIAM J. Appl. Math. **32**, 826–834 (1977)

15. Hoffmann, F., Kaufmann, M., Kriegel, K.: The art gallery theorem for polygons with holes. In: Proceedings of Foundations of Computer Science (FOCS 1991), pp. 39–48 (1991)

16. Kahn, J., Klawe, M.M., Kleitman, D.J.: Traditional galleries require fewer watchmen. SIAM J. Algebraic Discrete Methods **4**(2), 194–206 (1983)

17. Katz, M.J., Mitchell, J.S.B., Nir, Y.: Orthogonal segment stabbing. Comput. Geom. **30**(2), 197–205 (2005)

18. Katz, M.J., Morgenstern, G.: Guarding orthogonal art galleries with sliding cameras. Int. J. Comput. Geom. Appl. **21**(2), 241–250 (2011)

19. Katz, M.J., Roisman, G.S.: On guarding the vertices of rectilinear domains. Comput. Geom. **39**(3), 219–228 (2008)

20. Kirkpatrick, D.G.: An $O(\lg \lg opt)$-approximation algorithm for multi-guarding galleries. Discrete Comput. Geom. **53**(2), 327–343 (2015)

21. Krohn, E., Nilsson, B.J.: Approximate guarding of monotone and rectilinear polygons. Algorithmica **66**(3), 564–594 (2013)

22. Lee, D.T., Lin, A.K.: Computational complexity of art gallery problems. IEEE Trans. Inf. Theory **32**(2), 276–282 (1986)

23. Lubiw, A.: Decomposing polygonal regions into convex quadrilaterals. In: Proceedings of the ACM Symposium on Computational Geometry (SoCG 1985), pp. 97–106 (1985)

24. Mehrabi, S.: Geometric optimization problems on orthogonal polygons: hardness results and approximation algorithms. Ph.D. thesis, University of Manitoba, Winnipeg, Canada, August 2015

25. O'Rourke, J.: The complexity of computing minimum convex covers for polygons. In: 20th Allerton Conference Communication, Control, and Computing, pp. 75–84 (1982)

26. O'Rourke, J.: Galleries need fewer mobile guards: a variation to Chvátal's theorem. Geom. Dedicata **14**, 273–283 (1983)

27. O'Rourke, J.: Art Gallery Theorems and Algorithms. The International Series of Monographs on Computer Science. Oxford University Press, New York (1987)

28. Schuchardt, D., Hecker, H.: Two NP-hard art-gallery problems for ortho-polygons. Math. Logic Q. **41**(2), 261–267 (1995)

29. Tamassia, R., Tollis, I.: A unified approach a visibility representation of planar graphs. Discrete Comput. Geom. **1**, 321–341 (1986)

30. Tomás, A.P.: Guarding thin orthogonal polygons is hard. In: Gąsieniec, L., Wolter, F. (eds.) FCT 2013. LNCS, vol. 8070, pp. 305–316. Springer, Heidelberg (2013). doi:10.1007/978-3-642-40164-0_29

Bundling Two Simple Polygons to Minimize Their Convex Hull

Jongmin Choi, Dongwoo Park, and Hee-Kap Ahn[✉]

Department of Computer Science and Engineering, POSTECH,
Pohang, South Korea
{icothos,dwpark,heekap}@postech.ac.kr

Abstract. Given two simple polygons P and Q in the plane, we study the problem of finding a placement φP of P such that φP and Q are disjoint in their interiors and the convex hull of their union is minimized. We present exact algorithms for this problem that use much less space than the complexity of the Minkowski sum of P and Q. When the orientation of P is fixed, we find an optimal translation of P in $O(n^2 m^2 \log n)$ time using $O(nm)$ space, where n and m ($n \geq m$) denote the number of edges of P and Q, respectively. When we allow reorienting P, we find an optimal rigid motion of P in $O(n^3 m^3 \log n)$ time using $O(nm)$ space. In both cases, we find an optimal placement of P using linear space at the expense of slightly increased running time. For two polyhedra in three dimensional space, we find an optimal translation in $O(n^3 m^3 \log n)$ time using $O(nm)$ space or in $O(n^3 m^3 (m + \log n))$ time using linear space.

1 Introduction

Given two simple polygons P and Q in the plane, we study the bundling problem of finding a placement φP of P such that $\varphi P \cup Q$ is contained in a smallest possible convex region while φP and Q are disjoint in their interiors. We consider minimizing the area of the convex hull of $\varphi P \cup Q$ under either translations or rigid motions.

The bundling problem is related to the *packing problem* in which the shape of a container is predefined (such as a disk, a square, or a rectangle) and we aim to find a smallest container for input objects while the input objects remain disjoint in their interiors. Packing problems have been studied for a long time. It dates back to 1611 when Kepler studied sphere packing in three-dimensional Euclidean space [10]. Sugihara et al. [14] lately studied a disk packing problem that finds a smallest enclosing circle containing a set of disks in the plane and proposed an $O(n^4)$-time heuristic algorithm, where n is the number of disks.

For a set of polygons in the plane, Milenkovic [12] studied the problem of packing them into a given axis-parallel rectangle under translations. He proposed an $O(n^{k-1} \log n)$-time algorithm using linear programming techniques,

This work was supported by the NRF grant 2011-0030044 (SRC-GAIA) funded by the government of Korea.

S.-H. Poon et al. (Eds.): WALCOM 2017, LNCS 10167, pp. 66–77, 2017.
DOI: 10.1007/978-3-319-53925-6_6

where k is the number of given polygons and n is the maximum complexity of each polygon. Later, Alt and Hurtado [5] studied an optimization version of the problem for two convex polygons with n vertices in total to minimize the area or the perimeter of the container rectangle. They presented a near linear time algorithm for translations and an $O(n^3)$-time algorithm for rigid motions. This problem is known to be NP-hard for arbitrary numbers of input polygons even if the polygons are rectangles [7].

In the bundling problem, there is no predefined shape of the container as long as it is convex and the goal is to find placements of input objects such that they are disjoint in their interiors and the convex hull of their union is minimized with respect to the area or the perimeter. For two convex polygons, an optimal translation and an optimal rigid motion of P can be found in $O(n)$ time using $O(n)$ space [11] and in $O(n^3)$ time using $O(n)$ space [15], respectively, where n is the total number of vertices of P and Q. In both cases, they use the property that in every optimal placement the polygons are in contact along their boundaries. Later, Ahn and Cheong presented a near linear time algorithm that returns a rigid motion achieving a factor $(1 + \varepsilon)$ to the optimum for two convex polygons under rigid motion [1]. For two convex polytopes in 3 or higher dimensional space, Ahn et al. [2,3] showed an example of two convex polytopes for which no optimal translation aligns them to be in contact. They presented an algorithm that returns an optimal translation for two convex d-polytopes with n vertices in total in $O(n^{\lceil \frac{d}{2} \rceil (d-3)+d})$ time using $O(n^{\lceil \frac{d}{2} \rceil (d-3)+d})$ space with respect to the volume or the surface area of their convex hull.

For two simple polygons P and Q with n and m vertices, respectively, in the plane, there always exists an optimal translation (and an optimal rigid motion) of P that aligns the polygons to be in contact. In other words, there is an optimal translation lying on the boundary of the *Minkowski sum* of $-P$ and Q. (We provide a formal proof for this claim in the paper.) The Minkowski sum of P and Q has complexity $\Theta(n^2 m^2)$ and it can be computed in $O(n^2 m^2 \log n)$ time [6,9].

Our Results. We present algorithms for this problem using much less space than the complexity of the Minkowski sum. When the orientation of P is fixed, we find an optimal translation of P in $O(n^2 m^2 \log n)$ time using $O(nm)$ space, where n and m $(n \geq m)$ denote the number of edges of P and Q, respectively. When we allow reorienting P, we find an optimal rigid motion of P in $O(n^3 m^3 \log n)$ time using $O(nm)$ space. In both cases, we find an optimal placement of P using only $O(n)$ space at the expense of slightly increased running time. For three dimensional space and two polyhedra, we find an optimal translation in $O(n^3 m^3 \log n)$ time using $O(nm)$ space. We summarize our results in Table 1.

Throughout the paper, we use $V(R)$ and $E(R)$ to denote the set of vertices and the set of edges of a polygon R, respectively. In three dimensional space, we use $F(R)$ to denote the set of faces of a polyhedron R. For a compact set S in the plane (or three dimensional space), we use $\mathsf{conv}(S)$ to denote the convex hull of S, and $\|\mathsf{conv}(S)\|$ to denote the area (volume) of $\mathsf{conv}(S)$. Note that we mainly focus on minimizing the area (volume) of the convex hull in the following.

Table 1. Time and space complexities of our algorithms.

Algorithm	Time	Space
Under translations in 2D	$O(n^2 m^2 \log n)$	$O(nm)$
	$O(n^2 m^2 (m + \log n))$	$O(n)$
Under rigid motions in 2D	$O(n^3 m^3 \log n)$	$O(nm)$
	$O(n^3 m^3 (m + \log n))$	$O(n)$
Under translations in 3D	$O(n^3 m^3 \log n)$	$O(nm)$
	$O(n^3 m^3 (m + \log n))$	$O(n)$

The similar algorithm works for minimizing the perimeter (surface area) since the perimeter (surface area) function is also piecewise linear under translations, and piecewise trigonometric under rigid motions.

2 Minimizing the Convex Hull Under Translations

Let P and Q denote two simple polygons in the plane with n and m ($n \geq m$) vertices, respectively. Let $P + \tau$ denote the translated copy of P by $\tau \in \mathbb{R}^2$, that is, $P + \tau = \{p + \tau \mid p \in P\}$. We explain how to find an optimal translation τ^* such that the area $\|\mathsf{conv}((P + \tau^*) \cup Q)\|$ is minimized while $P + \tau^*$ and Q are disjoint in their interiors. It is known that the area function is convex if the polygons are allowed to intersect [1]. We will show how to find an optimal translation using this convexity.

For two simple polygons P and Q, we call an edge of $\mathsf{conv}(P \cup Q)$ a *bridge* if it has an endpoint at a vertex of P and an endpoint at a vertex of Q.

Lemma 1. *Any two disjoint simple polygons in the plane have zero or two bridges in their convex hull.*

Proof. The vertices of $\mathsf{conv}(P \cup Q)$ are from $V(P)$ and $V(Q)$ and each bridge has one endpoint at a vertex of $V(P)$ and the other at a vertex of $V(Q)$. Therefore, there are even number (including 0) of bridges on $\mathsf{conv}(P \cup Q)$.

Assume to the contrary that there are more than two bridges on the convex hull. Let p_i, q_i for $1 \leq i \leq 4$ be four bridges in $\mathsf{conv}(P \cup Q)$ for $p_i \in V(P)$ and $q_i \in V(Q)$. Since P is connected, $\mathsf{conv}(P \cup Q) \setminus P$ consists of at most four connected components. One connected component has at most two q_i's for which $p_i q_i$ is a bridge. Since Q is connected, at least one boundary chain of Q connecting two q_i's crosses P. This implies that the polygons overlap in their interiors. See the Fig. 1. ⊡

For a pair of a vertex $v \in V(P)$ and an edge $e \in E(Q)$ (or a vertex $v \in V(Q)$ and an edge $e \in E(P)$), let $T_{\mathsf{ve}}(v, e)$ denote the set of all translations τ of P that align v on e. We call such a pair a *vertex-edge pair*. For a pair of a vertex $v \in V(P)$ and an edge $h \in E(\mathsf{conv}(Q)) \setminus E(Q)$ (or a vertex $v \in V(Q)$ and an edge

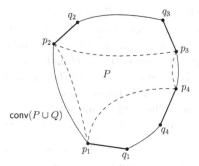

Fig. 1. Four bridges p_1q_1, p_2q_2 and p_3q_3 and p_4p_4 on the boundary of $\mathsf{conv}(P \cup Q)$. It is impossible to connect the four q_i's in a closed curve in $\mathsf{conv}(P \cup Q)$ without crossing P.

$h \in E(\mathsf{conv}(P)) \setminus E(P))$, let $T_{\mathsf{vh}}(v,h)$ denote the set of all translations τ of P that align v on h. We call such a pair a *vertex-hull-edge pair*. Note that $T_{\mathsf{ve}}(v,e)$ (and $T_{\mathsf{vh}}(v,h)$) is connected and contained in a line in the translation space $\mathbb{R}^2$. If $T_{\mathsf{ve}}(v,e)$ (and $T_{\mathsf{vh}}(v,h)$) is not degenerate, the line is uniquely defined. Let $\ell(v,e)$ denote the supporting line of $T_{\mathsf{ve}}(v,e)$. Then $\ell(v,e)$ intersects other supporting lines determined by vertex-edge pairs and vertex-hull-edge pairs. We call the intersection of two supporting lines a *double-contact*.

Lemma 2. *There is an optimal translation τ^* and a vertex-edge pair (v,e) such that τ^* lies in $T_{\mathsf{ve}}(v,e)$.*

Proof. Assume that there is an optimal translation τ such that $P+\tau$ and Q are apart. If there is no bridge in $\mathsf{conv}((P+\tau)\cup Q)$, then either $P+\tau$ is contained in a connected component of $\mathsf{conv}(Q)\setminus Q$ or Q is contained in a connected component of $\mathsf{conv}(P) \setminus P$. In both cases, we translate $P + \tau$ along a direction parallel to the convex-hull edge of the component without increasing the area until the two polygons become in contact with a vertex-edge pair.

If there are two bridges in $\mathsf{conv}((P + \tau) \cup Q)$, then we can always translate P in a direction parallel to one of the bridges such that the area of the convex hull decreased, which contradicts to the optimality of τ.   □

By using the two lemmas above, we give a characterization of an optimal placement.

Lemma 3. *There is an optimal placement τ^* that lies at an intersection of two supporting lines in the translation space $\mathbb{R}^2$. Moreover, τ^* lies in $T_{\mathsf{ve}}(v,e)$ of a vertex-edge pair (v,e).*

Proof. Consider two disjoint simple polygons P and Q that are in contact at a vertex v of P and an edge e of Q. By Lemma 1, there are zero or two bridges, say p_lq_l and p_rq_r ($p_l, p_r \in V(P)$ and $q_l, q_r \in V(Q)$), in the convex hull of P and Q. Imagine that P is translated by τ while P and Q are in contact with the vertex-edge pair (v,e). If there is no bridge, the area function $f(\tau) = \|\mathsf{conv}((P+\tau)\cup Q)\|$

is constant. Otherwise, the function is determined by the quadrilateral $p_l p_r q_r q_l$ because the area of each component of $\mathsf{conv}((P+\tau) \cup Q) \setminus p_l p_r q_r q_l$ remains the same during the translation as long as the combinatorial structure of $\mathsf{conv}((P+\tau) \cup Q)$ remains the same. Moreover, the combinatorial structure changes when passing $T_{\mathsf{vh}}(v', h)$. Note that the area of quadrilateral $p_l p_r q_r q_l$ changes linearly as τ changes. So the area function is non-increasing in one of the two directions parallel to $\ell(v, e)$. Therefore the area is minimized at the intersection of two supporting lines one of which is determined by a vertex-edge pair. ⊡

Based on the lemma above, our algorithm searches, for each supporting line determined by a vertex-edge pair, the intersections with other supporting lines, and finds an intersection realizing the minimum area as follows. For each vertex-edge pair (v, e), we compute the intersections on $\ell(v, e)$ with other supporting lines and sort them along $\ell(v, e)$ in $O(nm \log n)$ time. Then we process the translations τ, each of which is at the intersection of $\ell(v, e)$ and other supporting line, one by one in order and compute $\|(P+\tau) \cup Q\|$ if $P+\tau$ and Q are disjoint in their interiors.

Disjointness Test. We check whether the two polygons are disjoint or not by counting the number c of the connected components in the overlap of the polygons for translations of P along $\ell(v, e)$ for a vertex-edge pair (v, e). We consider the following four types of events that occur during the translation of P along $\ell(v, e)$. For ease of description, we enumerate the event types with respect to vertices of P in the following. There are another four event types with respect to vertices of Q, and they can be defined analogously.

- Type (a): A convex vertex v of P enters into Q and a new connected component appears in the overlap locally around v. See Fig. 2(a).
- Type (b): A convex vertex v of P leaves Q and a connected component of the overlap disappears locally around v. See Fig. 2(b).
- Type (c): A reflex vertex v of P leaves Q and a connected component is subdivided into two connected components in the overlap around v. See Fig. 2(c).
- Type (d): A reflex vertex v of P enters into Q and two distinct connected components in the overlap are merged into one around v. See Fig. 2(d).

The counter c is initialized to 0 in the beginning. During the translation along the supporting line determined by a vertex-edge pair, it increases by 1 for each event of type (a) or (c), and decreases by 1 for each event of type (b) or (d). Since the number of connected components in the overlap of $P+\tau$ and Q changes only for events of the four types above, $P+\tau$ and Q are disjoint if and only if $c = 0$.

For a vertex-edge pair $T_{\mathsf{ve}}(v, e)$ for any v and e, it may contain a double-contact that aligns the two polygons to be disjoint in their interiors, which we call *a feasible double-contact*. We apply the concept of the convolution of the two polygons [8,13] to find all feasible double-contacts as follows. A *state* x is a pair of a position $\dot{x}$ and a direction $\boldsymbol{x}$. A *move* is a set of states with constant direction and position varying along a line segment parallel to the direction. A *turn* is a set of states with constant position and direction varying along an arc of the circle of directions. A *polygonal trip* is a continuous sequence of moves and turns. A polygonal trip is closed if it starts and ends at the same state. A *polygonal tracing* is a collection of closed polygonal trips.

Given a simple polygon P, we denote by $\hat{P}$ the counter-clockwise tracing of the boundary of P. Then the convolution of $\hat{P}$ and $\hat{Q}$ is defined as follows. If p and q are states in $\hat{P}$ and $\hat{Q}$, respectively, having the same direction $\boldsymbol{p} = \boldsymbol{q}$, the state $c = (\dot{p} + \dot{q}, \boldsymbol{p})$ is a state of the convolution. Denote it as $*$.

Every vertex-edge pair containing a feasible double-contact is covered by the convolution. Therefore, it is sufficient to test the disjointness of the two polygons on the convolution by following lemma.

Lemma 4. *The convolution of two polygons contains all vertex-edge pairs containing a feasible double-contact.*

Proof. The *winding number* of a point x with respect to $-\hat{P} * \hat{Q}$ is the number of connected components in $(-P + x) \cap Q$ [8]. Thus, the Minkowski sum $-P \oplus Q$ is the region of $-\hat{P} * \hat{Q}$ such that the winding number of a point $x \in -\hat{P} * \hat{Q}$ is non-zero. By the definition of a vertex-edge pair, a vertex-edge pair containing no feasible double-contact is a region with non-zero winding number. From this, we know that the convolution $-\hat{P} * \hat{Q}$ covers every vertex-edge pair containing a feasible double-contact. ☐

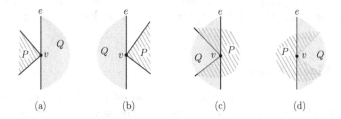

(a)   (b)   (c)   (d)

Fig. 2. Four types of events with respect to vertices of P that occur during the translation of P to the right.

Evaluation of the Area. To evaluate $\|\mathsf{conv}((P + \tau) \cup Q)\|$, we obtain a canonical triangulation T_τ of $\|\mathsf{conv}((P+\tau)\cup Q)\|$ as follows. Choose a point c in the interior of Q and connecting c to all the vertices of $\mathsf{conv}((P + \tau) \cup Q)$ by line segments. Then the area of convex hull $\|\mathsf{conv}((P + \tau) \cup Q)\|$ is the sum of the area of $\triangle \in T_\tau$. To evaluate the area, we maintain an area formula for each triangle of T_τ and their sum.

Let τ_1 and τ_2 be two translations corresponding to two adjacent double-contacts on a supporting line of a vertex-edge pair. Assume that we have just processed τ_1 and we are about to process τ_2 and we have the set of area formula for the triangles in T_{τ_1} and the sum of the formula. The translation of P from $P+\tau_1$ to $P+\tau_2$ may cause a change to the convex hull: either an edge is split into two edges or two adjacent edges merge into one. This makes a constant number of changes in T_{τ_1}: a triangle disappears and a new triangle appears. (See Fig. 3). We compute T_{τ_2} of triangles with their formula and the sum of the formula from T_{τ_1} in $O(1)$ time using $O(n)$ space.

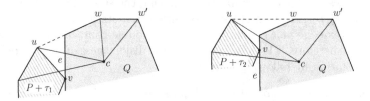

Fig. 3. When polygon $P + \tau_1$ moves to $P + \tau_2$, two triangles, cuw and cww', disappear and new triangle cuw' appears.

As mentioned above, the edges of the convolution of $\hat{P}$ and $\hat{Q}$ cover all vertex-edge pairs containing a feasible double-contact. We can compute an optimal solution by using the convolution in $O(k^2 \log n)$ time using $O(k)$ space, where k is the size of convolution. k is shown to be $O(nm)$ [13].

Theorem 1. *Let P and Q be simple polygons with n and m vertices with $n \geq m$, respectively. We can compute a translation $\tau \in \mathbb{R}^2$ that minimizes $\|\mathsf{conv}((P + \tau) \cup Q)\|$ satisfying $\mathsf{int}(P + \tau) \cap \mathsf{int}(Q) = \emptyset$ in $O(n^2 m^2 \log n)$ time using $O(nm)$ space.*

2.1 Using Linear Space

Now we show how to find an optimal translation using only linear space, at the expense of slightly increased time complexity. The key idea is to compute the double-contacts on a supporting line determined by a vertex-edge pair one by one along the line and evaluate the area of the convex hull, instead of computing all the intersections at once.

Finding the Next Intersection. Again, let $\ell(v, e)$ denote the supporting line determined by a vertex-edge pair (v, e). We compute the double-contacts on $\ell(v, e)$ one by one as follows. We maintain a min-heap H consisting of $O(n)$ double-contacts. Initially, it contains, for each vertex u of P, the first intersection of $\ell(v, e)$ among the supporting lines of (u, e') pairs for all $e' \in E(Q)$. It also contains, for each edge f of P, the first intersection of $\ell(v, e)$ among the supporting lines of (v', f) pairs for all $v' \in V(Q)$.

Our algorithm gets the first intersection τ (determined by, say (u, f)) from H, determines whether $P + \tau$ and Q are disjoint in their interiors, and computes the area of $\mathsf{conv}((P + \tau) \cup Q)$ if they are disjoint in their interiors. Then it frees the space used for processing the intersection and inserts the next intersection (with respect to u or to f) to H. The intersection can be computed in $O(m)$ time by selection algorithm. Then we check whether a double-contact is in convolution or not by disjointness test when we insert it to the heap, which takes $O(\log n)$ time. Thus, we can search an optimal double-contact in convolution in $O(k^2(m+\log n))$ time.

Theorem 2. *Let P and Q be simple polygons with n and m vertices with $n \geq m$, respectively. We can compute a translation $\tau \in \mathbb{R}^2$ that minimizes $\|\text{conv}((P + \tau) \cup Q)\|$ satisfying $\text{int}(P + \tau)) \cap \text{int}(Q) = \emptyset$ in $O(n^2 m^2(m + \log n))$ time using $O(n)$ space.*

3   Minimizing the Convex Hull Under Rigid Motions

In this section, we allow reorienting P and show how to find an optimal rigid motion $\rho \in \mathbb{R}^2 \times [0, 2\pi)$ which minimizes $\|\text{conv}(\rho P \cup Q)\|$.

For two vertex-edge pairs (v, e) and (v', e') of $v, v' \in V(P) \cup V(Q)$ and $e, e' \in E(P) \cup E(Q)$, let $t(ve, v'e')$ denote the set of all rigid motions ρ of P that align v on e and v' on e' simultaneously. Similarly, for a vertex-edge pairs (v, e) and a vertex-hull-edge pair (v', h) of $v, v' \in V(P) \cup V(Q)$, $e \in E(P) \cup E(Q)$, and $h \in E(\text{conv}(P)) \cup E(\text{conv}(Q))$, let $t(ve, v'h)$ denote the set of all rigid motions ρ of P that align v on e and v' on h simultaneously. Note that $t(ve, v'e')$ and $t(ve, v'h)$ might be empty. Let $\ell(ve, v'e')$ denote the supporting curve defined by a nonempty set $t(ve, v'e')$ in the motion space. See Fig. 4. The supporting curve can be represented by a trigonometric function as follows.

The supporting curve $\ell(ve, v'e')$ determined by $T_{ve}(v, e)$ and $T_{ve}(v', e')$ can be represented by one of two types of trigonometric functions depending on whether v and v' are from the same polygon or not. Let v_l and v_r be the endpoints of edge e. Similarly, let v_l' and v_r' be the endpoints of edge e'. Let $t = \frac{|v_l v|}{|v_l v_r|}$, where $|uu'|$ is the distance between two points u and u'. Assume that e is parallel to x-axis for the sake of convenience.

Consider the case that v and v' are from the same polygon. We use $x(u)$ and $y(u)$ to denote the x- and y-coordinate of a point u. Then v' has x-coordinate $t + |vv'| \cos(\theta)$ and y-coordinate $|vv'| \sin(\theta)$, where θ is the angle between vv_r and vv'. We can derive $kx(v_l') + (1 - k)x(v_r') = t + |vv'| \cos(\theta)$ and $ky(v_l') + (1 - k)y(v_r') = |vv'| \sin(\theta)$, where k is a real, $0 \leq k \leq 1$. Then, we derive the following formula by dispelling the parameter k,

$$t = x(v_r) + \frac{x(v_l) - x(v_r)}{y(v_l) - y(v_r)}(|vv'| \sin(\theta + \theta_v)) - |vv'| \cos(\theta + \theta_v).$$

Next, consider the case that v and v' are from different polygons. Let θ_l and θ_r be the angle between vv_r and vv_l' and the angle between vv_r and vv_r', respectively. Then v_l' has x-coordinate $t + |vv_l'| \cos(\theta + \theta_l)$ and y-coordinate $|vv_l'| \sin(\theta + \theta_l)$, and v_r' has x-coordinate $t + |vv_r'| \cos(\theta + \theta_r)$ and y-coordinate $|vv_r'| \sin(\theta + \theta_r)$. Then, we derive the following formula similarly to the above case,

$$t = x(v) - \frac{|vv_l'| \cos(\theta + \theta_l) - |vv_r'| \cos(\theta + \theta_r)}{|vv_l'| \sin(\theta + \theta_l) - |vv_r'| \sin(\theta + \theta_r)}(y(v) - |vv_r'| \sin(\theta + \theta_r)).$$

The supporting curve $\ell(ve, v'h)$ corresponding to $t(ve, v'h')$ is also defined analogously.

Our algorithm works similarly to the one in Sect. 2. We compute, for each vertex-edge pair (v, e), the supporting curves with other vertex-edge pairs or vertex-hull-edge pairs. Let C_{ve} denote the set of the supporting curves determined by (v, e). Next, we search, for each supporting curve in C_{ve}, the intersections with other supporting curves in C_{ve}. We do this for each vertex-pair (v, e). Then we evaluate a local optimal on each supporting curve. Note that the area function is also trigonometric. Hence, this function is derived by additions and multiplications of the coordinates of the vertices, and an optimal rigid motion may lie on any place of the supporting curve. (Under translations, we evaluate only the intersections of two supporting lines.) The intersections of two supporting curves may cause a change to the combinatorial structure of convex hull or to the disjointness of two polygons. Thus we update the formula of the area of convex hull and the disjointness at intersections of two supporting lines if necessary.

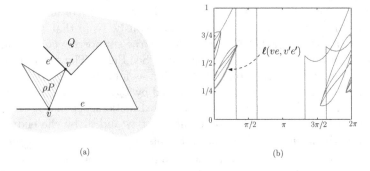

Fig. 4. (a) A vertex v lies on an edge e and a vertex v' lies on an edge e' simultaneously. (b) Two vertex-edge pairs (v, e) and $(v'e')$ determine the supporting curve $\ell(ve, v'e')$.

The number of vertex-edge pairs is $O(nm)$. For a vertex-edge pair (v, e), the number of supporting curves in C_{ve} is bounded by the number of other vertex-edge pairs or vertex-hull-edge pairs, which is $O(nm)$. Any two supporting curves cross each other at most $O(1)$ times since each curve is represented by a trigonometric function, whose domain is a set $\{x \mid 0 \le x < 2\pi\}$. Thus, we conclude the following theorem.

Theorem 3. *Let P and Q be simple polygons with n and m vertices with $n \ge m$, respectively. We can compute a rigid motion ρ that minimizes $\|conv(\rho P \cup Q)\|$ satisfying $int(\rho P) \cap int(Q) = \emptyset$ in $O(n^3 m^3 \log n)$ time using $O(nm)$ space.*

By using an approach similar to the one in Sect. 2.1, we can find a rigid motion that minimizes the area of the convex hull using only linear space, at the expense of slightly increased time complexity.

Theorem 4. *Let P and Q be simple polygons with n and m vertices with $n \geq m$, respectively. We can compute a rigid motion ρ that minimizes $\|conv(\rho P \cup Q)\|$ satisfying $int(\rho P) \cap int(Q) = \emptyset$ in $O(n^3 m^3 (m + \log n))$ time using $O(n)$ space.*

4   Extension to Three Dimensional Space

One may think that Lemma 3 extends to three dimensional space naturally. However, this does not work even for two convex polyhedra in 3-dimension. Ahn et al. [3] showed two convex polytopes whose unique optimal translation separates them apart. Thus, for an optimal translation τ^*, $P + \tau^*$ and Q are either apart or in contact. In the case that $P + \tau^*$ and Q are apart, we find the optimal translation by any algorithm for minimizing the volume of their convex hull [4].

Therefore, we focus on the problem where the two polyhedra are supposed to be in contact with each other. There are three types of contact pairs – vertex-facet pairs (v, f), vertex-hull-facet pairs (v, h), and edge-edge pairs (e, e) – each of which defines a set of translations in the 3-dimensional translation space, denoted by $T_{vf}(v, f)$, $T_{vh}(v, h)$, $T_{ee}(e, e')$, respectively. Here, a hull facet is either a facet of $F(conv(P)) \setminus F(P)$ or a facet of $F(conv(Q)) \setminus F(Q)$, where $F(R)$ denotes the set of facets of a polyhedron R. Clearly, a nonempty set $T_{vf}(v, f)$ (or $T_{vh}(v, h)$) forms a polygon in the space $\mathbb{R}^3$, which is a translate of $f - v$ (or $h - v$) if $v \in V(P)$, or a translate of $-f + v$ (or $-h + v$) otherwise. A nonempty set $T_{ee}(e, e')$ forms a parallelogram $e \oplus (-e')$ in the 3-dimensional translation space.

Lemma 5. *If there is an optimal translation that aligns two polytopes to be in contact, it always lies at the intersection of three supporting planes of translation polygons determined by contact pairs. Moreover, one of the three supporting planes must be determined by a vertex-facet or edge-edge pair.*

Proof. It suffices to prove that the volume function $f(\tau) = \|conv((P + \tau) \cup Q)\|$ is piecewise linear along an arbitrary line. Assume that P moves along a line parallel to the x-axis, that is, for $t \in \mathbb{R}$, $\tau = P + (t, 0, 0)$. We will show that the function $\omega(t) = \|conv((P + (t, 0, 0)) \cup Q)\|$ is piecewise linear. Let P_k be the intersection of a polyhedron P and a plane $z = k$. Then ω can be expressed as the sum of $conv((P + \tau) \cup Q)_k$ over all $k \in \mathbb{R}$. Each of the intersections can be interpreted by the area function of the convex hull of two convex polygons which is piecewise linear [1]. The sum of piecewise linear functions is also piecewise linear, and therefore, ω is piecewise linear. $\qquad\square$

Every two supporting planes meet along a line unless they are parallel. Thus, we consider each intersection line of two supporting planes and compute the intersections of the line with other supporting planes. Let $H(v, f)$ denote the supporting plane of $T_{vf}(v, f)$.

We check whether the two polytopes are disjoint or not in a way similar to the disjointness test in Sect. 2.

Evaluation of the Volume. The volume of $\mathsf{conv}((P + \tau) \cup Q)$ can be evaluated by subdividing it into a set of tetrahedra as follows. First, triangulate each facet of $\mathsf{conv}((P + \tau) \cup Q)$. Let T_τ be the set of those triangles on the boundary $\partial\mathsf{conv}((P+\tau)\cup Q)$. Next, choose a point c in the interior of Q and connect c to every vertices of $\mathsf{conv}((P+\tau)\cup Q)$ with edges. For each triangle $\triangle \in T_\tau$, let $\triangle^+$ be the tetrahedron with base $\triangle$ and apex c. Thus, the volume of $\mathsf{conv}((P+\tau)\cup Q)$ is the sum of $\|\triangle^+\|$ for $\triangle \in \partial\mathsf{conv}((P + \tau) \cup Q)$. Hence, the volume can be computed in $O(\mathrm{card}(T_\tau)) = O(n)$ time, where $\mathrm{card}(T_\tau)$ is the cardinality of T_τ. We show that evaluating the volume for each intersection is computed in $O(1)$ time by amortized analysis exploiting coherence as follows.

Exploiting Coherence. Let τ and τ' be translations corresponding to two adjacent intersections on an intersection line ℓ of two supporting planes. Assume that we have just processed τ and we are about to process τ'. We maintain T_τ, a set of formulas each of which represents the area function $\|\triangle^+\|$ for a $\triangle \in T_\tau$, and sum of these functions which is the volume formula of $\|\mathsf{conv}((P + \tau) \cup Q)\|$.

Assume that τ' is the intersection of ℓ with a supporting plane determined by a vertex-facet contact (v, f). The translation τ' from τ causes a change to the convex hull of the two polyhedra. We update $T_{\tau'}$ from T_τ and their formulas of $\triangle^+ \in T_{\tau'}$ accordingly as follows. A vertex v of $P+\tau'$ lies on the supporting plane of f so that the triangles defined on f disappear (and their tetrahedra) and the triangles defined by the edges connecting the boundary vertices of f and v appear at τ'. This implies that the number of triangles that disappear or appear at the supporting plane does not exceed the number of boundary vertices of facet f. We update $T_{\tau'}$ by removing all triangles that disappear and adding all triangles that appear and then update the formulas for the volumes. This can be done in $O(N_f)$ time, where N_f denotes the number of boundary vertices of f. The sum of N_f for $f \in F(P)\cup F(Q)$ is bounded by $2 \times (|V(P)| \cdot |F(Q)| + |V(Q)| \cdot |F(P)|)$. Thus, we can evaluate the volume in $O(1)$ time per each intersection.

The number of supporting plane is $O(nm)$. Thus, the number of the intersection lines determined by two supporting planes is $O(n^2m^2)$. Each intersection line intersects $O(nm)$ supporting planes. Thus, we conclude following theorems.

Theorem 5. *Let P and Q be polyhedra with n and m vertices with $n \geq m$, respectively. We can compute a placement $\tau \in \mathbb{R}^3$ that minimizes $\|\mathsf{conv}((P + \tau) \cup Q)\|$ satisfying $\mathit{int}(P + \tau) \cap \mathit{int}(Q) = \emptyset$ in $O(n^3m^3 \log n)$ time using $O(nm)$ space.*

Theorem 6. *Let P and Q be polyhedra with n and m vertices with $n \geq m$, respectively. We can compute a placement $\tau \in \mathbb{R}^3$ that minimizes $\|\mathsf{conv}((P + \tau) \cup Q)\|$ satisfying $\mathit{int}(\varphi P) \cap \mathit{int}(Q) = \emptyset$ in $O(n^3m^3(m + \log n))$ time using $O(n)$ space.*

References

1. Ahn, H.K., Cheong, O.: Aligning two convex figures to minimize area or perimeter. Algorithmica **62**, 464–479 (2012)
2. Ahn, H.K., Abardia, J., Bae, S.W., Cheong, O., Dann, S., Park, D., Shin, C.S.: The minimum convex container of two convex polytopes under translations, submitted manuscript
3. Ahn, H.K., Bae, S.W., Cheong, O., Park, D., Shin, C.S.: Minimum convex container of two convex polytopes under translations. In: Proceedings of the 26th Canadian Conference on Computational Geometry (CCCG 2014) (2014)
4. Ahn, H.K., Brass, P., Shin, C.S.: Maximum overlap and minimum convex hull of two convex polyhedra under translations. Comput. Geom. **40**(2), 171–177 (2008). http://www.sciencedirect.com/science/article/pii/S0925772107000909
5. Alt, H., Hurtado, F.: Packing convex polygons into rectangular boxes. In: Akiyama, J., Kano, M., Urabe, M. (eds.) JCDCG 2000. LNCS, vol. 2098, pp. 67–80. Springer, Heidelberg (2001). doi:10.1007/3-540-47738-1_5
6. de Berg, M., Cheong, O., van Kreveld, M., Overmars, M.: Computational Geomtry, Algorithms and Applications, 3rd edn. Springer, Berlin (2008)
7. Daniels, K., Milenkovic, V.: Multiple translational containment, part I: an approximation algorithm. Algorithmica **19**, 148–182 (1997)
8. Guibas, L., Ramshaw, L., Stolfi, J.: A kinetic framework for computational geometry. In: Proceedings of the 24th Annual Symposium on Foundations of Computer Science (FOCS 1983), pp. 100–111. IEEE (1983)
9. Kaul, A., O'Connor, M.A., Srinivasan, V.: Computing Minkowski sums of regular polygons. In: Proceedings of the 3rd Canadian Conference on Computational Geometry (CCCG 1991), pp. 74–77 (1991)
10. Kepler, J.: Vom sechseckigen Schnee, Ostwalds Klassiker der Exakten Wissenschaften, vol. 273. Akademische Verlagsgesellschaft Geest & Portig K.-G., Leipzig, strena seu de Nive sexangula, Translated from the Latin and with an introduction and notes by Dorothea Goetz (1987)
11. Lee, H.C., Woo, T.C.: Determining in linear time the minimum area convex hull of two polygons. IIE Trans. **20**(4), 338–345 (1988)
12. Milenkovic, V.: Rotational polygon containment and minimum enclosure using robust 2D constructions. Comput. Geom.: Theory Appl. **13**, 3–19 (1999)
13. Ramkumar, G.: An algorithm to compute the Minkowski sum outer-face of two simple polygons. In: Proceedings of the 12th Annual Symposium on Computational Geometry (SoCG 1996), pp. 234–241. ACM (1996)
14. Sugihara, K., Sawai, M., Sano, H., Kim, D.S., Kim, D.: Disk packing for the estimation of the size of a wire bundle. Jpn. J. Ind. Appl. Math. **21**, 259–278 (2004)
15. Tang, K., Wang, C.C.L., Chen, D.Z.: Minimum area convex packing of two convex polygons. Int. J. Comput. Geom. Appl. **16**(1), 41–74 (2006)

Combinatorial Optimization

Tangle and Maximal Ideal

Koichi Yamazaki[✉]

Faculty of Science and Technology, Gunma University,
1-5-1 Tenjin-cho, Kiryu, Gunma, Japan
koichi@cs.gunma-u.ac.jp

Abstract. Tangle is a dual notion of the graph parameter branch-width. The notion of tangle can be extended to connectivity systems. Ideal is an important notion that plays a foundational role in ring, set, and order theories. Tangle and (maximal) ideal are defined by axiomatic systems, and they have some axioms in common. In this paper, we define ideals on connectivity systems. Then, we address the relations between tangles and maximal ideals on connectivity systems. We demonstrate that a tangle can be considered as a non-principle maximal ideal on a connectivity system.

Keywords: Tangle · Ideal · Submodularity

1 Introduction

In this paper, we study the relations between two axiomatic systems: tangles and maximal ideals. The notion of tangle was first introduced by Robertson and Seymour in [18] for (hyper)graphs as a dual notion of a graph parameter branch-width. This was later extended to matroids (connectivity systems) [6, 7,11]. In addition, Oum and Seymour introduced a relaxed notion of tangle in [15], called a *loose tangle*, and showed that there exists a tangle of order k if and only if there exists a loose tangle of order k. The notion of an ideal is wide-ranging; it can be found in various contexts, such as ring theory (ring), set theory (boolean algebra), and order theory (lattice, poset), and is defined in their different contexts. Ideals discussed in this paper are those that appear in the context of set theory (see cf. [21]).

1.1 Motivation and Contribution: From the Perspective of Ideals

For an underling set X, a family $\mathscr{I} \subset 2^X$ is an *ideal* on X if it satisfies the following axioms:

(IH) $A, B \subseteq X$, $A \subset B$, and $B \in \mathscr{I} \implies A \in \mathscr{I}$.
(IU) $\forall A, B \in \mathscr{I}$, $A \cup B \in \mathscr{I}$.
(IW) $X \notin \mathscr{I}$.

© Springer International Publishing AG 2017
S.-H. Poon et al. (Eds.): WALCOM 2017, LNCS 10167, pp. 81–92, 2017.
DOI: 10.1007/978-3-319-53925-6_7

Intuitively speaking, ideals can be regarded as a family of "small sets" such as "virtually empty sets", "negligibly small sets", or sets that do not contain a "specified something that is important". For example, (IH) indicates that if B does not contain something important, say x, then any subset A of B also does not contain x. (IU) tells us that if A and B do not contain x, then $A \cup B$ also does not contain x. (IW) guarantees that the underling set X does contain x. In previous studies, ideals have usually been considered on the power set 2^X for an underlying (usually infinite) set X. The study of ideals for a finite underlying sets has not yet received significant attention. In this paper, roughly speaking instead of 2^X we consider ideals on $\mathscr{S}_k = \{A \subset X \mid f(A) \leq k\}$ for a submodular function f on a *finite* set X and a positive integer k. We call a set $A \subseteq X$ *small* if $f(A) \leq k$, and we will refer to such ideals as *ideals on the connectivity system* (X, f). That is, we extend the notion of an ideal on 2^X to a notion of one on $\mathscr{S}_k$ as follows:

(TB) $\forall A \in \mathscr{I}, f(A) \leq k$.
(TH) $A, B \subseteq X, \overline{A \subset B, B \in \mathscr{I}, \text{ and } f(A) \leq k} \implies A \in \mathscr{I}$.
(TU) $A, B \in \mathscr{I}, \overline{f(A \cup B) \leq k} \implies \overline{A \cup B \in \mathscr{I}}$.
(IW) $X \notin \mathscr{I}$.

The only differences between the definitions of ideals on 2^X and on $\mathscr{S}_k$ are the underlined parts. To the authors' knowledge, no previous study has addressed such ideals on $\mathscr{S}_k$. Ideals on 2^X should take all sets in 2^X into consideration, while for ideals on $\mathscr{S}_k$ only the small sets in 2^X should be considered. It should be noted that this extension is indeed a natural generalization of the usual notion of an ideal, because it holds that $2^X = \mathscr{S}_k$ for sufficiently large k if f is submodular. In this paper, we essentially address the following questions regarding maximal ideals:

– A maximal ideal $\mathscr{I}_a$ (on 2^X) can be obtained from an element $a \in X$ by setting $\mathscr{I}_a = \{A \subset X \mid a \notin A\}$. Such a maximal ideal is called *principle (generated by a)*, and it is well known that if the underlying set X is finite then all maximal ideals on 2^X are principle. (This does not hold in general for infinite underlying sets, cf., Fréchet filter.) Namely, there is no non-principle maximal ideal on 2^X for a *finite* underlying set X. The first question is whether this also holds for ideals on $\mathscr{S}_k$.
– It is also widely known that an ideal $\mathscr{I}$ (on 2^X) is maximal if and only if for any $A \in 2^X$, exactly one of A and $\overline{A}(= X \backslash A)$ is in $\mathscr{I}$, where X is not necessarily finite in this case (see, e.g., Sect. 7 in [21], Subsect. 2.2). This fact is frequently found in proofs of the *ultrafilter lemma*. For the case that X is finite, this is obvious by the fact mentioned in describing the first question. The second question is the following: Does it hold for any ideal $\mathscr{I}$ on $\mathscr{S}_k$ that $\mathscr{I}$ is maximal if and only if for any $A \in \mathscr{S}_k$ exactly one of A and $\overline{A}$ is in $\mathscr{I}$?

In this paper, we show that the answer to first question is no and the answer to second question is yes.

1.2   Motivation and Contribution: From the Perspective of Tangles

As mentioned above, Robertson and Seymour introduced tangles as follows [18]:

(TB) $\forall A \in \mathscr{T}, f(A) \leq k$.
(TE) $A \subseteq X, f(A) \leq k \implies$ either $(A \in \mathscr{T})$ or $(\overline{A} \in \mathscr{T})$.
(TC) $A, B, C \in \mathscr{T} \implies A \cup B \cup C \neq X$.
(TL) $\forall a \in X, X \backslash \{a\} \notin \mathscr{T}$.

We often call the axiom (TE) the *excluded middle* axiom. A set $A \subseteq X$ is called *k-small* (or simply small when k is clear) if $f(A)$ is at most k. Oum and Seymour introduced loose tangles as follows [15]:

(LL) $A \subseteq X, |A| \leq 1$, and $f(A) \leq k \implies A \in \mathscr{L}$.
(LSU) $A, B \in \mathscr{L}, C \subseteq A \cup B$, and $f(C) \leq k \implies C \in \mathscr{L}$.
(IW) $X \notin \mathscr{L}$.

For the following reasons, we suspect that there may be some connections between tangles and maximal ideals:

- The excluded middle axiom (TE) and the characterization of a maximal ideal (on 2^X) in the second question above have essentially the same form.
- There are some similarities between loose tangles and ideals (To see this, consult the proof of Theorem 5 in [15], especially k-branched sets.)
- The notion of a loose tangle is a relaxation of the notion of a tangle.

In this paper, we clarify these connections. More precisely, we show the following (see Fig. 1):

- The notion of a loose tangle coincides with that of an ideal on a connectivity system (Theorem 1).
- The notion of a tangle coincides with that of an ideal on a connectivity system satisfying the excluded middle axiom (TE) (Theorem 2).
- Any maximal ideal on a connectivity system satisfies the axiom (TE) (Theorem 3). This means that an ideal $\mathscr{I}$ on a connectivity system satisfies the axiom (TE) if and only if $\mathscr{I}$ is a maximal ideal on the connectivity system.

From the second and third properties above, we can conclude that the notion of a tangle coincides with that of a maximal ideal on connectivity systems. Considering this connection between tangles and ideals, the definition of a tangle can be interpreted as follows:

(TB) "$\forall A \in \mathscr{T}, f(A) \leq k$" means that we only have to consider *small* sets.
(TE) "$A \subseteq X, f(A) \leq k \implies$ either $(A \in \mathscr{T})$ or $(\overline{A} \in \mathscr{T})$" represents the maximality.
(TC) "$A, B, C \in \mathscr{T} \implies A \cup B \cup C \neq X$" indicates that the tangle is an ideal.
(TL) "$\forall a \in X, X \backslash \{a\} \notin \mathscr{T}$" means that the tangle is non-principle.

From this interpretation, we can see that the axioms indicates a non-principle maximal ideal on a connectivity system and that this definition is very natural and reasonable.

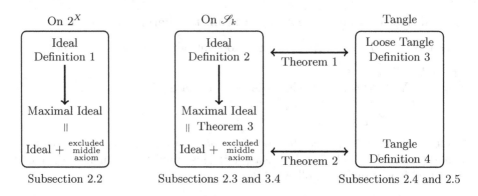

Fig. 1. Relations between theorems and definitions in this paper.

1.3 Relation Between Our Results and Duality Theorems

Many duality theorems have been proved decompositions (such as path, tree, and branch decompositions) and dual combinatorial objects (such as blockages, brambles, and tangles), [1–4,7,9,10,14,17–20]. Based on the results in this paper, some duality theorem can be obtained by considering the duality between ideals and filters. In fact, the complement of a tangle can be thought of as an ultrafilter as indicated in [8,9,12]. In this paper, we will not address any such duality theorems, but rather we consider the relation between maximality and the excluded middle axiom in ideals on connectivity systems. However the proofs in this paper rely heavily upon the techniques used and developed in the proving the duality theorems, especially those used in [11,15,18].

The organization of the remainder of this paper is as follows. We present definitions and some known results regarding ideals and tangles in Sect. 2. Then, we present our results in Sect. 3. (We defer the proofs of some of these to the appendix.) Section 4 describes the conclusions and directions for further research.

2 Definitions and Known Results

In this section, we present some definitions and known results relating to tangles and ideals that we will be required in this paper.

2.1 Submodular Functions and Connectivity Systems

A function f is called *symmetric submodular* if it satisfies the symmetry property that $\forall A \subseteq X$, $f(A) = f(\overline{A})$, and the submodularity property that $\forall A, B \subseteq X$, $f(A) + f(B) \geq f(A \cap B) + f(A \cup B)$. For an integer $k > 0$, a set $A \subset X$ is called *k small* (or simply small) if it holds that $f(A) \leq k$. Next, we provide some useful inequalities that will be used in our proofs.

Lemma 1. *A symmetric submodular function f satisfies the following inequalities (see cf. [11]):*

1. $\forall A \subseteq X,\ f(A) \geq f(\emptyset)$.
2. $f(A) + f(B) \geq f(A \backslash B) + f(B \backslash A)$.

Proof. (1): $f(A) + f(A) = f(A) + f(\overline{A}) \geq f(A \cup \overline{A}) + f(A \cap \overline{A}) = f(X) + f(\emptyset) = f(\emptyset) + f(\emptyset)$. (2): $f(A) + f(B) = f(A) + f(\overline{B}) \geq f(A \cup \overline{B}) + f(A \cap \overline{B}) = f(B \backslash A) + f(A \backslash B) = f(B \backslash A) + f(A \backslash B)$.

We call the pair (X, f) a *connectivity system*.

2.2   Ideals and Maximal Ideals on a Boolean Algebra

Definition 1. *Let X be an underlying set. A collection $\mathscr{I} \subseteq 2^X$ is called an ideal (on a boolean algebra) if $\mathscr{I}$ satisfies the following axioms:*

(IH) $A, B \subseteq X$, $A \subset B$, and $B \in \mathscr{I} \implies A \in \mathscr{I}$.
(IU) $\forall A, B \in \mathscr{I}$, $A \cup B \in \mathscr{I}$.
(IW) $X \notin \mathscr{I}$.

An ideal $\mathscr{I}$ (on a boolean algebra) is *maximal* if there is no $\mathscr{I}'$ properly containing $\mathscr{I}'$. Maximal ideals have several characterizations. For example, an ideal $\mathscr{I}$ with an underlying set X is *maximal* if and only if

– there is an element $a \in X$ such that $A \in \mathscr{I}$ if and only if $a \notin A$. This characterization holds when X is finite. Such an $\mathscr{I}$ is called the *principle ideal generated by a*.
– for any $A \subseteq 2^X$ either $A \in \mathscr{I}$ or $\overline{A} \in \mathscr{I}$ holds (i.e., the excluded middle axiom).
– for any partition (X_1, X_2, X_3) of X, *exactly* one of $\{X_1, X_2, X_3\}$ does not belong to $\mathscr{I}$ (see Corollary 1.6 in [13]).
– there exists a two valued measure m such that $m(A) = 0$ if and only if $A \in \mathscr{I}$ (see cf. 6.1 in Sect. 6 in [21], Lemma 3.1 in [13]).

2.3   Ideals and Maximal Ideals on Connectivity Systems

In Subsect. 2.2, we recall the definition of an ideal on *a boolean algebra*. In this subsection, we introduce the definition of an ideal on a *connectivity system*.

Definition 2. *Let (X, f) be a connectivity system and $k > 0$ an integer. A collection $\mathscr{I} \subseteq 2^X$ is called an ideal on (X, f) of order $k + 1$ if $\mathscr{I}$ satisfies the following axioms:*

(TB) $\forall A \in \mathscr{I}$, $f(A) \leq k$.
(TH) $A, B \subseteq X$, $A \subset B$, $B \in \mathscr{I}$, and $f(A) \leq k \implies A \in \mathscr{I}$.
(TU) $A, B \in \mathscr{I}$, $f(A \cup B) \leq k \implies A \cup B \in \mathscr{I}$.
(IW) $X \notin \mathscr{I}$.

We will refer to the ideal $\mathscr{I}$ as an ideal on (the connectivity system) (X, f) of order $k + 1$.

An ideal $\mathscr{I}$ is called *maximal* (on a connectivity system) if there exists no $\mathscr{I}'$ that properly contains $\mathscr{I}$. A natural question arises here: can a maximal ideal on a connectivity system be characterized by the excluded middle axiom (TE)? In Subsect. 3.4, we show that the answer to this is yes (see Theorem 3). Another natural question is that of whether it holds that for an ideal $\mathscr{I}$ on a connectivity system, $\mathscr{I}$ is maximal if and only if $\mathscr{I}$ is prime. Here, an ideal $\mathscr{P}$ on a connectivity system is *prime* if $\mathscr{P}$ satisfies the following additional axiom (IP): For all A, B with $f(A) \leq k$ and $f(B) \leq k$, if $A \cap B \in \mathscr{P}$, then it holds that either $A \in \mathscr{P}$ or $B \in \mathscr{P}$. We demonstrate the equivalence in the next lemma.

Lemma 2. *Let $\mathscr{I} \neq \emptyset$ be an ideal on a connectivity system (X, f). Then, $\mathscr{I}$ satisfies the excluded middle axiom (TE) if and only if $\mathscr{I}$ is prime.*

Proof. First, we show that $\mathscr{I}$ satisfies (TE) $\implies$ $\mathscr{I}$ is prime. Let A and B be sets such that $f(A) \leq k$ and $f(B) \leq k$. We will demonstrate that $A, B \notin \mathscr{I}$ implies that $A \cap B \notin \mathscr{I}$. Because $A, B \notin \mathscr{I}$ and $f(A), f(B) \leq k$, we have that $\overline{A}, \overline{B} \in \mathscr{I}$ by (TE). If $k < f(\overline{A} \cup \overline{B}) = f(\overline{A \cap B}) = f(A \cap B)$, then by (TB) we have that $A \cap B \notin \mathscr{I}$, as desired. Hence, we may assume that $f(\overline{A} \cup \overline{B}) \leq k$. Then, it follows that $\overline{A \cap B} = \overline{A} \cup \overline{B} \in \mathscr{I}$ by (TU). Hence, by (TE) we have that $A \cap B \notin \mathscr{I}$.

Next, we show that $\mathscr{I}$ is prime $\implies$ $\mathscr{I}$ satisfies (TE). As $\mathscr{I} \neq \emptyset$, there exists a set $A \in \mathscr{I}$. Hence, from Lemma 1, $f(\emptyset) \leq k$, which implies that $\emptyset \in \mathscr{I}$ by (TH). Let Y be a subset of X such that $f(Y) \leq k$. As $Y \cap \overline{Y} = \emptyset \in \mathscr{I}$, it follows from definition of prime that either $Y \in \mathscr{I}$ or $\overline{Y} \in \mathscr{I}$. If both are in $\mathscr{I}$, then it holds that $X \in \mathscr{I}$ by (TU) and the fact that $f(X) = f(\emptyset) \leq k$, but this contradicts (IW). Thus, exactly one of these must hold. $\square$

In Subsect. 3.4, we will prove Theorem 3, which essentially states that $\mathscr{I}$ is maximal if and only if $\mathscr{I}$ satisfies the excluded middle axiom (TE). By combining Lemma 2 with Theorem 3, we will obtain the equivalence.

2.4 Loose Tangles

Definition 3. *For a connectivity system (X, f), a collection $\mathscr{L} \subseteq 2^X$ is called a loose tangle of order $k + 1$ if $\mathscr{L}$ satisfies the following axioms [15]:*

(LL) $A \subseteq X$, $|A| \leq 1$, and $f(A) \leq k \implies A \in \mathscr{L}$.
(LSU) $A, B \in \mathscr{L}$, $C \subseteq A \cup B$, and $f(C) \leq k \implies C \in \mathscr{L}$.
(IW) $X \notin \mathscr{L}$.

The definition appears to be similar to that of an ideal on a connectivity system. In fact, we will show that a loose tangle corresponds to an ideal on a connectivity system and that a tangle corresponds to a maximal ideal on a connectivity system. It is known that there exists no loose tangle of order $k + 1$ if and only if branch-width of (X, f) is at most k (see Theorem 5 in [15]). That is, there exists a tangle of order k if and only if there exists a loose tangle of order k. As we will see, this corresponds to the fact that for a connectivity system, there exists a maximal ideal of order k if and only if there exists an ideal of order k.

2.5 Tangles

Definition 4. *Let X be an underlying set, and let f be a symmetric submodular function. A collection $\mathscr{T} \subseteq 2^X$ on the connectivity system (X, f) is called a tangle of order $k + 1$ if $\mathscr{T}$ satisfies the following axioms [11,18]:*

(TB) $\forall A \in \mathscr{T}, f(A) \leq k.$
(TE) $A \subseteq X, f(A) \leq k \implies$ *either $(A \in \mathscr{T})$ or $(\overline{A} \in \mathscr{T})$.*
(TC) $A, B, C \in \mathscr{T} \implies A \cup B \cup C \neq X.$
(TL) $\forall a \in X, X \backslash \{a\} \notin \mathscr{T}.$

It is known that the tangle admits the following alternative definition [11]:

(TB) $\forall A \in \mathscr{T}, f(A) \leq k.$
(TE) $A \subseteq X, f(A) \leq k \implies$ either $(A \in \mathscr{T})$ or $(\overline{A} \in \mathscr{T})$.
(TH) $A, B \subseteq X, A \subset B, B \in \mathscr{T}$, and $f(A) \leq k \implies A \in \mathscr{T}.$
(T3P) $\forall$ partitions $(X_1, X_2, X_3), \exists 1 \leq i \leq 3$ s.t. $X_i \notin \mathscr{T}.$
(TL) $\forall a \in X, X \backslash \{a\} \notin \mathscr{T}.$

Because $\{A \mid f(A) \leq k - 1\} \subseteq \{A \mid f(A) \leq k\}$, if there exists a tangle of order k then there also exists a tangle of order $k - 1$. For a connectivity system (X, f), the maximum integer k for which there exists a tangle of order k on (X, f) is denoted by $tn_f(X)$ (or simply $tn(X)$), and is called the *tangle number* of (X, f). It is known that $tn_f(X)$ is equal to the branch-width of (X, f) (see Theorem 3.2 in [11]). Suppose that there exists an element $b \in X$ such that $f(\{b\}) > k$. In this case, there trivially exists a tangle. In fact, the family $\{A \mid b \notin A, f(A) \leq k\}$ is a tangle. Usually we are not interested in such trivial case. We revisit this in Subsect. 3.3.

3 Results

In this section, we first provide a new definition of a loose tangle that is similar to the definition of an ideal on a connectivity system. Next, we give a new definition of a tangle that constitutes the definition of an ideal on a connectivity system with an additional axiom (TE). Finally, we show that in connectivity systems, an ideal satisfying (TE) coincides with a maximal ideal.

3.1 A New Definition of a Loose Tangle

Theorem 1. *A collection $\mathscr{L} \subseteq 2^X$ is a loose tangle of order $k + 1$ if and only if $\mathscr{L}$ satisfies the following axioms:*

(TB) $\forall A \in \mathscr{L}, f(A) \leq k.$
(TH) $A, B \subseteq X, A \subset B, B \in \mathscr{L}$, and $f(A) \leq k \implies A \in \mathscr{L}.$
(TU) $A, B \in \mathscr{L}, f(A \cup B) \leq k \implies A \cup B \in \mathscr{L}.$
(LL) $\forall A \subseteq X, |A| \leq 1, f(A) \leq k \implies A \in \mathscr{L}.$
(IW) $X \notin \mathscr{L}.$

Proof. Let us recall the following:

(LSU) $A, B \in \mathscr{L}$, $C \subseteq A \cup B$, and $f(C) \leq k \implies C \in \mathscr{L}$.

Because the original definition of a loose tangle also satisfies the axiom (TB), the only difference between the original definition of a loose tangle and the axiomatic system stated in the theorem is that (LSU) is replaced with (TH) and (TU) (see Definition 3).

We first show that (LSU) $\implies$ (TU) and (TH). It is clear that (LSU) implies (TU). To show that (LSU) $\implies$ (TH), take $\emptyset (\in \mathscr{L})$ as A in (LSU). Then, for any $C \subseteq B(\in \mathscr{L})$ with $f(C) \leq k$, we have that $C \in \mathscr{L}$, as required.

Next, we show that (TU) and (TH) $\implies$ (LSU) by adopting the technique used in [16]. Let $A, B \in \mathscr{L}$, $C \subseteq A \cup B$, and $f(C) \leq k$. (Note that $f(A), f(B) \leq k$ by (TB).) Furthermore, let W be a set such that $C \subseteq W \subseteq A \cup B$ that minimizes $f(W)$. Note that by the choice of W, we have that $f(W) \leq f(C) \leq k$ and $f(W) \leq f(Y \cup W)$ for any $Y \subseteq A \cup B$. From this, it follows that $f(Y) + f(W) \geq f(Y \cap W) + f(Y \cup W) \geq f(Y \cap W) + f(W)$. Thus, it holds that $f(Y) \geq f(Y \cap W)$ for any $Y \subseteq A \cup B$ (we will refer to this as (‡)).

We now demonstrate that $W \in \mathscr{L}$, from which it follows from (TH) that $C \in \mathscr{L}$, as desired. By taking Y in (‡) as A, we have that $f(A \cap W) \leq f(A) \leq k$. Thus, because $A \in \mathscr{L}$, we have that $A \cap W \in \mathscr{L}$ by (TH). Similarly, by taking Y in (‡) as B, we have that $B \cap W \in \mathscr{L}$. Then, because $f(W) \leq k$ and $W = W \cap (A \cup B) = (A \cap W) \cup (B \cap W)$, we have that $W \in \mathscr{L}$ by (TU). $\square$

3.2   A New Definition of a Tangle

In this subsection, we give a characterization of tangles from the viewpoint of ideal.

Theorem 2. *A collection $\mathscr{T} \subseteq 2^X$ is a tangle of order $k + 1$ if and only if $\mathscr{T}$ satisfies the following axioms:*

(TB) $\forall A \in \mathscr{T}$, $f(A) \leq k$.
(TE) $A \subseteq X$, $f(A) \leq k \implies$ either $(A \in \mathscr{T})$ or $(\overline{A} \in \mathscr{T})$.
(TH) $A, B \subseteq X$, $A \subseteq B$, $B \in \mathscr{T}$, and $f(A) \leq k \implies A \in \mathscr{T}$.
(TU) $A, B \in \mathscr{T}$, $f(A \cup B) \leq k \implies A \cup B \in \mathscr{T}$.
(LL) $\forall A \subseteq X$, $|A| \leq 1$, $f(A) \leq k \implies A \in \mathscr{T}$.

Proof. Because (LL) together with (TE) can be replaced with (TL), the difference between the original definition of a tangle and the axiomatic system stated in the theorem is that (TC) is replaced with (TH) and (TU) (see Definition 4). Hence, it is sufficient to show that (TC) holds if and only if both (TH) and (TU) hold, under assumption that (TB) $\wedge$ (TE) $\wedge$ (TL) holds.

First, we show that (TC) $\implies$ (TH). Note that, since $f(\emptyset) \leq f(U), \forall U \subseteq X$ by Lemma 1, and $f(A) \leq k$, we have that $f(\emptyset) \leq f(A) \leq k$. Hence, by (TE), it holds that either $\emptyset \in \mathscr{T}$ or $X \in \mathscr{T}$. Assume to the contrary that there are sets A and B such that $B \in \mathscr{T}$, $A \subseteq B$, $f(A) \leq k$, and $A \notin \mathscr{T}$. Then, by the assumption and (TE), we have that $\overline{A} \in \mathscr{T}$, and thus, $\emptyset \cup B \cup \overline{A} = X$ (we will refer to this as ($\sharp$)) from $A \subseteq B$. On the other hand, it clearly holds that $X \cup B \cup \overline{A} = X$ (we will refer to this as ($\flat$)). Therefore, if $\emptyset \in \mathscr{T}$ then (TC) does not hold from ($\sharp$). On the other hand, if $X \in \mathscr{T}$ then (TC) also does not hold from ($\flat$).

Next, we show that (TC) $\implies$ (TU). Suppose to the contrary that there exist sets A and B such that $A, B \in \mathscr{T}$, $f(A \cup B) \leq k$, and $A \cup B \notin \mathscr{T}$. From the fact that $f(A \cup B) \leq k$, (TE) implies that $\overline{A \cup B} \in \mathscr{T}$. As $A, B \in \mathscr{T}$, it holds that $A \cup B \cup (\overline{A \cup B}) = X$, which contradicts (TC).

Finally, we show that (TH$^{\flat}$), (TU) $\implies$ (TC). Assume to the contrary that there exist $A, B, C \in \mathscr{T}$ such that $A \cup B \cup C = X$. Then, choose A, B, and C to minimize $|A \cap B| + |B \cap C| + |C \cap A|$. For such A, B, C, we now claim that (A, B, C) is a partition; that is, $|A \cap B| + |B \cap C| + |C \cap A| = 0$. We show this through the same technique as is used in Lemma 3.1 in [11]. Suppose this does not hold. Then, with loss of generality, we may assume that $|A \cap B| \geq 1$. Furthermore, from $f(A) + f(B) \geq f(A \backslash B) + f(B \backslash A)$ and (TB), it follows that $f(A \backslash B) \leq k$ or $f(B \backslash A) \leq k$. With loss of generality, we may assume that $f(A \backslash B) \leq k$. From $f(A \backslash B) \leq k$ and $A \backslash B \subseteq A$, (TH) implies that $A \backslash B \in \mathscr{T}$. Clearly $(A \backslash B) \cup B \cup C = X$, so we have:

$$|A \cap B| \quad + |B \cap C| + \quad |C \cap A| \quad >$$
$$|(A \backslash B) \cap B| + |B \cap C| + |C \cap (A \backslash B)|.$$

This is because $|A \cap B| \geq 1$, while $|(A \backslash B) \cap B| = 0$ and $|C \cap (A \backslash B)| \leq |C \cap A|$. However, this contradicts the minimality of $|A \cap B| + |B \cap C| + |C \cap A|$. Now, from the demonstrated fact that (A, B, C) is a partition, we have that $C = \overline{A \cup B}$. Notice that $f(A \cup B) = f(\overline{A \cup B}) = f(C) \leq k$. Then, because $A, B \in \mathscr{T}$, and $f(A \cup B) \leq k$, it follows that $A \cup B \in \mathscr{T}$ by (TU). Hence, by (TE) we have that $C = \overline{A \cup B} \notin \mathscr{T}$, which is a contradiction.

3.3   Non-principle: The Role of the Axioms (TL) and (LL)

First, note that the combination of (TL) and (TE) plays the same role as (LL). Consider a sufficiently large k_ℓ such that $f(A) \leq k_\ell$ for any $A \subseteq X$. Note that in this case, the family of small sets $\mathscr{S}_k = \{A \subset X \mid f(A) \leq k\}$ coincides with 2^X. It is obvious that in this case there exists a branch decomposition of width k_ℓ. Thus, by the duality theorem there exists no tangle for k_ℓ. (For details regarding branch decomposition and the duality theorem, see [15].) This nonexistence result for tangles can be explained in terms of maximal ideals as follows. By setting $k := k_\ell$, the new definition of a maximal ideal on a connectivity system (i.e., Theorem 2) becomes essentially the same as the definition of a maximal ideal on a boolean algebra *together with the condition that* $\{a\} \in \mathscr{T}$ *for any* $a \in X$. Note that this condition comes from the axiom (LL) and also that we have (IW), because of the fact that $\emptyset \in \mathscr{T}$ and (TE). The condition imposes that the maximal ideal cannot be principle. However, mentioned in the introduction, any maximal ideal on 2^X for a finite underlying set X must be principle. Therefore, we may conclude that there is no tangle for k_ℓ.

Now, consider a sufficiently small k_s such that there is an element $b \in X$ with $f(\{b\}) > k_s$. Then, the family $\{A \mid b \notin A, f(A) \leq k_s\}$ satisfies the axioms in Theorem 2; that is, there exists a tangle for k_s. It should be noted that this family can be considered as a principle ideal generated by b. As mentioned briefly in Subsect. 2.5, this case is not interesting for discussing the tangle number (i.e., branch width). Indeed, the tangle number coincides with the largest integer k for which there exists a non-principle maximal ideal on the connectivity system (X, f).

Recall that there exists no non-principle maximal ideal on 2^X for a finite underlying set X. In contrast, there can exist non-principle maximal ideals on the small sets $\mathscr{I}_k$, even for a finite underlying set X. This disparity is attributed to the fact that $A_1 \cup A_2 \cup \cdots \cup A_\ell \in \mathscr{I}$ may not hold in general, even when $A_i \in \mathscr{I}$ for each $1 \leq i \leq \ell$ and $f(A_1 \cup A_2 \cup \cdots \cup A_\ell) \leq k$.

3.4   A Characterization of Maximal Ideals on Connectivity Systems

It is known that for every proper ideal $\mathscr{I}_0$ on a boolean algebra there exists a maximal ideal containing $\mathscr{I}_0$ (see cf. [21]). In this subsection, we show that the same holds true for ideals on connectivity systems by proving Theorem 3. Owing to considerations of space, we omit the full proof in the proceedings version. The proof relies heavily on results in [11,18].

Theorem 3. *Let $\mathscr{M}$ be a maximal ideal of order $k + 1$ on a connectivity system (X, f). Then, there exists no $Y \subseteq X$ with $f(Y) \leq k$ such that $Y \notin \mathscr{M}$ and $\overline{Y} \notin \mathscr{M}$.*

The outline of the proof is as follows: Suppose to the contrary that there is a set Y such that $f(Y) \leq k$ and neither Y nor $\overline{Y}$ is in $\mathscr{M}$. We choose such a Y that minimizes $f(Y)$. Let us consider the closure $\mathscr{M}_Y$ ($\mathscr{M}_{\overline{Y}}$, resp.) of $\mathscr{M} \cup \{Y\}$ ($\mathscr{M} \cup \{\overline{Y}\}$, resp.) under the following operations.

(TU) operation. If $A, B \in \mathscr{M}_Y$ ($A, B \in \mathscr{M}_{\overline{Y}}$, resp.) and $f(A \cup B) \leq k$, then $A \cup B \in \mathscr{M}_Y$ ($A \cup B \in \mathscr{M}_{\overline{Y}}$, resp.),
(TH) operation. If $B \in \mathscr{M}_Y$ ($B \in \mathscr{M}_{\overline{Y}}$, resp.), $A \subseteq B$, and $f(A) \leq k$, then $A \in \mathscr{M}_Y$ ($A \in \mathscr{M}_{\overline{Y}}$, resp.).

Note that $\mathscr{M}_Y$ and $\mathscr{M}_{\overline{Y}}$ satisfy (TB), (TU), and (TH). From the maximality of $\mathscr{M}$, $\mathscr{M}_Y$ and $\mathscr{M}_{\overline{Y}}$ both violate the axiom (IW); that is, both contain X.

We first construct two ternary trees T_Y and $T_{\overline{Y}}$. To have that $X \in \mathscr{M}_Y$, there must exist sets $S_0, S_1 \in \mathscr{M}_Y$ from which X can be derived by (TU), because $X \notin \mathscr{M} \cup \{Y\}$. Then, for S_0 there must exist sets $S_{00}, S_{01} \in \mathscr{M}_Y$ from which S_0 can be derived by (TU) or (TH). That is, if S_0 is obtained by (TU), then there exist sets $S_{00}, S_{01} \in \mathscr{M}_Y$ such that $S_0 = S_{00} \cup S_{01}$. If S_0 is obtained by (TH), then there must exist sets $S_{00}, S_{01}, W \in \mathscr{M}_Y$ such that $S_0 \subseteq W = S_{00} \cup S_{01}$. Note that in either case, it holds that $S_0 \subseteq S_{00} \cup S_{01}$.

By recursively repeating this *decompose operation*, eventually we obtain a *binary* tree such that each leaf corresponds to a member of $\mathscr{M} \cup \downarrow Y$, where

$\downarrow Y := \mathcal{M} \cup \{Y' \mid Y' \subseteq Y, f(Y') \leq k\}$. Hence, each leaf is contained in a member of $\mathcal{M} \cup \{Y\}$. From this binary tree, the ternary tree T_Y can be naturally obtained by deleting the root X and adding an edge e between S_0 and S_1.

The rest of proof is quite similar to that of Theorem 3.3 in [11]: Next, we associate sets to the incidences in the trees. That is, we define functions α_Y and $\alpha_{\overline{Y}}$ from which we obtain two exact tree-labelings (T_Y, α_Y) and $(T_{\overline{Y}}, \alpha_{\overline{Y}})$. Then, we modify the function α_Y ($\alpha_{\overline{Y}}$, resp.) so that (T_Y, β_Y) and $(T_{\overline{Y}}, \beta_{\overline{Y}})$ can be merged, where β_Y ($\beta_{\overline{Y}}$, resp.) is the modified function of α_Y ($\alpha_{\overline{Y}}$, resp.). In this modification step, the submodularity of the function f plays an important role. Finally, we merge these into an exact tree-labeling over $\mathcal{M}$, from which we derive a contradiction.

4   Conclusions and Further Research

In the paper, we have demonstrated that a tangle can be considered as a non-principle maximal ideal on a connectivity system, while a loose tangle can be regarded as a non-principle ideal. Although we have not discussed tangles from the point of view of filters (i.e., that is dual to ideals), this viewpoint is important. Intuitively speaking, filters can be thought of as a family of "big sets" which could include "essentially the whole set" or sets that contain a "specified something that is important." For example, the duality theorems mentioned in Subsect. 1.3 yield a game-theoretical interpretation (see e.g. [1,5,19]). The "something important" for filters may be regarded as being a "robber" in the game-theoretical interpretation.

Ultrafilters and partitions would play an important role in the research of tangles. An ultrafilter is a notion that is dual to that of a maximal ideal. A connection between ultrafilters and tangles has been suggested in [8,9,12]. The study of brambles (i.e., a dual notion to that of tree-width) and tangles has already been approached from the view point of partitions (e.g., [2,14]). Ultrafilters have a high affinity with partitions. In fact, in [13], Leinster brings a little-known characterization of ultrafilters to light: an ultrafilter is characterized by partitions into three subsets. In terms of ideals, this characterization is restated as follows: $\mathscr{I}$ is a maximal ideal on 2^X if and only if for any partition of X with at most three parts (X_1, X_2, X_3) (where X_i can be $\emptyset$), exactly one of X_1, X_2, and X_3 does not belong $\mathscr{I}$. We conjecture the following.

Conjecture 1. $\mathscr{I}$ is a maximal ideal on $\mathscr{S}_k = \{A \subset X \mid f(A) \leq k\}$ if and only if for any partition of X with at most three parts (X_1, X_2, X_3) such that $f(X_i) \leq k$ for all $i \in \{1, 2, 3\}$, exactly one of X_1, X_2, and X_3 does not belong $\mathscr{I}$.

It is easy to prove the direction from maximal ideal to partition. The other direction, in particular showing that (TH) holds, has not yet been proved.

Acknowledgements. This work was supported by JSPS KAKENHI Grant Number 15K00007.

References

1. Adler, I.: Games for width parameters and monotonicity. arXiv preprint arXiv:0906.3857 (2009)
2. Amini, O., Mazoit, F., Nisse, N., Thomassé, S.: Submodular partition functions. Discrete Math. **309**(20), 6000–6008 (2009)
3. Bellenbaum, P., Diestel, R.: Two short proofs concerning tree-decompositions. Comb. Probab. Comput. **11**(06), 541–547 (2002)
4. Bienstock, D., Robertson, N., Seymour, P., Thomas, R.: Quickly excluding a forest. J. Comb. Theory Ser. B **52**(2), 274–283 (1991)
5. Bonato, A., Yang, B.: Graph searching and related problems. In: Pardalos, P.M., Du, D.-Z., Graham, R.L. (eds.) Handbook of Combinatorial Optimization, pp. 1511–1558. Springer, Heidelberg (2013)
6. Clark, B., Whittle, G.: Tangles, trees, and flowers. J. Comb. Theory Ser. B **103**(3), 385–407 (2013)
7. Dharmatilake, J.: A min-max theorem using matroid separations. Matroid Theory Contemp. Math. **197**, 333–342 (1996)
8. Diestel, R.: Ends and tangles. arXiv preprint arXiv:1510.04050v2 (2015)
9. Diestel, R., Oum, S.: Unifying duality theorems for width parameters in graphs and matroids. I. Weak and strong duality. arXiv preprint arXiv:1406.3797 (2014)
10. Fomin, F., Thilikos, D.: On the monotonicity of games generated by symmetric submodular functions. Discrete Appl. Math. **131**(2), 323–335 (2003)
11. Geelen, J., Gerards, B., Robertson, N., Whittle, G.: Obstructions to branch-decomposition of matroids. J. Comb. Theory Ser. B **96**(4), 560–570 (2006)
12. Grohe, M.: Tangled up in blue (a survey on connectivity, decompositions, and tangles). arXiv preprint arXiv:1605.06704 (2016)
13. Leinster, T.: Codensity and the ultrafilter monad. Theory Appl. Categories **28**(13), 332–370 (2013)
14. Lyaudet, L., Mazoit, F., Thomassé, S.: Partitions versus sets: a case of duality. Eur. J. Comb. **31**(3), 681–687 (2010)
15. Oum, S., Seymour, P.: Testing branch-width. J. Comb. Theory Ser. B **97**(3), 385–393 (2007)
16. Oum, S., Seymour, P.: Corrigendum to our paper "testing branch-width" (2009)
17. Reed, B.: Tree width and tangles: a new connectivity measure and some applications. Surv. Comb. **241**, 87–162 (1997)
18. Robertson, N., Seymour, P.: Graph minors. X. Obstructions to tree-decomposition. J. Comb. Theory Ser. B **52**(2), 153–190 (1991)
19. Seymour, P., Thomas, R.: Graph searching and a min-max theorem for tree-width. J. Comb. Theory Ser. B **58**(1), 22–33 (1993)
20. Seymour, P., Thomas, R.: Call routing and the ratcatcher. Combinatorica **14**(2), 217–241 (1994)
21. Sikorski, R.: Boolean Algebras. Springer, Heidelberg (1969)

A Width Parameter Useful for Chordal and Co-comparability Graphs

Dong Yeap Kang[1], O-joung Kwon[2(✉)], Torstein J.F. Strømme[3], and Jan Arne Telle[3]

[1] Department of Mathematical Sciences, KAIST, Daejeon, South Korea
dynamical@kaist.ac.kr
[2] Logic and Semantics, TU Berlin, Berlin, Germany
ojoungkwon@gmail.com
[3] Department of Informatics, University of Bergen, Bergen, Norway
torstein.stromme@ii.uib.no, Jan.Arne.Telle@uib.no

Abstract. In 2013 Belmonte and Vatshelle used mim-width, a graph parameter bounded on interval graphs and permutation graphs that strictly generalizes clique-width, to explain existing algorithms for many domination-type problems, known as LC-VSVP problems. We focus on chordal graphs and co-comparability graphs, that strictly contain interval graphs and permutation graphs respectively. First, we show that mim-width is unbounded on these classes, thereby settling an open problem from 2012. Then, we introduce two graphs $K_t \boxminus K_t$ and $K_t \boxminus S_t$ to restrict these graph classes, obtained from the disjoint union of two cliques of size t, and one clique of size t and one independent set of size t respectively, by adding a perfect matching. We prove that $(K_t \boxminus S_t)$-free chordal graphs have mim-width at most $t-1$, and $(K_t \boxminus K_t)$-free co-comparability graphs have mim-width at most $t - 1$. From this, we obtain several algorithmic consequences, for instance, while DOMINATING SET is NP-complete on chordal graphs, it can be solved in time $\mathcal{O}(n^t)$ on chordal graphs where t is the maximum among induced subgraphs $K_t \boxminus S_t$ in the given graph. We also show that classes restricted in this way have unbounded rank-width which validates our approach.

In the second part, we generalize these results to bigger classes. We introduce a new width parameter sim-width, special induced matching-width, by making only a small change in the definition of mim-width. We prove that chordal and co-comparability graphs have sim-width at most 1. Since DOMINATING SET is NP-complete on chordal graphs, an XP algorithm parameterized only by sim-width would imply $P = NP$. Therefore, to apply the algorithms for domination-type problems mentioned above, we parameterize by both sim-width w and a further parameter t, which is the smallest value such that the input has no induced minor isomorphic to $K_t \boxminus S_t$ or $K_t \boxminus S_t$. We show that such graphs have mim-width at most $8(w + 1)t^3$ and that the resulting algorithms for domination-type problems have runtime $n^{\mathcal{O}(wt^3)}$, when the decomposition tree is given.

The first author is supported by TJ Park Science Fellowship of POSCO TJ Park Foundation. The second author is supported by the European Research Council (ERC) under the European Union's Horizon 2020 research and innovation programme (ERC consolidator grant DISTRUCT, agreement No 648527).

S.-H. Poon et al. (Eds.): WALCOM 2017, LNCS 10167, pp. 93–105, 2017.
DOI: 10.1007/978-3-319-53925-6_8

1   Introduction

Graph width parameters like tree-width and clique-width have been studied for many years, and their algorithmic use has been steadily increasing. In 2012 Vatshelle introduced mim-width[1] which is a parameter with even stronger modelling power than clique-width. This parameter is defined using branch decompositions over the vertex set with the cut function computing the maximum induced matching of the bipartite graph obtained by removing the edges in both parts. Well-known graph classes have bounded mim-width, e.g. interval graphs and permutation graphs have mim-width 1 while their clique-width can be quadratic in the number of vertices [21]. Thus for mim-width k an XP algorithm - runtime $n^{f(k)}$ - is already very interesting. XP algorithms based on mim-width were used by Belmonte and Vatshelle [3] and Bui-Xuan et al. [5] to give a common explanation for the existence of polynomial-time algorithms on many well-known graph classes like interval graphs and permutation graphs, for LC-VSVP problems - Locally Checkable Vertex Subset and Vertex Partitioning problems. We define the class of LC-VSVP problems formally in Sect. 4. In this paper, we extend these algorithms for LC-VSVP problems to subclasses of chordal graphs and co-comparability graphs, that strictly contain interval graphs and permutation graphs respectively. We also show that mim-width is unbounded on general chordal and co-comparability graphs.

The LC-VSVP problems include the class of domination-type problems known as (σ, ρ)-problems, whose intractability on chordal graphs is well known [10]. For two subsets of non-negative numbers σ and ρ, a set S of vertices is called a (σ, ρ)-dominating set if for every vertex $v \in S$, $|S \cap N(v)| \in \sigma$, and for every $v \notin S$, $|S \cap N(v)| \in \rho$, where $N(v)$ denotes the set of neighbors of v. Golovach and Kratochvíl [10] showed that for chordal graphs, the problem of deciding if a graph has a (σ, ρ)-dominating set is NP-complete if σ and ρ are such that there exists at least one chordal graph containing more than one such set. Golovach et al. [11] extended these results to the parameterized setting, showing that the existence of a (σ, ρ)-dominating set of size k, and at most k, are W[1]-complete problems when parameterized by k for any pair of finite sets σ and ρ.

Fig. 1. $K_5 \boxminus S_5$ and $K_5 \boxminus K_5$.

In this paper we apply a different parametrization to solve these problems efficiently on chordal graphs, and on co-comparability graphs. We introduce

---

[1] Introduced formally by Vatshelle in [21], but implicitly used by Belmonte and Vatshelle in [3].

two graphs $K_t \boxminus K_t$ and $K_t \boxminus S_t$, which are obtained from the disjoint union of two cliques of size t, and one clique of size t and one independent set of size t respectively, by adding a perfect matching. See Fig. 1. We prove that $(K_t \boxminus S_t)$-free chordal graphs have mim-width at most $t-1$, and $(K_t \boxminus S_t)$-free co-comparability graphs have linear mim-width at most $t-1$. These results comprise newly discovered graph classes of bounded mim-width, as already $(K_3 \boxminus S_3)$-free chordal graphs contain all interval graphs, and $(K_3 \boxminus K_3)$-free co-comparability graphs contain all permutation graphs. In particular, previous known classes including interval and permutation graphs have bounded linear mim-width, while $(K_t \boxminus S_t)$-free chordal graphs in general have unbounded linear mim-width. By applying algorithms of Bui-Xuan et al. [5] we obtain the following.

Theorem 1. *Given an n-vertex $(K_t \boxminus S_t)$-free chordal graph or an n-vertex $(K_t \boxminus K_t)$-free co-comparability graph, we can solve any LC-VSVP problem in time $n^{O(t)}$, including all (σ, ρ)-domination problems for finite or co-finite σ, ρ.*

For example, MINIMUM DOMINATING SET is solved in time $O(n^{3t+4})$ and q-COLORING in time $O(qn^{3qt+4})$. In the second part of this paper we generalize these results to larger classes. We also ask the question - what algorithmic use can we make of a width parameter with even stronger modelling power than mim-width? We define the parameter *sim-width* (special induced matching-width) by making only a small change in the definition of mim-width, simply requiring that a special induced matching across a cut of the graph cannot contain edges between two vertices on the same side of the cut. See Sect. 2 for the precise definitions. The linear variant of sim-width will be called *linear sim-width*. We show that graphs of bounded sim-width are closed under taking induced minors, and the modelling power of sim-width is strictly stronger than mim-width.

Theorem 2. *Chordal graphs have sim-width at most 1 while split graphs have unbounded mim-width, and a branch-decomposition of sim-width at most 1 can be found in polynomial time. Co-comparability graphs have linear sim-width at most 1 but unbounded mim-width, and a linear branch-decomposition of sim-width at most 1 can be found in polynomial time.*

This confirms a conjecture of Vatshelle and Belmonte from 2012 [3,21] that chordal graphs and co-comparability graphs have unbounded mim-width[2]. See Fig. 2 for an inclusion diagram of some well-known graph classes. We conjecture that circle graphs and weakly chordal graphs, also have constant sim-width.

The sim-width parameter thus meets our goal of having stronger modelling power than mim-width, but stronger modelling power automatically implies weaker analytic power, i.e. fewer problems will have FPT or XP algorithms when parameterized by sim-width. As an example, for problems like MINIMUM DOMINATING SET which are NP-complete on chordal graphs [4], we cannot expect an XP algorithm parameterized by sim-width, i.e. with runtime $|V(G)|^{f(\mathrm{simw}(G))}$,

---

[2] Our result appeared on arxiv June 2016. In August 2016 a similar result by Mengel, developed independently, also appeared on arxiv [15].

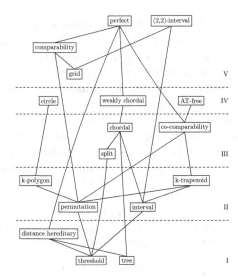

Fig. 2. Inclusion diagram of some well-known graph classes. (I) Classes where clique-width and rank-width are constant. (II) Classes where mim-width is constant. (III) Classes where sim-width is constant. (IV) Classes where it is unknown if sim-width is constant. (V) Classes where sim-width is unbounded.

even if we are given a branch-decomposition, unless $P = NP$. Thus, for the algorithmic use of sim-width we must either strongly restrict the problems we consider, or we must put a further restriction on the input graphs. In this paper we take the latter approach, and apply the same graphs $K_t \boxminus K_t$ and $K_t \boxminus S_t$ as we did earlier for chordal and co-comparability graphs, but now we disallow them as induced minors, which is natural as the resulting classes are then closed under induced minors.

Theorem 3. *Every graph with sim-width w and no induced minor isomorphic to $K_t \boxminus K_t$ and $K_t \boxminus S_t$ has mim-width at most $8(w+1)t^3 - 1$.*

In a further result we show that we can also exclude these graphs as an induced subgraph to bound the mim-width, but in that case, we need to use Ramsey's theorem, and the bound will in general become worse.

Combining again with the algorithms of [5] we get the following.

Theorem 4. *Given an n-vertex graph G having no induced minor isomorphic to $K_t \boxminus K_t$ and $K_t \boxminus S_t$, with a branch-decomposition of sim-width w, we can solve any LC-VSVP problem in time $O(n^{O(wt^3)})$.*

2   Preliminaries

We use standard graph terminology. Here we define only a few terms, see the Appendix for a full list. A tree is called *subcubic* if every internal node has exactly

3 neighbors. A tree T is called a *caterpillar* if contains a path P where for every vertex in T either it is in P or has a neighbor on P. A graph is called *chordal* if it contains no induced subgraph isomorphic to a cycle of length 4 or more. An ordering $v_1, \ldots, v_n$ of the vertex set of a graph G is called a *co-comparability ordering* if for every triple i, j, k with $i < j < k$, v_j has a neighbor in each path from v_i to v_k avoiding v_j. A graph is a *co-comparability graph* if it admits a co-comparability ordering. A graph is *H-free* if it contains no induced subgraph isomorphic to H. A set of edges $\{v_1 w_1, v_2 w_2, \ldots, v_m w_m\}$ of G is called an *induced matching* in G if there are no other edges in $G[\{v_1, \ldots, v_m, w_1, \ldots, w_m\}]$. For a vertex partition (A, B) of a graph G, we denote by $G[A, B]$ the bipartite graph on the bipartition (A, B) where for $a \in A, b \in B$, a and b are adjacent in $G[A, B]$ if and only if they are adjacent in G.

Let G be a graph. We define functions $\mathrm{cutrk}_G, \mathrm{mimval}_G, \mathrm{simval}_G$ from $2^{V(G)}$ to $\mathbb{Z}$ such that

- $\mathrm{cutrk}_G(A)$ is the rank of the bipartite-adjacency matrix of $G[A, V(G) \setminus A]$ where the rank is computed over the binary field,
- $\mathrm{mimval}_G(A)$ is the maximum size of an induced matching of $G[A, V(G) \setminus A]$,
- $\mathrm{simval}_G(A)$ is the maximum size of an induced matching $\{a_1 b_1, \ldots, a_m b_m\}$ in G where $a_1, \ldots, a_m \in A$ and $b_1, \ldots, b_m \in V(G) \setminus A$.

A pair (T, L) of a subcubic tree T and a function L from $V(G)$ to the set of leaves of T is called a *branch-decomposition*. For each edge e of T, let (A_1^e, A_2^e) be the vertex partition of G where T_1^e, T_2^e are the two connected components of $T - e$, and for each $i \in \{1, 2\}$, A_i^e is the set of all vertices in G mapped to leaves contained in T_i^e. We call it the vertex partition of G associated with e. For a branch-decomposition (T, L) of a graph G and an edge e in T and a function $f : 2^{V(G)} \rightarrow \mathbb{Z}$, the *width* of e with respect to f, denote by $f_{(T,L)}(e)$, is define as $f(A_1^e)$ where (A_1^e, A_2^e) is the vertex partition associated with e. The *width* of (T, L) with respect to f is the maximum width over all edges in T.

The *rank-width*, *mim-width*, and *sim-width* of a graph G are the minimum widths over all their branch-decompositions with respect to $\mathrm{cutrk}_G, \mathrm{mimval}_G$, and simval_G and denote by $\mathrm{rw}(G), \mathrm{mimw}(G)$, and $\mathrm{simw}(G)$, respectively. If T is a subcubic caterpillar tree, then (T, L) is called a *linear branch-decomposition*. The *linear mim-width* and *linear sim-width* of a graph G are the minimum widths over all their linear branch-decompositions with respect to mimval_G and simval_G, and denote by $\mathrm{lmimw}(G)$ and $\mathrm{lsimw}(G)$, respectively. By definitions, we have $\mathrm{simw}(G) \leq \mathrm{mimw}(G) \leq \mathrm{rw}(G)$ for every graph G.

3   Mim-Width of Chordal and Co-comparability Graphs

We show that *chordal graphs* and *co-comparability graphs* have sim-width at most 1, but have unbounded mim-width. Belmonte and Vatshelle [3] showed that chordal graphs either do not have constant mim-width or it is NP-complete to find such a decomposition. We strengthen their result. We further show that

$(K_t \boxminus S_t)$-free chordal graphs have mim-width at most $t - 1$, and similarly, $(K_t \boxminus S_t)$-free co-comparabililty graphshave mim-width at most $t - 1$.

We use the fact that chordal graphs admit a tree-decomposition whose bags are maximal cliques.

Proposition 1. *Given a chordal graph, one can output a branch-decomposition of sim-width at most 1 in polynomial time. Moreover for every positive integer t, given a $(K_t \boxminus S_t)$-free chordal graph, one can output a mim-decomposition of width at most $t - 1$ in polynomial time.*

Proof. To prove both statements, we construct a certain branch-decomposition explicitly. Let G be a chordal graph. We may assume that G is connected. We compute a tree-decomposition $(F, \mathcal{B} = \{B_t\}_{t \in V(F)})$ of G where every bag induces a maximal clique of G. It is known that such a decomposition can be computed in polynomial time; for instance, see [16]. Let us choose a root node r of F.

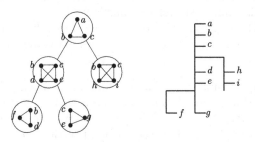

Fig. 3. Constructing a branch-decomposition (T, L) of a chordal graph G of sim-width at most 1 from its tree-decomposition.

We construct a tree (T, L) from F as follows. We attach a leaf r' to the root node r and regard it as the parent of r and let $B_r := \emptyset$. For every $t \in V(F)$ with its parent t', we subdivide the edge tt' into a path $t v_1^t \cdots v_{|B_t \setminus B_{t'}|}^t t'$ with $|B_t \setminus B_{t'}|$ internal nodes, and for each internal node q, we attach a leaf q' and assign those leaves to the vertices of $B_t \setminus B_{t'}$ in any order as the images of L. Then remove r'. For every $t \in V(F)$ with its children $t_1, \ldots, t_m$ we remove t and introduce a path $w_1^t w_2^t \cdots w_m^t$. If t is a leaf, then we just remove it. We add an edge $w_1^t v_1^t$, and for each $i \in \{1, \ldots, m\}$, add an edge $w_i^t v_{|B_{t_i} \setminus B_t|}^{t_i}$. See Fig. 3 for an illustration of the construction.

Let T' be the resulting tree, and we obtain a tree T from T' by smoothing all nodes of degree 2. Let (T, L) be the resulting branch-decomposition. Remark that F has $\mathcal{O}(|V(G)|)$ many nodes, and thus we can construct (T, L) in linear time. We claim that (T, L) has sim-width at most 1. We prove a stronger result that for every edge e of T with a partition (A, B) associated with e, either $N_G(A) \cap B$ or $N_G(B) \cap A$ is a clique.

Claim. Let e be an edge of T and let (A, B) be a partition of $V(G)$ associated with e. Then either $N_G(A) \cap B$ or $N_G(B) \cap A$ is a clique.

Proof. We prove for T', which is the tree before smoothing. We may assume that both end nodes of e are internal nodes of T'. There are 4 types of e; (1) $e = v_i^t v_{i+1}^t$ for some t and i, (2) $e = w_1^t v_1^t$ for some t, (3) $e = v_{|B_{t_i} \setminus B_t|}^t w_i^t$ where t_i is a child of t, (4) $e = w_i^t w_{i+1}^t$ for some t and i. Suppose $e = v_i^t v_{i+1}^t$ for some t and i, and let t' be the parent of t. We may assume that A corresponds to the part consisting of descendants of v_i^t. It is not difficult to check that for every $v \in A \setminus B_t$, $N_G(v) \cap B \subseteq B_t$. Furthermore, for $v \in A \cap B_t$, we have $N_G(v) \cap B \subseteq B_t$, because v is contained in $B_t \setminus B_{t'}$ by construction. Thus, $N_G(A) \cap B$ is a subset of B_t which is a clique. We can similarly prove for Cases 2 and 3.

We assume $e = w_i^t w_{i+1}^t$ for some t and i. Without loss of generality, we assume that $B_t \subseteq B$. We can observe that for every $v \in A$, $N_G(v) \cap B \subseteq B_t$. Thus $N_G(A) \cap B$ is a clique, as required.   □

We prove that If G is $(K_t \boxminus S_t)$-free, then (T, L) has mim-width at most $t-1$. We may assume $t \geq 2$. We show that for every edge e or T, $\mathrm{mimval}_{(T,L)}(e) \leq t-1$. Suppose for contradiction that there is an edge e of T and a partition (A, B) associated with e, where $\mathrm{mimval}_G(A) \geq t$. We may assume that both end nodes of e are internal nodes of T. By **Claim**, one of $N_G(A) \cap B$ and $N_G(B) \cap A$ is a clique. Without loss of generality we assume $N_G(B) \cap A$ is a clique C. If there is an induced matching $\{a_1 b_1, \ldots, a_t b_t\}$ in $G[A, B]$ where $a_1, \ldots, a_t \in A$, then we have $a_1, \ldots, a_t \in V(C)$. Furthermore there are no edges between vertices in $\{b_1, \ldots, b_t\}$, otherwise, it creates an induced C_4. Thus, we have an induced subgraph isomorphic to $K_t \boxminus S_t$, which contradicts to our assumption. We conclude that (T, L) has mim-width at most $t - 1$.   □

We now prove the lower bound on the mim-width of general chordal graphs. We in fact show this for the class of split graphs that is a subclass of chordal graphs. A *split* graph is a graph that can be partitioned into two vertex sets C and I where C is a clique and I is an independent set.

Proposition 2. *For every large enough n, there is a split graph on n vertices having mim-width at least $\sqrt{\log_2 \frac{n}{2}}$.*

Proof. The full proof is given in the Appendix. Let $m \geq 10000$ be an integer and let $n := m + (2^m - 1)$. Let G be a split graph on the vertex partition (C, I) where C is a clique of size m, I is an independent set of size $2^m - 1$, and all vertices in I have pairwise distinct non-empty neighborhoods on C. Let (T, L) be a branch-decomposition of G. It is well known that there is an edge of T inducing a balanced vertex partition whose each part has at most $\frac{2n}{3}$ vertices. Let (A_1, A_2) be such a partition, and without loss of generality, we may assume that $|A_1 \cap C| \geq |A_2 \cap C|$, and thus we have $\frac{m}{2} \leq |A_1 \cap C| \leq m$. Note that $|A_2 \cap I| > \frac{n}{3} - m \geq \frac{2^m - 2m - 1}{3} \geq 2^{m-3}$. Since $|A_2 \cap C| < \frac{m}{2}$ and $m \geq 8$, there are at least $\frac{2^{m-3}}{2^{\frac{m}{2}}} \geq 2^{\frac{m}{2}-3}$ vertices in $A_2 \cap I$ that have pairwise distinct neighbors on $A_1 \cap C$. Let $I' \subseteq A_2 \cap I$ be the set of such vertices.

By the Sauer-Shelah lemma [18,19], if $|I'| \geq |A_1 \cap C|^k$ for some k, then there will be an induced matching of size k between $A_1 \cap C$ and I' in $G[A_1, A_2]$. We choose $k := \sqrt{m}$. As $m \geq 10000$, we can deduce that $\frac{m}{2} - 3 \geq \sqrt{m} \log_2 m$. Therefore, we have $|I'| \geq 2^{\frac{m}{2}-3} \geq m^{\sqrt{m}} \geq |A_1 \cap C|^{\sqrt{m}}$, and there is an induced matching of size $\sqrt{m}$ between $A_1 \cap C$ and I' in $G[A_1, A_2]$. This shows that the mim-width of G is at least $\sqrt{m} \geq \sqrt{\log_2 \frac{n}{2}}$.   □

We observe similar properties for co-comparability graphs.

Proposition 3. *Given a co-comparability graph, one can output a linear branch-decomposition of linear sim-width at most 1 in polynomial time. Moreover for every positive integer t, given a $(K_t \boxminus K_t)$-free co-comparability graph, one can output a linear mim-decomposition of width at most $t - 1$ in polynomial time.*

Proof. We use a result by McConnell and Spinrad [14] that one can output a co-comparability ordering of a co-comparability graph in polynomial time. This ordering provides a linear branch-decomposition of sim-width at most 1, and mim-width at most $t - 1$ when the given graph is $K_t \boxminus K_t$-free.   □

Proposition 4. *For every large enough n, there is a co-comparability graph on n vertices having mim-width at least $\sqrt{\frac{n}{12}}$.*

4   Algorithms for LC-VSVP Problems

We describe algorithmic applications for restricted subclasses of chordal and co-comparability graphs described in Sect. 3. Telle and Proskurowski [20] classified a class of problems called *Locally Checkable Vertex Subset and Vertex Partitioning problems*, which is a subclass of MSO_1 problems. These problems generalize problems like MINIMUM DOMINATING SET and q-COLORING.

Let σ, ρ be finite or co-finite subsets of natural numbers. For a graph G and $S \subseteq V(G)$, we call S a (σ, ρ)-*dominating set* of G if (1) for every $v \in S$, $|N_G(v) \cap S| \in \sigma$, and (2) for every $v \in V(G) \setminus S$, $|N_G(v) \cap S| \in \rho$. For instance, a $(0, \mathbb{N})$-set is an independent set as there are no edges inside of the set, and we do not care about the adjacency between S and $V(G) \setminus S$. Another example is that a $(\mathbb{N}, \mathbb{N}^+)$-set is a dominating set as we require that for each vertex in $V(G) \setminus S$, it has at least one neighbor in S. The class of *locally checkable vertex subset problems* consist of finding a minimum or maximum (σ, ρ)-dominating set in an input graph G, and possibly on vertex-weighted graphs.

For a positive integer q, a $(q \times q)$-matrix D_q is called a *degree constraint matrix* if each element is either a finite or co-finite subset of natural numbers. A partition $\{V_1, V_2, \ldots, V_q\}$ of the vertex set of a graph G is called a D_q-*partition* if for every $i, j \in \{1, \ldots, q\}$ and $v \in V_i$, $|N_G(v) \cap V_j| \in D_q[i, j]$. For instance, if we take a matrix D_q where all diagonal entries are 0, and all other entries are $\mathbb{N}$, then a D_q-partition is a partition into q independent sets, which corresponds to a q-coloring of the graph. The class of *locally checkable vertex partitioning problems* consist of deciding if G admits a D_q-partition.

All these problems will be called *Locally Checkable Vertex Subset and Vertex Partitioning problems*, shortly LC-VSVP problems. As shown in [5] the runtime solving an LC-VSVP problem by dynamic programming relates to the finite or co-finite subsets of natural numbers used in its definition. The following function d is central. Let $d(\mathbb{N}) = 0$ and for every finite or co-finite set $\mu \subseteq \mathbb{N}$, let $d(\mu) = 1 + \min(\max\{x \in \mathbb{N} : x \in \mu\}, \max\{x \in \mathbb{N} : x \notin \mu\})$. For example, for MINIMUM DOMINATING SET and q-COLORING problems we plug in $d = 1$ because $\max(d(\mathbb{N}), d(\mathbb{N}^{+})) = 1$ and $\max(d(0), d(\mathbb{N})) = 1$.

Theorem 5 (Belmonte and Vatshelle [3] and Bui-Xuan et al. [5]). *Given an n-vertex graph and its branch-decomposition (T, L) of mim-width w we solve any (σ, ρ)-vertex subset problem with $d = \max(d(\sigma), d(\rho))$ in time $\mathcal{O}(n^{3dw+4})$, and solve any D_q-vertex partitioning problem with $d = \max_{i,j} d(D_q[i, j])$ in time $\mathcal{O}(qn^{3dwq+4})$.*

Combining Theorem 5 with Propositions 1 and 3 we get the following.

Corollary 1. *Given an n-vertex chordal graph having no induced subgraph isomorphic to $K_t \boxminus S_t$, or co-comparability graph with no induced subgraph isomorphic to $K_t \boxminus K_t$, we solve any (σ, ρ)-vertex subset problem with $d = \max(d(\sigma), d(\rho))$ in time $\mathcal{O}(n^{3dt+4})$, any D_q-vertex partitioning problem with $d = \max_{i,j} d(D_q[i, j])$ in time $\mathcal{O}(qn^{3dtq+4})$.*

DOMINATING SET is NP-complete on chordal graphs [4], but for fixed t, it can be solved in polynomial on $(K_t \boxminus S_t)$-free chordal graphs. Also, WEIGHTED DOMINATING SET is NP-complete on co-comparability graphs [13], but for every fixed t, it can be solved in polynomial time on $(K_t \boxminus K_t)$-free co-comparability graphs.

5   Extending to Graphs of Bounded Sim-Width

In Sect. 3, we proved that graphs of sim-width at most 1 contain all chordal and co-comparability graphs. A classical result on chordal graphs is that the problem of finding a minimum dominating set in a chordal graph is NP-complete [4]. So, even for this kind of locally-checkable problem, we cannot expect efficient algorithms on graphs of sim-width at most w. Therefore, to obtain a meta-algorithm for graphs of bounded sim-width encompassing many locally-checkable problems, we must impose some restrictions. We approach this problem in a way analogous to what has previously been done in the realm of rank-width [8].

Complete graphs have rank-width at most 1, but they have unbounded tree-width. Fomin et al. [8] showed that the tree-width of a K_r-minor free graph is bounded by $c \cdot \mathrm{rw}(G)$ where c is a constant depending on r. This can be utilized algorithmically, to get a result for graphs of bounded rank-width when excluding a fixed minor, as the class of problems solvable in FPT time is strictly larger when parameterized by tree-width than rank-width [12].

We will do something similar by focusing on the distinction between mim-width and sim-width. However, K_r-minor free graphs are too strong, as one can show that on K_r-minor free graphs, the tree-width of a graph is also bounded by some constant factor of its sim-width [7].

Instead of using minors, we exclude certain graphs as induced subgraphs or induced minors. The induced minor operation is rather natural because sim-width does not increase when taking induced minors; the proof is given in the Appendix. We observe that indeed $K_t \boxminus K_t$ and $K_t \boxminus S_t$ are essential graphs to bound mim-width from graphs of bounded sim-width.

Proposition 5. *Every graph with sim-width w and no induced minor isomorphic to $K_t \boxminus K_t$ and $K_t \boxminus S_t$ has mim-width at most $8(w+1)t^3 - 1$.*

Note that we can also exclude these graphs as an induced subgraph to bound mim-width, but in that case, we need to use the Ramsey's theorem, and the bound on the exponent becomes again an exponential in w and t. We denote by $R(k, \ell)$ the Ramsey number, which is the minimum integer satisfying that every graph with at least $R(k, \ell)$ vertices contains either a clique of size k or an independent set of size ℓ. By Ramsey's Theorem [17], $R(k, \ell)$ exists for every pair of positive integers k and ℓ.

Proposition 6. *Every graph with sim-width w and no induced subgraph isomorphic to $K_t \boxminus K_t$ and $K_t \boxminus S_t$ has mim-width at most $R(R(w+1, t), R(t, t))$.*

Propositions 1 and 3 can be seen as special cases of Proposition 6 because chordal graphs do not have any $K_t \boxminus K_t$ for $t \geq 2$, and co-comparability graphs do not have any $K_t \boxminus S_t$ for $t \geq 3$. Remark that Belmonte et al. [2, Corollary 1] discussed that the Ramsey number can be a polynomial function of k and ℓ if the underlying graphs are some special classes such as chordal graphs, interval graphs, proper interval graphs, comparability graphs, co-comparability graphs, and permutation graphs. While comparability graphs have unbounded sim-width, all other graphs have either bounded mim-width, or sim-width 1. To have more applications of Proposition 6, it is interesting to see whether some graphs with constant sim-width admit a polynomial function for the Ramsey number.

We prove Proposition 5. Notice that the optimal bound of Theorem 6 has been slightly improved by Fox [9], and then by Balogh and Kostochka [1].

Theorem 6 (Duchet and Meyniel [6]). *For positive integers k and n, every n-vertex graph contains either an independent set of size k or a K_t-minor where $t \geq \frac{n}{2k-1}$.*

Proof (of Proposition 5). Let G be a graph with sim-width w and no induced minor isomorphic to $K_t \boxminus K_t$ and $K_t \boxminus S_t$. Let (T, L) be a branch-decomposition of G of width w with respect to the simval$_G$ function. We claim that for each $e \in E(T)$, mimval$_{(T,L)}(e) \leq 8(w+1)t^3 - 1$.

Let $e \in E(T)$, and (A, B) be the vertex partition of G associated with e. Suppose for contradiction that there is an induced matching $\{v_1 w_1, \ldots, v_m w_m\}$ in $G[A, B]$ where $v_1, \ldots, v_m \in A$, $w_1, \ldots, w_m \in B$, and $m \geq 8(w+1)t^3$. Let f be the function from $\{v_1, \ldots, v_m\}$ to $\{w_1, \ldots, w_m\}$ such that $f(v_i) = w_i$ for each $i \in \{1, \ldots, m\}$. As $m \geq 8(w+1)t^3$, by Theorem 6, the subgraph $G[\{v_1, \ldots, v_m\}]$ contains either an independent set of size $2(w+1)t$, or a K_{2t^2}-minor.

If $G[\{v_1, \ldots, v_m\}]$ contains a K_{2t^2}-minor, then there exist pairwise disjoint subsets $S_1, \ldots, S_{2t^2}$ of $\{v_1, \ldots, v_m\}$ such that for each $i \in \{1, \ldots, 2t^2\}$, $G[S_i]$ is connected, and for two distinct integers $i, j \in \{1, \ldots, 2t^2\}$, there is an edge between S_i and S_j. In this case, for each $i \in \{1, \ldots, 2t^2\}$, we choose a representative d_i in each $f(S_i)$ and contract S_i to a vertex c_i. Let G' be the resulting graph. Then $G'[\{c_1, \ldots, c_{2t^2}\}, \{d_1, \ldots, d_{2t^2}\}]$ is an induced matching of size t, and $\{c_1, \ldots, c_{2t^2}\}$ is a clique in G'. We can do the same procedure for the set $\{d_1, \ldots, d_{2t^2}\}$, and by Theorem 6, the subgraph $G'[\{d_1, \ldots, d_{2t^2}\}]$ contains either an independent set of size t, or a K_t-minor. In both cases, one can observe that G' contains an induced minor isomorphic to $K_t \boxminus K_t$ or $K_t \boxminus S_t$, contradiction.

Assume that $G[\{v_1, \ldots, v_m\}]$ contains an independent set $\{c_1, \ldots, c_{2(w+1)t}\}$, and for each $i \in \{1, \ldots, 2(w+1)t\}$, let $d_i := f(c_i)$. Then by Theorem 6, $G[\{d_1, \ldots, d_{2(w+1)t}\}]$ contains either an independent set of size $w + 1$ or a K_t-minor. In the former case, we obtain an induced matching of size $w + 1$, contradicting to the assumption that $\mathrm{simval}_{(T,L)}(e) \leq w$. In the latter case, we obtain an induced minor isomorphic to $K_t \boxminus S_t$, contradiction. We conclude $\mathrm{mimval}_{(T,L)}(e) \leq 8(w+1)t^3$. $\qquad\square$

We can prove Proposition 6 in a similar manner by replacing the application of Theorem 6 with the Ramsey's Theorem to find an exact clique or an independent set. We extend Corollary 1 for general classes of graphs.

Corollary 2. *Let G be a given n-vertex graph with a branch-decomposition of sim-width w.*

- *If G has no induced minor isomorphic to $K_t \boxminus K_t$ or $K_t \boxminus S_t$ and $t' := 8(w + 1)t^3$, then we can solve any (σ, ρ)-vertex subset problem in time $\mathcal{O}(n^{3dt'+4})$ where $d = \max(d(\sigma), d(\rho))$, and solve any D_q-vertex partitioning problem in time $\mathcal{O}(qn^{3dt'q+4})$ where $d = \max_{i,j} d(D_q[i, j])$.*
- *If G has no induced subgraph isomorphic to $K_t \boxminus K_t$ or $K_t \boxminus S_t$ and $t' := R(R(w + 1, t), R(t, t))$, then we can solve any (σ, ρ)-vertex subset problem in time $\mathcal{O}(n^{3dt'+4})$ where $d = \max(d(\sigma), d(\rho))$, and solve any D_q-vertex partitioning problem in time $\mathcal{O}(qn^{3dt'q+4})$ where $d = \max_{i,j} d(D_q[i, j])$.*

6   Concluding Remarks

We showed that every LC-VSVP problem can be solved in XP time parameterized by t on $(K_t \boxminus S_t)$-free chordal graphs and $(K_t \boxminus K_t)$-free co-comparability graphs. We further generalized this to every graph with sim-width at most w and having no induced minor isomorphic to $K_t \boxminus S_t$ or $K_t \boxminus K_t$ has mim-width at most $8(w+1)t^3$, by showing that every LC-VSVP problem can be solved in time $n^{\mathcal{O}(wt^3)}$ on such n-vertex graphs, when its branch-decomposition is given.

It would be interesting to find more classes having constant sim-width, but unbounded mim-width. We propose some possible classes.

Question 1. *Do weakly chordal graphs, AT-free graphs, or circle graphs have constant sim-width?*

We showed that DOMINATING SET can be solved in time $n^{\mathcal{O}(t)}$ on $(K_t \boxminus S_t)$-free chordal graphs, but we could not obtain an FPT algorithm. We ask whether it is W[1]-hard or not. This may be a right direction to show that DOMINATING SET is W[1]-hard parameterized by mim-width, which is open.

Question 2. *Is* DOMINATING SET *on chordal graphs W[1]-hard parameterized by the maximum t such that it has an $K_t \boxminus S_t$ induced subgraph?*

References

1. Balogh, J., Kostochka, A.: Large minors in graphs with given independence number. Discret. Math. **311**(20), 2203–2215 (2011)
2. Belmonte, R., Heggernes, P., van't Hof, P., Rafiey, A., Saei, R.: Graph classes and Ramsey numbers. Discret. Appl. Math. **173**, 16–27 (2014)
3. Belmonte, R., Vatshelle, M.: Graph classes with structured neighborhoods and algorithmic applications. Theor. Comput. Sci. **511**, 54–65 (2013)
4. Booth, K.S., Johnson, J.H.: Dominating sets in chordal graphs. SIAM J. Comput. **11**(1), 191–199 (1982)
5. Bui-Xuan, B.M., Telle, J.A., Vatshelle, M.: Fast dynamic programming for locally checkable vertex subset and vertex partitioning problems. Theor. Comput. Sci. **511**, 66–76 (2013)
6. Duchet, P., Meyniel, H.: On Hadwiger's number and the stability number. In: Bollobás, B. (ed.) Graph Theory Proceedings of the Conference on Graph Theory. North-Holland Mathematics Studies, vol. 62, pp. 71–73. North-Holland, Amsterdam (1982)
7. Fomin, F.V., Golovach, P., Thilikos, D.M.: Contraction obstructions for treewidth. J. Comb. Theory Ser. B **101**(5), 302–314 (2011)
8. Fomin, F.V., Oum, S., Thilikos, D.M.: Rank-width and tree-width of H-minor-free graphs. Eur. J. Comb. **31**(7), 1617–1628 (2010)
9. Fox, J.: Complete minors and independence number. SIAM J. Discret. Math. **24**(4), 1313–1321 (2010)
10. Golovach, P., Kratochvíl, J.: Computational complexity of generalized domination: a complete dichotomy for chordal graphs. In: Brandstädt, A., Kratsch, D., Müller, H. (eds.) WG 2007. LNCS, vol. 4769, pp. 1–11. Springer, Heidelberg (2007). doi:10.1007/978-3-540-74839-7_1
11. Golovach, P.A., Kratochvíl, J., Suchý, O.: Parameterized complexity of generalized domination problems. Discret. Appl. Math. **160**(6), 780–792 (2012)
12. Hliněný, P., Oum, S., Seese, D., Gottlob, G.: Width parameters beyond tree-width and their applications. Comput. J. **51**(3), 326–362 (2008)
13. Maw-Shang, C.: Weighted domination of cocomparability graphs. Discret. Appl. Math. **80**(2), 135–148 (1997)
14. McConnell, R.M., Spinrad, J.P.: Linear-time modular decomposition and efficient transitive orientation of comparability graphs. In: Proceedings of the Fifth Annual ACM-SIAM Symposium on Discrete Algorithms SODA 1994, pp. 536–545. Society for Industrial and Applied Mathematics, Philadelphia (1994)
15. Mengel, S.: Lower bounds on the mim-width of some perfect graph classes. Preprint arXiv.org/abs/1608.01542 (2016)
16. Naor, J., Naor, M., Schäffer, A.A.: Fast parallel algorithms for chordal graphs. SIAM J. Comput. **18**(2), 327–349 (1989)

17. Ramsey, F.P.: On a problem of formal logic. Proc. Lond. Math. Soc. **30**(s2), 264–286 (1930)
18. Sauer, N.: On the density of families of sets. J. Comb. Theory Ser. A **13**(1), 145–147 (1972)
19. Shelah, S.: A combinatorial problem; stability and order for models and theories in infinitary languages. Pac. J. Math. **41**(1), 247–261 (1972)
20. Telle, J.A., Proskurowski, A.: Algorithms for vertex partitioning problems on partial k-trees. SIAM J. Discret. Math. **10**(4), 529–550 (1997)
21. Vatshelle, M.: New width parameters of graphs. Ph.D. thesis, University of Bergen (2012)

Byzantine Gathering in Networks
with Authenticated Whiteboards

Masashi Tsuchida$^{(\boxtimes)}$, Fukuhito Ooshita, and Michiko Inoue

Nara Institute of Science and Technology, Ikoma, Japan
tsuchida.masashi.td8@is.naist.jp

Abstract. We propose an algorithm for the gathering problem of mobile agents in Byzantine environments. Our algorithm can make all correct agents meet at a single node in $O(fm)$ time (f is the upper bound of the number of Byzantine agents and m is the number of edges) under the assumption that agents have unique ID and behave synchronously, each node is equipped with an authenticated whiteboard, and f is known to agents. Since the existing algorithm achieves gathering without a whiteboard in $\tilde{O}(n^9\lambda)$ time, where n is the number of nodes and λ is the length of the longest ID, our algorithm shows an authenticated whiteboard can significantly reduce the time for the gathering problem in Byzantine environments.

1 Introduction

Background. Distributed systems, which are composed of multiple computers (nodes) that can communicate with each other, have become larger in scale recently. This makes it complicated to design distributed systems because developers must maintain a huge number of nodes and treat massive data communication among them. As a way to mitigate the difficulty, (mobile) agents have attracted a lot of attention [2]. Agents are software programs that can autonomously move from a node to a node and execute various tasks in distributed systems. In systems with agents, nodes do not need to communicate with other nodes because agents themselves can collect and analyze data by moving around the network, which simplifies design of distributed systems. In addition, agents can efficiently execute tasks by cooperating with other agents. Hence many works study algorithms to realize cooperation among multiple agents.

The gathering problem is a fundamental task to realize cooperation among multiple agents. The goal of the gathering problem is to make all agents meet at a single node within a finite time. By achieving gathering, all agents can communicate with each other at the single node.

Related Works. The gathering problem has been widely studied in literature [10,12]. Most studies aim to clarify solvability of the gathering problem in various environments, and, if it is solvable, they aim to clarify costs (e.g., time, number

---

This work was supported by JSPS KAKENHI Grant Numbers 26330084 and 15H00816. The full version of this paper is provided in [15].

S.-H. Poon et al. (Eds.): WALCOM 2017, LNCS 10167, pp. 106–118, 2017.
DOI: 10.1007/978-3-319-53925-6_9

of moves, and memory space) required to achieve gathering. To do this, many studies have been conducted under various environments such that assumptions on synchronization, anonymity, randomized behavior, topology, and presence of node memory (whiteboard) are different. Table 1 summarizes some of the results.

Table 1. Gathering of synchronous agents with unique IDs in arbitrary graphs (n is the number of nodes, l is the length of the smallest ID of agents, τ is the maximum difference among activation times of agents, m is the number of edges, λ is the length of the longest ID of agents, f is the upper bound of the number of Byzantine agents).

	Byzantine	Whiteboard	Time complexity
[5]	None	None	$\tilde{O}(n^5\sqrt{\tau l} + n^{10}l)$
[9]	None	None	$\tilde{O}(n^{15} + l^3)$
[14]	None	None	$\tilde{O}(n^5 l)$
Trivial algorithm	None	Non-authenticated	$O(m)$
[6]	Weak	None	$\tilde{O}(n^9 \lambda)$
[1,6]	Strong	None	Exponential
Trivial extension of [6]	Weak	Authenticated	$O(n^5 \lambda)$
Proposed algorithm	Weak	Authenticated	$O(fm)$

For environments such that no whiteboard exists (i.e., agents cannot leave any information on nodes), many deterministic algorithms to achieve gathering of two agents have been proposed. Note that these algorithms can be easily extended to a case of more than two agents [9]. If agents do not have unique IDs, they cannot achieve gathering for some symmetric graphs. Therefore some works [5,9,14] assume unique IDs and achieve gathering for any graph. Dessmark et al. [5] proposed an algorithm that realizes gathering in $\tilde{O}(n^5\sqrt{\tau l} + n^{10}l)$ time for any graph, where n is the number of nodes, l is the length of the smaller ID of agents, and τ is the difference between activation times of two agents. Kowalski and Malinowski [9] and Ta-Shma and Zwick [14] improved the time complexity to $\tilde{O}(n^{15} + l^3)$ and $\tilde{O}(n^5 l)$ respectively, which are independent of τ. On the other hand, some works [3,4,8] studied the case that agents have no unique IDs. In this case, gathering is not solvable for some graphs and initial positions of agents. So the works proposed algorithms only for solvable graphs and initial positions. They proposed memory-efficient gathering algorithms for trees [3,8] and arbitrary graphs [4].

If whiteboard exists on each node, the time required for gathering can be significantly reduced. For example, when agents have unique IDs, they can write their IDs into whiteboards on their initial nodes. Agents can collect all the IDs by traversing the network [13], and thus they can achieve gathering by moving to the initial node of the agent with the smallest ID. This trivial algorithm achieves gathering in $O(m)$ time, where m is the number of edges. On the other hand, when agents have no unique IDs, gathering is not trivial even if they use whiteboard and randomization. Ooshita et al. [11] clarified the relationship between solvability of randomized gathering and termination detection in ring networks with whiteboard.

Recently some works [1,6] have considered gathering in the presence of Byzantine agents, which can behave arbitrarily. They modeled agents controlled by crackers or corrupted by software errors as Byzantine agents. These works assume agents have unique IDs, behave synchronously, and cannot use whiteboard. They consider two types of Byzantine agents. While a weakly Byzantine agent can make arbitrary behavior except falsifying its ID, a strongly Byzantine agent can make arbitrary behavior including falsifying its ID. Dieudonné et al. [6] proposed algorithms to achieve gathering in arbitrary graphs against weakly Byzantine agents and strongly Byzantine agents, both when the number of nodes n is known and when it is unknown. For weakly Byzantine agents, when n is known, they proposed an algorithm that achieves gathering in $4n^4 \cdot P(n, \lambda)$ time, where $P(n, l)$ is the time required for gathering of two correct agents (l is the length of the smaller ID) and λ is the length of the longest ID among all agents. Since two agents can meet in $P(n, l) = \tilde{O}(n^5 l)$ time [14], the algorithm achieves gathering in $\tilde{O}(n^9 \lambda)$ time. For weakly Byzantine agents, when n is unknown, they also proposed a polynomial-time algorithm. However, for strongly Byzantine agents, they proposed only exponential-time algorithms. Bouchard et al. [1] minimized the number of correct agents required to achieve gathering for strongly Byzantine agents, however the time complexity is still exponential.

Our Contributions. The purpose of this work is to reduce the time required for gathering by using whiteboard on each node. However, if Byzantine agents can erase all information on whiteboard, correct agents cannot see the information and thus whiteboard is useless. For this reason, we assume that an authentication function is available on the system and this provides authenticated whiteboard. In authenticated whiteboard, each agent is given a dedicated area to write information. In other words, each agent can write information to the dedicated area and cannot write to other areas. Regarding read operations, each agent can read information from all areas on the whiteboard. In addition, we assume, by using the authentication function, each agent can write information with signature that guarantees the writer and the writing node.

No gathering algorithms have been proposed for environments with whiteboard in the presence of Byzantine agents. However, since two agents can meet quickly by using authenticated whiteboard, the time complexity of an algorithm in [6] can be reduced. More specifically, each agent can explore the network in $O(m)$ time by the depth-first search (DFS), and after the first exploration it continues to explore the network in $O(n)$ time for each exploration. By applying this to Dessmark's algorithm [5], two agents can meet in $P(n, l) = O(nl)$ time. Thus, for weakly Byzantine agents, agents can achieve gathering in $O(n^5 \lambda)$ time.

In this work, we propose a new algorithm to achieve gathering in shorter time. Similarly to [6], we assume agents have unique IDs and behave synchronously. When at most f weakly Byzantine agents exist and f is known to agents, our algorithm achieves gathering in $O(fm)$ time by using authenticated whiteboard.

That is, our algorithm significantly reduces the time required for gathering by using authenticated whiteboard. To realize this algorithm, we newly propose a technique to simulate message-passing algorithms by agents. Our algorithm overcomes difficulty of Byzantine agents by simulating a Byzantine-tolerant consensus algorithm [7]. This technique is general and not limited to the gathering problem, and hence it can be applied to other problems of agents.

2 Preliminaries

A Distributed System and Mobile Agents. A distributed system is modeled by a connected undirected graph $G = (V, E)$, where V is a set of nodes and E is a set of edges. The number of nodes is denoted by $n = |V|$. When $(u, v) \in E$ holds, u and v are adjacent. A set of adjacent nodes of node v is denoted by $N_v = \{u | (u, v) \in E\}$. The degree of node v is defined as $d(v) = |N_v|$. Each edge is labeled locally by function $\lambda_v : \{(v, u) | u \in N_v\} \rightarrow \{1, 2, \cdots, d(v)\}$ such that $\lambda_v(v, u) \neq \lambda_v(v, w)$ holds for $u \neq w$. We say $\lambda_v(v, u)$ is a port number (or port) of edge (v, u) on node v.

Each node does not have a unique ID. Each node has whiteboard where agents can leave information. Each agent is assigned a dedicated writable area in the whiteboard, and the agent can write information only to that area. On the other hand, each agent can read information from all areas (including areas of other agents) in whiteboard.

Multiple agents exist in a distributed system. The number of agents is denoted by k, and a set of agents is denoted by $A = \{a_1, a_2, \cdots, a_k\}$. Each agent has a unique ID, and the length of the ID is $O(\log k)$ bits. The ID of agent a_i is denoted by ID_i. Each agent knows neither n nor k.

Each agent is modeled as a state machine (S, δ). The first element S is the set of agent states, where each agent state is determined by values of variables in its memory. The second element δ is the state transition function that decides the behavior of an agent. The input of δ is the current agent state, the content of the whiteboard in the current node, and the incoming port number. The output of δ is the next agent state, the next content of the whiteboard, whether the agent stays or leaves, and the outgoing port number if the agent leaves.

Agents move in synchronous rounds. That is, the time required for each correct agent to move to the adjacent node is identical. In the initial configuration, each agent is inactive and stays at an arbitrary node. Some agents spontaneously become active and start the algorithm. When active agent a_i encounters inactive agent a_j at some node v, agent a_i can make a_j active. In this case, a_j starts the algorithm before a_i executes the algorithm at v.

Each agent a_i can sign a value x that guarantees its ID ID_i and its current node v. That is, any agent identifies an ID of the signed agent and whether it is signed at the current node or not from the signature. We assume a_i can use signature function $Sign_{i,v}(x)$ at v and we denote the output of $Sign_{i,v}(x)$ by $\langle x \rangle : (ID_i, v)$. Each agent a_i can compute $Sign_{i,v}(x)$ for value x at v, however cannot compute $Sign_{j,w}(x)$ for either $j \neq i$ or $w \neq v$. Therefore, it is guaranteed that signed value $\langle x \rangle : (ID_i, v)$ is created by a_i at v. For signed value

$x = \langle value \rangle : (id_1, v_1) : (id_2, v_2) : \cdots : (id_j, v_j)$, the output of $Sign_{i,v}(x)$ is denoted by $\langle value \rangle : (id_1, v_1) : (id_2, v_2) : \cdots : (id_j, v_j) : (ID_i, v)$. In this paper, when an algorithm treats a signed value, it first checks the validity of signatures and ignores the signed value if it includes wrong signatures. We omit this behavior from descriptions, and assume all signatures of every signed value are valid.

Byzantine agents may exist in a distributed system. Each Byzantine agent behaves arbitrarily without being synchronized with other agents. However, each Byzantine agent cannot change its ID. In addition, even if agent a_i is Byzantine, a_i cannot compute $Sign_{j,v}(x)(j \neq i)$ for value x, and therefore a_i cannot create $\langle x \rangle : (ID_j, v)$ for $j \neq i$. We assume the number of Byzantine agents is at most $f(< k)$ and f is known to each agent.

The Gathering Problem. The gathering problem is a problem to make all correct agents meet at a single node and declare termination. In the initial configuration, each agent stays at an arbitrary node and multiple agents can stay at a single node. If an agent declares termination, it never works after that.

To evaluate the performance of the algorithm, we consider the time required for all agents to declare termination after some agent starts the algorithm. We assume the time required for a correct agent to move to the adjacent node is one unit time, and we ignore the time required for local computation.

3   A Byzantine-Tolerant Consensus Algorithm for Message-Passing Systems [7]

In this section, we explain a Byzantine-tolerant consensus algorithm in [7] that will be used as building blocks in our algorithm.

3.1   A Message-Passing System

The consensus algorithm is proposed in a fully-connected synchronous message-passing system. That is, we assume that processes form a complete network. We assume the number of processes is k and denote a set of processes by $P = \{p_1, p_2, \ldots, p_k\}$. Each process has a unique ID, and the ID of p_i is denoted by ID_i. All processes execute an algorithm in synchronous phases. In the 0-th (or initial) phase, every process computes locally and sends messages (if any). In the r-th phase ($r > 0$), every process receives messages, computes locally, and sends messages (if any). If process p_i sends a message to process p_j in the r-th phase, p_j receives the message at the beginning of $(r + 1)$-th phase.

Similarly to Sect. 2, each process p_i has signature function $Sign_i(x)$. The output of $Sign_i(x)$ is denoted by $\langle x \rangle : ID_i$, and only p_i can compute $Sign_i(x)$.

Some Byzantine processes may exist in the message-passing system. Byzantine processes can behave arbitrarily. But even if p_i is Byzantine, p_i cannot compute $Sign_j(x)$ $(j \neq i)$ for value x. We assume the number of Byzantine processes is at most $f < k$ and f is known to each process.

3.2   A Byzantine-Tolerant Consensus Algorithm

In this subsection, we explain a Byzantine-tolerant consensus algorithm in [7]. In the consensus algorithm, each process p_i is given at most one value x_i as its input. If p_i is not given an input value, we say $x_i =\perp$. The goal of the consensus algorithm is to agree on the set of all input values. Of course, some Byzantine processes behave arbitrarily and forge inconsistent input values. However, by the consensus algorithm in [7], all correct agents can agree on the same set $X \supseteq X_c$, where X_c is a set of all values input by correct processes.

We show the details of the consensus algorithm. Each process p_i has one variable $p_i.W$ to keep a set of input values, and initially $p_i.W = \emptyset$ holds. The algorithm consists of $f + 2$ phases (from the 0-th phase to $(f + 1)$-th phase). After processes terminate, they have the same values in W.

In the 0-th phase, if p_i is given an input value $x_i(\neq \perp)$, process p_i broadcasts $Sign_i(x_i) = \langle x_i \rangle : ID_i$ to all processes and adds x_i to variable $p_i.W$. If p_i is not given an input value, it does not do anything.

In the r-th phase $(1 \leq r \leq f + 1)$, p_i receives all messages (or signed values) broadcasted in $(r - 1)$-th phase. After that, for every received message, process p_i checks its validity. We say message $t = \langle x \rangle : id_1 : id_2 : \cdots : id_y$ is valid if and only if t satisfies all the following conditions.

1. The number y of signatures in t is equal to r.
2. All signatures in t are distinct.
3. Message t does not contain p_i's signature.
4. Value x is not in $p_i.W$.

If message $t = \langle x \rangle : id_1 : id_2 : \cdots : id_y$ is valid, p_i broadcasts $Sign_i(t) = \langle x \rangle : id_1 : id_2 : \cdots : id_y : ID_i$ to all processes (if $r \leq f$) and adds x to variable $p_i.W$.

For this algorithm, the following theorem holds.

Theorem 1. [7] *After all processes terminate, all the following holds.*

1. *For any correct process p_i, $x_i \in p_i.W$ holds if $x_i \neq \perp$.*
2. *For any two correct processes p_i and p_j, $p_i.W = p_j.W$ holds.*

4   Our Algorithm

4.1   Overview

First, we give an overview of our algorithm. When agent a_i starts the algorithm, a_i leaves its starting information to whiteboard at its initial node v. The starting information includes ID_i, and consequently it can notify other agents that a_i starts at v. After that, a_i explores the network and collects starting information of all agents. If no Byzantine agent exists, all agents collect the same set of starting information, and thus all agents can meet at a single node by visiting the node where the agent with the smallest ID leaves the starting information.

However, when some Byzantine agent exists, it can write and delete its starting information repeatedly so that only a subset of agents see the information.

This implies some agents may obtain a set of starting information different from others and thus may fail to achieve gathering.

To overcome this difficulty, our algorithm makes all correct agents agree on the same set of starting information at each node. That is, letting $a_i.X_v$ be the set of starting information that a_i obtains at node v, we guarantee that $a_i.X_v = a_j.X_v$ holds for any two correct agents a_i and a_j. In addition, we also guarantee that, if correct agent a_c starts at v, then $a_i.X_v$ contains a_c's starting information and $a_i.X_w(w \neq v)$ does not contain a_c's starting information. We later explain the details of this procedure.

After that, each agent a_i can obtain $a_i.X_{all} = \bigcup_{v \in V} a_i.X_v$, and clearly $a_i.X_{all} = a_j.X_{all}$ holds for any two correct agents a_i and a_j. Consequently each agent a_i can compute the same gathering node based on $a_i.X_{all}$ as follows. First a_i removes all duplicated starting information from $a_i.X_{all}$ because a Byzantine agent may leave its starting information at several nodes. After that, a_i finds the starting information of the agent with the smallest ID and selects the node with the starting information as the gathering node. By this behavior, all correct agents can meet at the same gathering node.

In the rest of this subsection, we explain the way to make all correct agents agree on the same set of starting information at each node. To realize this, our algorithm uses a Byzantine-tolerant consensus algorithm in Sect. 3. At each node, agents simulate the consensus algorithm and then agree on the same set. However, since the consensus algorithm is proposed for synchronous message-passing systems, we need additional synchronization mechanism. We realize this by using the depth-first search (DFS).

DFS and Round Synchronization. The DFS is a well-known technique to explore a graph. In the DFS, an agent continues to explore a port as long as it visits a new node. If the agent visits an already visited node, it backtracks to the previous node and explores another unexplored port. If no unexplored port exists, the agent backtracks to the previous node again. By repeating this behavior, each agent can visit all nodes in $2m$ unit times, where m is the number of edges. Note that, since each agent can realize the DFS by using only its dedicated area on whiteboard, Byzantine agents cannot disturb the DFS of correct agents.

To simulate the consensus algorithm, we realize round synchronization of agents by the DFS. More specifically, we guarantee that, before some agent a_i makes the r-th visit to v, all agents finish the $(r-1)$-th visit to v. To realize this, each agent a_i executes the following procedure in addition to the DFS.

– If a_i finds an inactive agent, a_i makes the agent active.
– Every time a_i completes a DFS, it waits for the same time as the exploration time. That is, a_i waits for $2m$ unit times after each DFS.

We define the r-th exploration period of a_i as the period during which a_i executes the r-th DFS exploration, and define the r-th waiting period of a_i as the period during which a_i waits after the r-th DFS exploration. In addition, we define the r-th round of a_i as the period from the beginning of the r-th exploration period to the end of the r-th waiting period. As shown in the Fig. 1, before some agent starts the r-th exploration period, every correct agent completes the $(r-1)$-th exploration period.

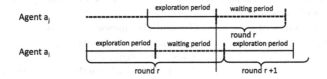

Fig. 1. Exploration and waiting periods.

Simulation of Consensus Algorithm. In the following, we explain the way to apply the consensus algorithm in Sect. 3. The goal is to make all correct agents agree on the same set of starting information at each node. To achieve this, we assume k virtual processes $v.p_1, v.p_2, \ldots, v.p_k$ exist at each node v and form a message-passing system in Sect. 3 (See Fig. 2). When agent a_i visits node v, it simulates $v.p_i$'s behavior of the consensus algorithm.

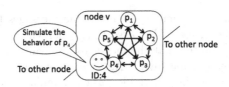

Fig. 2. Virtual processes.

In the consensus algorithm on node v, each virtual process decides its input value as follows. If a_i starts the algorithm at v, the input of virtual process $v.p_i$ is the starting information of a_i. Otherwise, the input of virtual process $v.p_i$ is not given. Thus, after completion of the consensus algorithm, all virtual processes at v agree on the same set X_v of starting information. From the property of the consensus algorithm, X_v contains starting information of all correct agents that start at v.

Next, we explain how to simulate the behaviors of virtual processes. Each agent a_i simulates the r-th phase of virtual process $v.p_i$ when a_i visits v for the first time in the exploration period of r-th round. Recall that, by the round synchronization, when some correct agent a_i starts the exploration period of the r-th round, all correct agents have already completed the exploration period of the $(r-1)$-th round. This implies, a_i can simulate the r-th phase of virtual process $v.p_i$ after all virtual processes complete the $(r-1)$-th phase.

To simulate $v.p_i$, a_i uses variables $v.wb[ID_i].T$ and $v.wb[ID_i].W$ in whiteboard of node v. We denote variable var in the dedicated area of a_i by $v.wb[ID_i].var$. Agent a_i uses $v.wb[ID_i].T$ to simulate communications among

---

Algorithm 1. main( )

1: —Variables in whiteboard of node v—
2: **var** $v.wb[ID_i].T$ and $v.wb[ID_i].W$
3: **var** $v.wb[ID_i].round$, $v.wb[ID_i].from\_port$, and $v.wb[ID_i].unexplored\_port$
4: —Variables of agent a_i—
5: **var** $a_i.node\_num = 0$ // count the number of nodes
6: **var** $a_i.all\_edge\_num = 0$ // count the number of edges
7: **var** $a_i.r = 0$ // keep the current round
8: **var** $a_i.W = \emptyset$ // collect a set of starting information
9: ————————————
10: $consensus()$
11: **for** $a_i.r = 1$ to $f + 1$ **do**
12:     $a_i.node\_num = 1$
13:     $a_i.all\_edge\_num = 0$
14:     $DFS(null)$
15:     **wait** $a_i.all\_edge\_num \times 2$
16: **end for**
17: Delete duplicated *candidate* from $a_i.W$
18: Move to a node where the minimum *candidate* in $a_i.W$ is written
19: Declare termination

---

virtual processes. That is, when $v.p_i$ sends some messages to other processes, a_i stores the messages in $v.wb[ID_i].T$ so that other virtual processes read the messages. Here, to guarantee that the messages are available on only node v, a_i stores $Sign_{i,v}(t)$ instead of message t. Agent a_i uses $v.wb[ID_i].W$ to memorize variables of $v.p_i$. By using these variables, a_i can simulate the r-th phase of $v.p_i$ as follows:

1. By reading from all variables $v.wb[id].T$ (for some id), a_i receives messages that virtual processes have sent to $v.p_i$ in the $(r-1)$-th phase.
2. From $v.p_i$'s variables stored in $v.wb[ID_i].W$ and messages received in 1, agent a_i simulates local computation of $v.p_i$'s r-th phase.
3. Agent a_i writes updated variables of $v.p_i$ to $v.wb[ID_i].W$. If $v.p_i$ sends some messages, a_i writes the messages with signatures to $v.wb[ID_i].T$.

Note that, since only agent a_i can update variables $v.wb[ID_i].T$ and $v.wb[ID_i].W$, agent a_i simulates the correct behavior of $v.p_i$ if a_i is correct. This implies that the simulated message-passing system contains at most f Byzantine processes. Consequently (correct) virtual processes can agree on the same set by the consensus algorithm that can tolerate at most f Byzantine processes. Thus correct agents can agree on the same set of starting information at v.

4.2   Details

The pseudo-code of the algorithm is given in Algorithms 1, 2, and 3. Due to limitation of space, the details of $main()$ and $DFS()$ are provided in the full

Algorithm 2. DFS(f_port)

```
 1: make an inactive agent active if such an agent exists at v
 2: if v.wb[ID_i].round ≠ a_i.r then
 3:     v.wb[ID_i].round = a_i.r
 4:     v.wb[ID_i].from_port = f_port
 5:     if f_port = null then
 6:         v.wb[ID_i].unexplored_port = {1, ..., d(v)}
 7:     else
 8:         v.wb[ID_i].unexplored_port = {1, ..., d(v)} \ {f_port}
 9:     end if
10:     a_i.node_num + +
11:     consensus()
12:     if a_i.r = f + 1 then
13:         for all candidate in v.wb[ID_i].W do
14:             a_i.W = a_i.W ∪ {(candidate, a_i.node_num)}
15:         end for
16:     end if
17:     while v.wb[ID_i].unexplored_port ≠ ∅ do
18:         x = min(v.wb[ID_i].unexplored_port)
19:         a_i.all_edge_num + +
20:         v.wb[ID_i].unexplored_port = v.wb[ID_i].unexplored_port \ {x}
21:         Go to the next node via port x
22:         DFS(Port number via which a_i enters the current node)
23:     end while
24:     Backtrack via port v.wb[ID_i].from_port. If it is null, do not move.
25: else
26:     v.wb[ID_i].unexplored_port = v.wb[ID_i].unexplored_port \ {f_port}
27:     Backtrack via port f_port. If it is null, do not move.
28: end if
```

version [15]. Simply put, functions $main()$ and $DFS()$ realize the DFS traversal of agent a_i. When a_i starts the algorithm, a_i executes $consensus()$ once to simulate the 0-th phase of virtual process $v.p_i$. After that, for each node v, a_i calls $consensus()$ to simulate the r-th phase of $v.p_i$ when it visits v for the first time during the r-th round.

Function $consensus()$ simulates the consensus algorithm in Sect. 3 by following the strategy in Sect. 4.1. In the 0-th round, a_i simulates the 0-th phase of the consensus algorithm. That is, a_i makes virtual process $v.p_i$ broadcast a signed value $Sign_{i,v}(x_i)$ if $v.p_i$ is given an input value x_i. Recall that $v.p_i$ is given starting information of a_i as an input if a_i starts at v. This means the simulation of the 0-th phase is required only for the initial node of a_i. In other words, a_i completes the 0-th round without exploring the network. Specifically, a_i adds $Sign_{i,v}(ID_i)$ to $v.wb[ID_i].T$ as its stating information, and adds ID_i to $v.wb[ID_i].W$ (lines 1 to 3).

In the r-th round (lines 4 to 11), a_i simulates the r-th phase of the consensus algorithm. To realize this, for every node v, a_i simulates the r-th phase of $v.p_i$

Algorithm 3. consensus( )

1: **if** $a_i.r = 0$ **then**
2:     $v.wb[ID_i].T = \{Sign_{i,v}(ID_i)\}$
3:     $v.wb[ID_i].W = \{ID_i\}$
4: **else**
5:     **for all** t such that $t \in v.wb[id].T$ for some id **do**
6:         **if** (t is valid) **then**
7:             $v.wb[ID_i].T = v.wb[ID_i].T \cup \{Sign_{i,v}(t)\}$
8:             $v.wb[ID_i].W = v.wb[ID_i].W \cup \{value(t)\}$
9:         **end if**
10:     **end for**
11: **end if**

---

when it visits v for the first time during the round. Specifically, for every message received by $v.p_i$, a_i checks its validity. Note that messages received by $v.p_i$ are stored in $\bigcup_{a_j \in A} v.wb[ID_j].T$. We say message $t = \langle x \rangle : (id_1, v_1) : (id_2, v_2) : \cdots : (id_y, v_y)$ is valid if and only if t satisfies all the following conditions, where we define $value(t) = x$ and $initial(t) = id_1$.

1. The number y of signatures in t is equal to r.
2. All signatures in t are distinct.
3. Message t does not contain a_i's signature.
4. $value(t)$ is not in $v.wb[ID_i].W$.
5. $value(t) = initial(t)$ holds.
6. All the y signatures are given at the current node.

Conditions 1–4 are identical to conditions in Sect. 3. Condition 5 is introduced to assure that value ID_i in messages is originated from a_i. Note that, since correct agent a_i can initially add $\langle ID_i \rangle : (ID_i, v)$ to $v.wb[ID_i].T$, every message t forwarded by correct agents satisfies $value(t) = initial(t)$. This implies condition 5 does not discard messages originated from and forwarded by correct agents, and consequently does not influence the simulation of correct processes. Condition 6 is introduced to assure that message t is generated at the current node. If t is valid, a_i adds $Sign_{i,v}(t)$ to $v.wb[ID_i].T$ to simulate broadcast of $Sign_{i,v}(t)$ by virtual process $v.p_i$. At the same time, a_i adds $value(t)$ to $v.wb[ID_i].W$.

In the $(f + 1)$-th round, all agents complete simulating the consensus algorithm. That is, $v.wb[ID_i].W = v.wb[ID_j].W$ holds for any two correct agents a_i and a_j. During the $(f + 1)$-th round, a_i collects contents in $v.wb[ID_i].W$ for all v by variable $a_i.W$ (lines 12 to 16 of $DFS()$). Recall that $v.wb[ID_i].W$ includes IDs of agents that start at v. When a_i memorizes $candidate \in v.wb[ID_i].W$, a_i memorizes it as a pair $(candidate, a_i.node\_num)$ to recognize the node later.

After that, a_i computes the gathering node from the collected information in $a_i.W$ (lines 17 to 18 in $main()$). Since IDs of Byzantine agents may appear more than once in $a_i.W$, a_i deletes all pairs from $a_i.W$ such that $candidate$ is duplicated. Then a_i finds the pair such that $candidate$ is the smallest, and it selects the node of the pair as the gathering node. Note that the pair includes *candidate* and $a_i.node\_num$. Hence a_i can move to the gathering node by executing the DFS until $a_i.node\_num$ becomes the same number as the pair (this procedure is omitted in *main*()).

Theorem 2. *Our algorithm solves the gathering problem in $O(fm)$ unit times.*

5 Summary

In this paper, we proposed a Byzantine-tolerant gathering algorithm for mobile agents in synchronous networks with authenticated whiteboards. In our algorithm, each agent first writes its starting information to the initial node, and then each agent executes a consensus algorithm so that every correct agent agrees on the same set of starting information. Once correct agents obtain the set, they can calculate the same gathering node. By this algorithm, all correct agents can achieve gathering in $O(fm)$ time. An important open problem is to develop a Byzantine-tolerant gathering algorithm in asynchronous networks with authenticated whiteboards. Since the consensus algorithm is proven to be unsolvable in asynchronous networks, we must consider other approaches.

References

1. Bouchard, S., Dieudonné, Y., Ducourthial, B.: Byzantine gathering in networks. In: Scheideler, C. (ed.) Structural Information and Communication Complexity. LNCS, vol. 9439, pp. 179–193. Springer, Heidelberg (2015). doi:10.1007/978-3-319-25258-2_13
2. Cao, J., Das, S.K.: Mobile Agents in Networking and Distributed Computing. Wiley-Interscience, Hoboken (2012)
3. Czyzowicz, J., Kosowski, A., Pelc, A.: Time versus space trade-offs for rendezvous in trees. Distrib. Comput. **27**(2), 95–109 (2014)
4. Czyzowicz, J., Kosowski, A., Pelc, A.: How to meet when you forget: log-space rendezvous in arbitrary graphs. Distrib. Comput. **25**(2), 165–178 (2012)
5. Dessmark, A., Fraigniaud, P., Kowalski, D.R., Pelc, A.: Deterministic rendezvous in graphs. Algorithmica **46**(1), 69–96 (2006)
6. Dieudonné, Y., Pelc, A., Peleg, D.: Gathering despite mischief. ACM Trans. Algorithms **11**(1), 1:1–1:28 (2014). Article 1
7. Dolev, D., Strong, H.R.: Authenticated algorithms for byzantine agreement. SIAM J. Comput. **12**(4), 656–666 (1983)
8. Fraigniaud, P., Pelc, A.: Delays induce an exponential memory gap for rendezvous in trees. ACM Trans. Algorithms **9**(2), 17 (2013)
9. Kowalski, D.R., Malinowski, A.: How to meet in anonymous network. Theor. Comput. Sci. **399**(1–2), 141–156 (2008)
10. Kranakis, E., Krizanc, D., Markou, E.: The Mobile Agent Rendezvous Problem in the Ring, 1st edn. Morgan and Claypool Publishers (2010). ISBN: 1608451364, 9781608451364
11. Ooshita, F., Kawai, S., Kakugawa, H., Masuzawa, T.: Randomized gathering of mobile agents in anonymous unidirectional ring networks. IEEE Trans. Parallel Distrib. Syst. **25**(5), 1289–1296 (2014)

12. Pelc, A.: Deterministic rendezvous in networks: a comprehensive survey. Networks **59**, 331–347 (2012)
13. Sudo, Y., Baba, D., Nakamura, J., Ooshita, F., Kakugawa, H., Masuzawa, T.: A single agent exploration in unknown undirected graphs with whiteboards. IEICE Trans. **98–A**(10), 2117–2128 (2015)
14. Ta-Shma, A., Zwick, U.: Deterministic rendezvous, treasure hunts, and strongly universal exploration sequences. ACM Trans. Algorithms **10**(3), 12:1–12:15 (2014)
15. Tsuchida, M., Ooshita, F., Inoue, M.: Byzantine gathering in networks with authenticated whiteboards. NAIST Information Science Technical report, NAIST-IS-TR2016001 (2016)

Generating All Patterns of Graph Partitions Within a Disparity Bound

Jun Kawahara[1]([✉]), Takashi Horiyama[2], Keisuke Hotta[3], and Shin-ichi Minato[4]

[1] Nara Institute of Science and Technology, Ikoma, Japan
jkawahara@is.naist.jp
[2] Saitama University, Saitama, Japan
horiyama@al.ics.saitama-u.ac.jp
[3] Bunkyo University, Chigasaki, Japan
khotta@shonan.bunkyo.ac.jp
[4] Hokkaido University, Sapporo, Japan
minato@ist.hokudai.ac.jp

Abstract. A balanced graph partition on a vertex-weighted graph is a partition of the vertex set such that the partition has k parts and the disparity, which is defined as the ratio of the maximum total weight of parts to the minimum one, is at most r. In this paper, a novel algorithm is proposed that enumerates all the graph partitions with small disparity. Experimental results show that five millions of partitions with small disparity for some graph with more than 100 edges can be enumerated within ten minutes.

1 Introduction

A k-balanced graph partitioning problem requires splitting a given graph into k connected components of almost equal size. This problem appears in several applications such as parallel computing, image analysis, floorplan design, and so on. Among them, designing social systems such as elections is an important application. Suppose that we would like to elect k representatives from a prefecture, which is composed of some cities. We divide the prefecture into k connected components to elect one representative for each component under the condition that any city must not be split and all the components are desired to be balanced. For this purpose, we represent the prefecture as a vertex-weighted graph, where its vertex corresponds to a city, two vertices are adjacent if and only if the corresponding cities have the common border, and the weight of a vertex means the population in the city. Our concern is a balanced partition on such a graph.

Nemoto and Hotta [12] proposed an integer programming-based algorithm that obtains the graph partition with the smallest *disparity*, that is defined as the maximum ratio of the weights of two connected components in the partition. Their aim of minimizing the disparity seems to make current electoral partitions more balanced in accordance with the requisition of a law in Japan [12]. However, from a practical point of view, we desire not the partition with simply the

© Springer International Publishing AG 2017
S.-H. Poon et al. (Eds.): WALCOM 2017, LNCS 10167, pp. 119–131, 2017.
DOI: 10.1007/978-3-319-53925-6_10

smallest disparity but one satisfying many other conditions derived from geographical, sociological, legal, and political requirements with moderately small disparity. It is quite hard that we impose such complex constraints on a solution in the integer programming-based algorithm and solve it.

In this paper, we adopt an enumeration algorithm-based approach to obtain desired partitions. Specifically, we propose an algorithm that generates all partitions within a given disparity for the input vertex-weighted graph. Needless to say, the number of ways for partitioning a graph exponentially increases as the graph size grows. For example, there are more than 10^{28} ways of partitioning an 8×8 grid graph with 64 vertices into at least two connected components. To treat a tremendous number of partitions, we use a compact data structure called the zero-suppressed decision diagram (ZDD) [11], which is a variant of binary trees for representing a family of sets. Our algorithm directly constructs a ZDD representing all the partitions without enumerating them explicitly, which implies that our algorithm runs significantly faster than algorithms that generate and output partitions one by one. We develop a novel technique that imposes the disparity constraint on resulting partitions. We also show some results on the complexity of our algorithms.

We mention why we enumerate a number of partitions: (i) We can count the number of partitions with every 0.1 disparity, and obtain a histogram showing them (examples are shown in Figs. 1 and 2). The histogram tells us whether there is room for improvement of the current partition in terms of the disparity. If there are a number of partitions whose disparity is much smaller than the current one, we can suggest adopting one of them. Such a histogram would be useful for practitioners. (ii) As mentioned above, it is difficult to obtain the partition simultaneously satisfying multiple conditions (including the disparity condition). Therefore, we first generate a million of partitions with small disparity by our proposed algorithm (not taking other conditions into consideration), and then one by one check whether each partition satisfies desired conditions, e.g., the slenderness of components, the amount of change from the current partition. Note that each of million partitions can be checked in a brute-force manner in a distributed environment. Since it will be a human who determines which partition is adopted, it is essential to narrow enumerated partitions with small disparity down from millions to dozens. Practitioners often would like to compare dozens of good partitions by hand.

The paper is organized as follows. In Sect. 2, we provide an explanation of our framework. Sect. 3 describes an algorithm that enumerates all the graph partitions on a given graph. We show how to enumerate partitions with only small disparity in Sect. 4. In Sect. 5, the results of numerical experiments are shown to confirm the effectiveness of our algorithms. We conclude the paper in Sect. 6.

Related work: There are several research adopting the approach of using ZDDs and enumeration algorithms. Inoue et al. [6] optimized the power loss for smart grid by enumerating all the rooted spanning forests of a given distribution network. The technique is also used for evacuation planning [14] and evaluating the network reliability [3,4]. Yoshinaka et al. [15] proposed algorithms for solving and enumerating some puzzle problems.

The special case where $k = 2$, the disparity is exactly one, and all the weights are one is known as the minimum bisection problem, which has been widely studied for decades [7]. In this problem, the objective function is not the disparity but the number of edges whose two endpoints belong to distinct connected components. Andreev and Räcke [1] introduced a (k, ν)-balanced graph partition, which is a graph partition such that each component must not have more than $\nu n/k$ vertices. Note that their definition of "balanced" is different from our definition of disparity. They proved that there is no polynomial time approximation algorithm unless P = NP when $k \geq 3$ and $\nu = 1$, and designed an approximation algorithm which attains $O(\log^{1.5} n)$ approximation ratio for fixed $\nu > 1$. Local search-based methods for graph partitioning have been proposed by King et al. [9].

2   Preliminaries

We explain some notation and definitions. An undirected vertex-weighted graph is defined as a triple (V, E, h), where V is a set of vertices, E is a set of edges, each of which is a vertex set having exactly two elements of V, and $h : V \to \mathbb{N}$ is a weighted function mapping a vertex into a natural number. Throughout this paper, we use the notation G, V, E and h for the input graph and assume that the input graph is simple. We let $m = |E|$ and $E = \{e_1, \ldots, e_m\}$. In this paper, when we use the phrase "subgraph of G," it always indicates a subgraph (V, E', h) of G for an edge set $E' \subseteq E$, that is, a subgraph whose vertex set coincides with V, and consequently we can identify a subgraph with the set of its edges.

In this section, we explain an algorithm, called the frontier-based search [8, 13], which constructs a directed acyclic graph compactly expressing all the subgraphs of a given graph such that the subgraphs satisfy specified conditions. For ease of explanation, we explain the frontier-based search for spanning trees as an example, that is, we describe how to construct the directed acyclic graph representing all the spanning trees on a given graph. Note that the frontier-based search can treat various subgraphs such as paths, matchings, covers besides spanning trees [8,10,15].

We describe the property of the directed acyclic graph $D = (N, A)$ that the frontier-based search constructs, which we call the ZDD. D has two special terminal nodes, called the 0-terminal and 1-terminal and denoted by $\mathbf{0}$ and $\mathbf{1}$, respectively, and has one node which has no incoming arc, called the root node and denoted by n_{root} (see figures in [8]). Each node has exactly two arcs, called the 0-arc and the 1-arc, and has a label e_i for some $i = 1, \ldots, m$. The root node has a label e_1. A node with label e_i points at one with e_{i+1} or a terminal. Each path from the root node to a node n_i with label e_i corresponds to a subgraph in the following manner: Let $P = \langle n_1, a_1, n_2, a_2, \ldots, n_{i-1}, a_{i-1}, n_i \rangle$ be a directed path on the ZDD such that $n_1 = n_{\text{root}}$, n_j is a node with label e_j and a_j is an arc of n_j for $j = 1, \ldots, i$. We define $E(P) = \{e_i \mid a_i \text{ is a 1-arc}\}$, and say that P *corresponds* to the subgraph $(V, E(P))$. We also define $\mathcal{G}(n_i) = \{(V, E(P')) \mid P' \text{ is a directed path from } n_{\text{root}} \text{ to } n_i\}$ and say that the node n_i *corresponds* to $\mathcal{G}(n_i)$. We interpret that the ZDD represents the set $\mathcal{G}(1)$ of subgraphs.

Now we describe how to construct the ZDD. First, we start from creating the root node n_{root} and its 0-arc and 1-arc. Then, we create nodes with label e_2 as the destinations of the arcs. Nodes of the ZDD are created in a breadth-first manner, that is, for each $i = 2, \ldots, m - 1$, we create nodes labeled e_i after all nodes labeled e_{i-1} are created. Note that a node n with label e_i corresponds to the set $\mathcal{G}(n)$ of subgraphs whose edge set consists of some of $e_1, \ldots, e_{i-1}$. For each node n, to efficiently decide whether subgraphs in $\mathcal{G}(n)$ have a cycle, we store a partition over vertices, denoted by $n.\text{comp}$, representing the connectivity of them into n. The partition $n.\text{comp}$ means that for vertices v and w, v and w belong to the same cell of the partition if and only if v and w belong to the same connected component for all the subgraphs in $\mathcal{G}(n)$. (Given a partition $\mathcal{V} = \{V_1, V_2, \ldots\}$ over V such that $V_j \subseteq V$ and $V_j \cap V_{j'} = \emptyset$ with $j \neq j'$, we call each V_j a cell of the partition $\mathcal{V}$.)

Consider the situation where we are creating a node n' with label e_{i+1} as the destination of the x-arc of a node n with label $e_i = \{v, w\}$ for $x = 0, 1$. If $x = 1$, n' corresponds to the set of subgraphs each of which is obtained by adding an edge e_i to a subgraph in $\mathcal{G}(n)$. Then, if v and w belong to the same cell of $n.\text{comp}$, a cycle is generated in the subgraphs, which means that there is no chance that spanning trees are completed from the subgraphs. In this case, we stop creating n' and let the x-arc of n point at the 0-terminal instead of n'.

The partition $n.\text{comp}$ of a node n is also used for ensuring that resulting subgraphs are connected. Consider the situation described above again (for $x = 0, 1$). Suppose that e_i has the last index among the edges incident to v. Then, if all the vertices in the connected component including v will never be connected with the other vertices, the connected component is isolated and the resulting subgraphs are not spanning trees. Such a situation occurs when all the vertices in the connected component including v are incident to edges only in $e_1, \ldots, e_i$. In this case, we let the x-arc of n point at the 0-terminal. More formally, for each $i = 1, \ldots, m - 1$, we define

$$F_i = \left(\bigcup_{j=1,\ldots,i} e_j \right) \cap \left(\bigcup_{j=i+1,\ldots,m} e_j \right),$$

called the *i-th frontier* F_i. We also define $F_0 = F_m = \emptyset$. If $v \in F_{i-1}$, $v \notin F_i$ and there is no vertex w on the frontier F_i such that v and w belong to the same cell of $n.\text{comp}$, it means that the connected component including v is never connected with other connected components even if some of $e_{i+1}, \ldots, e_m$ are added.

The decision described above needs the information of the connectivity of only vertices in F_{i-1} when e_i is processed. Therefore, we restrict the domain of $n.\text{comp}$ to F_{i-1}. For nodes n' and n'', we say that n' and n'' are *identical* if n' and n'' have the same label and $n'.\text{comp} = n''.\text{comp}$ (on F_{i-1}). When we create a

---

Algorithm 1. CONSTRUCTZDD

---

1 $N_1 \leftarrow \{n_{\text{root}}\}$. $N_i \leftarrow \emptyset$ for $i = 2, \ldots, m + 1$.
2 **for** $i \leftarrow 1$ **to** m **do**
3   **foreach** $n \in N_i$ **do**
4   **foreach** $x \in \{0, 1\}$ **do**   // process for the 0/1-arc
5              $n' \leftarrow$ MAKENEWNODE(n, i, x) // returns a new node or the
                       0/1-terminal
6   **if** $n' \neq 0, 1$ **then**   // n' is neither 0 nor 1.
7   **if** *there exists a node* $n'' \in N_{i+1}$ *s.t.* n'' *is identical to* n' **then**
8                      $n' \leftarrow n''$
9   **else**
10                     $N_{i+1} \leftarrow N_{i+1} \cup \{n'\}$
11   Create the x-arc of n and make it point at n'.

---

node n' as the destination of an arc of n, if there is a node n'' which is identical to n', we stop creating n' and make the arc of n point at n'' instead of n'.

When n has a label $e_m = \{v, w\}$, if v and w are in the same cell of $n.\texttt{comp}$, the destination of the 0-arc of n is the 1-terminal because it means that a spanning tree is completed, and that of the 1-arc of n is the 0-terminal because a cycle occurs. Otherwise (v and w are in distinct cells of $n.\texttt{comp}$), the destination of the 0-arc of n is the 0-terminal because the connected component including v and that including w are not connected, and that of the 1-arc of n is the 1-terminal because two remaining connected components are connected by adding e_m.

3   Constructing a ZDD Representing All Graph Partitioning

In this section, we propose a novel algorithm for constructing the ZDD representing all the partitions of a given graph. We describe the algorithm by modifying the frontier-based search for spanning trees explained in Sect. 2. We define a *partition of a graph* as a partition over the vertex set of the graph such that the subgraph induced by each cell of the partition is connected. Note that in the frontier-based search framework, a subgraph needs to be represented as a set of edges. Although a partition of a graph can be represented as a spanning forest by regarding its connected components as cells of a partition, its representation is not unique. Therefore, we consider a subgraph such that for vertices $u, w \in V$ which belong to the same connected component in the subgraph, the subgraph must have the edge $\{v, w\}$ if the original graph has $\{v, w\}$, and call a *partition subgraph*. It is clear that there is one-to-one correspondence between partition subgraphs of a graph and partitions of the graph.

We sometimes would like to specify the number of cells of each partition we generate. We use notation $K \subseteq \{1, \ldots, |V|\}$ to specify a set of the possible numbers of cells of partitions. Let $|\mathcal{V}|$ denote the number of cells of a partition $\mathcal{V}$ of V. We define

$$\mathcal{P}(K) = \{(V, E') \mid \mathcal{V} \text{ is a partition of } V, \ |\mathcal{V}| \in K, \{v, w\} \in E' \text{ if and only if}$$
$$v \text{ and } w \text{ are in the same cell of } \mathcal{V}\}.$$

Our goal in this section is to design the frontier-based search for constructing the ZDD representing $\mathcal{P}(K)$ for any integer set $K \subseteq \{1, \ldots, |V|\}$.

First, we describe how to count the number of connected components. Recall that in the frontier-based search for spanning trees, when creating a new node, we decide whether an isolated connected component occurs in the subgraphs corresponding to the node. Therefore, we can count the number of isolated connected components by storing the number into each node. For a node n, we use variable $n.\mathtt{cc}$ to count it. More formally, consider the situation where we are creating a node n' with label e_{i+1} as the destination of an arc of a node n with label $e_i = \{v, w\}$. For each $u \in \{v, w\}$, if e_i has the last index among the edges incident to u, and there is no vertex $x \in F_i$ such that $x \neq u$ and x and u belong to the same cell of $n.\mathtt{comp}$, we increment $\mathtt{cc}$. In other words, $n'.\mathtt{cc} \leftarrow n.\mathtt{cc} + a$, where a is the number of vertices $u \in \{v, w\}$ such that the above condition holds. These processes are carried out in Line 18–19 in Algorithm 2, which is called from Algorithm 1. If $n'.\mathtt{cc} > \max(K)$, the node can be pruned, that is, we let the destination of the arc point at the 0-terminal instead of n (see Line 20–21) because the number of isolated connected components exceeds $\max(K)$ even if some of edges $e_{i+1}, \ldots, e_m$ are added. After the last edge is processed, that is, in case of $i = m$, we check whether $\mathtt{cc} \in K$. If so, we let the terminal be 1, otherwise 0 (Line 25–28).

Next, we describe how to ensure that each connected component of resulting subgraphs composes an induced subgraph. For a node n with a label $e_i = \{v, w\}$, if v and w are in the same cell of $n.\mathtt{comp}$, the edge $\{v, w\}$ must be adopted because v and w have already been in the same connected component. In this case, we let the destination of the 0-arc of n point at the 0-terminal (Line 13–14). If v and w are in distinct cells of $n.\mathtt{comp}$, we cannot immediately decide whether v and w are in the same connected component at the end of processing edges. If we decide not to adopt $e_i = \{v, w\}$, two components including v and w are not allowed to be connected in the future. Therefore, we need to remember which components are not allowed to be connected. For this purpose, we introduce a variable $n.\mathtt{fps}$, called a *forbidden pair set*, to remember such pairs of connected components. For connected components C and C' in $n.\mathtt{comp}$, $\{C, C'\} \in n.\mathtt{fps}$ means that C and C' must not be connected in the future. (Note that we identify a connected component with a cell of a partition over vertices.) For a vertex u, let C_u be a connected component in $n.\mathtt{comp}$. When we adopt $e_i = \{v, w\}$, if $C_v \neq C_w$ and $\{C_v, C_w\} \in n.\mathtt{fps}$, we let the destination of the 1-arc of n point at the 0-terminal (Line 8–9). Otherwise, we update the connected components in $n.\mathtt{fps}$, that is, we replace all C_v's and C_w's with $C_v \cup C_w$ (Line 10–11). If we decide not to adopt $e_i = \{v, w\}$, add the pair $\{C_v, C_w\}$ to $\mathtt{fps}$ (Line 15–16).

Algorithm 2. MAKENEWNODE(n, i, x) for partition subgraphs

---

1 Let $e_i = \{v, w\}$.
2 Copy n to n'.
3 **foreach** $u \in \{v, w\}$ *such that* $u \notin F_{i-1}$ **do**   // u is entering the frontier.
4    $\lfloor$ n'.comp $\leftarrow n'$.comp $\cup \{\{u\}\}$ // add the singleton set $\{u\}$ to n'.comp
5 Let C_v and C_w be the vertex set containing v and w in n'.comp, respectively.
6 **if** $x = 1$ **then**
7    $|$ n'.comp $\leftarrow (n'$.comp $\setminus \{C_v\} \setminus \{C_w\}) \cup \{C_v \cup C_w\}$ // The two components become connected.
8    $|$ **if** $C_v \neq C_w$ *and* $\{C_v, C_w\} \in n'$.fps **then**   // C_v and C_w must not be merged
9    $|$ $\lfloor$ **return** 0
10   $|$ **else**
11   $|$ $\lfloor$ Replace all C_v's and C_w's in n'.fps with $C_v \cup C_w$.

12 **else**
13   $|$ **if** $C_v = C_w$ **then**   // $\{v, w\}$ must be adopted because v and w are included in the connected component
14   $|$ $\lfloor$ **return** 0
15   $|$ **else**
16   $|$ $\lfloor$ n'.fps $\leftarrow n'$.fps $\cup \{\{C_v, C_w\}\}$

17 **foreach** $u \in \{v, w\}$ *such that* $u \notin F_i$ **do**   // u is leaving the frontier.
18   $|$ **if** $\{u\} \in n'$.comp **then**   // u is in an isolated connected component
19   $|$ $|$ n'.cc $\leftarrow n'$.cc $+ 1$
20   $|$ $|$ **if** n'.cc $> \max(K)$ **then**   // pruning
21   $|$ $|$ $\lfloor$ **return** 0
22   $|$ Remove u from n'.comp.
23   $\lfloor$ Remove $\{\{u\}, X\}$ for any $X \in n'$.comp from n'.fps.
24 **if** $i = m$ **then**   // The processing of the last edge has been done.
25   $|$ **if** n'.cc $\in K$ **then**
26   $|$ $\lfloor$ **return** 1   // All the constraints are satisfied.
27   $|$ **else**
28   $|$ $\lfloor$ **return** 0

29 **return** n'

In the frontier-based search for partition subgraphs, we say that n and n' are *identical* if n and n' have the same label, n.comp $= n'$.comp, n.cc $= n'$.cc and n.fps $= n'$.fps. The CONSTRUCTZDD function of the frontier-based search for partition subgraphs is almost the same as that for spanning trees. The only different thing is the MAKENEWNODE functions called from CONSTRUCTZDD. The MAKENEWNODE function for partition subgraphs is shown in Algorithm 2.

3.1 Complexity of Our Algorithm

In this subsection we discuss the complexity of our algorithm. First, we estimate it for general graphs. In this subsection, n denotes the number of vertices in G. The cardinality of the forbidden pair set in a ZDD node affects the number of nodes in a ZDD and the computation time.

Lemma 1. *In the process of the ZDD construction, the cardinality of the forbidden pair set in any ZDD node is at most $O(n^2)$.*

This estimation is tight because there is a case in which the cardinality of a forbidden pair set is $\Omega(n^2)$ (for example, in case of complete graphs).

Next, we consider the case of planar graphs. We define $f = \max_i |F_i|$, where $|F_i|$ is the cardinality of F_i. Let C_j be the j-th Catalan number, defined by $\frac{1}{j+1}\binom{2j}{j}$, and G_j be the number of graphs having j nodes on a circle without crossing edges [5], which is $2^j \sum_{0 \le v \le (j-1)/2} (-1)^v \frac{1 \cdot 3 \cdots (2j - 2v - 5)}{v!(j-1-2v)!} 3^{j-1-2v} 2^{-v-2}$ [2].

Theorem 1. *When the input graph is planar, the algorithm constructs the ZDD within $O(mnC_fG_f)$ time.*

The proof of Theorem 1 is omitted due to the space limitation. Note that for planar graphs, there exists an edge ordering such that f is bounded by $O(\sqrt{n})$ [13].

4   Enumerating All the Graph Partitions with a Small Disparity

In this section, we construct the ZDD representing all the partition subgraphs which have exactly k cells within disparity r for fixed k and r. For a partition subgraph, if we fix k and r, all the weights of connected components of the subgraph must range between some values. Let a_i denote the weight of the i-th smallest connected component for $i = 1, \ldots, k$ and $S = \sum_{v \in V} h(v) (= \sum_{i=1}^{k} a_i)$. Note that $a_k/a_1 \le r$ must hold. Since $S = \sum_{i=1}^{k} a_i \ge (k-1)a_1 + a_k \ge (k-1)a_k/r + a_k$ holds, we have $a_k \le rS/(r + (k-1))$. Similarly, since $S = \sum_{i=1}^{k} a_i \le a_1 + (k-1)a_k \le a_1 + r(k-1)a_1$ holds, we have $a_1 \ge S/(r(k-1) + 1)$. Letting $U(k, r) = rS/(r + (k-1))$ and $L(k, r) = S/(r(k-1) + 1)$, we have $L(k, r) \le a_i \le U(k, r)$ for all $i = 1, \ldots, k$. We sometimes drop k and r from $U(k, r)$ and $L(k, r)$ and simply denote them by U and L if it is clear from the context.

We now design the frontier-based search for partition subgraphs within disparity r by modifying that for $\mathcal{P}(\{k\})$ described in Sect. 3. Our idea is to store the current weight of each connected component on the frontier into a node. For a node n with label e_i, we introduce a variable $n.\texttt{weight}$. For a connected component C on F_{i-1}, $n.\texttt{weight}[C]$ represents the total weight of the connected component including C on V. For a node n with label $e_i = \{v, w\}$, if e_i is the first edge incident to v, we set $\texttt{weight}[\{v\}] \leftarrow h(v)$. Then, if we decide to adopt e_i and $C_v \ne C_w$, $\texttt{weight}[C_v \cup C_w] \leftarrow \texttt{weight}[C_v] + \texttt{weight}[C_w]$. At this time, if weight$[C_v \cup C_w]$ exceeds U, we prune the resulting node. When a connected component C becomes isolated, if weight$[C] < L$, then we also prune the resulting node.

Since the condition that $L \leq a_i \leq U$ for all i is not a sufficient condition, to ensure that the disparity does not exceed r, we also store the maximum and the minimum weights of isolated connected components into a node. For a node n with label $e_i = \{v, w\}$, we store such values into n.maxw and n.minw, respectively. We set n_{root}.maxw $\leftarrow 0$ and n_{root}.minw $\leftarrow \infty$ as the initial values. When creating a node n' as the destination of an arc of n, we update maxw and minw as follows. If $v \in F_{i-1}$, $v \notin F_i$, $\{v\} \in n$.comp, and n.maxw $< n$.weight$[\{v\}]$, then we let n'.maxw $\leftarrow n$.weight$[\{v\}]$. Similarly, if $v \in F_{i-1}$, $v \notin F_i$, $\{v\} \in n$.comp, and n.minw $> n$.weight$[\{v\}]$, then we let n'.minw $\leftarrow n$.weight$[\{v\}]$. After we update n.maxw and/or n.minw, if n.maxw$/n$.minw $> r$ holds, we prune the resulting node.

The identicalness of nodes is decided by weight, maxw and minw in addition to other variables described in the previous sections. At the end of processing all the edges, we obtain the ZDD representing all the partition subgraphs within disparity r. Two nodes cannot be merged if weight, maxw or minw of the two nodes is different, which causes an increase of nodes in the constructed ZDD, whereas pruning a node occurs if it is decided that the weight of an isolated connected component is not between L and U, which causes a decrease of nodes.

5 Experimental Results

In this section, we show the results of experiments about the computation time of our algorithms. All experiments in this paper have been carried out on a machine with Intel Xeon E5-2630 (2.30 GHz) CPU and 128 GB memory (Linux Centos 6.6). We have implemented the algorithms in C++ and compiled them by gcc with the -O3 optimization option.

The input graphs we use in the experiments have been created from the maps of Japan's prefectures. As we have already described in the introduction, a vertex of the input graph corresponds to a city, two vertices are adjacent if the corresponding cities have the common border, and the weight of the graph corresponds to the number of residents living in the city. The number of residents for each graph is given in the 2015 census in Japan, and that of connected components for each graph is specified by an election low of Japan. The properties of the input graphs are shown in Table 1. The order of edges in each graph is determined by a heuristic algorithm so that the frontier size is small.

First, we have carried out the algorithm in Sect. 3, which enumerates all the partition subgraphs without the condition of disparity. Table 2 shows the computation time of the algorithm, the number of nodes in the constructed ZDD and the number of solutions for each graph. We have succeeded in enumerating about 4.8×10^{30} partition subgraphs on G_{Osa} in 0.34 s. Note that once a ZDD is obtained, the number of subgraphs represented by the ZDD can be easily computed by a dynamic programming algorithm. We also show the results for $\ell \times \ell$ grid graphs dividing them into two and three connected components in Table 3.

Table 1. Property of the input graphs

Pref. name	Graph name	# of vertices	# of edges	# of components
Fukui	G_{Fuk}	17	26	2
Kyoto	G_{Kyo}	36	78	6
Niigata	G_{Nii}	38	81	6
Ibaragi	G_{Iba}	44	95	7
Hyogo	G_{Hyo}	49	107	13
Kanagawa	G_{Kan}	58	135	20
Osaka	G_{Osa}	69	161	21

Table 2. Enumeration of all the partitions

Graph	Time (sec.)	# of ZDD nodes	# of solutions
G_{Fuk}	0.07	719	86
G_{Kyo}	0.21	44,209	44,063,998,545
G_{Nii}	0.20	44,540	386,618,915,837
G_{Iba}	0.20	53,018	63,497,174,378,978
G_{Hyo}	2.27	1,211,157	5,882,276,420,292,896,537
G_{Kan}	0.48	167,530	15,178,369,667,784,648,635,562,083
G_{Osa}	0.34	626,192	4,893,281,393,039,250,022,519,012,101,206

Next, using the algorithm in Sect. 4 we have conducted enumerating all the partition subgraphs whose disparity is at most r for G_{Kyo}, G_{Iba}, G_{Kan} and G_{Osa}. Due to the space limitation, we show the results only for G_{Iba} and G_{Osa} in Tables 4 and 5, respectively. Columns "L" and "U" mean the lower and upper bounds shown in the previous section, respectively. "Time" is the time of constructing the ZDD. "# of Solutions" is the number of partition subgraphs with disparity at most r, and "# of ZDD Nodes" is that of nodes of the constructed ZDD. "N/A" indicates that the computation has failed due to out of memory. We also show the histograms describing the numbers of partitions with every 0.1

Table 3. Enumeration of all the partitions on $\ell \times \ell$ grid graphs

ℓ	Time (sec.) for $k = 2$	# of solutions for $k = 2$	Time (sec.) for $k = 3$	# of solutions for $k = 3$
4	0.11	627	0.13	10,830
5	0.19	16,213	0.20	709,351
6	0.41	1,123,743	0.49	99,699,033
7	2.25	221,984,391	3.17	34,719,687,359
8	22.95	127,561,384,993	34.31	32,128,580,602,967
9	256.90	215,767,063,451,331	388.81	82,102,610,820,820,733
10	2844.84	1,082,828,220,389,781,579	4457.78	593,301,237,469,990,370,097

Table 4. Enumerating all the partition subgraphs with disparity r for G_{Iba}.

r	L	U	Time	# of ZDD nodes	# of solutions
1.1	383,928	452,063	56.98	8,907,949	135,158
1.2	355,836	486,310	593.31	236,736,206	5,106,426
1.3	331,574	519,619	1,733.53	1,376,840,028	21,434,502
1.4	310,410	552,027	N/A	–	–

Table 5. Enumerating all the partition subgraphs with disparity r for G_{Osa}.

r	L	U	Time	# of ZDD nodes	# of solutions
1.1	384,300	460,797	0.04	70	0
1.2	353,556	500,316	1.20	746,969	0
1.3	327,366	539,464	N/A	–	–

and 0.05 disparity for G_{Kyo} and G_{Iba} in Figs. 1 and 2, respectively. The value $\hat{x}$ of the x-axis represents the number of partitions between $\hat{x} - 0.1$ (or $\hat{x} - 0.05$) and $\hat{x}$.

For G_{Kyo}, G_{Iba} and G_{Kan}, we succeeded in generating all the partition subgraphs with disparity $r = 2.0$, $r = 1.3$ and $r = 1.4$, respectively. For G_{Osa}, we failed to construct the ZDD even for $r = 1.3$. It is known that the smallest disparity of the partition of G_{Osa} is 1.330 [12].

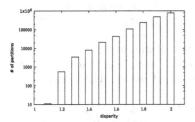

Fig. 1. Disparity histogram for G_{Kyo}. The y-axis is log scale.

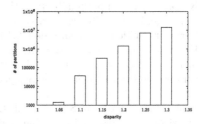

Fig. 2. Disparity histogram for G_{Iba}. The y-axis is log scale.

6   Conclusion

We have proposed a novel algorithm that enumerates all the partitions of a graph. As shown in the experiments, our algorithm enables us to generate millions of partitions with only small disparity. Generated partitions are represented as ZDDs, which save much memory space than explicitly storing them. Moreover, the ZDD representation brings about great benefits for applications because there are dozens of useful ZDD features, e.g., taking set operations, filtering by conditions, counting, and random sampling. We believe that by the frontier-based search, we can enumerate graph partitions having various conditions other than the disparity condition directly.

Acknowledgment. The authors would like to thank Dr. Toshiki Saitoh, Dr. Norihito Yasuda and Dr. Ryo Yoshinaka for their valuable comments. This work was partly supported by JSPS KAKENHI 15H05711, 24106007 and 15K00008.

References

1. Andreev, K., Räcke, H.: Balanced graph partitioning. Theor. Comput. Syst. **39**(6), 929–939 (2006)
2. Flajolet, P., Noy, M.: Analytic combinatorics of non-crossing configurations. Discrete Math. **204**(1), 203–229 (1999)
3. Hardy, G., Lucet, C., Limnios, N.: K-terminal network reliability measures with binary decision diagrams. IEEE Trans. Reliab. **56**(3), 506–515 (2007)
4. Imai, H., Sekine, K., Imai, K.: Computational investigations of all-terminal network reliability via BDDs. IEICE Trans. Fundam. Electron. Commun. Comput. Sci. **E82-A**, 714–721 (1999)
5. OEIS Foundation Inc.: The on-line encyclopedia of integer sequences (2011). http://oeis.org/A054726
6. Inoue, T., Takano, K., Watanabe, T., Kawahara, J., Yoshinaka, R., Kishimoto, A., Tsuda, K., Minato, S., Hayashi, Y.: Distribution loss minimization with guaranteed error bound. IEEE Trans. Smart Grid **5**(1), 102–111 (2014)
7. Karpinski, M.: Approximability of the minimum bisection problem: an algorithmic challenge. In: Diks, K., Rytter, W. (eds.) MFCS 2002. LNCS, vol. 2420, pp. 59–67. Springer, Heidelberg (2002). doi:10.1007/3-540-45687-2_4
8. Kawahara, J., Inoue, T., Iwashita, H., Minato, S.: Frontier-based search for enumerating all constrained subgraphs with compressed representation. Hokkaido University, Division of Computer Science, TCS Technical reports TCS-TR-A-13-76 (2014)
9. King, D.M., Jacobson, S.H., Sewell, E.C.: Efficient geo-graph contiguity and hole algorithms for geographic zoning and dynamic plane graph partitioning. Math. Program. **149**(1–2), 425–457 (2015). http://dx.doi.org/10.1007/s10107-014-0762-4
10. Knuth, D.E.: The Art of Computer Programming, Vol. 4A. Combinatorial Algorithms, Part 1, 1st edn. Addison-Wesley Professional, Boston (2011)
11. Minato, S.: Zero-suppressed BDDs for set manipulation in combinatorial problems. In: Proceedings of the 30th ACM/IEEE Design Automation Conference, pp. 272–277 (1993)
12. Nemoto, T., Hotta, K.: On the limits of the reduction in population disparity between single-member election districts in Japan. Jpn. J. Electoral Stud. **20**, 136–147 (2005). (in Japanese)
13. Sekine, K., Imai, H., Tani, S.: Computing the Tutte polynomial of a graph of moderate size. In: Staples, J., Eades, P., Katoh, N., Moffat, A. (eds.) ISAAC 1995. LNCS, vol. 1004, pp. 224–233. Springer, Heidelberg (1995). doi:10.1007/BFb0015427

14. Takizawa, A., Takechi, Y., Ohta, A., Katoh, N., Inoue, T., Horiyama, T., Kawahara, J., Minato, S.: Enumeration of region partitioning for evacuation planning based on ZDD. In: Proceedings of 11th International Symposium on Operations Research and its Applications in Engineering, Technology and Management (ISORA 2013), pp. 1–8 (2013)
15. Yoshinaka, R., Saitoh, T., Kawahara, J., Tsuruma, K., Iwashita, H., Minato, S.: Finding all solutions and instances of numberlink and slitherlink by ZDDs. Algorithms **5**(2), 176–213 (2012). http://www.mdpi.com/1999-4893/5/2/176/

Graph Drawing

An Experimental Study on the Ply Number
of Straight-Line Drawings

Felice De Luca[1]([✉]), Emilio Di Giacomo[1], Walter Didimo[1], Stephen Kobourov[2],
and Giuseppe Liotta[1]

[1] Università degli Studi di Perugia, Perugia, Italy
felice.deluca@studenti.unipg.it,
{emilio.digiacomo,walter.didimo,giuseppe.liotta}@unipg.it
[2] University of Arizona, Tucson, USA
kobourov@cs.arizona.edu

Abstract. The *ply number* of a drawing is a new criterion of interest for
graph drawing. Informally, the ply number of a straight-line drawing of
a graph is defined as the maximum number of overlapping disks, where
each disk is associated with a vertex and has a radius that is half the
length of the longest edge incident to that vertex. This paper reports
the results of an extensive experimental study that attempts to estimate
correlations between the ply numbers and other aesthetic quality metrics for a graph layout, such as stress, edge-length uniformity, and edge
crossings. We also investigate the performances of several graph drawing
algorithms in terms of ply number, and provides new insights on the
theoretical gap between lower and upper bounds on the ply number of
k-ary trees.

1 Introduction

Graphs occur naturally in many domains: from sociology and biology, to software
engineering and transportation. When the vertices and edges of the given graph
have no inherent geographical locations, graph layout algorithms are used to
try to capture the underlying relationships in the data (see, e.g., [8,10,22,33]).
In order to make the graph layout readable for the user, such algorithms are
designed to optimize several quality metrics, like minimizing the number of edge
crossings, striving for uniform edge lengths, or maximizing the vertex angular
resolution [8].

Force-directed methods are among the most flexible and popular graph layout
algorithms [24]. They tend to compute drawings that are aesthetically pleasing,
exhibit symmetries, and contain no, or a few, edge crossings when the graph
is planar. Classic examples include the spring layout method of Eades [12] and

Research supported in part by the MIUR project AMANDA "Algorithmics for MAssive and Networked DAta", prot. 2012C4E3KT_001. Work on this problem began
at the NII Shonan Meeting *Big Graph Drawing: Metrics and Methods*, Jan. 12–15,
2015. We thank M. Kaufmann and his staff in the University of Tübingen for sharing
their code to compute ply number. We also thank A. Wolff and F. Montecchiani for
many useful discussions.

© Springer International Publishing AG 2017
S.-H. Poon et al. (Eds.): WALCOM 2017, LNCS 10167, pp. 135–148, 2017.
DOI: 10.1007/978-3-319-53925-6_11

the algorithm of Fruchterman and Reingold [16], both of which rely on spring forces, similar to those in Hooke's law. In these methods, there are repulsive forces between all nodes and attractive forces between nodes that are adjacent. Alternatively, forces between the nodes can be computed based on their graph theoretic distances, determined by the lengths of shortest paths between them. For instance, the algorithm of Kamada and Kawai [23] uses spring forces proportional to the graph theoretic distances. In general, force-directed methods define an objective function which maps each graph layout into a number in $\mathbb{R}^+$ representing the *energy* of the layout. This function is defined in such a way that low energies correspond to layouts in which adjacent nodes are near some pre-specified distance from each other, and in which non-adjacent nodes are well-spaced. The notion of the energy of the system is related to the notion of *stress* in multidimensional scaling [26]. A careful look into stress-based layout algorithms and force-directed algorithms shows that despite some similarities, they optimize a spectrum of different functions. For example, methods such as Kamada-Kawai optimize long graph distances, while Noack's LinLog layout [27] optimizes short graph distances. A later study by Chen and Buja explores further the differences between the energy models [5].

Recently, a new parameter, called *ply number*, has been proposed as a quality metric for graph layouts [9], partly inspired by the fact that real road networks have small values of such a parameter [13]. Roughly speaking, a drawing has a small ply number if some, suitably defined, regions of influence of the vertices in the drawing are well spread out. More precisely, let Γ be a straight-line drawing of a graph. For each vertex $v \in \Gamma$, let C_v be the *open* disk centered at v whose radius r_v is half the length of the longest edge incident to v. Denote by S_q the set of disks C_v sharing a point $q \in \mathbb{R}^2$. The *ply number* of Γ is defined as $\mathsf{pn}(\Gamma) = \max_{q \in \mathbb{R}^2} |S_q|$. In other words, the ply number of Γ is the maximum number of disks C_v mutually intersecting in Γ (see, e.g., Fig. 1).

The ply number $\mathsf{pn}(G)$ of a graph G is the minimum ply number over all straight-line drawings of G. Computing the ply number of a given graph is NP-hard [9]. Figure 2 shows two drawings of the same graph. Intuitively, the drawing to the left is more readable than the drawing to the right and in fact the ply number of the left drawing is significantly smaller than the ply number of the right drawing.

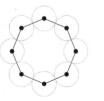

Fig. 1. (a) A drawing with ply number 1. (b) A drawing with ply number 2.

Our Contribution. While preliminary theoretical results about computing drawings with low ply number have already appeared [1,9], our work is an experimental study whose main goals are: (*i*) to shed more light on the quality of drawings computed by some of the most popular algorithms, and in particular by different types of force-directed methods; (*ii*) to investigate whether the ply number of a drawing can be actually regarded as a quality metric, which possibly

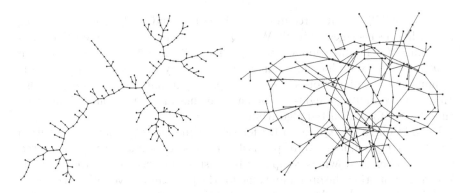

Fig. 2. Two drawings of the same graph with ply number 3 (left) and 12 (right).

encompasses other popular metrics; (*iii*) to guide further theoretical studies of the combinatorial properties of drawings with low ply number. Specifically, our experiments involve several graph layout algorithms and several graph families, and we establish a correlation between the ply number and some classical quality metrics like stress and edge length uniformity. Additionally, we give some insights about the known theoretical gap between lower and upper bounds for the ply number of k-ary trees.

The paper is structured as follows. Sections 2–4 provide details about our experimental questions, setting, and procedures. The results of our study are presented and discussed in Sect. 5. Conclusions and future research directions are given in Sect. 6.

2   Experimental Questions

As mentioned in the introduction, our experiment has the following main objectives: (*i*) to shed more light on the quality of drawings computed by some of the most popular algorithms, with particular on force-directed methods; (*ii*) to investigate whether the ply number of a drawing can be regarded as a quality metric; (*iii*) to guide further theoretical studies of the combinatorial properties of drawings with low ply number. We pose the following experimental questions:

Q1. *How good are the layouts computed by different drawing algorithms in terms of ply number?*

Q2. *How close is the ply number of drawings produced by existing algorithms to the ply number of the input graph (i.e., to the optimum value)?*

Q3. *Does the ply number correlate with some other commonly used quality metrics?*

Q4. *Can we establish empirical upper bounds on the ply number of k-ary trees?*

Questions **Q1–Q3** are concerned with objective (*i*) and (*ii*), while **Q4** is relevant for objective (*iii*). Below, we discuss the motivation behind each question in more detail.

The possibility that force-directed algorithms indirectly optimize the ply number has been suggested in [9]: With Q1 we compare force-directed algorithms based on different force models to experimentally investigate this hypothesis. In [9] it is observed that non-planar drawings may have significantly smaller ply number than planar ones. Hence, for planar graphs, we also consider algorithms that compute straight-line planar drawings in comparison with drawings computed by force-directed algorithms.

In addition to Q1, Question Q2 focuses on the quality of layout algorithms in terms of ply number. More precisely, it aims to estimate the gap between the ply number of drawings computed by existing algorithms and the optimum. However, computing the optimum value for the ply number over all drawings of a graph is NP-hard [9], and the (worst-case) optimum value of ply number is known only for simple graph families, like paths, cycles, binary trees, and caterpillars (whose optimum is either 1 or 2). We then restrict Q2 to these families.

Question Q3 is more focused on understanding whether the ply number can be used as a quality metric for graph layouts, which possibly encompasses several other popular quality measures. We are mainly interested in three measures that are among the most used in graph visualization and that we expect to affect (positively or negatively) the ply number more than others: number of crossings, stress, and edge-length uniformity. It is worth remarking that the number of crossings is widely adopted to evaluate the quality of graph layouts, especially for graphs of small and medium size (see, e.g., [21,29,30,35]). Studying the correlation between ply number and crossings, is further motivated by the fact that, as observed above, the ply number is sometimes reduced at the expense of edge crossings. The *stress* of a graph layout captures how well the realized geometric distances between pairs of vertices reflect their graph-theoretic distances in the graph; in the standard formulation, all edges are assumed to have about the same length; thus stress is related to edge-length uniformity (see Sect. 3). Recent studies give some evidence that reducing the stress of a graph layout is correlated with improved aesthetics [11,25]. Studying the correlation between ply number, stress, and edge-length uniformity is also motivated by the fact that in a drawing with ply number one, all edges have the same length [9].

Question Q4 is motivated by objective (*iii*) and arises from theoretical results on the ply number of k-ary trees. Namely, it is known that every 2-ary (i.e., binary) tree has ply number at most two [9], while the ply number of 10-ary trees is not bounded by a constant [1]. What happens for values of k in the range [3, 9] is an interesting theoretical question. With Q4 we experimentally investigate this question.

3   Experimental Setting

In order to answer Questions Q1–Q4, we selected different graph datasets, algorithms, and measures. In what follows we describe each of these experimental components.

Graph Datasets. To understand whether the experimental results are influenced by the structure of the graph we considered several graph families. All graphs are of small or medium size, expressed as the number of their vertices. In some cases, the size and the number of instances used for each graph family depends on the type of question we want to answer (see Sect. 5 for details). We used the following datasets:

Trees. Generated with uniform probability distribution using Prüfer sequences [28].

Planar. Connected simple planar graphs, generated with the OGDF library [6].

General. Connected simple graphs, generated with uniform probability distribution.

Scale-free. Scale-free graphs, generated according to the Barabási-Albert model [2].

Caterpillars. Each caterpillar of n vertices is generated by first creating a path (spine of the caterpillar) of length $k \in [\frac{n}{4}, \frac{n}{2}]$ (randomly chosen), and attaching each remaining vertex to a randomly selected vertex of the spine.

Paths, Cycles. For each desired number of vertices n, there is only one (unlabeled) path and one (unlabeled) cycle of n vertices.

k-ary Trees. Rooted trees where each node has either 0 or k children. Each tree is generated by starting with a single vertex and then creating k children of a randomly selected leaf, until the desired number n of vertices is achieved. When n cannot be obtained, we use a value close to it.

Table 1 shows which datasets are used to answer each question. Note that the datasets Caterpillars, Paths, Cycles, and 2-ary Trees are explicitly used to answer Q2. All these families (except Cycles) are special cases of trees and we do not use them to answer Q1 and Q3. The dataset of k-ary Trees is explicitly designed to answer Q4. See Sect. 4 for more details.

Table 1. Table summarizing which datasets are used to answer each question.

	Q1	Q2	Q3	Q4
Trees	✓		✓	
Planar	✓		✓	
General	✓		✓	
Scale-free	✓		✓	
Caterpillars		✓		
Paths		✓		
Cycles		✓		
k-ary Trees ($k = 2$)		✓		
k-ary Trees ($k = 3, 6, 9$)				✓

Algorithms. Among the many force-directed algorithms, we considered some of the most popular ones [24]. We used the following algorithms, available in OGDF:

FR. This algorithm is based on the Fruchterman-Reingold model [16], an improvement of the seminal algorithm by Eades [12]. Vertices are viewed as equally-charged electrical particles and edges act similar to springs; electrical charges cause repulsion between vertices and springs cause attraction. It also introduces a temperature function, which reduces the displacement of the vertices as the layout becomes better.

GEM. This algorithm is proposed by Frick et al. [15]. It is a variant of the FR algorithm, which adds several new heuristics to improve the convergence, including local temperatures, gravitational forces, and the detection of rotations and oscillations.

KK. This algorithm is described by Kamada-Kawai [23]. Unlike FR and GEM, it aims to compute a layout where the geometric distance between two vertices equals their graph-theoretic distance in the graph. The energy function minimized by this algorithm is therefore a type of stress function.

SM. This technique is proposed by Gansner et al. [17]. It minimizes a stress function similar to that proposed by KK, which can be minimized more efficiently via majorization.

FM3. The fast multipole multilevel method of Hachul and Jünger [18] is among the most effective force-directed algorithms in the literature [19].

We also used the following algorithm, available in Gephi [3]:

LL. This is a force-directed algorithm based on the LinLog energy model proposed by Noack [27]. It is specifically conceived to emphasize clusters in the graph.

For instances of Planar, Trees, and 2-ary Trees, we also considered planar straight-line drawing algorithms, still using the implementations in OGDF. The algorithms are:

PL. This is an improved version of the planar straight-line drawing algorithm proposed by Chrobak and Kant [7], based on the shift algorithm of de Fraysseix et al. [14].

TR. The tree layout algorithm of Buchheim et al. [4] is an efficient version of Walker's algorithm [34], which in turn is an extension of the Reingold-Tilford algorithm for rooted binary trees [31].

Measures. We considered four measures: ply number (PN), number of crossings (CR), stress (ST), and edge-length uniformity (EU). Let Γ be a straight-line drawing of a graph $G = (V, E)$. EU corresponds to the normalized standard deviation of the edge length, i.e.:

$$\mathrm{EU}(\Gamma) = \sqrt{\sum_{e \in E} \frac{(l_e - l_{avg})^2}{|E| l_{avg}^2}},$$

where l_e is the length of edge e and l_{avg} is the average length of the edges. The stress of Γ is defined as:

$$\mathrm{ST}(\Gamma) = \sum_{i,j \in V} w_{ij}(\| p_i - p_j \| - d_{ij})^2,$$

where $w_{ij} = d_{ij}^{-2}$, p_i and p_j are the positions of i and j in Γ, and d_{ij} is the graph theoretic distance of i and j in G.

4   Experimental Procedures

In the following we describe the different experimental procedures executed to answer each question. In Sect. 5 we present the results and summarize the main findings.

Procedure for Q1. We drew each instance of each dataset with all the algorithms (clearly, PL has been used only for Planar and TR only for Trees), and measured the ply number of each drawing. Both for Trees and for Planar, we generated 10 instances for each fixed number of vertices $n \in \{50, 100, 150, 200, \ldots, 450\}$: the average density of the planar graphs is 1.75. In the General dataset we generated 10 instances for each pair $\langle n, d \rangle$, where $n \in \{50, 100, 150, 200, \ldots, 450\}$ is still the number of vertices of the graph and $d \in \{1.5, 2.5\}$ is its edge density. Indeed, we want also understand to which degree the ply number is influenced by the density of the graph. In the Scale-free dataset we generated 10 instances for each pair $\langle n, d \rangle$, where $n \in \{50, 100, 150, 200, \ldots, 450\}$ and $d \in \{2, 3\}$ (the graph generator that we used required to specify an integer number as edge density [20]).

Procedure for Q2. To answer Q2 we need to compare the ply number of the drawings computed by the various algorithms with the ply number of the input graph (i.e., the optimum value). Hence, we considered families of graphs whose the (worst-case optimal) ply number is known. In particular, we considered the Paths and Cycles instances, which have ply number one, and the Caterpillars and 2-ary Trees instances, whose ply number is at most two (see [9]). For each $n \in \{50, 100, 150, 200, \ldots, 450\}$, we generated a single instance in Paths and Cycles, and 5 instances in the Caterpillars and 2-ary Trees datasets. For each instance, we computed 10 different drawings with the algorithms KK, FM3, SM (those with better performances based on the results of Q1). We then took the minimum value of ply number over all the drawings of each instance.

Procedure for Q3. We took a representative instance for each sample (i.e., size or size and density) of each dataset. As reported in Table 1, we used the same datasets as for Q1. For each representative instance we computed a series of 60 different drawings and on this series we measured Spearman's rank correlation coefficient ρ [32] between the ply number and all the other quality metrics described in Sect. 3, i.e., ST, CR, and EU. The series of drawings are produced by running 10 times each of the 6 force-directed algorithms, varying the initial layout every time.

Procedure for Q4. We generated a k-ary tree for each pair $\langle n, k \rangle$, where $n = \{100, 150, 200, \ldots, 950, 1000, 2000, 3000, 4000\}$ is the desired number of vertices and $k \in \{3, 6, 9\}$. The choice for the values of k is motivated by the fact that we want to experimentally understand if we can empirically establish a constant upper bound to the ply number of k-trees for $2 < k < 10$. For each instance, we measured the ply number of a drawing computed with SM, which turned out to be the best performing algorithm for this measure, according to the experiment for Question Q1.

5   Experimental Results and Findings

We first report the experimental data, by presenting tables and charts. Then, we list and discuss the main findings.

Results for Q1. Figure 3 reports the average ply number for each sample (number of vertices) of drawings computed by the same algorithm. For Trees and Planar we observed a similar trend. The algorithms that give the lowest values of the ply number are SM, KK, and FM3 (with KK and SM that always have almost identical values and FM3 that is slightly better for larger planar graphs). FR produces drawings with ply number higher than the previous three algorithms, although it has a similar trend. GEM and LL produces drawings with ply number much higher than the other force-directed algorithms, with GEM that becomes worse than LL as n grows. For Trees the TR algorithm has quite good performances (between FM3 and FR), while the algorithm PL (for the Planar dataset) produces the drawings with the worse values of ply, thus confirming that planar drawings often have higher ply number than non-planar ones. For the General and Scale-free datasets, we have a similar situation. The main difference is that GEM performances do not worsen as fast as in the cases of Trees and Planar (in particular, it is always better than LL). For the Scale-free dataset, the values computed by KK and SM increase, as n grows, more than those of FR and FM3. Thus, for larger values of n they are outperformed by FR and FM3, and approach the performances of GEM.

Results for Q2. For paths and cycles, all three algorithms compute drawings with ply number 2, i.e., just one unit larger than the optimal value, for instances up to 250. For larger sizes, FM3 still computes drawing with ply number 2, while KK and SM produce drawings with ply number 3. It is worth saying that all the drawings computed by FM3 have ply number 2 with the only exception of one drawing of the cycle of size 450. This is consistent with the fact that FM3 (a multilevel algorithm) is less affected by the initial position of the vertices. The maximum value of ply for drawings produced by KK and SM is 6 and 4, respectively. Concerning the binary trees, FM3 has the worst performance with an average ply number ranging from 3 to 5.2; the other two algorithms have almost the same values of the average ply number, ranging from 2 to 4.2. Also, FM3 and KK tend to have larger differences between the (average) maximum and the (average) minimum ply number than SM.

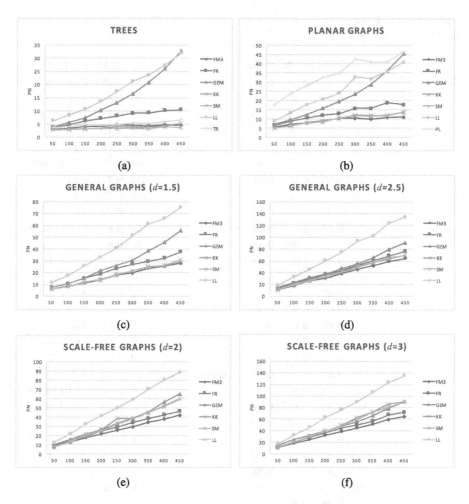

Fig. 3. Average ply number for (a) Trees; (b) Planar; (c) General with density $d = 1.5$; (d) General with density $d = 2.5$; (e) Scale-free with density $d = 2$; (f) Scale-free with density $d = 3$. The x-axis reports the number of vertices.

For caterpillars, instead, the three algorithms have similar performances. The average ply number is around 3.5. Also in this case SM is the algorithm with the smallest difference between maximum and minimum error.

Results for Q3. Table 2 shows, for each instance, the values of the Spearman's rank correlation coefficient ρ. We have high values of correlation (i.e., $\rho \geq 0.7$) between ply number and stress for almost every instance. The exceptions are larger scale-free graphs, for which in most cases we have a moderate correlation (i.e., $0.3 \leq \rho < 0.7$). The correlation between PN and EU is high/moderate in all cases. High values of correlation are obtained for the smaller instances of each dataset, with the only exception of scale-free graphs, where there is a high correlation for all sizes. Concerning PN and CR, we have (high/moderate)

Table 2. Correlation coefficient ρ between PN and ST, CR, EU. The values in bold indicate a strong correlation ($\rho \geq 0.7$).

	n	PN, ST	PN, CR	PN, EU		n	PN, ST	PN, CR	PN, EU
Trees	50	**0.92**	**0.88**	**0.86**	General ($d = 2.5$)	50	**0.80**	0.19	**0.89**
	100	**0.90**	**0.84**	**0.89**		100	**0.80**	0.21	**0.72**
	150	**0.94**	**0.92**	**0.78**		150	**0.78**	0.20	0.62
	200	**0.97**	**0.89**	**0.87**		200	**0.79**	0.04	0.54
	250	**0.96**	**0.91**	0.69		250	**0.78**	0.47	0.58
	300	**0.97**	**0.92**	0.65		300	**0.71**	−0.34	0.63
	350	**0.93**	**0.90**	0.62		350	**0.70**	−0.19	0.59
	400	**0.96**	**0.90**	0.61		400	**0.73**	−0.13	0.55
	450	**0.95**	**0.96**	0.51		450	**0.72**	−0.19	0.54
Planar	50	**0.92**	**0.80**	**0.76**	Scale-free ($d = 2$)	50	**0.86**	0.29	**0.74**
	100	**0.91**	**0.83**	**0.86**		100	**0.83**	0.08	**0.80**
	150	**0.90**	0.66	**0.89**		150	0.32	−0.08	**0.82**
	200	**0.77**	**0.87**	**0.71**		200	0.57	0.33	**0.72**
	250	**0.87**	**0.83**	**0.81**		250	0.38	0.27	**0.90**
	300	**0.93**	**0.80**	**0.80**		300	0.21	0.09	**0.90**
	350	**0.80**	**0.91**	**0.74**		350	0.48	0.53	**0.82**
	400	**0.77**	**0.93**	0.67		400	0.45	0.60	**0.84**
	450	**0.76**	**0.84**	0.58		450	0.50	0.54	**0.80**
General ($d = 1.5$)	50	**0.88**	0.39	**0.87**	Scale-free ($d = 3$)	50	**0.72**	0.15	**0.91**
	100	**0.92**	**0.80**	**0.82**		100	0.67	−0.04	**0.88**
	150	**0.89**	**0.81**	**0.86**		150	0.54	−0.32	**0.87**
	200	**0.87**	**0.83**	**0.89**		200	0.47	−0.16	**0.87**
	250	**0.85**	0.60	**0.84**		250	0.30	0.13	**0.90**
	300	**0.85**	**0.73**	**0.73**		300	0.18	0.27	**0.95**
	350	**0.84**	**0.85**	**0.85**		350	0.14	0.36	**0.96**
	400	**0.85**	**0.79**	0.49		400	0.29	0.22	**0.92**
	450	**0.88**	**0.86**	0.59		450	0.30	0.41	**0.90**

correlation only for trees, planar graphs, and for general graphs with density 1.5. For denser general graphs and for scale-free graphs there is little correlation between PN and CR.

Results for Q4. For the sizes of the trees that we considered, we always observed progressively increasing ply numbers, even for $k = 3$, and the ply number function does not exhibit any asymptotic trend towards a constant upper bound.

We now summarize the main findings. We denote by Fi the finding concerned with Question Qi ($i = 1, \ldots, 4$).

F1. Algorithms designed to minimize stress and edge-length uniformity (like SM and KK) compute drawings with smaller values of ply number. This behavior confirms the intuition that low ply number is related to stress and edge uniformity optimization (see also F3). Also multilevel algorithms (like

FM3) have good performances and are more stable on denser graphs. We think this is a consequence of their coarsening phase, which indirectly tends to evenly distribute the vertices in the plane, thus producing drawings with good edge length uniformity, independently of the original placement of the nodes. We also observed a good behavior of FR on denser graphs. Force-directed algorithms whose energy model is conceived to highlight clusters, such as LL, tend to produce drawings with high ply numbers, as they give rise to very different edge lengths in the same drawing.

F2. The best performing algorithms in terms of ply number very often generate drawings whose ply number is close to the optimum for graphs like paths, cycles, caterpillars, and binary trees. Hence, they can be considered good heuristics to compute graph layouts with minimum ply number, at least for these simple graph families.

F3. There is a strong correlation between ply and stress, and a strong/moderate correlation between ply and edge-length uniformity. For planar graphs and low density graphs, the correlation between ply and crossings is also observed, while ply is definitely non-correlated with edge crossings on denser graphs and, in particular, on scale-free graphs. Overall, these data indicate that ply number can be often regarded as a unifying quality metric, which encompasses at least stress and edge length uniformity. For very sparse graphs, it also encompasses edge crossings. Note that, the correlation between ply number and stress does not always imply that low ply number equals low stress.

F4. We could not observe any asymptotic trend of the ply number towards a constant upper bound for k-ary trees ($k \in \{3, 6, 9\}$). This indicates that the ply number for such graphs is likely unbounded, which should be confirmed by a theoretical proof.

6   Conclusions and Future Work

Our graph datasets and the data collected in the different experiments are publicly available at http://www.felicedeluca.com/ply/. These data answer, or partially answer, several of our initial questions and raise new interesting questions for further study.

We remark that, as in many experimental studies, ours has some limitations and should be interpreted in the context of the specific datasets, layout algorithms, and measurements used. For example, a more complete picture can be obtained with a more diverse set of graphs. In particular, computing the ply number of a drawing in a reliable way requires high arithmetic precision, which is computationally expensive. This limited the sizes of graphs that we could consider. Further, we mostly used algorithms available in OGDF, considering only one other algorithm. Comparing a wider spectrum of algorithms might help to identify what type of stress-minimization and energy minimization functions are best suited to minimize the ply number.

Our study provides answers several of the questions that we asked and also suggests several natural research directions that remain to be explored:

- Can new layout algorithms be developed that directly optimize ply number? Modifying stress-based methods such as Kamada-Kawai would be difficult, but perhaps force-directed methods such as Fruchterman-Reingold can be augmented with additional forces to separate overlapping disks.
- Considering more carefully the variations in the exact functions used in different stress-based methods and force-directed methods and their impact on optimizing ply might lead to better insights about how to compute low-ply layouts.
- There is growing evidence that different types of graph layout algorithms are suited to different types of graphs. A cognitive study could consider the impact of minimizing ply number, compared to the impact of minimizing edge crossings and other aesthetic criteria that are not correlated with the ply number.
- More experiments can be performed to look for possible correlations between ply and other quality metrics, such as angular resolution and drawing symmetries.
- Further experiments may help to identify graph families with constant or unbounded ply number. Experimental data can then be formally verified. In particular, our experiments indicate that k-ary trees, with $k \in [3, 9]$ have unbounded ply number; proving this would close a gap in our theoretical knowledge.

References

1. Angelini, P., Bekos, M.A., Bruckdorfer, T., Hančl, J., Kaufmann, M., Kobourov, S., Symvonis, A., Valtr, P.: Low ply drawings of trees. In: Hu, Y., Nöllenburg, M. (eds.) GD 2016. LNCS, vol. 9801, pp. 236–248. Springer, Heidelberg (2016). doi:10.1007/978-3-319-50106-2_19
2. Barabási, A.L., Albert, R.: Emergence of scaling in random networks. Science **286**(5439), 509–512 (1999)
3. Bastian, M., Heymann, S., Jacomy, M.: Gephi: an open source software for exploring and manipulating networks (2009). http://www.aaai.org/ocs/index.php/ICWSM/09/paper/view/154
4. Buchheim, C., Jünger, M., Leipert, S.: Drawing rooted trees in linear time. Softw.: Pract. Exp. **36**(6), 651–665 (2006)
5. Chen, L., Buja, A.: Stress functions for nonlinear dimension reduction, proximity analysis, and graph drawing. J. Mach. Learn. Res. **14**(1), 1145–1173 (2013)
6. Chimani, M., Gutwenger, C., Jünger, M., Klau, G.W., Klein, K., Mutzel, P.: The open graph drawing framework (OGDF). In: Tamassia, R. (ed.) Handbook of Graph Drawing and Visualization, pp. 543–569. CRC Press, Boca Raton (2013)
7. Chrobak, M., Kant, G.: Convex grid drawings of 3-connected planar graphs. Int. J. Comput. Geom. Appl. **07**(03), 211–223 (1997)
8. Di Battista, G., Eades, P., Tamassia, R., Tollis, I.G.: Graph Drawing. Prentice Hall, Upper Saddle River (1999)
9. Di Giacomo, E., Didimo, W., Hong, S.H., Kaufmann, M., Kobourov, S.G., Liotta, G., Misue, K., Symvonis, A., Yen, H.C.: Low ply graph drawing. In: IISA 2015, pp. 1–6. IEEE (2015)

10. Didimo, W., Liotta, G.: Mining graph data. In: Cook, D.J., Holder, L.B. (eds.) Graph Visualization and Data Mining, pp. 35–64. Wiley, Hoboken (2007)
11. Dwyer, T., Lee, B., Fisher, D., Quinn, K.I., Isenberg, P., Robertson, G.G., North, C.: A comparison of user-generated and automatic graph layouts. IEEE Trans. Vis. Comput. Graph. 15(6), 961–968 (2009)
12. Eades, P.: A heuristics for graph drawing. Congressus Numerantium 42, 146–160 (1984)
13. Eppstein, D., Goodrich, M.T.: Studying (non-planar) road networks through an algorithmic lens. In: GIS 2008, pp. 1–10. ACM (2008)
14. de Fraysseix, H., Pach, J., Pollack, R.: How to draw a planar graph on a grid. Combinatorica 10(1), 41–51 (1990)
15. Frick, A., Ludwig, A., Mehldau, H.: A fast adaptive layout algorithm for undirected graphs (extended abstract and system demonstration). In: Tamassia, R., Tollis, I.G. (eds.) GD 1994. LNCS, vol. 894, pp. 388–403. Springer, Heidelberg (1995). doi:10.1007/3-540-58950-3_393
16. Fruchterman, T.M.J., Reingold, E.M.: Graph drawing by force-directed placement. Softw.: Pract. Exp. 21(11), 1129–1164 (1991)
17. Gansner, E.R., Koren, Y., North, S.: Graph drawing by stress majorization. In: Pach, J. (ed.) GD 2004. LNCS, vol. 3383, pp. 239–250. Springer, Heidelberg (2005). doi:10.1007/978-3-540-31843-9_25
18. Hachul, S., Jünger, M.: Drawing large graphs with a potential-field-based multilevel algorithm. In: Pach, J. (ed.) GD 2004. LNCS, vol. 3383, pp. 285–295. Springer, Heidelberg (2005). doi:10.1007/978-3-540-31843-9_29
19. Hachul, S., Jünger, M.: Large-graph layout algorithms at work: an experimental study. J. Graph Algorithms Appl. 11(2), 345–369 (2007)
20. Hagberg, A.A., Schult, D.A., Swart, P.J.: Exploring network structure, dynamics, and function using networkX. In: Varoquaux, G., Vaught, T., Millman, J. (eds.) Proceedings of the 7th Python in Science Conference, Pasadena, CA, USA, pp. 11–15 (2008)
21. Huang, W., Hong, S.H., Eades, P.: Effects of sociogram drawing conventions and edge crossings in social network visualization. J. Graph Algorithms Appl. 11(2), 397–429 (2007)
22. Jünger, M., Mutzel, P. (eds.): Graph Drawing Software. Springer, Heidelberg (2003)
23. Kamada, T., Kawai, S.: An algorithm for drawing general undirected graphs. Inf. Process. Lett. 31(1), 7–15 (1989)
24. Kobourov, S.G.: Force-directed drawing algorithms. In: Tamassia, R. (ed.) Handbook of Graph Drawing and Visualization, pp. 383–408. CRC Press, Boca Raton (2013)
25. Kobourov, S.G., Pupyrev, S., Saket, B.: Are crossings important for drawing large graphs? In: Duncan, C., Symvonis, A. (eds.) GD 2014. LNCS, vol. 8871, pp. 234–245. Springer, Heidelberg (2014). doi:10.1007/978-3-662-45803-7_20
26. Kruskal, J.B., Wish, M.: Multidimensional Scaling. Sage Press, Thousand Oaks (1978)
27. Noack, A.: Energy models for graph clustering. J. Graph Algorithms Appl. 11(2), 453–480 (2007)
28. Prüfer, H.: Neuer beweis eines satzesüber permutationen. Arch. Math. Phys. 27, 742–744 (1918)
29. Purchase, H.C.: Effective information visualisation: a study of graph drawing aesthetics and algorithms. Interact. Comput. 13(2), 147–162 (2000)

30. Purchase, H.C., Carrington, D.A., Allder, J.A.: Empirical evaluation of aesthetics-based graph layout. Empir. Softw. Eng. **7**(3), 233–255 (2002)
31. Reingold, E.M., Tilford, J.S.: Tidier drawings of trees. IEEE Trans. Softw. Eng. **SE-7**(2), 223–228 (1981)
32. Spearman, C.: The proof and measurement of association between two things. Am. J. Psychol. **15**(1), 72–99 (1904)
33. Tamassia, R. (ed.): Handbook of Graph Drawing and Visualization. CRC Press, Boca Raton (2013)
34. Walker, J.Q.: A node-positioning algorithm for general trees. Softw.: Pract. Exp. **20**(7), 685–705 (1990)
35. Ware, C., Purchase, H.C., Colpoys, L., McGill, M.: Cognitive measurements of graph aesthetics. Inf. Vis. **1**(2), 103–110 (2002)

Complexity Measures for Mosaic Drawings

Quirijn W. Bouts[✉], Bettina Speckmann, and Kevin Verbeek

Department of Mathematics and Computer Science, TU Eindhoven,
Eindhoven, The Netherlands
{q.w.bouts,b.speckmann,k.a.b.verbeek}@tue.nl

Abstract. Graph Drawing uses a well established set of complexity measures to determine the quality of a drawing, most notably the area of the drawing and the complexity of the edges. For contact representations the complexity of the shapes representing vertices also clearly contributes to the complexity of the drawing. Furthermore, if a contact representation does not fill its bounding shape completely, then also the complexity of its complement is visually salient.

We study the complexity of contact representations with variable shapes, specifically mosaic drawings. Mosaic drawings are drawn on a tiling of the plane and represent vertices by *configurations*: simply-connected sets of tiles. The complement of a mosaic drawing with respect to its bounding rectangle is also a set of simply-connected tiles, the *channels*.

We prove that simple mosaic drawings without channels may require $\Omega(n^2)$ area. This bound is tight. If we use only straight channels, then outerplanar graphs with k ears may require $\Omega(\min(nk, n^2/k))$ area. This bound is partially tight: we show how to draw outerplanar graphs with k ears in $O(nk)$ area with L-shaped vertex configurations and straight channels. Finally, we argue that L-shaped channels are strictly more powerful than straight channels, but may still require $\Omega(n^{7/6})$ area.

1 Introduction

Graph Drawing uses a well established set of complexity measures to determine the quality of a drawing (see, for example, the overview by Purchase [9]). For node-link drawings of planar graphs the arguably most prominent measures are the area of the drawing and the complexity (number of bends) of the edges. In addition to the classic node-link drawings, there are also other well-established drawings styles, most notably *contact representations*. Here the vertices of a graph are represented by a variety of possible shapes and the edges are implied by point or side contacts between these shapes. In certain settings the shapes are fixed: for example circles in Koebes theorem [7] or rectangles in rectangular duals and rectangular cartograms [8,10,11]. In other scenarios there is a certain variability in the shapes: for example, when using rectilinear polygons for rectilinear cartograms [1,3] or mosaic tiles and pixels [2,4] to compose shapes.

Contact representations do not draw edges explicitly and hence the complexity of the edges is not a valid quality measure. However, in scenarios where the shapes representing vertices are not fixed, the complexity of these shapes

© Springer International Publishing AG 2017
S.-H. Poon et al. (Eds.): WALCOM 2017, LNCS 10167, pp. 149–160, 2017.
DOI: 10.1007/978-3-319-53925-6_12

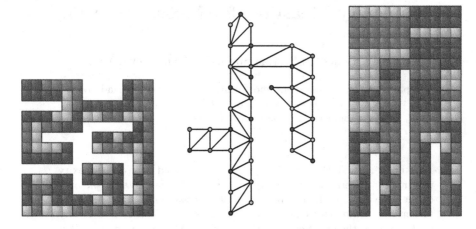

Fig. 1. A mosaic drawing with complex channels (left) of an outerplanar graph (middle). The same graph drawn with straight channels (right).

clearly contributes to the visual quality of the drawing. For example, there are several sequences of papers which strive to represent (weighted) planar triangulated graphs with rectilinear polygons of the lowest possible complexity, which is 8 in both the weighted [1] and in the unweighted case [5].

Contact representations are said to be *proper* if the shapes corresponding to vertices form a partition of the bounding shape. For example, a rectangular dual is a proper contact representation, since it consists of a partition of a rectangle into rectangles. Similarly, certain outerplanar graphs have a proper touching triangle representation [6], which consists of a partition of a triangle into triangles. If a contact representation is not proper, then the complexity of its complement with respect to an appropriate bounding shape is also visually salient.

We propose to study the complexity of contact representations with variable shapes. Specifically, we focus on mosaic drawings, which were recently introduced by Cano *et al.* [4]. The same drawing style was independently described by Alam *et al.* [2] as pixel and voxel (in 3D) drawings. Mosaic drawings are drawn on a tiling of the plane and represent vertices by so-called *configurations*: simply-connected sets of tiles (see Fig. 1). The complement of a simple mosaic drawing with respect to its bounding rectangle is also a set of simply-connected tiles, which we call *channels*. Figure 1 shows two different representations of an outerplanar graph. In both cases the vertices have complexity 6 (rectangles and L-shapes), but the complexity of the channels differs substantially. We would like to argue that a higher complexity of the channels increases the visual complexity of the drawing and hence is likely to be an impediment to the user. However, more complex channels allow us to significantly reduce the area of a drawing. We hence investigate the trade-offs between these two competing quality measures.

Mosaic Drawings. Following Cano *et al.* [4] we define mosaic drawings for *planar triangulated graphs*. Mosaic drawings are drawn on a *tiling* T of the plane. For a given set $S \subseteq T$ of tiles we define the *tile dual* of S to be the graph which has a vertex for each tile in S and an edge connecting two vertices if and only if the two corresponding tiles share a side. A *configuration* is a set of tiles with a connected tile dual. A configuration C is *simple* if the tiles in C are simply connected. The *boundary* of a simple configuration C is a simple rectilinear polygon and the *complexity* of a configuration is the number of vertices of its boundary (see, for example, Fig. 2 which consists of complexity 4 rectangular and complexity 6 L-shaped configurations).

Two configurations C_1 and C_2 are *adjacent* if and only if they contain tiles $t_1 \in C_1$ and $t_2 \in C_2$ such that t_1 and t_2 share a side. A *mosaic drawing* $D_T(G)$ of a planar triangulated graph $G = (V, E)$ on T represents every vertex $v \in V$ by a simple configuration $C(v)$ of tiles from T. Two configuration $C(u)$ and $C(v)$ representing vertices u and v are adjacent if and only if $(u, v) \in E$.

Simple Mosaic Drawings. Cano *et al.* [4] define a mosaic drawing to be *simple* if (*i*) the union of its configurations is simply connected (there are no holes in the interior of the mosaic drawing). Here we extend this definition by requiring in addition that in a simple mosaic drawing (*ii*) two adjacent configurations share exactly one contiguous piece of boundary (two configurations do not touch two or more times) and (*iii*) whenever four tiles meet in a point at least two adjacent tiles belong to the same configuration (this might be the outer configuration).

We focus on regular tilings, and specifically, the square tiling and hence denote mosaic drawings simply by $D(G)$. We construct mosaic drawings within a bounding rectangle R. The area of a drawing $D(G)$ is the number of tiles inside R. The complement of a drawing $D(G)$ with respect to R is a set of simple configurations, which we call *channels* (see Fig. 2). The boundary of a simple mosaic drawing $D(G)$ is a simple rectilinear polygon as

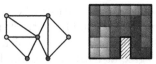

Fig. 2. Simple mosaic drawing $D(G)$ with one straight channel (shaded).

well. The boundary of $D(G)$ naturally divides into maximal straight segments, which we refer to as the *boundary segments* of $D(G)$. Finally, note that a simple mosaic drawing $D(G)$ induces an embedding of G.

Results and Organization. In Sect. 2 we first introduce some additional definitions and notation. Then we argue that any simple mosaic drawing of a maximal outerplanar graph is *natural*, that is, it follows the unique outerplanar embedding of an outerplanar graph. Hence, in the remainder of this paper, we assume that the embedding of each outerplanar graph is fixed to be outerplanar.

If we allow general channels of arbitrary complexity, then Alam *et al.* [2] show that each triangulated outerplanar graph has a simple mosaic drawing in $O(n)$ area, which is trivially tight. Note that they do not make any assumptions on the existence or absence of channels.

In Sect. 3 we consider mosaic drawing without channels, that is, mosaic drawings which are proper contact representations of a given triangulated graph G with n vertices. Alam *et al.* [2] prove that there are k-outerplanar graphs such that any mosaic drawing of these graphs requires $\Omega(kn)$ area. We strengthen this result by constructing outerplanar (that is, 1-outerplanar) graphs, such that any mosaic drawing of these graphs *without channels* requires $\Omega(n^2)$ area. More specifically, there exist outerplanar graphs with k ears, such that any mosaic drawing of these graphs has either $\Omega(n^2/k^2)$ area or total channel complexity $\Omega(k)$. This bound is tight, since the algorithm by Chiang *et al.* [5] can be used to construct mosaic drawings without channels in $O(n^2)$ area.

In Sect. 4 we consider *straight channels*. In this setting we prove that outerplanar graphs with k ears may require $\Omega(\min(nk, n^2/k))$ area. This bound is partially tight: we show how to draw outerplanar graphs with k ears in $O(nk)$ area with L-shaped vertex configurations and straight channels. Finally, in Sect. 5 we show that L-shaped channels are strictly more powerful than straight channels, but may still require $\Omega(n^{7/6})$ area.

2   Preliminaries

Outer-Path. Let G be a maximal outerplanar graph and let G^* be its weak dual. If G^* is a path then we call G an *outerpath*. The *length* of an outerpath is the number of vertices of G^*. A so-called *ear* of G is a triangle which is dual to a vertex of degree 1 in G^*. Each ear has exactly one vertex of degree 2, its so-called *tip*. Every outerpath has exactly two vertices of degree 2, the tips of its two ears. These two tips naturally divide the vertices of G into two consecutive sequences of vertices, the *upper* and the *lower sequence* of the outerpath G. In our lower bound constructions we use a particular type of outerpath, namely a so-called *outerzigzag*, which is also known as a triangle strip. Its vertices have degree at most 4. The number of vertices in the upper and the lower sequence of an outerzigzag differ by at most one.

Natural Mosaic Drawings. A mosaic drawing is *natural* if it follows the unique outerplanar embedding of a maximal outerplanar graph, where each configuration is adjacent to the outer face in the same order as in the outerplane embedding (or its reverse). We prove below that we can restrict ourselves to natural mosaic drawings of maximal outerplanar graphs.

Lemma 1. *In any simple mosaic drawing $D(G)$ of a maximal outerplanar graph G, every configuration must be adjacent to the outer face.*

Proof. For the sake of contradiction, assume that a configuration $C(v)$ is not adjacent to the outer face. We perform a case analysis on the degree of v in G. Since G is maximally outerplanar, the degree of v is at least two. If v has at least 3 neigh-

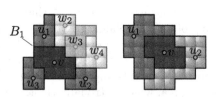

bors u_1, u_2, and u_3, then the boundary of $C(v)$ contains three parts: the shared boundaries $B_i = C(v) \cap C(u_i)$ for $i = 1, 2, 3$. Now consider the part of the boundary of $C(v)$ between B_1 and B_2 (see figure left). Since $D(G)$ has no point contacts and $C(v)$ is not adjacent to the outer face, there must be a sequence of vertices $u_1 = w_1, w_2, \ldots, w_k = u_2$ such that $C(w_i)$ is adjacent to $C(w_{i+1})$ for $1 \leq i < k$. The same holds for the parts of the boundary of $C(v)$ between B_2

and B_3, and B_3 and B_1. Hence G contains a K_4 $\{v, u_1, u_2, u_3\}$ as a minor, which contradicts the outerplanarity of G.

If v has two neighbors u_1 and u_2, then the boundary of $C(v)$ can be partitioned into two parts $B_1 = C(v) \cap C(u_1)$ and $B_2 = C(v) \cap C(u_2)$. This directly implies that $D(G)$ has a point contact or that $C(u_1) \cap C(u_2)$ is not contiguous (see figure right). This is not allowed in a simple mosaic drawing. □

Lemma 2. *Every simple mosaic drawing $D(G)$ of a maximal outerplanar graph G is natural.*

Proof. Lemma 1 shows that every configuration $C(v)$ must be adjacent to the outer face. Furthermore, the shared boundary between a configuration $C(v)$ and the outer face must be contiguous, since otherwise $C(v)$ forms a cut of $D(G)$, implying that v is a cut vertex of G. Now let $v_1, \ldots, v_n$ be the order of vertices implied by the unique outerplane embedding of G. Furthermore, let B_i be the shared boundary between $C(v_i)$ and the outer face. Note that, by definition, a set of consecutive vertices $\{v_i, v_{i+1}\}$ cannot be a cut of G. If B_i and B_{i+1} are not consecutive along the boundary of $D(G)$, then $C(v_i) \cup C(v_{i+1})$ forms a cut of $D(G)$. Thus, $D(G)$ must be natural. □

3 Mosaic Drawings Without Channels

In this section we construct a family of maximal outerplanar graphs that require $\Omega(n^2)$ area to be drawn if we do not allow channels. More precisely we prove: to obtain sub-quadratic area, the number of boundary segments on the boundary of $D(G)$ must be proportional to the number of ears of G.

Lemma 3. *Any mosaic drawing $D(G)$ of an outerzigzag G of size n, where all vertex configurations are adjacent to the same or two consecutive boundary segments, requires $\Omega(n)$ width and $\Omega(n)$ height.*

Proof. Let $v_1, \ldots, v_n$ be the vertices of G in order along the outer face, and let $v_{n/2}$ be the tip of an ear of G. Now consider a horizontal and a vertical ray starting anywhere in $C(v_{n/2})$ and pointing away from the respective boundary

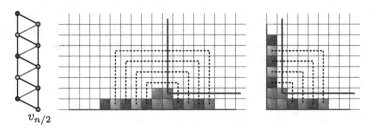

Fig. 3. $D(G)$ requires $\Omega(n)$ width and $\Omega(n)$ height since the horizontal and vertical lines must be crossed $\Omega(n)$ times. The dashed lines indicate the adjacencies which need to be satisfied to complete the drawing.

segment(s). These lines must both cross all $\Omega(n)$ (disjoint) regions of the form $C(v_i) \cup C(v_{n-i})$ ($1 \leq i < n/2$) before leaving $D(G)$ (see Fig. 3). Thus, $D(G)$ must have $\Omega(n)$ width and $\Omega(n)$ height.   □

A *k-comb* ($k \geq 3$) is an outerplanar graph with n vertices and k ears, where each outerpath incident to an ear (a *leg*) is an outerzigzag with $\Theta(n/k)$ vertices, and all legs are on the same side with respect to the remaining part of the graph (see Fig. 4).

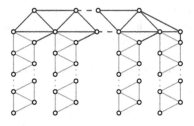

Lemma 4. *Any mosaic drawing of a k-comb G with n vertices requires $2k$ boundary segments or $\Omega(n/k)$ width and $\Omega(n/k)$ height.*

Fig. 4. A k-comb with k legs. (orange) (Color figure online)

Proof. If all the vertex configurations of one of the legs are adjacent to the same or two consecutive boundary segments of $D(G)$, then $D(G)$ requires $\Omega(n/k)$ height and $\Omega(n/k)$ width by Lemma 3. Otherwise, the vertex configurations of a single leg must be adjacent to at least 3 consecutive (the embedding is fixed) boundary segments of $D(G)$. Since two consecutive legs can share at most one of the boundary segments, we need at least 2 boundary segments for each leg, resulting in a total of $2k$ boundary segments.   □

Corollary 1. *Any mosaic drawing without channels of a 3-comb with n vertices requires $\Omega(n^2)$ area.*

4   Straight Channels

We now consider mosaic drawings with straight channels. We first argue that outerplanar graphs with k ears may require $\Omega(\min(nk, n^2/k))$ area. To prove this lower bound, we consider *one-sided* mosaic drawings: every channel must have a tile adjacent to the bottom of R, and every vertex configuration must have a tile adjacent to a channel or the bottom of R. These restrictions apply only to the bottom of R. One-sided mosaic drawings can otherwise have an arbitrary boundary.

Lemma 5. *Let G be an outerplanar graph with n vertices and k ears such that a one-sided mosaic drawing $D(G)$ of G with certain channel restrictions requires $\Omega(f(n, k))$ area. Then there exists an outerplanar graph G' with $O(n)$ vertices and $O(k)$ ears such that a mosaic drawing $D(G')$ of G' with the same channel restrictions also requires $\Omega(f(n, k))$ area.*

Proof. We construct G' by combining 5 copies of G, attaching them to a triangulated pentagon as illustrated in the figure. Since R has only 4 sides, there must be a copy of G such that all corresponding vertex configuration have a tile adjacent to the same side of R in $D(G')$. These vertex configurations thus form a one-sided mosaic drawing of G, possibly by rotating the drawing.   □

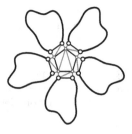

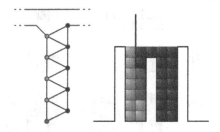

Fig. 5. The red line segment has to be crossed $\Omega(n)$ times. (Color figure online)

Lemma 6. *There exists an outerplanar graph G with n vertices and k ears such that any one-sided mosaic drawing $D(G)$ with only straight channels requires $\Omega(\min(nk, n^2/k))$ area.*

Proof. Let G be the graph obtained by attaching another leg (outerzigzag) of size $n/2$ to a $(k-1)$-comb with $n/2$ vertices. Note that the other legs have size $\Theta(n/k)$. We will argue that any one-sided drawing $D(G)$ of G has width $\Omega(\min(k, n/k))$ and height $\Omega(n)$.

Consider the leg of size $n/2$. Let $v_1, \ldots, v_{n/2}$ be the vertices of the leg in order along the outer face and assume w.l.o.g. that the corresponding configurations occur counter-clockwise along the boundary of $D(G)$. Since the drawing has only straight channels, the tiles of $C(v_1), \ldots, C(v_{n/2})$ adjacent to the boundary of $D(G)$ must have non-decreasing x-coordinates. Let $v_{n/4}$ be the tip of the ear of the long leg, and let t be the last tile of $C(v_{n/4})$ adjacent to the boundary (see Fig. 5). If we draw a vertical line segment up from t, then this line segment must be crossed by all $\Omega(n)$ (disjoint) regions of the form $C(v_i) \cup C(v_{n/2-i})$. This directly implies that $D(G)$ must have height $\Omega(n)$.

Next we consider the other legs. Lemma 3 implies that we need $\Omega(k)$ boundary segments, or $D(G)$ has width at least $\Omega(n/k)$. Since every channel can add at most 4 boundary segments, we need at least $\Omega(k)$ channels in the first case, which require a tile at the bottom of $D(G)$ each. Thus the width of $D(G)$ is at least $\Omega(\min(k, n/k))$, implying a total area of at least $\Omega(\min(nk, n^2/k))$. □

Drawing Algorithm. We now show how to draw outerplanar graphs with k ears in $O(nk)$ area with L-shaped vertex configurations and straight channels. Our algorithm is incremental. It starts with a single edge on the boundary of G and repeatedly draws a vertex attached to the endpoints of an existing edge.

An edge in this construction is called *open* if we still need to add a vertex to it. That is, an edge is open if and only if it is the initial edge or an internal edge of G. In every step we arbitrarily choose an open edge to extend with a vertex.

An open edge (a, b) is represented in $D(G)$ in two possible ways, as shown at the top of Fig. 6: one of the two vertex configurations is an L-shape and either touches the other configuration from the side (Case (I)) or lies on top of the other configuration (Case (II)). The horizontally mirrored case, where $C(b)$ is an L-shape instead of $C(a)$, is also possible, but completely symmetric. Therefore we will consider only the cases shown in the figure. Furthermore note that the

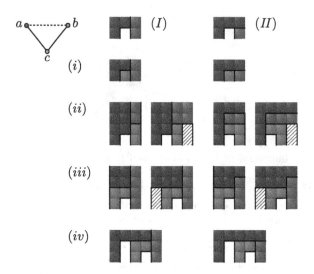

Fig. 6. The cases of the incremental drawing algorithm.

vertex configurations can still extend further as indicated by the open borders in the figure. In particular, both $C(a)$ and $C(b)$ can be L-shapes. However, such extensions are not relevant for the different cases.

We now show how to extend the drawing when a vertex c is added to an open edge (a, b). There are four cases (see Fig. 6):

(i) **(a, c) and (c, b) are not open.** In both Case (I) and (II) we simply fill up the remaining hole with $C(c)$.

(ii) **(a, c) is open and (c, b) is not open.** In Case (I) we place $C(c)$ below $C(b)$ and shift $C(a)$ down such that it forms a new Case (I) with $C(c)$. In Case (II) we extend $C(b)$ into the hole and place $C(c)$ below it as an L-shape pointing to the left. The open edge (a, c) then forms the symmetric version of Case (I).

(iii) **(a, c) is not open and (c, b) is open.** In both Case (I) and (II) we extend $C(a)$ into the hole and place $C(c)$ below it as an L-shape pointing to the right. The open edge (c, b) then forms a new Case (I).

(iv) **(a, c) and (c, b) are open.** In both Case (I) and (II) we put $C(c)$ in the hole as an L-shape pointing to the right. Then the open edge (a, c) forms a new Case (II) and the open edge (c, b) forms a new Case (I).

In the above construction Case (iv) is a special case. We call a vertex c in Case (iv) a *splitter*. The configuration of a splitter can and sometimes must extend a channel downwards. It must do so before it is no longer part of an open edge. Therefore, some cases must be handled slightly differently when a or b is a splitter. This is shown on the right side of each case in Fig. 6 whenever this is relevant. Furthermore, splitters are part of two open edges simultaneously. Thus, in the independent construction of the two open edges, its configuration may get a different height. This can be fixed by extending the construction of one of the open edges vertically until the height of the configuration of the splitter matches. This way the configuration of the splitter will not have any additional complexity or incorrect adjacencies.

Lemma 7. *The above algorithm computes a simple mosaic drawing $D(G)$ that correctly represents G.*

Proof. By construction the drawing $D(G)$ contains all adjacencies in G. Thus, we need to argue that $D(G)$ does not have any adjacencies not in G. We use the following invariant: for every open edge (a, b) the left side of the configuration of a is on the boundary of $D(G)$ or a is a splitter. The same holds for b and the right side. This invariant is maintained throughout all cases and hence no unwanted adjacencies are introduced when adding a vertex c. Therefore, $D(G)$ correctly represents G. □

Theorem 1. *For every maximal outerplanar graph G with n vertices and k ears there is a mosaic drawing $D(G)$ with the following properties:*

1. *$D(G)$ has $O(k)$ straight channels.*
2. *$D(G)$ has $O(nk)$ area.*
3. *All vertex configurations of $D(G)$ are L-shaped or rectangular.*

Proof. First note that only splitters introduce (straight) channels (one each). Since every open edge must correspond to a unique ear, and the number of open edges increases for every splitter (Case (iv)), there can be at most k channels. For the complexity of the vertex configurations, note that the complexity of the configurations never increases when adding a vertex. Therefore the complexity of a configuration can be at most its initial complexity: rectangular or L-shaped. To argue the area of $D(G)$, we show that $D(G)$ has $O(k)$ width and $O(n)$ height. The width only increases in Case (iv) (by 2 columns) or when a channel is introduced (by 1 column), which both involves a splitter. Every splitter is involved in exactly one Case (iv) and can introduce a channel only once. Therefore, the width of $D(G)$ is linear in the number of splitters, which is $O(k)$. To argue the height we use a compaction argument. Consider two rows of $D(G)$ such that no vertex configuration has a horizontal boundary on the horizontal line separating the two rows. Then we can replace these two rows by a single row without changing adjacencies and without increasing the complexities of the vertex configurations. Now, since every line separating two rows must contain a horizontal boundary of a vertex configuration, and there are at most $3n$ such horizontal boundaries (vertex configurations are at most L-shapes), the height of $D(G)$ is $O(n)$. □

5 L-Shaped Channels

In this section we prove a lower bound on the area of mosaic drawings with L-shaped channels. We first show that the lower bound construction for straight channels cannot directly be extended to L-shaped channels. For this we define *generalized k-combs*: k-combs where legs can have different sizes. Note that the example used for the lower bound in Lemma 6 is a generalized k-comb.

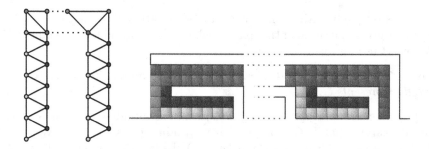

Fig. 7. One-sided drawing of a generalized k-comb in linear space.

Lemma 8. *Every generalized k-comb G with n vertices allows a one-sided mosaic drawing $D(G)$ with L-shaped channels of $O(n)$ area.*

Proof. By using one L-shaped channel per leg which almost immediately bends to the left, a complete leg can be drawn with $O(1)$ height (see Fig. 7). We can then draw all legs next to each other to obtain a mosaic drawing of $O(n)$ width. Care has to be taken when drawing the first and last leg of the comb to ensure all vertex configurations are adjacent to a channel originating from the bottom. Finally, one big L-shaped channel spanning the entire drawing is used to border the "spine" of the generalized k-comb, resulting in a drawing of $O(n)$ area.   □

We now give a more general result on mosaic drawings with channels of constant complexity, using the following trivial observation.

Observation 2. *A channel of complexity c adds c boundary segments to a drawing $D(G)$.*

To prove the following lower bounds, we need a special type of outerplanar graph: a *(k, r)-signpost*. A (k, r)-signpost is like a k-comb, but with two important differences: (1) the legs are connected on alternating sides of the "spine", and (2) every leg is an r-comb. The graph dual of a (k, r)-signpost is shown in Fig. 8.

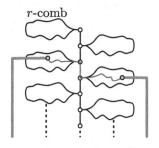

Fig. 8. The graph dual of a (k, r)-signpost.

Lemma 9. *There exists some constant r such that any one-sided mosaic drawing $D(G)$ with constant complexity channels of a (k, r)-signpost G with n vertices must have width and height $\Omega(\min(k, n/k))$.*

Proof. If the channels of $D(G)$ can have complexity c, then we require that $r > c/2$. Since r is constant, the legs of the r-combs in a (k, r)-signpost have size $\Omega(n/k)$. By Lemma 4 we know that each r-comb either requires width and height $\Omega(n/k)$, or is adjacent to $2r$ boundary segments of $D(G)$. In the latter case, since $2r > c$, every representation of an r-comb must contain a tile adjacent to the bottom of R. Therefore, the union of the configurations of two consecutive r-combs on opposite sides of the "spine" of the (k,r)-signpost separates $D(G)$ into two parts (see Fig. 8). Since a (k,r)-signpost contains $\Omega(k)$ of such pairs of consecutive r-combs, $D(G)$ must have width and height $\Omega(k)$. Combining both cases results in the claimed lower bound of $\Omega(\min(k,n/k))$ for the width and height of $D(G)$. $\qquad\square$

We can now prove lower bounds for mosaic drawings with L-shaped channels. In the following, let the x-coordinate of an L-shaped channel be the x-coordinate of its vertical leg.

Lemma 10. *Let $D(G)$ be a one-sided mosaic drawing of an outerzigzag G on n vertices with m L-shaped channels, width w, and height h. Then $dm + w + h = \Omega(n)$, where d is the difference between the x-coordinates of the rightmost and leftmost channels.*

Proof. Let $v_1, \ldots, v_n$ be the vertices of G in order along the outer face, and let $v_{n/2}$ be the tip of an ear of G. Starting from a tile of $C(v_{n/2})$ we draw a y-monotone rectilinear path P in $D(G)$ until it escapes $D(G)$ using as few horizontal segments as possible (see figure). Note that such a path exists and can have at most m horizontal segments. Further,

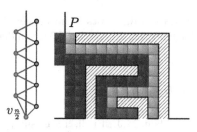

P must be crossed by all $\Omega(n)$ (disjoint) regions of the form $C(v_i) \cup C(v_{n-i})$, which implies that the length of P is $\Omega(n)$. The sum of the lengths of the vertical segments of P is at most h. The horizontal segments of P have length at most d, except the highest segment, which can have length w. We directly obtain that $dm + w + h = \Omega(n)$. $\qquad\square$

Lemma 11. *Any one-sided mosaic drawing $D(G)$ with only L-shaped channels of a k-comb G with $k = n^{1/3}$ requires $\Omega(n^{2/3})$ width or height.*

Proof. Let w and h be the width and height of $D(G)$, respectively. For the ith leg of G, let L_i be the set of channels adjacent to the configurations of leg i. Note that the set L_i must be consecutive along the boundary of $D(G)$. Furthermore, let $m_i = |L_i|$ and let d_i be the difference between the x-coordinates of the rightmost and leftmost channels in L_i. By Lemma 10 we have that $d_i m_i + w + h = \Omega(n^{2/3})$ for each leg. Furthermore, $d_i \geq m_i$ and $\sum_i d_i \leq w$, since $D(G)$ is one-sided. If $d_i m_i = \Omega(n^{2/3})$ for all i, then $d_i = \Omega(n^{1/3})$ for all i. As a result, $w \geq \sum_i d_i = n^{1/3}\Omega(n^{1/3}) = \Omega(n^{2/3})$. Otherwise we obtain that $w + h = \Omega(n^{2/3})$, which directly implies the claimed result. $\qquad\square$

Theorem 3. *There exists an outerplanar graph G with n vertices such that any mosaic drawing $D(G)$ with only L-shaped channels requires $\Omega(n^{7/6})$ area.*

Proof. Let G_1 be a $(\sqrt{n},3)$-signpost with $n/2$ vertices, and let G_2 be a $(n^{1/3})$-comb with $n/2$ vertices. We construct G by attaching G_1 to G_2 similarly to the construction in Lemma 5. Lemma 9 implies that the width and height of $D(G)$ are both $\Omega(\sqrt{n})$. Furthermore, Lemma 11 implies that the width or height of $D(G)$ is $\Omega(n^{2/3})$. As a result, $D(G)$ has area $\Omega(n^{7/6})$. Although we have only argued this for a one-sided mosaic drawing now, this result also holds for a general mosaic drawing with L-shaped channels due to Lemma 5. $\square$

6   Conclusions and Open Problems

We investigated the trade-offs between two complexity measures for mosaic drawings of outerplanar graphs: channel complexity and area. Both measures clearly contribute to the quality of mosaic drawings. Several intriguing open questions remain, such as: is there a non-trivial upper bound for the area of simple mosaic drawings using only L-shaped channels? And do more complex channels allow for drawings with linear area?

Acknowledgments. The authors are supported by the Netherlands Organisation for Scientific Research (NWO) under grant numbers 639.023.208 (Q.B. & B.S.) and 639.021.541 (K.V.).

References

1. Alam, M.J., Biedl, T., Felsner, S., Kaufmann, M., Kobourov, S.G., Ueckerdt, T.: Computing cartograms with optimal complexity. Discret. Comput. Geom. **50**(3), 784–810 (2013)
2. Alam, M.J., Bläsius, T., Rutter, I., Ueckerdt, T., Wolff, A.: Pixel and voxel representations of graphs. In: Di Giacomo, E., Lubiw, A. (eds.) GD 2015. LNCS, vol. 9411, pp. 472–486. Springer, Heidelberg (2015). doi:10.1007/978-3-319-27261-0_39
3. de Berg, M., Mumford, E., Speckmann, B.: On rectilinear duals for vertex-weighted plane graphs. Discret. Math. **309**(7), 1794–1812 (2009)
4. Cano, R.G., Buchin, K., Castermans, T., Pieterse, A., Sonke, W., Speckmann, B.: Mosaic drawings and cartograms. Comput. Graph. Forum **34**(3), 361–370 (2015)
5. Chiang, Y.T., Lin, C.C., Lu, H.I.: Orderly spanning trees with applications. SIAM J. Comput. **34**(4), 924–945 (2005)
6. Fowler, J.J.: Strongly-connected outerplanar graphs with proper touching triangle representations. In: Wismath, S., Wolff, A. (eds.) GD 2013. LNCS, vol. 8242, pp. 155–160. Springer, Heidelberg (2013). doi:10.1007/978-3-319-03841-4_14
7. Koebe, P.: Kontaktprobleme der konformen Abbildung. Berichte über die Verhandlungen der Sächsischen Akademie der Wissenschaften zu Leipzig. Math.-Phys. Klasse **88**, 141–164 (1936)
8. Koźmiński, K., Kinnen, E.: Rectangular duals of planar graphs. Networks **5**(2), 145–157 (1985)
9. Purchase, H.C.: Metrics for graph drawing aesthetics. J. Vis. Lang. Comput. **13**(5), 501–516 (2002)
10. Raisz, E.: The rectangular statistical cartogram. Geogr. Rev. **24**(2), 292–296 (1934)
11. Ungar, P.: On diagrams representing maps. J. Lond. Math. Soc. **s1−28**(3), 336–342 (1953)

Fast Optimal Labelings for Rotating Maps

Rafael G. Cano$^{(\boxtimes)}$, Cid C. de Souza, and Pedro J. de Rezende

Institute of Computing, University of Campinas, Campinas, Brazil
{rgcano,cid,rezende}@ic.unicamp.br

Abstract. We study a dynamic labeling problem on rotating maps, i.e., maps that allow for continuous rotations. As the map is rotated, labels must remain horizontally aligned. Rotations may cause labels that were previously disjoint to overlap. For each label, we must determine a set of active ranges (i.e., angular ranges during which the label is visible) such that at any rotation angle all active labels are disjoint. The objective is to maximize the sum of the angular length of all active ranges. We prove a number of properties of optimal solutions which allow us to significantly reduce the size of an integer programming model from the literature. We report the results of several experiments using two existing benchmarks with 180 real-world instances. We obtained reductions of over 100 times in the number of variables and constraints of the model. The compact formulation solved all but 5 instances to optimality in under a minute.

1 Introduction

Labeling problems are well-studied in the optimization literature. The input consists of a map and several points of interest. Each point is associated with a textual label that identifies or describes it. Labels must be placed on the map close to their associated point and without overlap. Classic problems usually aim at producing static maps to be used in print. However, with the recent dissemination of navigation systems and digital maps, a new branch of labeling problems has arisen. These are referred to as *dynamic labeling problems* since they are inserted in a context that allows the user to interact with the map.

Here, we focus on rotating maps, i.e., maps that permit continuous rotations. Each label is anchored at a point that is specified in the input. We assume that each anchor point lies on the boundary of its associated label. As the map is rotated, all labels remain horizontally aligned and anchored at their associated points (see Fig. 1). This may cause labels that were previously disjoint to overlap. We handle such cases by omitting one or more labels at specific angular intervals. Therefore, we must determine, for each label, a set of *active ranges* (i.e., maximal angular intervals during which the label is visible) such that, at any rotation angle, all active labels are disjoint. The goal is to maximize the total angular length of the active ranges of all labels for a full rotation of the map.

Supported by Fundação de Amparo à Pesquisa do Estado de São Paulo (FAPESP), grant 2012/00673-2, and Conselho Nacional de Desenvolvimento Científico e Tecnológico (CNPq), grants #155498/2016-9, #311140/2014-9 and #304727/2014-8.

S.-H. Poon et al. (Eds.): WALCOM 2017, LNCS 10167, pp. 161–173, 2017.
DOI: 10.1007/978-3-319-53925-6_13

Fig. 1. Rotations may create overlaps that previously did not exist.

Been et al. [1] observed that naive (dynamic) labeling algorithms may cause unpleasant visual effects. Good maps must not allow labels to pop or jump during monotonous navigation. In rotating maps, this happens if a label has too many active ranges. In such cases, labels appear to flicker on the screen as the map is rotated. Gemsa et al. [2,3] solve this issue by letting each label have only a limited number of active ranges. They refer to this setting as the kR-*model*, where k denotes the maximum number of active ranges allowed per label.

In some applications, the occlusion of anchor points might be undesirable, since they may convey useful information even if their associated labels are inactive. This motivates the following distinction. If two labels ℓ and ℓ' intersect at some rotation angle α, we say that they have a *soft conflict* at α. If, in addition, ℓ contains ℓ''s anchor point, then ℓ also has a *hard conflict* with ℓ' at α. Note that soft conflicts define a symmetric relation, whereas hard conflicts (generally) do not. One may optionally require that hard conflicts be avoided.

Gemsa et al. [2] were the first to study this problem. They prove that it is NP-hard to find an optimal solution in the 1R-model with hard conflicts. They also provide efficient approximation algorithms. The first Integer Linear Programming (ILP) formulation for this problem was given by the same authors in [3]. They experimented with an extensive benchmark of real-world instances, several of which could only be addressed heuristically.

Our Contribution: we prove a number of properties of optimal solutions that allow us to significantly reduce the ILP formulation proposed in [3]. We obtained reductions of up to 100 times in the number of variables and constraints of the model. The compact formulation was able to optimally solve 178 out of 180 instances. Moreover, we found 12 optimal solutions that were previously unknown. The text is organized as follows. Section 2 presents the said ILP formulation. Section 3 proves several properties of optimal solutions. Section 4 discusses some implementation details and reports our computational results.

2   Model

Let $L = \{\ell_1, \ldots, \ell_n\}$ be a set of n rectangular, axis-aligned, non-overlapping, closed labels in the plane. Let $P = \{p_1, \ldots, p_n\}$ be a set of points in the plane such that each p_i corresponds to the anchor of label ℓ_i and lies on its boundary. Given two labels ℓ and ℓ', the *conflict set* $C(\ell, \ell')$ is defined as the set of all rotation angles at which ℓ and ℓ' overlap, i.e., $C(\ell, \ell') = \{\alpha \in [0, 2\pi) \mid \ell \text{ and } \ell'$ have a conflict at $\alpha\}$. A *conflict range* is a maximal contiguous range in $C(\ell, \ell')$. We refer to the endpoints of maximal conflict ranges as *conflict events*.

Let E be an ordered set that contains all conflict events and, additionally, the values 0 and 2π. For $0 \leq j \leq |E| - 1$, we denote by e_j the j-th element of E. An interval between two consecutive elements of E is an *atomic interval*. For $0 \leq j \leq |E| - 2$, we denote by $E[j]$ the j-th atomic interval, i.e., the interval between e_j and e_{j+1}. Given the circular nature of the problem, in all further references, indices of conflict events and atomic intervals should be taken modulo E and $|E| - 1$, respectively. Gemsa et al. [2,3] showed that there always exists an optimal solution in which all active ranges correspond to unions of atomic intervals. This allowed the authors to present the following ILP formulation.

For $1 \leq i \leq n$ and $0 \leq j \leq |E| - 2$ we define two sets of binary variables x_i^j and b_i^j. Variable x_i^j has value 1 if label ℓ_i is active during $E[j]$ and 0 otherwise. Variable b_i^j takes value 1 when ℓ_i is inactive during $E[j-1]$ and active during $E[j]$. The objective function to be maximized is $\sum_{i=1}^{n} \sum_{j=0}^{|E|-2} x_i^j \cdot |E[j]|$, where $|E[j]|$ represents the angular length of $E[j]$. The following constraints must hold:

$$x_i^j - b_i^j \leq x_i^{j-1} \qquad\qquad 1 \leq i \leq n, 0 \leq j \leq |E| - 2; \quad (1)$$

$$\sum_{j=0}^{|E|-2} b_i^j \leq k \qquad\qquad 1 \leq i \leq n; \quad (2)$$

$$x_i^j + x_k^j \leq 1 \qquad 0 \leq j \leq |E| - 2 \text{ and } \forall\, \ell_i, \ell_k \text{ in conflict on } E[j]. \quad (3)$$

Inequalities (1) guarantee that if a new active range starts at $E[j]$ (i.e., if $x_i^{j-1} = 0$ and $x_i^j = 1$), then b_i^j must be set to 1. Inequalities (2) limit the number of active ranges of each label to at most k. Inequalities (3) prevent pairs of overlapping labels from being simultaneously active. In the text that follows, we will refer to inequalities (3) as *conflict inequalities*. Finally, hard conflicts can be incorporated by simply fixing the appropriate activity variables at zero.

3   Properties of Optimal Solutions

Our goal is to show how to reduce the ILP from [3] without affecting its correctness. We first need a few auxiliary results. In order to simplify the discussion that follows, we assume that all conflict events occur at distinct rotation angles and, thus, all atomic intervals have positive length. We also assume that L is defined as a set of non-overlapping labels, so no conflicts exist at rotation angle 0. It should be noted that the techniques described here would work even if these assumptions are not satisfied.

Let $G[j]$ be the *conflict graph* for the j-th atomic interval, i.e., an undirected graph with a vertex for each label and edges indicating pairs of labels that are in conflict on $E[j]$. Conflict event e_{j+1} either starts or ends a conflict range, so the transition from $G[j]$ to $G[j+1]$ consists of either the insertion or the deletion of a single edge, respectively. These transitions cause labels to change their state in the solution. Our results derive from the observation that conflict events usually have a somewhat local effect, often leaving the state of distant labels unchanged.

We formalize this notion by analyzing the connected components of conflict graphs. Given a connected component C of $G[j]$, we say that C is *unaffected* on atomic interval $E[j]$ if the edge inserted or deleted in the transition from $G[j-1]$ to $G[j]$ was not incident to any vertices of C. Note that this implies that C is also a connected component of $G[j-1]$. If C does not satisfy the previous condition, we say that it is *affected* on $E[j]$.

We now define two basic operations that we will use extensively to manipulate solutions in our proofs. Given a set of labels $S \subseteq L$, a solution (x, b) to the ILP from Sect. 2, and an atomic interval $E[\gamma]$, a *backward copy operation* (denoted BACKWARD-COPY$(S, (x, b), E[\gamma])$) creates a new solution $(\widehat{x}, \widehat{b})$ in which the state of each label in S is copied from atomic interval $E[\gamma]$ to $E[\gamma-1]$. More formally, BACKWARD-COPY$(S, (x, b), E[\gamma])$ constructs $(\widehat{x}, \widehat{b})$ by assigning:

$$\widehat{x}_i^j = \begin{cases} x_i^\gamma \text{ if } \ell_i \in S \text{ and} \\ \qquad j = \gamma - 1 \\ x_i^j \text{ otherwise;} \end{cases} \qquad \widehat{b}_i^j = \begin{cases} \max(0, \ \widehat{x}_i^j - \widehat{x}_i^{j-1}) \text{ if } \ell_i \in S \text{ and } j = \gamma - 1 \\ 0 \qquad\qquad\qquad \text{ if } \ell_i \in S \text{ and } j = \gamma \\ b_i^j \qquad\qquad\qquad \text{ otherwise.} \end{cases}$$

Analogously, a *forward copy operation* (denoted FORWARD-COPY$(S, (x, b), E[\gamma])$) copies the state of each label in S from atomic interval $E[\gamma]$ to $E[\gamma+1]$:

$$\widehat{x}_i^j = \begin{cases} x_i^\gamma \text{ if } \ell_i \in S \text{ and} \\ \qquad j = \gamma + 1 \\ x_i^j \text{ otherwise;} \end{cases} \qquad \widehat{b}_i^j = \begin{cases} \max(0, \ \widehat{x}_i^j - \widehat{x}_i^{j-1}) \text{ if } \ell_i \in S \text{ and } j = \gamma + 2 \\ 0 \qquad\qquad\qquad \text{ if } \ell_i \in S \text{ and } j = \gamma + 1 \\ b_i^j \qquad\qquad\qquad \text{ otherwise.} \end{cases}$$

The reader may want to keep in mind that in the following proofs each operation that employs variable values of the ILP solution has a corresponding effect on the geometry of the label conflicts. Proposition 1 shows that solutions produced by either of the two operations always satisfy inequalities (1) and (2).

Proposition 1. *Let $S \subseteq L$ be a set of labels and (x, b) be a feasible solution to the ILP from Sect. 2. Then, for any atomic interval $E[\gamma]$, $(\widehat{x}, \widehat{b}) =$ BACKWARD-COPY$(S, (x, b), E[\gamma])$ and $(x', b') =$ FORWARD-COPY$(S, (x, b), E[\gamma])$ satisfy inequalities (1) and (2).*

Proof. It is easy to verify that, by construction, inequalities (1) are always satisfied. As for inequalities (2), note that a copy operation can only extend an existing active range or merge two active ranges. In both cases, the number of active ranges does not increase, and the result follows. □

Next, we deal with conflict inequalities. Given a connected component C of a conflict graph, denote by $L(C)$ the set of labels associated with the vertices of C.

Proposition 2. *Let C be an unaffected component of a conflict graph $G[\gamma]$. Let (x, b) be a feasible solution to the ILP from Sect. 2. Then $(\widehat{x}, \widehat{b}) =$ BACKWARD-COPY$(L(C), (x, b), E[\gamma])$ and $(x', b') =$ FORWARD-COPY$(L(C), (x, b), E[\gamma - 1])$ satisfy all conflict inequalities.*

Proof. Since C is an unaffected component of $G[\gamma]$, a conflict involving the labels in $L(C)$ exists on $E[\gamma - 1]$ if and only if it also exists on $E[\gamma]$. Thus, copying the state of these labels from $E[\gamma]$ to $E[\gamma - 1]$ (or vice versa) cannot violate any conflict inequalities.   □

Proposition 3. *Let C be an affected component of a conflict graph $G[\gamma]$. Let (x, b) be a feasible solution to the ILP from Sect. 2. Let $(\widehat{x}, \widehat{b})$ = BACKWARD-COPY$(L(C), (x, b), E[\gamma])$ and (x', b') = FORWARD-COPY$(L(C), (x, b), E[\gamma - 1])$. Let ℓ_t and ℓ_u be the two labels that caused conflict event e_γ. Then:*

1. *If e_γ starts a conflict range, $\widehat{x}$ satisfies all conflict inequalities and x' may violate at most one, namely $x_t'^\gamma + x_u'^\gamma \le 1$;*
2. *If e_γ ends a conflict range, x' satisfies all conflict inequalities and $\widehat{x}$ may violate at most one, namely $\widehat{x}_t^{\gamma-1} + \widehat{x}_u^{\gamma-1} \le 1$.*

Proof. We prove the result for $\widehat{x}$. The proof for x' is analogous. The only values that may differ between x and $\widehat{x}$ are the ones for $E[\gamma-1]$, so it suffices to analyze conflicts for that interval. Note that the only pair of labels that may have a conflict either on $E[\gamma - 1]$ or $E[\gamma]$ (but not both) is ℓ_t and ℓ_u. All other conflicts are present on $E[\gamma - 1]$ if and only if they are also present on $E[\gamma]$. Let ℓ_i and ℓ_k be two labels that have a conflict on $E[\gamma-1]$. If neither ℓ_i nor ℓ_k belongs to $L(C)$, $\widehat{x}_i^{\gamma-1} = x_i^{\gamma-1}$ and $\widehat{x}_k^{\gamma-1} = x_k^{\gamma-1}$. Thus, the corresponding conflict inequality must be satisfied by $\widehat{x}$ (otherwise x would not be feasible). If both ℓ_i and ℓ_k belong to $L(C)$, $\widehat{x}_i^{\gamma-1} = x_i^\gamma$ and $\widehat{x}_k^{\gamma-1} = x_k^\gamma$. Thus, $\widehat{x}_i^{\gamma-1} + \widehat{x}_k^{\gamma-1} > 1$ implies that this conflict is not present on $E[\gamma]$ (otherwise x would not be feasible). This can only happen if $\{\ell_i, \ell_k\} = \{\ell_t, \ell_u\}$ and, specifically, if e_γ ends a conflict range.

Now, suppose $\ell_i \in L(C)$ and $\ell_k \notin L(C)$. Note that if a conflict exists between ℓ_i and ℓ_k on $E[\gamma - 1]$, then either (i) this conflict is also present on $E[\gamma]$ or (ii) $\{\ell_i, \ell_k\} = \{\ell_t, \ell_u\}$ and e_γ ends a conflict range. Case (i) implies $\ell_k \in L(C)$ because the vertices for ℓ_i and ℓ_k would be connected by an edge in $G[\gamma]$. Case (ii) is precisely the conflict that may be violated, so the proof is complete.   □

The results in Propositions 1–3 are summarized in Lemmas 1 and 2. We actually prove slightly stronger results that will be useful in our remaining proofs.

Lemma 1 (Backward Copy Lemma). *Let C be a component of a conflict graph $G[\gamma]$. Let (x, b) be a feasible solution to the ILP from Sect. 2. Let $Z \subseteq L(C)$ be a (possibly empty) set of labels that are inactive on $E[\gamma - 1]$, i.e., $\forall \ell_i \in Z$, $x_i^{\gamma-1} = 0$. Let $(\widehat{x}, \widehat{b}) = $ BACKWARD-COPY$(L(C) \setminus Z, (x, b), E[\gamma])$. Let ℓ_t and ℓ_u be the two labels that caused conflict event e_γ. Then:*

1. *If C is unaffected on $E[\gamma]$, $(\widehat{x}, \widehat{b})$ is feasible;*
2. *If C is affected on $E[\gamma]$ and e_γ starts a conflict range, $(\widehat{x}, \widehat{b})$ is feasible;*
3. *If C is affected on $E[\gamma]$ and e_γ ends a conflict range, $(\widehat{x}, \widehat{b})$ may violate at most one inequality, namely $\widehat{x}_t^{\gamma-1} + \widehat{x}_u^{\gamma-1} \le 1$.*

Proof. If $Z = \emptyset$, the result follows directly from Propositions 1–3. Otherwise, by Proposition 1, $(\widehat{x}, \widehat{b})$ satisfies all inequalitie (1) and (2). Now, consider the solution $(x', b') = \text{BACKWARD-COPY}(L(C), (x, b), E[\gamma])$. Note that if two values differ between $\widehat{x}$ and x', the one in $\widehat{x}$ must be 0. This implies $\widehat{x} \le x'$ (where the comparison is taken between corresponding entries). Thus, every conflict inequality satisfied by x' must also be satisfied by $\widehat{x}$, and the result follows.   □

Lemma 2 (Forward Copy Lemma). *Let C be a component of a conflict graph $G[\gamma]$. Let (x, b) be a feasible solution to the ILP from Sect. 2. Let $Z \subseteq L(C)$ be a (possibly empty) set of labels that are inactive on $E[\gamma]$, i.e., $\forall\, \ell_i \in Z$, $x_i^\gamma = 0$. Let $(\widehat{x}, \widehat{b}) = \text{FORWARD-COPY}(L(C) \setminus Z, (x, b), E[\gamma - 1])$. Let ℓ_t and ℓ_u be the two labels that caused conflict event e_γ. Then:*

1. *If C is unaffected on $E[\gamma]$, $(\widehat{x}, \widehat{b})$ is feasible;*
2. *If C is affected on $E[\gamma]$ and e_γ ends a conflict range, $(\widehat{x}, \widehat{b})$ is feasible;*
3. *If C is affected on $E[\gamma]$ and e_γ starts a conflict range, $(\widehat{x}, \widehat{b})$ may violate at most one inequality, namely $\widehat{x}_t^\gamma + \widehat{x}_u^\gamma \le 1$.*

Proof. Analogous to the proof of the Backward Copy Lemma.   □

The next results show that, for optimal solutions, the number of active labels on consecutive intervals cannot vary much. Given a solution (x, b) and a component C of an atomic interval $E[\gamma]$, we define

$$\delta_C^\gamma(x) = \sum_{\ell_i \in L(C)} x_i^\gamma - \sum_{\ell_i \in L(C)} x_i^{\gamma-1},$$

i.e., $\delta_C^\gamma(x)$ represents the variation in the number of active labels in $L(C)$ from atomic interval $E[\gamma - 1]$ to $E[\gamma]$.

Lemma 3. *Let C be a component of a conflict graph $G[\gamma]$. Let (x^*, b^*) be an optimal solution to the ILP from Sect. 2. Then:*

1. *If C is unaffected on $E[\gamma]$, $\delta_C^\gamma(x^*) = 0$;*
2. *If C is affected on $E[\gamma]$ and e_γ starts a conflict range, $\delta_C^\gamma(x^*) \in \{-1, 0\}$;*
3. *If C is affected on $E[\gamma]$ and e_γ ends a conflict range, $\delta_C^\gamma(x^*) \in \{0, 1\}$.*

Proof. Case 1. Suppose $\delta_C^\gamma(x^*) \ne 0$, i.e., $\sum_{\ell_i \in L(C)} x_i^{*\gamma-1} \ne \sum_{\ell_i \in L(C)} x_i^{*\gamma}$. Consider the solutions provided by $\text{BACKWARD-COPY}(L(C), (x^*, b^*), E[\gamma])$ and $\text{FORWARD-COPY}(L(C), (x^*, b^*), E[\gamma - 1])$. By the two Copy Lemmas, both solutions are feasible. One of these alternatives gives a solution with a higher number of active labels during either $E[\gamma - 1]$ or $E[\gamma]$. Thus (x^*, b^*) cannot be optimal.

Case 2. Let ℓ_t and ℓ_u be the two labels that caused conflict event e_γ. Initially, suppose $\delta_C^\gamma(x^*) \le -2$. We must have $x_t^{*\gamma} = 0$ or $x_u^{*\gamma} = 0$ (otherwise x^* would not be feasible). Suppose w.l.g. $x_u^{*\gamma} = 0$. Let $(\widehat{x}, \widehat{b}) = \text{FORWARD-COPY}(L(C) \setminus \{\ell_u\}, (x^*, b^*), E[\gamma - 1])$. Since $\widehat{x}_u^\gamma = 0$, the Forward Copy Lemma guarantees that $(\widehat{x}, \widehat{b})$ is feasible. Moreover, $\delta_C^\gamma(\widehat{x}) \ge -1$, so $(\widehat{x}, \widehat{b})$ has more visible labels than (x^*, b^*) on $E[\gamma]$, and hence attains a better objective value. Consequently, (x^*, b^*) cannot be optimal. Now, suppose $\delta_C^\gamma(x^*) \geq 1$. By the Backward Copy Lemma, $(\widehat{x}, \widehat{b}) = \text{BACKWARD-COPY}(L(C), (x^*, b^*), E[\gamma])$ is feasible. Also, it has more visible labels on $E[\gamma - 1]$ than (x^*, b^*). Thus, (x^*, b^*) cannot be optimal.

Case 3. Analogous to the proof of Case 2.   □

In our next proofs, it will be necessary to execute copy operations sequentially on consecutive atomic intervals. We indicate this by passing a list of intervals as parameters to BACKWARD- and FORWARD-COPY. Formally, given a set of labels $S \subseteq L$, a solution (x, b) and a list of consecutive atomic intervals $Q = \langle E[j_1], E[j_1 + 1], \ldots, E[j_2] \rangle$, BACKWARD-COPY$(S, (x, b), E[j_2], \ldots, E[j_1 + 1], E[j_1])$ produces a list of solutions by setting $(x[0], b[0]) = (x, b)$ and, for $0 \leq \lambda \leq |Q| - 1$, $(x[\lambda+1], b[\lambda+1]) = \text{BACKWARD-COPY}(S, (x[\lambda], b[\lambda]), E[j_2 - \lambda])$. The last solution is returned as the result. FORWARD-COPY$(S, (x, b), E[j_1], E[j_1 + 1], \ldots, E[j_2])$ is defined analogously (note, however, that the order of the intervals is reversed).

The following definition will also be useful. Let C be a component of a conflict graph $G[\gamma]$. The *lifespan* of C (denoted $\mathcal{LS}(C)$) is a maximal set of consecutive intervals $\{E[j_1], E[j_1 + 1], \ldots, E[\gamma], \ldots, E[j_2 - 1], E[j_2]\}$ that contains $E[\gamma]$ and such that C is a component of $G[j_1], \ldots, G[\gamma], \ldots, G[j_2]$. Here, we consider that two components are distinct even if only their edge sets differ. This implies that, if C is affected in some conflict graph $G[j]$, then it cannot be a component of $G[j - 1]$. Therefore, C is unaffected on $E[j_1 + 1], \ldots, E[\gamma], \ldots, E[j_2]$.

Moreover, suppose C is also unaffected on $E[j_1]$. Then, C must be a component of $E[j_1 - 1]$, and since $\mathcal{LS}(C)$ is maximal, $E[j_1 - 1] \in \mathcal{LS}(C)$. This can only be true if $E[j_1 - 1] = E[j_2]$ (recall that interval indices are taken modulo $|E| - 1$), which implies $\mathcal{LS}(C)$ contains all atomic intervals and C is unaffected in all of them. Due to the fact that all labels are disjoint at rotation angle 0, C must consist of a single isolated label that can be made visible over the full map rotation. We assume that such a trivial case is treated in a preprocessing step. Thus, in the remainder of this text, we consider that C is affected on $E[j_1]$ and unaffected on every other interval of its lifespan.

We can now present our main results. We say that a solution is *harmonious* if for every label ℓ_i of each unaffected component C of $G[j]$ $(0 \leq j \leq |E| - 2)$ we have $x_i^{j-1} = x_i^j$ and $b_i^j = 0$. In other words, in a harmonious solution, a label ℓ_i may change its state during the j-th atomic interval only if the component to which it belongs in $G[j]$ is affected.

Theorem 1. *The ILP from Sect. 2 always has a harmonious optimal solution.*

Proof. Let (x^*, b^*) be an optimal solution. For each $\ell_i \in L$ and for $0 \leq j \leq |E| - 2$, we assume $b_i^{*j} = \max(0, x_i^{*j} - x_i^{*j-1})$, i.e., b_i^{*j} is set to 1 only if strictly necessary to satisfy inequalities (1). Note that, if (x^*, b^*) does not satisfy this condition, there exists a label ℓ_i and an interval $E[j]$ such that $b_i^{*j} = 1$ and $\max(0, x_i^{*j} - x_i^{*j-1}) = 0$. But, then, setting b_i^{*j} to 0 does not violate inequalities (1) and, since this reduces its value, inequalities (2) must also remain satisfied.

Now, suppose that for some atomic interval $E[\gamma]$ there exists a label ℓ_i that belongs to an unaffected component C of $G[\gamma]$. In addition, suppose that the state of ℓ_i changed in (x^*, b^*) on $E[\gamma]$, i.e., $x_i^{*\gamma-1} \neq x_i^{*\gamma}$. Let $\mathcal{LS}(C) = \{E[j_1], \ldots, E[\gamma], \ldots, E[j_2]\}$. Let $(\widehat{x}, \widehat{b}) = \text{BACKWARD-COPY}(L(C), (x^*, b^*), E[j_2], \ldots, E[\gamma], \ldots, E[j_1 + 1])$. By the Backward Copy Lemma, $(\widehat{x}, \widehat{b})$ is feasible. Also, by Lemma 3, the copy operations do not change the number of visible labels on any atomic interval, so $(\widehat{x}, \widehat{b})$ is optimal. Finally, for all labels $\ell_i \in L(C)$ and all intervals $E[j] \in \mathcal{LS}(C)$ on which C is unaffected, $\widehat{x}_i^{j-1} = \widehat{x}_i^j$ and $\widehat{b}_i^j = 0$.

To complete the proof, note that, if $(\widehat{x}, \widehat{b})$ is still not harmonious, the process described here for component C can be repeated independently for any other unaffected component. Once this is done for all unaffected components of all atomic intervals, the resulting solution will be harmonious. $\square$

Theorem 1 shows that there always exists an optimal solution in which the state of labels in all unaffected components remains unchanged. Next, we provide a similar result for affected components. Let $\mathcal{A}$ be a set containing all affected components of all atomic intervals, i.e., $\mathcal{A} = \{C \mid C$ is an affected component of some $G[j], 0 \leq j \leq |E| - 2\}$. We remark that it is possible that two affected components of two distinct conflict graphs have exactly the same set of vertices and edges. However, since they do not belong to the same conflict graph, they are represented by distinct elements in $\mathcal{A}$.

Let $C \in \mathcal{A}$ be an affected component of a conflict graph $G[j_1]$. Let $\mathcal{LS}(C) = \{E[j_1], E[j_1 + 1], \ldots, E[j_2]\}$ be C's lifespan. Since $\mathcal{LS}(C)$ is maximal, conflict event e_{j_2+1} affects at least one vertex of C, i.e., the transition from $G[j_2]$ to $G[j_2 + 1]$ either inserts or deletes an edge incident to a vertex of C. Graph $G[j_2 + 1]$ may have one or two affected components, both distinct from C. Let $D \in \mathcal{A}$ be an affected component of $G[j_2 + 1]$. We say that C is a *parent* of D and, conversely, D is a *child* of C. Alternatively, the parents of D are precisely the components of $G[j_2]$ that contain the vertices of D. Here, two observations are in order. First, note that a component can have up to two parents and two children. Second, we do not intend to establish a partial order on the elements of $\mathcal{A}$ (this is not possible due to the circular nature of the problem). We only wish to determine how components interact with each other to form new ones.

We want to show that there always exists an optimal solution in which the state of most labels in a well-chosen set of affected components does not change. Let $C \in \mathcal{A}$ be an affected component of $G[j]$. Let ℓ_t and ℓ_u be the two labels that caused conflict event e_j. Generally, we cannot guarantee that the state of ℓ_t and ℓ_u will remain the same (actually, it often changes). However, the other labels in $L(C)$ may keep their previous state. Theorem 2 formalizes this result. We define a component graph $G_{\mathcal{A}}$ as a simple undirected graph with a vertex v_C for each component $C \in \mathcal{A}$ and an edge $\{v_C, v_D\}$ for each pair of components $C, D \in \mathcal{A}$ such that C is either a parent or a child of D.

Theorem 2. *Let $\mathcal{I} \subseteq \mathcal{A}$ be a set of components such that their associated vertices form an independent set in $G_{\mathcal{A}}$. There always exists a harmonious optimal solution (x^*, b^*) to the ILP from Sect. 2 such that the following is satisfied for all elements of $\mathcal{I}$: if $C \in \mathcal{I}$ is an affected component of $G[\gamma]$, and ℓ_t and ℓ_u are* *the two labels that caused conflict event e_γ, then for each $\ell_i \in L(C) \setminus \{\ell_t, \ell_u\}$ we have $x_i^{*\gamma-1} = x_i^{*\gamma}$ and $b_i^{*\gamma} = 0$.*

Proof. Let (x^*, b^*) be a harmonious optimal solution and suppose it does not satisfy the desired condition. Let $C \in \mathcal{I}$ be an affected component of $G[\gamma]$, and ℓ_t and ℓ_u be the two labels that caused conflict event e_γ. We will show that there exists a harmonious optimal solution $(\widehat{x}, \widehat{b})$ in which the state of the labels in $L(C) \setminus \{\ell_t, \ell_u\}$ is the same on $E[\gamma - 1]$ and $E[\gamma]$. We will do that by executing copy operations on (x^*, b^*), either from $E[\gamma - 1]$ to $E[\gamma]$ or from $E[\gamma]$ to $E[\gamma - 1]$. By Lemma 3, if e_γ starts a conflict range, $\delta_C^\gamma(x^*) \in \{-1, 0\}$, otherwise, $\delta_C^\gamma(x^*) \in \{0, 1\}$. We treat these four cases separately. The feasibility of all solutions considered here follows from the two Copy Lemmas. We omit further references for brevity.

Case 1: e_γ starts a conflict range and $\delta_C^\gamma(x^*) = -1$. We must have $x_t^{*\gamma} = 0$ or $x_u^{*\gamma} = 0$. Suppose w.l.o.g. $x_u^{*\gamma} = 0$. Let $(x', b') = \text{FORWARD-COPY}(L(C) \setminus \{\ell_u\}, (x^*, b^*), E[\gamma - 1])$. Since $x_u'^\gamma = 0$, (x', b') is feasible. Also, $\delta_C^\gamma(x') \geq -1$, so it has the same objective value as (x^*, b^*) and is optimal. However, we modified some variables for $E[\gamma]$, so (x', b') may not be harmonious (especially if C is unaffected on $E[\gamma + 1]$). Let $\{E[\gamma], E[\gamma + 1], \ldots, E[j_2]\}$ be the lifespan of C. Solution $(\widehat{x}, \widehat{b}) = \text{FORWARD-COPY}(L(C), (x', b'), E[\gamma], E[\gamma + 1], \ldots, E[j_2 - 1])$ is feasible and harmonious (recall that C is unaffected on $E[\gamma + 1], \ldots, E[j_2]$). Finally, by Lemma 3, it is also optimal.

Case 2: e_γ starts a conflict range and $\delta_C^\gamma(x^*) = 0$. Let $(x', b') = \text{BACKWARD-COPY}(L(C), (x^*, b^*), E[\gamma])$. Since $\delta_C^\gamma(x') = 0$, (x', b') is optimal, but (possibly) not harmonious. Let $D \in \mathcal{A}$ be a parent of C. Let $\{E[j_1], E[j_1 + 1], \ldots, E[\gamma - 1]\}$ be the lifespan of D. Solution $(\widehat{x}, \widehat{b}) = \text{BACKWARD-COPY}(L(D), (x', b'), E[\gamma - 1], E[\gamma - 2], \ldots, E[j_1 + 1])$ is feasible (because D is unaffected on $E[j_1 + 1], \ldots, E[\gamma - 1]$) and, by Lemma 3, optimal. If C has only one parent, it is also harmonious. Otherwise, let D' be the other parent. Labels ℓ_t and ℓ_u belong to distinct components in $G[\gamma - 1]$ (otherwise, there would be a single parent), and $L(C) = L(D) \cup L(D')$. This means that the first backward copy operation changed the state of all labels in $L(C)$, but the others only addressed the labels in $L(D)$. Thus, we must also apply backward copy operations to the labels in $L(D')$ on the intervals that belong to the lifespan of D'. Once this is done, the solution obtained will be both harmonious and optimal.

Case 3: e_γ ends a conflict range and $\delta_C^\gamma(x^*) = 0$. Analogous to Case 1.
Case 4: e_γ ends a conflict range and $\delta_C^\gamma(x^*) = 1$. Analogous to Case 2.

In all four cases, we obtain a harmonious optimal solution in which the state of the labels in $L(C)$ remains unchanged (except for ℓ_t and ℓ_u), so it satisfies the desired condition for component C. It remains to be shown that this procedure can be executed independently for all components in $\mathcal{I}$. Observe that we needed to execute copy operations on the intervals that belong to C's lifespan (Cases 1 and 3) and also to its parents' lifespan (Cases 2 and 4). Therefore, if we apply the same procedure to C's parents, we might overwrite some variables that had already been set. However, we can execute it independently for other components as long as we do not pick a parent and its child. This is equivalent to choosing components that form an independent set in $G_\mathcal{A}$, so the proof is complete.   □

4 Implementation and Experiments

Implementation. The results from Sect. 3 can be used to significantly reduce the size of the ILP model. Let $\mathcal{I} \subseteq \mathcal{A}$ be a set of components that form an independent set in $G_\mathcal{A}$. Let C be a component of conflict graph $G[\gamma]$ and let ℓ_i be a label in $L(C)$. If C is unaffected on $E[\gamma]$, Theorem 1 allows us to add the following equality constraints to the model: $x_i^{\gamma-1} = x_i^\gamma$ and $b_i^\gamma = 0$. Otherwise, if C is affected on $E[\gamma]$ and $C \in \mathcal{I}$, Theorem 2 allows us to add the same constraints, except if ℓ_i is one of the labels that caused conflict event e_γ.

When it is the case that this pair of equalities is included in the model, two variables may be removed: b_i^γ and either $x_i^{\gamma-1}$ or x_i^γ (but not both). Additionally, some constraints might become redundant. Clearly, $x_i^\gamma - b_i^\gamma \le x_i^{\gamma-1}$ is always satisfied and can be removed. Also, some conflict constraints become unnecessary. If ℓ_i has a conflict with ℓ_k on both $E[\gamma - 1]$ and $E[\gamma]$, and if $x_k^{\gamma-1} = x_k^\gamma$, then we only need the conflict constraint for either $E[\gamma - 1]$ or $E[\gamma]$.

In our implementation, we used CPLEX to solve the ILPs. We tested two approaches to reduce the size of the formulations. Initially, we loaded all variables and constraints to the solver, including the equality constraints we just described. Next, we let CPLEX's presolve procedure remove all unnecessary variables and constraints. This option led to satisfactory results for models with less than a million variables and constraints, which are usually associated with instances that have up to 250 labels. However, for larger instances, CPLEX needed a prohibitive amount of memory and most of the computation time was spent presolving, and not actually solving the model. Therefore, we implemented our own lightweight presolve procedure, taking advantage of the fact that we know a *priori* the main simplifications to be performed. We then loaded only the reduced model to CPLEX. Our procedure is able to presolve even the largest instances in less than a second, so we used it in all experiments.

Finally, regarding the choice of set $\mathcal{I}$, observe that, although any independent set in $G_\mathcal{A}$ leads to a correct model, some sets might allow more variables to be removed than others. In particular, given a component $C \in \mathcal{A}$, it is easy to calculate how many variables will be removed if C is included in $\mathcal{I}$. We would like to select the set that leads to the deletion of the maximum possible number of variables. Therefore, we select $\mathcal{I}$ by solving a weighted maximum independent set problem on $G_\mathcal{A}$. We formulated this problem as an ILP and used CPLEX to solve it. Because $G_\mathcal{A}$ is sparse, the time necessary to solve this auxiliary model is approximately four orders of magnitude smaller than the time needed for the actual labeling problem. Thus, it is acceptable to solve this problem optimally.

Experiments. We now evaluate the performance of the reduced ILP formulation, obtained after performing the simplifications just described. Experiments were run on an Intel Xeon CPU X3470, 2.93 GHz, with 8 GB RAM. Integer programs were solved with CPLEX 12.5.1 using traditional search with a single thread.

The code was written in C++ and compiled with gcc 5.4.0. We considered two problem variants: one that only handles soft conflicts, and another that takes both soft and hard conflicts into account. We experimented with models 1R, 2R and 3R (recall that model kR allows at most k active ranges per label).

We used two sets of maps constructed from real-world data and made available by Gemsa et al. [3]. The first set was built from data on cities from six countries (France, Germany, United Kingdom, Italy, Japan and USA) using map scales of 65pixel $\hat{=}$ 20 km, 50 km and 100 km. The second set is based on data from restaurants in four cities (Berlin, London, New York and Paris) and uses map scales of 65pixel $\hat{=}$ 20 m, 50 m, and 100 m. The two benchmarks have a total of 30 input maps, yielding $2 \cdot 3 \cdot 30 = 180$ distinct instances.

As in [3], we attempt to decompose each input map before building the ILP models. We construct a conflict graph for a full rotation of the map, i.e., each label is represented by a vertex and two labels are connected by an edge if they have a conflict at any rotation angle. Clearly, connected components in this graph can be solved separately. In all experiments, we allowed a time limit of one hour per component before interrupting the resolution of the hardest problems.

In Fig. 2 we plot, for each input map, the number of variables (left) and constraints (right) in the original (abscissa) and in the reduced formulation (ordinate). The reductions with and without hard conflicts differ for the same map, so we also distinguish data points by conflict type. In all cases, a notable reduction can be observed. The most impressive case occurred for Paris at scale 100 m, with a reduction of over 99% of variables and constraints for both soft and hard conflicts. On average, we eliminated 86.6% of the variables and 80.2% of the constraints from the original model when using only soft conflicts. With hard conflicts, these values are 86.2% and 88.1% respectively. In general, these percentages increase with the size of the largest connected component.

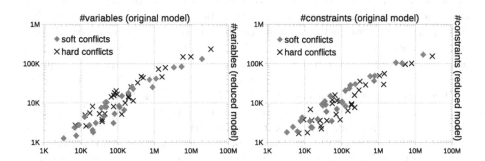

Fig. 2. Variables and constraints in the original and reduced formulations.

Table 1 shows running times for the original model (T_o) and for the reduced formulations obtained by applying Theorem 1 alone (T_1) and Theorems 1 and 2 together ($T_{1,2}$). We also give the ratios T_o/T_1 and $T_o/T_{1,2}$. Additionally, for each instance we report the total number of labels (n), the number of labels in the

Table 1. Experimental results. Times are given in seconds.

Map	n (n_{lcc})	kR	s|h	%var	%con	T_o	T_1	$T_{1,2}$	T_o/T_1	$T_o/T_{1,2}$
USA-100	288 (213)	2R	s	97.2	95.3	3228	44	29	73.9	111.9
France-100	69 (69)	1R	s	93.0	90.1	1614	26	10	61.0	162.4
Berlin-100	2628 (416)	3R	s	98.5	97.7	4231	4	2	1073.7	1940.6
London-50	1620 (175)	1R	s	94.7	92.0	–	90	34	–	–
Paris-100	2141 (952)	3R	s	99.3	99.0	–	40	13	–	–
London-50	1620 (175)	2R	h	94.4	95.4	97	2	1	49.7	122.2
London-100	1457 (371)	2R	h	97.5	98.2	773	3	1	292.9	581.5
USA-100	288 (213)	1R	h	97.2	97.9	540	1	1	568.7	1039.0
Berlin-100	2628 (416)	1R	h	98.5	98.7	1869	3	2	716.0	1099.2
Paris-100	2141 (952)	3R	h	99.3	99.5	–	6	3	–	–

largest connected component (n_{lcc}), and the fraction of variables (%var) and constraints (%con) eliminated from the original formulation. Column s|h indicates whether the soft or the hard conflict model is considered. Some instances could not be solved by the original formulation within the given time limit. With soft conflicts, for instances London-50 1R and Paris-100 3R the solver was halted while still leaving duality gaps of 3.55% and 50.59%, respectively. With hard conflicts, for instance Paris-100 3R there remained a gap of 43.75%.

As expected, the size reductions in the ILPs had a direct effect on execution times and the more compact formulation was faster for all instances. At scales 20 km and 20 m, components are very small and both versions of the ILP performed relatively well. However, at other scales, the reduced model performed significantly better. The best speedup occurred for the Berlin map at scale 100 m with the 3R soft-conflict model, for which the reduced formulation was 1940 times faster. Only two instances could not be solved by the reduced formulation, namely London and Paris, at scale 100 m, with the 1R soft-conflict model. In both cases, there was a gap of 0.03% when the time limit was reached. For the same instances, the original model left gaps of 30.78% and 51.27% (respectively).

The times presented in columns T_1 and $T_{1,2}$ of Table 1 allow us to assess the individual contribution of Theorems 1 and 2 to the final results. Reductions that come from unaffected components clearly produce the most impressive speed-ups. There are two main reasons for this. Firstly, Theorem 1 allows us to remove all variables associated with unaffected components, whereas Theorem 2 is restricted to subsets of affected components that form independent sets in G_A. Moreover, no more than two components of any conflict graph may be affected. Nevertheless, by using Theorem 2, running times can be reduced by up to an additional factor of three.

Concluding Remarks. The reduced model had improvements of up to three orders of magnitude in running times. All but two instances were solved within the time limit, and all but five in less than a minute. We solved components with up to 952 labels and found 12 optimal solutions that were previously unknown.

References

1. Been, K., Daiches, E., Yap, C.: Dynamic map labeling. IEEE Trans. Vis. Comput. Graph. **12**(5), 773–780 (2006)
2. Gemsa, A., Nöllenburg, M., Rutter, I.: Consistent labeling of rotating maps. J. Comput. Geom. **7**(1), 308–331 (2016)
3. Gemsa, A., Nöllenburg, M., Rutter, I.: Evaluation of labeling strategies for rotating maps. ACM J. Exp. Algorithmics **21**(1), 1.4:1–1.4:21 (2016)

Graph Algorithms I

Recognizing Simple-Triangle Graphs by Restricted 2-Chain Subgraph Cover

Asahi Takaoka[✉]

Department of Information Systems Creation, Kanagawa University,
Rokkakubashi 3-27-1, Kanagawa-ku, Kanagawa 221-8686, Japan
takaoka@jindai.jp

Abstract. A simple-triangle graph (also known as a PI graph) is the intersection graph of a family of triangles defined by a point on a horizontal line and an interval on another horizontal line. The recognition problem for simple-triangle graphs was a longstanding open problem, and recently a polynomial-time algorithm has been given (SIAM J. Discrete Math. 29(3):1150–1185, 2015). Along with the approach of this paper, we show a simpler recognition algorithm for simple-triangle graphs. To do this, we provide a polynomial-time algorithm to solve the following problem: Given a bipartite graph G and a set F of edges of G, find a 2-chain subgraph cover of G such that one of two chain subgraphs has no edges in F.

Keywords: Chain cover · Graph sandwich problem · PI graphs · Simple-triangle graphs · Threshold dimension 2 graphs

1 Introduction

Let L_1 and L_2 be two horizontal lines in the plane with L_1 above L_2. A *simple-triangle graph* is the intersection graph of a family of triangles spanned by a point on L_1 and an interval on L_2. That is, a simple undirected graph is called a simple-triangle graph if there is such a triangle for each vertex and two vertices are adjacent if and only if the corresponding triangles have a nonempty intersection. See Fig. 1(a) and (b) for example. Simple-triangle graphs are also known as *PI graphs* [3,5], where *PI* stands for *Point-Interval*. Simple-triangle graphs were introduced in [5] as a generalization of both interval graphs and permutation graphs. Simple-triangle graphs are also known as a proper subclass of trapezoid graphs [5,6], another generalization of interval graphs and permutation graphs.

Recently, the graph isomorphism problem for trapezoid graphs has shown to be isomorphism-complete [23] (that is, polynomial-time equivalent to the problem for general graphs). Since the problem can be solved in linear time for interval graphs [13] and for permutation graphs [4], it has become an interesting question to give the structural characterization of graph classes lying strictly between permutation graphs and trapezoid graphs or between interval graphs and trapezoid graphs [25]. Although a lot of research has been done for interval

© Springer International Publishing AG 2017
S.-H. Poon et al. (Eds.): WALCOM 2017, LNCS 10167, pp. 177–189, 2017.
DOI: 10.1007/978-3-319-53925-6_14

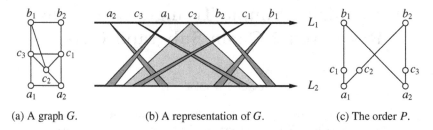

(a) A graph G.   (b) A representation of G.   (c) The order P.

Fig. 1. A simple-triangle graph G, an intersection representation of G, and the Hasse diagram of the linear-interval order P obtained from G.

graphs, for permutation graphs, and for trapezoid graphs (see [22] for example), there are few results for simple-triangle graphs [2,3,5]. It is only recently that a polynomial-time recognition algorithm have been given [17,18].

The recognition algorithm first reduces the recognition problem to the *linear-interval cover* problem. The algorithm then reduces the linear-interval cover problem to *gradually mixed* formulas, a tractable subclass of 3-satisfiability (3SAT). Finally, the algorithm solves the gradually mixed formulas by reducing it to 2-satisfiability (2SAT), which can be solved in linear time (see [1] for example). The total running time of the algorithm is $O(n^2\bar{m})$, where n and $\bar{m}$ are the number of vertices and non-edges of the given graph, respectively.

In this paper, we introduce the *restricted 2-chain subgraph cover* problem as a generalization of the linear-interval cover problem. Then, we show that our problem is directly reducible to 2SAT. This result does not improve the running time, but it can simplify the previous algorithm for the recognition of simple-triangle graphs.

1.1   Linear-Interval Cover

In this section, we briefly describe the linear-interval cover problem and the reduction to it from the recognition problem for simple-triangle graphs. See [18] for the details. We first show that the recognition of simple-triangle graphs is reducible to that of linear-interval orders in $O(n^2)$ time, where n is the number of vertices of the given graph. A *partial order* is a pair $P = (V, \prec)$, where V is a finite set and $\prec$ is a binary relation on V that is irreflexive and transitive. Partial orders are represented by *transitively oriented graphs*, which are directed graphs such that if $u \rightarrow v$ and $v \rightarrow w$, then $u \rightarrow w$ for any three vertices u, v, w of the graphs.

There is a correspondence between partial orders and the intersection graphs of geometric objects spanned between two horizontal lines L_1 and L_2 [9]. A partial order $P = (V, \prec)$ is called a *linear-interval order* [2,3] if for each element $v \in V$, there is a triangle T_v spanned by a point on L_1 and an interval on L_2 such that $u \prec v$ if and only if T_u lies completely to the left of T_v for any two elements $u, v \in V$. See Fig. 1(b) and (c) for example.

For a graph $G = (V, E)$, the graph $\overline{G} = (V, \overline{E})$ is called the *complement* of G, where $uv \in \overline{E}$ if and only if $uv \notin E$ for any pair of vertices $u, v \in V$. We can obtain a linear-interval order from a simple-triangle graph G by giving a transitive orientation to the complement $\overline{G}$ of G. The complement $\overline{G}$ might have some different transitive orientations, but the following theorem states that any transitive orientation of $\overline{G}$ gives a linear-interval order if G is a simple-triangle graph. A property of partial orders is said to be a *comparability invariant* if either all orders obtained from the same graph have that property or none have that property.

Theorem 1 [3]. *Being a linear-interval order is a comparability invariant.*

Many algorithms have been proposed for transitive orientation, including a linear-time one [16]. Since the complement of a graph can be obtained in $O(n^2)$ time, the recognition of simple-triangle graphs is reducible to that of linear-interval orders in $O(n^2)$ time.

We then show that the recognition of linear-interval orders is reducible to the linear-interval cover problem in $O(n^2)$ time, where n is the number of elements of the given partial orders. Let $P = (V, \prec)$ be a partial order with $V = \{v_1, v_2, \ldots, v_n\}$, and let $V' = \{v'_1, v'_2, \ldots, v'_n\}$. The *domination bipartite graph* $C(P) = (V, V', E)$ of P is defined such that $v_i v'_j \in E$ if and only if $v_i \prec v_j$ in P [14]. We also define that $E_0 = \{v_i v'_i \mid v_i \in V\}$. The *bipartite complement* of $C(P)$ is the bipartite graph $\widehat{C(P)} = (V, V', \hat{E})$, where $\hat{E}$ is the set of non-edges between the vertices of V and V', that is, $v_i v'_j \in \hat{E}$ if and only if $v_i v'_j \notin E$ for any vertices $v_i \in V$ and $v'_j \in V'$. By definition, we have $E_0 \subseteq \hat{E}$.

Let $2K_2$ denote a graph consisting of four vertices u_1, u_2, v_1, v_2 with two edges $u_1 v_1, u_2 v_2$. A bipartite graph $G = (U, V, E)$ is called a *chain graph* [26] if it has no $2K_2$ as an induced subgraph. Equivalently, a bipartite graph G is a chain graph if and only if there is a linear ordering $u_1, u_2, \ldots, u_n$ on U (or V) such that $N_G(u_1) \subseteq N_G(u_2) \subseteq \ldots \subseteq N_G(u_n)$, where $N_G(u)$ is the set of vertices adjacent to u in G. A *chain subgraph* of G is a subgraph of G that has no induced $2K_2$. A bipartite graph $G = (U, V, E)$ is said to be *covered* by two chain subgraphs $G_1 = (U, V, E_1)$ and $G_2 = (U, V, E_2)$ if $E = E_1 \cup E_2$ (we note that in general, E_1 and E_2 are not disjoint), and the pair of chain subgraphs (G_1, G_2) is called a *2-chain subgraph cover* of G. For a partial order P, a 2-chain subgraph cover (G_1, G_2) of $\widehat{C(P)}$ is called a *linear-interval cover* if G_1 has no edges in E_0.

Theorem 2 [18]. *A partial order P is linear-interval order if and only if $\widehat{C(P)}$ has a linear-interval cover.*

The *linear-interval cover* problem asks whether $\widehat{C(P)}$ has a linear-interval cover. Since $C(P)$ and $\widehat{C(P)}$ can be obtained in $O(n^2)$ time from a partial order P, the recognition of linear-interval orders is reducible to the linear-interval cover problem in $O(n^2)$ time.

1.2 Restricted 2-Chain Subgraph Cover

As a generalization of the linear-interval cover problem, we consider the following restricted problem for 2-chain subgraph cover.

RESTRICTED 2-CHAIN SUBGRAPH COVER
Instance: A bipartite graph $G = (U, V, E)$ and a set F of edges of G.
Question: Find a 2-chain subgraph cover (G_1, G_2) of G
          such that G_1 has no edges in F.

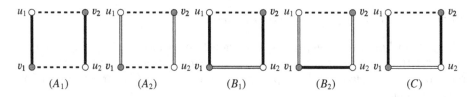

Fig. 2. Forbidden configurations. Solid lines and gray solid lines denote edges in E_r and E_b, respectively. Dashed lines denote non-edges in $\hat{E}$, and double lines denote edges in F.

Notice that G_2 has all the edges in F. Let $\hat{E}$ be the set of edges of the bipartite complement $\hat{G}$ of G. Let $m = |E|$, $\hat{m} = |\hat{E}|$, and $f = |F|$. The following is our main result.

Theorem 3. *The restricted 2-chain subgraph cover problem can be solved in* $O(m\hat{m} + \min\{m^2, \hat{m}(\hat{m} + f)\})$ *time.*

In the rest of this section, we describe the outline of our algorithm. The details are shown in Sect. 2. Two edges e and e' of a bipartite graph $G = (U, V, E)$ is said to be *in conflict in G* if the vertices of e and e' induce a $2K_2$ in G. An edge $e \in E$ is said to be *committed* if there is another edge $e' \in E$ such that e and e' are in conflict in G, and said to be *uncommitted* otherwise. Let E_c be the set of committed edges of G, and let E_u be the set of uncommitted edges of G.

Suppose G has a 2-chain subgraph cover (G_1, G_2) such that G_1 has no edges in F. If two edges $e, e' \in E$ are in conflict in G, then e and e' may not belong to the same chain subgraph. Therefore, each committed edge in E_c belongs to either G_1 or G_2. We refer to the committed edges of G_1 as *red* edges and the committed edges of G_2 as *blue* edges. Let E_r and E_b be the set of red edges and blue edges, respectively, and we call (E_r, E_b) the *bipartition* of E_c. Notice that $F \subseteq E_b \cup E_u$ since E_r has no edges in F. Hence, we assume without explicitly stating it in the rest of this paper that all the committed edges in F are in E_b. We can also see that the bipartition (E_r, E_b) does not have the following *forbidden configurations* (see Fig. 2).

- Configuration (A_1) [resp., (A_2)] consists of four vertices $u_1, u_2 \in U$ and $v_1, v_2 \in V$ with edges $u_1v_1, u_2v_2 \in E_r$ [resp., $u_1v_1, u_2v_2 \in E_b$] and non-edges $u_1v_2, u_2v_1 \in \hat{E}$, that is, u_1v_1 and u_2v_2 are in conflict in G;
- Configuration (B_1) [resp., (B_2)] consists of four vertices $u_1, u_2 \in U$ and $v_1, v_2 \in V$ with edges $u_1v_1, u_2v_2 \in E_r$ [resp., $u_1v_1, u_2v_2 \in E_b$], a non-edge $u_1v_2 \in \hat{E}$, and an edge $u_2v_1 \in E_b$ [resp., $u_2v_1 \in E_r$];
- Configuration (C) consists of four vertices $u_1, u_2 \in U$ and $v_1, v_2 \in V$ with edges $u_1v_1, u_2v_2 \in E_r$, a non-edge $u_1v_2 \in \hat{E}$, and an edge $u_2v_1 \in F$.

Our algorithm construct a bipartition (E_r, E_b) of E_c that does not have some forbidden configurations. A bipartition of E_c is called (A, C)-*free* if it has neither configuration (A_1), (A_2), nor (C). A bipartition of E_c is called (A, B, C)-*free* if it has neither configuration (A_1), (A_2), (B_1), (B_2), nor (C).

Theorem 4. *A bipartite graph G has a 2-chain subgraph cover (G_1, G_2) such that G_1 has no edges in F if and only if E_c has an (A, C)-free bipartition.*

The outline of our algorithm is as follows.

Step 1: Partition the set E_c of committed edges into an (A, C)-free bipartition (E_r, E_b) by solving 2SAT.

Step 2: From the (A, C)-free bipartition (E_r, E_b) of E_c, compute an (A, B, C)-free bipartition (E'_r, E'_b) of E_c by swapping some edges between E_r and E_b.

Step 3: From the (A, B, C)-free bipartition (E'_r, E'_b) of E_c, compute a desired 2-chain subgraph cover of G by adding some uncommitted edges into E'_r and E'_b.

We will show in Sects. 2.1, 2.2, and 2.3 that **Step 1**, **Step 2**, and **Step 3** can be done in $O(\min\{m^2, \hat{m}(\hat{m} + f)\})$ time, $O(m\hat{m})$ time, and linear time, respectively.

1.3 Related Work

A bipartite graph $G = (U, V, E)$ is said to be *covered* by k subgraphs $G_i = (U, V, E_i)$, $1 \le i \le k$, if $E = E_1 \cup E_2 \cup \cdots \cup E_k$. A k-*chain subgraph cover* problem asks whether a given bipartite graph can be covered by k chain subgraphs. The k-chain subgraph cover problem is NP-complete if $k \ge 3$, while it is polynomial-time solvable if $k \le 2$ [26].

The 2-chain subgraph cover problem is closely related to some recognition problems; more precisely, they can be efficiently reduced to the 2-chain subgraph cover problem. They are the recognition problems for threshold dimension 2 graphs on split graphs [11, 19], circular-arc graphs with clique cover number 2 [10, 21], 2-directional orthogonal ray graphs [20, 24], and trapezoid graphs [14]. Other related problems and surveys can be found in Chap. 8 of [15] and Sect. 13.5 of [22].

As far as we know, there are two approaches for the 2-chain subgraph cover problem and the other related problems. One approach is shown in [14, 21], which reduces the 2-chain subgraph cover problem to the recognition of 2-dimensional partial orders. This approach is used in the fastest known algorithm [14] with a running time of $O(n^2)$, where n is the number of vertices of the given graph. Another approach can be found in [10, 11, 19]. They show that a bipartite graph $G = (U, V, E)$ has a 2-chain subgraph cover if and only if the *conflict graph* $G^* = (V^*, E^*)$ of G is bipartite, where $V^* = E$ and two edges e and e' in E are adjacent in G^* if e and e' are in conflict in G. We note that the algorithm in this paper is based on the latter approach.

In Sect. 8.6 of [15], the following problem is considered for recognizing threshold dimension 2 graphs: Given a bipartite graph G and a pair (F_1, F_2) of edge sets, find a 2-chain subgraph cover (G_1, G_2) of G such that G_1 and G_2 have every edge in F_1 and F_2, respectively. We call such a problem the *extension problem* for 2-chain subgraph cover. We emphasize that the extension problem is not a generalization of our restricted 2-chain subgraph cover problem since in the extension problem, G_1

and G_2 are allowed to have all the uncommitted edges of G. As shown in [15], this problem can be solved in polynomial time by reducing it to some variation of the recognition problem for 2-dimensional partial orders. We note that this variation can be stated as the problem of extending a partial orientation of a permutation graph to a 2-dimensional partial order [12].

2 Algorithm

2.1 Partitioning Edges

A *2CNF formula* is a Boolean formula in conjunctive normal form with at most two literals per clause. In this section, we construct a 2CNF formula ϕ such that ϕ is satisfiable if and only if G has an (A, C)-free bipartition of E_c. The construction of ϕ is as follows:

- Assign the Boolean variable x_e to each committed edge $e \in E_c$;
- Add the clause (x_e) for each edge $e \in F \cap E_c$;
- For each pair of two edges e and e' in E_c, add the clauses $(x_e \vee x_{e'})$ and $(\overline{x_e} \vee \overline{x_{e'}})$ to ϕ if e and e' are in conflict in G;
- For each pair of two edges e and e' in E_c, add the clause $(x_e \vee x_{e'})$ to ϕ if the vertices of e and e' induce a path of length 3 whose middle edge is in F (see the forbidden configuration (C) in Fig. 2).

Then, we obtain the bipartition (E_r, E_b) of E_c from a truth assignment τ of the variables as follows:

- $x_e = 0$ in $\tau \iff e \in E_r$ (or $x_e = 1$ in $\tau \iff e \in E_b$).

It is obvious that a truth assignment τ satisfies ϕ if and only if the corresponding bipartition of E_c is (A, C)-free and all the committed edges in F are in E_b.

The 2CNF formula ϕ has at most m Boolean variables. We can also see that ϕ has at most $f + 2 \cdot \min\{m^2, \hat{m}(\hat{m} + f)\}$ clauses since ϕ has at most two clauses for each pair of two edges in E_c or for each pair of a non-edge in $\hat{E}$ and an edge in F. Then, ϕ can be obtained in $O(\min\{m^2, \hat{m}(\hat{m} + f)\})$ time from G and F. Since a satisfying truth assignment of a 2CNF formula can be computed in linear time (see [1] for example), we have the following.

Lemma 1. *An (A, C)-free bipartition of E_c can be computed in $O(\min\{m^2, \hat{m}(\hat{m} + f)\})$ time.*

2.2 Swapping Edges

In this section, we show an $O(m\hat{m})$-time algorithm to transform a given (A, C)-free bipartition (E_r, E_b) of E_c into an (A, B, C)-free bipartition (E'_r, E'_b) of E_c. For a non-edge $uv \in \hat{E}$, we define that

$$H_r = \{u'v' \in E_r \mid uv', u'v \in E_b\};$$
$$H_b = \{u'v' \in E_b \mid uv', u'v \in E_r\};$$
$$H = H_r \cup H_b.$$

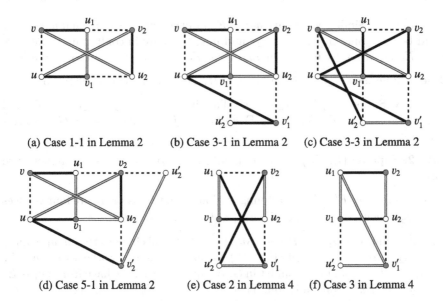

(a) Case 1-1 in Lemma 2   (b) Case 3-1 in Lemma 2   (c) Case 3-3 in Lemma 2

(d) Case 5-1 in Lemma 2   (e) Case 2 in Lemma 4   (f) Case 3 in Lemma 4

Fig. 3. Illustrating the proof of cases in Lemmas 2 and 4. Lines denote the same type of edges as in Fig. 2.

In other words, H_r is the set of red edges of all configurations (B_2) having non-edge uv, and H_b is the set of blue edges of all configurations (B_1) having non-edge uv. Between E_r and E_b, we swap all edges in H to obtain another bipartition (E'_r, E'_b) of E_c, that is, we define that

$$E'_r = (E_r \setminus H_r) \cup H_b;$$
$$E'_b = (E_b \setminus H_b) \cup H_r.$$

Since (E_r, E_b) is (A, C)-free, we have $F \cap H = \emptyset$. Hence, all the committed edges in F remain blue in the new bipartition (E'_r, E'_b). Notice that by swapping the edges, we remove all the configurations (B_1) and (B_2) having non-edge $uv \in \hat{E}$. We claim that the swapping generates no forbidden configurations.

Lemma 2. *No edges in H is an edge of any forbidden configurations of the new bipartition (E'_r, E'_b) of E_c.*

Proof. We assume that the new bipartition (E'_r, E'_b) has some configuration with at least one edge in H, and obtain a contradiction.

Case 1: Suppose (E'_r, E'_b) has a configuration (A_1), that is, there are four vertices u_1, v_1, u_2, v_2 with $u_1v_1, u_2v_2 \in E'_r$ and $u_1v_2, u_2v_1 \in \hat{E}$.

Case 1-1: Suppose $u_1v_1 \in H$ and $u_2v_2 \notin H$. This implies that $u_1v_1 \in E_b$ and $u_2v_2, uv_1, u_1v \in E_r$. See Fig. 3(a). We have $uv_2 \in E$, for otherwise $uv_1 \in E_r$ and $u_2v_2 \in E_r$ would be in conflict in G. Since uv_2 and $u_1v \in E_r$ are in conflict in G, we have $uv_2 \in E_b$. Similarly, we have $u_2v \in E$, for otherwise $u_1v \in E_r$ and $u_2v_2 \in E_r$ would be in conflict in G. Since u_2v and $uv_1 \in E_r$ are in conflict in G, we have $u_2v \in E_b$. However, we have from $uv_2, u_2v \in E_b$ that $u_2v_2 \in H_r$, a contradiction.

Case 1-2: Suppose $u_2v_2 \in H$ and $u_1v_1 \notin H$. This case is symmetric to Case 1-1.

Case 1-3: Suppose $u_1v_1, u_2v_2 \in H$. This implies that $u_1v_1, u_2v_2 \in E_b$ and $uv_1, u_1v, uv_2, u_2v \in E_r$, but $u_1v \in E_r$ and $uv_2 \in E_r$ are in conflict in G, a contradiction.

Case 2: Suppose (E_r', E_b') has a configuration (A_2). This case is symmetric to Case 1.

Case 3: Suppose (E_r', E_b') has a configuration (B_1), that is, there are four vertices u_1, v_1, u_2, v_2 with $u_1v_1, u_2v_2 \in E_r'$, $u_2v_1 \in E_b'$, and $u_1v_2 \in \hat{E}$.

Case 3-1: Suppose $u_1v_1 \in H$ and $u_2v_2, u_2v_1 \notin H$. This implies that $u_1v_1, u_2v_1 \in E_b$ and $u_2v_2, uv_1, u_1v \in E_r$. See Fig. 3(b). We have $u_2v \in E$, for otherwise $u_1v \in E_r$ and $u_2v_2 \in E_r$ would be in conflict in G. If $u_2v \in E_r$, then we have from $uv_1 \in E_r$ that $u_2v_1 \in H_b$, a contradiction. Therefore, $u_2v \in E_b \cup E_u$. Since $u_2v_1 \in E_b$, there is an edge $u_2'v_1' \in E_r$ such that u_2v_1 and $u_2'v_1'$ are in conflict in G, that is, $u_2v_1', u_2'v_1 \in \hat{E}$. We have $uv_1' \in E$, for otherwise $uv_1 \in E_r$ and $u_2'v_1' \in E_r$ would be in conflict in G. Since uv_1' and $u_2v \in E_b \cup E_u$ are in conflict in G, we have $uv_1' \in E_r$ and $u_2v \in E_b$. Then, we have $uv_2 \in E$, for otherwise $uv_1' \in E_r$ and $u_2v_2 \in E_r$ would be in conflict in G. Since uv_2 and $u_1v \in E_r$ are in conflict in G, we have $uv_2 \in E_b$. However, we have from $uv_2, u_2v \in E_b$ that $u_2v_2 \in H_r$, a contradiction.

Case 3-2: Suppose $u_2v_2 \in H$ and $u_1v_1, u_2v_1 \notin H$. This case is symmetric to Case 3-1.

Case 3-3: Suppose $u_2v_1 \in H$ and $u_1v_1, u_2v_2 \notin H$. This implies that $u_1v_1, u_2v_2, u_2v_1 \in E_r$ and $uv_1, u_2v \in E_b$. See Fig. 3(c). Since $u_2v_1 \in E_r$, there is an edge $u_2'v_1' \in E_b$ such that u_2v_1 and $u_2'v_1'$ are in conflict in G, that is, $u_2v_1', u_2'v_1 \in \hat{E}$. We have $uv_1' \in E$, for otherwise $uv_1 \in E_b$ and $u_2'v_1' \in E_b$ would be in conflict in G. Since uv_1' and $u_2v \in E_b$ are in conflict in G, we have $uv_1' \in E_r$. Then, we have $uv_2 \in E$, for otherwise $uv_1' \in E_r$ and $u_2v_2 \in E_r$ would be in conflict in G. If $uv_2 \in E_b$, then we have from $u_2v \in E_b$ that $u_2v_2 \in H_r$, a contradiction. Therefore, $uv_2 \in E_r \cup E_u$. Similarly, we have $u_2'v \in E$, for otherwise $u_2v \in E_b$ and $u_2'v_1' \in E_b$ would be in conflict in G. Since $u_2'v$ and $uv_1 \in E_b$ are in conflict in G, we have $u_2'v \in E_r$. Then, we have $u_1v \in E$, for otherwise $u_1v_1 \in E_r$ and $u_2'v \in E_r$ would be in conflict in G. Since u_1v and $uv_2 \in E_r \cup E_u$ are in conflict in G, we have $u_1v \in E_b$ and $uv_2 \in E_r$. However, we have from $uv_1 \in E_b$ that $u_1v_1 \in H_r$, a contradiction.

Case 3-4: Suppose $u_1v_1, u_2v_2 \in H$ and $u_2v_1 \notin H$. We have a contradiction as Case 1-3.

Case 3-5: Suppose $u_1v_1, u_2v_1 \in H$ and $u_2v_2 \notin H$. This implies that $u_1v_1 \in H_b$ and $u_2v_1 \in H_r$, but it follows that $uv_1 \in E_r$ from $u_1v_1 \in H_b$ and $uv_1 \in E_b$ from $u_2v_1 \in H_r$, a contradiction.

Case 3-6: Suppose $u_2v_2, u_2v_1 \in H$ and $u_1v_1 \notin H$. This case is symmetric to Case 3-5.

Case 3-7: Suppose $u_1v_1, u_2v_2, u_2v_1 \in H$. We have a contradiction as Case 3-5.

Case 4: Suppose (E'_r, E'_b) has a configuration (B_2). This case is symmetric to Case 3.

Case 5: Suppose (E'_r, E'_b) has a configuration (C), that is, there are four vertices u_1, v_1, u_2, v_2 with $u_1v_1, u_2v_2 \in E'_r$, $u_1v_2 \in \hat{E}$, and $u_2v_1 \in F$. Since the bipartition (E_r, E_b) is (A, C)-free, we have $u_2v_1 \notin H$.

Case 5-1: Suppose $u_1v_1 \in H$ and $u_2v_2 \notin H$. This implies that $u_1v_1 \in E_b$ and $u_2v_2, uv_1, u_1v \in E_r$. See Fig. 3(d). We have $uv_2 \in E$, for otherwise the vertices u, v_1, u_2, v_2 would induce a configuration (C). Since uv_2 and $u_1v \in E_r$ are in conflict in G, we have $uv_2 \in E_b$. Similarly, we have $u_2v \in E$, for otherwise $u_1v \in E_r$ and $u_2v_2 \in E_r$ would be in conflict in G. If $u_2v \in E_r$, then the vertices u, v_1, u_2, v would induce a configuration (C). Therefore, $u_2v \in E_b \cup E_u$. Since $u_2v_2 \in E_r$, there is an edge $u'_2v'_2 \in E_b$ such that u_2v_2 and $u'_2v'_2$ are in conflict in G, that is, $u_2v'_2, u'_2v_2 \in \hat{E}$. We have $uv'_2 \in E$, for otherwise $u'_2v'_2 \in E_b$ and $uv_2 \in E_b$ would be in conflict in G. Since uv'_2 and $u_2v \in E_b \cup E_u$ are in conflict in G, we have $uv'_2 \in E_r$ and $u_2v \in E_b$. However, we have from $uv_2, u_2v \in E_b$ that $u_2v_2 \in H_r$, a contradiction.

Case 5-2: Suppose $u_2v_2 \in H$ and $u_1v_1 \notin H$. This case is symmetric to Case 5-1.

Case 5-3: Suppose $u_1v_1, u_2v_2 \in H$. We have a contradiction as Case 1-3.

Since all the cases above lead to contradictions, we conclude that the new bipartition (E'_r, E'_b) has no forbidden configurations with an edge in H. □

It follows from Lemma 2 that continuing in this way for each non-edge in $\hat{E}$, we can obtain an (A, B, C)-free bipartition of E_c. Since the set H can be computed in $O(m)$ time for each non-edge in $\hat{E}$, the overall running time is $O(m\hat{m})$.

Lemma 3. *From a given (A, C)-free bipartition of E_c, an (A, B, C)-free bipartition of E_c can be computed in $O(m\hat{m})$ time.*

2.3 Adding Edges

In this section, we claim that a given (A, B, C)-free bipartition (E_r, E_b) of E_c can be extended in linear time into a 2-chain subgraph cover (G_1, G_2) of G such that G_1 has no edges in F. We first show the following.

Lemma 4. *The subgraph of G induced by $E_b \cup E_u$ is a chain graph.*

Proof. We show that no $2K_2$ is in the subgraphs of G induced by $E_b \cup E_u$.

Case 1: Suppose $u_1v_1, u_2v_2 \in E_b \cup E_u$ and $u_1v_2, u_2v_1 \in \hat{E}$. It is obvious that $u_1v_1, u_2v_2 \notin E_u$, but $u_1v_1, u_2v_2 \in E_b$ implies that the vertices u_1, v_1, u_2, v_2 induce a configuration (A_2), a contradiction.

Case 2: Suppose $u_1v_1, u_2v_2 \in E_b \cup E_u$, $u_1v_2 \in \hat{E}$, and $u_2v_1 \in E \setminus (E_b \cup E_u)$. Since $u_2v_1 \in E \setminus (E_b \cup E_u) = E_r$, there is an edge $u'_2v'_1 \in E_b$ such that u_2v_1 and $u'_2v'_1$

are in conflict in G, that is, $u_2'v_1, u_2v_1' \in \hat{E}$. See Fig. 3(e). We have $u_1v_1' \in E$, for otherwise the vertices u_1, v_1, u_2', v_1' would induce a configuration in Case 1. Since u_1v_1' and $u_2v_2 \in E_b \cup E_u$ are in conflict in G, we have $u_2v_2 \in E_b$. Similarly, we have $u_2'v_2 \in E$, for otherwise the vertices u_2, v_2, u_2', v_1' would induce a configuration in Case 1. Since $u_2'v_2$ and $u_1v_1 \in E_b \cup E_u$ are in conflict in G, we have $u_1v_1 \in E_b$, but then the vertices u_1, v_1, u_2, v_2 induce a configuration (B_2), a contradiction.

Case 3: Suppose $u_1v_1, u_2v_2 \in E_b \cup E_u$ and $u_1v_2, u_2v_1 \in E \setminus (E_b \cup E_u)$. Since $u_2v_1 \in E \setminus (E_b \cup E_u) = E_r$, there is an edge $u_2'v_1' \in E_b$ such that u_2v_1 and $u_2'v_1'$ are in conflict in G, that is, $u_2'v_1, u_2v_1' \in \hat{E}$. See Fig. 3(f). We have $u_1v_1' \notin \hat{E} \cup E_r$, for otherwise the vertices u_1, v_1, u_2', v_1' would induce a configuration in Case 1 or Case 2. However, $u_1v_1' \in E_b \cup E_u$ implies that the vertices u_2, v_2, u_1, v_1' induce a configuration in Case 2, a contradiction.

Since all the cases above lead to contradictions, we conclude that the subgraph of G induced by $E_b \cup E_u$ has no $2K_2$, and it is a chain subgraph of G. □

We next show that E_r can be extended into a chain graph in $G-F$, the subgraph of G obtained by removing all the edges in F. To do this, we consider the following problem: Given a graph H and a set M of edges of H, find a chain subgraph C of H containing all edges in M. This problem is called the *chain graph sandwich problem*, and the chain graph C is called a *chain completion* of M in H. Although the chain graph sandwich problem is NP-complete, it can be solved in linear time if H is a bipartite graph [7]. The chain graph sandwich problem on bipartite graphs is closely related to the threshold graph sandwich problem [8, 19] (see also Sect. 1.5 of [15]), and in the proof of Lemma 5, we will use an argument similar to that used in the literature.

Let $H = (U, V, E)$ be a bipartite graph, let $\hat{E}$ be the set of edges of the bipartite complement $\hat{H}$ of H, and let $k \geq 2$. A set of k distinct vertices $u_0, u_1, \ldots, u_{k-1}$ in U and k distinct vertices $v_0, v_1, \ldots, v_{k-1}$ in V is called an *alternating cycle of M relative to H* if $u_iv_i \in \hat{E}$ and $u_{i+1}v_i \in M$ for any $i, 0 \leq i < k$ (indices are modulo k). Note that an alternating cycle of M with lengh 4 relative to H is exactly a $2K_2$ of M in H.

Lemma 5. *Let M be a set of edges in a bipartite graph H.*

- *The set M of edges has a chain completion in H if and only if there are no alternating cycles of M relative to H.*
- *The chain completion of M in H can be computed in $O(n + m)$ time.*

Proof. The proof is in Appendix. The details of the algorithm are also shown in [7]. □

Then, we show that E_r has a chain completion in $G - F$.

Lemma 6. *There are no alternating cycles of E_r relative to $G - F$.*

Proof. We first prove that there are no alternating cycles of E_r with length 4 relative to $G - F$, that is, no two edges in E_r are in conflict in $G - F$. Since the bipartition (E_r, E_b) does not have a configuration (A_1) or (C), it is enough to show that

(E_r, E_b) has no configuration consisting of four vertices u_1, v_1, u_2, v_2 with edges $u_1v_1, u_2v_2 \in E_r$ and $u_1v_2, u_2v_1 \in F$. Suppose (E_r, E_b) has such a configuration. Since $u_1v_1 \in E_r$, there is an edge $u_1'v_1' \in E_b$ such that u_1v_1 and $u_1'v_1'$ are in conflict in G, that is, $u_1v_1', u_1'v_1 \in \hat{E}$. We have $u_2v_1' \in E$, for otherwise $u_2v_1 \in F$ and $u_1'v_1' \in E_b$ would be in conflict in G (recall that $F \subseteq E_b \cup E_u$). If $u_2v_1' \in E_r$, then the vertices u_1, v_1, u_2, v_1' would induce a configuration (C), a contradiction. Therefore, $u_2v_1' \in E_b \cup E_u$. Similarly, since $u_2v_2 \in E_r$, there is an edge $u_2'v_2' \in E_b$ such that u_2v_2 and $u_2'v_2'$ are in conflict in G, that is, $u_2v_2', u_2'v_2 \in \hat{E}$. We have $u_1v_2' \in E$, for otherwise $u_1v_2 \in F$ and $u_2'v_2' \in E_b$ would be in conflict in G. Since u_1v_2' and $u_2v_1' \in E_b \cup E_u$ are in conflict in G, we have $u_1v_2' \in E_r$ and $u_2v_1' \in E_b$. However, the vertices u_2, v_2, u_1, v_2' induce a configuration (C), a contradiction. Thus, there are no alternating cycles of E_r with length 4 relative to $G - F$.

We now suppose that there are an alternating cycle of E_r with length grater than 4 relative to $G - F$. Let AC be such an alternating cycle with minimal length, and let $u_0, v_0, u_1, v_1, \ldots u_{k-1}, v_{k-1}$ be the consecutive vertices of AC with $u_iv_i \in \hat{E} \cup F$ and $u_{i+1}v_i \in E_r$ for any i, $0 \leq i < k$ (indices are modulo k).

We claim that AC has no edges in F. Suppose $u_1v_1 \in F$. We have $u_2v_0 \in E$, for otherwise the vertices u_2, v_1, u_1, v_0 would induce a configuration (C). If $u_2v_0 \in E_r$, then the vertices $u_0, v_0, u_2, v_2, \ldots u_{k-1}, v_{k-1}$ form a shorter alternating cycle of E_r relative to $G - F$, contradicting the minimality of AC. Therefore, $u_2v_0 \in E_b \cup E_u$. Since $u_1v_0 \in E_r$, there is an edge $u_1'v_0' \in E_b$ such that u_1v_0 and $u_1'v_0'$ are in conflict in G, that is, $u_1'v_0, u_1v_0' \in \hat{E}$. Similarly, since $u_2v_1 \in E_r$, there is an edge $u_2'v_1' \in E_b$ such that u_2v_1 and $u_2'v_1'$ are in conflict in G, that is, $u_2'v_1, u_2v_1' \in \hat{E}$. The edges $u_1'v_0'$ and $u_2'v_1'$ are not the same edge, for otherwise $u_1'v_0' \in E_b$ and $u_2v_0 \in E_b \cup E_u$ would be in conflict in G. Then, the vertices $u_1, v_1, u_2', v_1', u_2, v_0, u_1', v_0'$ form an alternating cycle of $E_b \cup E_u$ relative to G (recall that $F \subseteq E_b \cup E_u$). It follows from Lemma 5 that $E_b \cup E_u$ does not induce a chain graph, contradicting Lemma 4. Thus, AC has no edges in F.

Recall that the length of AC is at least 6, and let $u_0, v_0, u_1, v_1, u_2, v_2$ denote the consecutive vertices of AC. Since AC has no edges in F, we have $u_0v_0, u_1v_1, u_2v_2 \in \hat{E}$ and $u_1v_0, u_2v_1 \in E_r$. We have $u_2v_0 \in E$, for otherwise $u_1v_0 \in E_r$ and $u_2v_1 \in E_r$ would be in conflict in G. If $u_2v_0 \in E_r$, then the vertices $u_0, v_0, u_2, v_2, \ldots u_{k-1}, v_{k-1}$ form a shorter alternating cycle of E_r relative to $G - F$, contradicting the minimality of AC. Therefore, $u_2v_0 \in E_b \cup E_u$. On the other hand, if $u_0v_1 \in \hat{E}$, then the vertices $u_0, v_1, u_2, v_2, \ldots u_{k-1}, v_{k-1}$ form a shorter alternating cycle of E_r relative to $G - F$, contradicting the minimality of AC. Therefore, $u_0v_1 \in E$. Since u_0v_1 and $u_1v_0 \in E_r$ are in conflict in G, we have $u_0v_1 \in E_b$. By similar arguments, we have $u_1v_2 \in E_b$. Then, we have $u_0v_2 \in E$, for otherwise $u_0v_1 \in E_b$ and $u_1v_2 \in E_b$ would be in conflict in G. Since u_0v_2 and $u_2v_0 \in E_b \cup E_u$ are in conflict in G, we have $u_0v_2 \in E_r$ and $u_2v_0 \in E_b$. This implies that the vertices u_1, v_0, u_2, v_1 induce a configurations (B_1), a contradiction.

Thus, we conclude that there are no alternating cycles of E_r relative to $G - F$. $\square$

Now, we have the following from Lemmas 5 and 6.

Lemma 7. *There is a chain completion of E_r in $G - F$, and it can be computed in linear time from E_r.*

Since every edge of G belongs to either E_r or $E_b \cup E_u$, G can be covered by the chain completion of E_r in $G - F$ and the chain subgraph of G induced by $E_b \cup E_u$. Thus, we have the following from Lemmas 4 and 7.

Lemma 8. *From a given (A, B, C)-free bipartition of E_c, a 2-chain subgraph cover (G_1, G_2) of G such that G_1 has no edges in F can be computed in linear time.*

3   Concluding Remarks

This paper provides an $O(m\hat{m} + \min\{m^2, \hat{m}(\hat{m} + f)\})$-time algorithm to solve the restricted 2-chain subgraph cover problem by reducing it to 2SAT. To do this, we show that the problem has a feasible solution if and only if there is an (A, C)-free bipartition of the set of committed edges of the given bipartite graph. This result implies a simpler recognition algorithm for simple-triangle graphs.

We finally note that for simple-triangle graphs, structure characterizations as well as the complexity of the graph isomorphism problem still remain open questions.

Acknowledgments. We are grateful to anonymous referees for careful reading and helpful comments. A part of this work was done while the author was in Tokyo Institute of Technology and supported by JSPS Grant-in-Aid for JSPS Fellows (26·8924).

References

1. Aspvall, B., Plass, M.F., Tarjan, R.E.: A linear-time algorithm for testing the truth of certain quantified boolean formulas. Inf. Process. Lett. **8**(3), 121–123 (1979)
2. Bogart, K.P., Laison, J.D., Ryan, S.P.: Triangle, parallelogram, and trapezoid orders. Order **27**(2), 163–175 (2010)
3. Cerioli, M.R., de Oliveira, F.S., Szwarcfiter, J.L.: Linear-interval dimension and PI orders. Electron. Notes Discrete Math. **30**, 111–116 (2008)
4. Colbourn, C.J.: On testing isomorphism of permutation graphs. Networks **11**(1), 13–21 (1981)
5. Corneil, D.G., Kamula, P.A.: Extensions of permutation and interval graphs. Congr. Numer. **58**, 267–275 (1987)
6. Dagan, I., Golumbic, M.C., Pinter, R.Y.: Trapezoid graphs and their coloring. Discrete Appl. Math. **21**(1), 35–46 (1988)
7. Dantas, S., de Figueiredo, C.M.H., Golumbic, M.C., Klein, S., Maffray, F.: The chain graph sandwich problem. Ann. Oper. Res. **188**(1), 133–139 (2011)
8. Golumbic, M.C., Kaplan, H., Shamir, R.: Graph sandwich problems. J. Algorithms **19**(3), 449–473 (1995)
9. Golumbic, M.C., Rotem, D., Urrutia, J.: Comparability graphs and intersection graphs. Discrete Math. **43**(1), 37–46 (1983)
10. Hell, P., Huang, J.: Two remarks on circular arc graphs. Graphs Comb. **13**(1), 65–72 (1997)

11. Ibaraki, T., Peled, U.: Sufficient conditions for graphs to have threshold number 2. Ann. Discrete Math. **11**, 241–268 (1981)
12. Klavík, P., Kratochvíl, J., Krawczyk, T., Walczak, B.: Extending partial representations of function graphs and permutation graphs. In: Epstein, L., Ferragina, P. (eds.) ESA 2012. LNCS, vol. 7501, pp. 671–682. Springer, Heidelberg (2012). doi:10.1007/978-3-642-33090-2_58
13. Lueker, G.S., Booth, K.S.: A linear time algorithm for deciding interval graph isomorphism. J. ACM **26**(2), 183–195 (1979)
14. Ma, T.H., Spinrad, J.P.: On the 2-chain subgraph cover and related problems. J. Algorithms **17**(2), 251–268 (1994)
15. Mahadev, N., Peled, U.: Threshold Graphs and Related Topics. Ann. Discrete Math., vol. 56. Elsevier Science B.V., Amsterdam (1995)
16. McConnell, R.M., Spinrad, J.P.: Modular decomposition and transitive orientation. Discrete Math. **201**(1–3), 189–241 (1999)
17. Mertzios, G.B.: The recognition of simple-triangle graphs and of linear-interval orders is polynomial. In: Bodlaender, H.L., Italiano, G.F. (eds.) ESA 2013. LNCS, vol. 8125, pp. 719–730. Springer, Heidelberg (2013). doi:10.1007/978-3-642-40450-4_61
18. Mertzios, G.B.: The recognition of simple-triangle graphs and of linear-interval orders is polynomial. SIAM J. Discrete Math. **29**(3), 1150–1185 (2015)
19. Raschle, T., Simon, K.: Recognition of graphs with threshold dimension two. In: STOC 1995, pp. 650–661. ACM, New York (1995)
20. Shrestha, A.M.S., Tayu, S., Ueno, S.: On orthogonal ray graphs. Discrete Appl. Math. **158**(15), 1650–1659 (2010)
21. Spinrad, J.P.: Circular-arc graphs with clique cover number two. J. Comb. Theory Ser. B **44**(3), 300–306 (1988)
22. Spinrad, J.P.: Efficient Graph Representations, Fields Institute Monographs, vol. 19. American Mathematical Society, Providence (2003)
23. Takaoka, A.: Graph isomorphism completeness for trapezoid graphs. IEICE Trans. Fundam. **98–A**(8), 1838–1840 (2015)
24. Takaoka, A., Tayu, S., Ueno, S.: Dominating sets and induced matchings in orthogonal ray graphs. IEICE Trans. Inf. Syst. **96–D**(11), 2327–2332 (2014)
25. Uehara, R.: The graph isomorphism problem on geometric graphs. Discrete Math. Theoret. Comput. Sci. **16**(2), 87–96 (2014)
26. Yannakakis, M.: The complexity of the partial order dimension problem. SIAM J. Algebraic Discrete Methods **3**(3), 351–358 (1982)

Tree-Deletion Pruning in Label-Correcting Algorithms for the Multiobjective Shortest Path Problem

Fritz Bökler[(✉)] and Petra Mutzel

Department of Computer Science, TU Dortmund, Dortmund, Germany
{fritz.boekler,petra.mutzel}@tu-dortmund.de

Abstract. In this paper, we consider algorithms for multi-objective shortest-path (MOSP) optimization in spatial decision making. We re-evaluate the basic strategies for label-correcting algorithms for the MOSP problem, i.e., node and label selection. In contrast to common believe, we show that—when carefully implemented—the node-selection strategy usually beats the label-selection strategy. Moreover, we present a new pruning method which is easy to implement and performs very well on real-world road networks. In this study, we test our hypotheses on artificial MOSP instances from the literature with up to 15 objectives and real-world road networks with up to almost 160,000 nodes. We also evaluate these algorithms on the problem of finding good power grid lines.

1 Introduction

In this paper we are concerned with one of the most famous problems from multiobjective optimization, the multiobjective shortest path (MOSP) problem. We are given a directed graph G, consisting of a finite set of nodes V and a set of directed arcs $A \subseteq V \times V$. We are interested in paths between a given *source node* s and a given *target node* t. Instead of a single-objective cost function, we are given an objective function c which maps each arc to a vector, i.e., $c : A \to \mathbb{Q}^d$ for $d \in \mathbb{N}$. The set of all directed paths from s to t in a given graph is called $\mathcal{P}_{s,t}$ and we assume that the objective function c is extended on these paths in the canonical way, i.e., for $p \in \mathcal{P}_{s,t} : c(p) := \sum_{a \in p} c(a)$, where $a \in p$ denotes the arcs in the path p.

In contrast to the single-objective case, where there exists only one unique optimal value, in the multiobjective case there usually does not exist a path minimizing all objectives at once. Thus, we are concerned with finding the *Pareto-front* of all s-t-paths, i.e., the minimal vectors of the set $c(\mathcal{P}_{s,t})$ with respect to the canonical componentwise partial order on vectors $\leq$. Moreover, we also want to find for each point y of the Pareto-front one representative path $p \in \mathcal{P}_{s,t}$,

F. Bökler—The author has been supported by the Bundesministerium für Wirtschaft und Energie (BMWi) within the research project "Bewertung und Planung von Stromnetzen" (promotional reference 03ET7505) and by DFG GRK 1855 (DOTS).

S.-H. Poon et al. (Eds.): WALCOM 2017, LNCS 10167, pp. 190–203, 2017.
DOI: 10.1007/978-3-319-53925-6_15

such that $c(p) = y$. Each such path is called *Pareto-optimal*. For more information on multiobjective path and tree problems we refer the reader to the latest survey [6].

It is long known that the Pareto-front of a MOSP instance can be of exponential size in the input [11]. Moreover, it has been recently shown in [3], that there does not exist an output-sensitive algorithm for this problem even in the case of $d = 2$ unless $\mathbf{P} = \mathbf{NP}$.

This study is motivated by the problem of finding good power grid lines. In Germany, in the process of the "Energiewende", or turnaround in energy policy, there are many energy producers in the north and many consumers in the south and west, while old producers in the south will be shut down in the next approximately 5 to 10 years. Thus, it is crucial to improve the power grid connectivity between these parts. Albeit, many power grid line projects get delayed by objections by the population and environmental organizations. In the research project "Bewertung und Planung von Stromnetzen" (assessment and planning of power grids) of the TU Dortmund (the spatial planning, mathematics and computer science departments), the power grid provider Amprion[1], and sponsored by the German Ministry for Economics and Energy (BMWi), we evaluate means to find good power grid line alternatives. This lead us to model the problem as a multiobjective optimization problem. The problem of finding among the Pareto-optimal solutions one solution to implement is not covered in this paper, but traditional methods like ELECTRE (cf. e.g. [9]) can be used on the solution set computed by our algorithms. See also [1] for more details on this application.

1.1 Previous Work

The techniques used for solving the MOSP problem are based on labeling algorithms. The majority of the literature is concerned with the biobjective case and we will not be concerned with algorithms for this special case. The latest computational study for more than 2 objectives is from 2009 [13] and compares 27 variants of labeling algorithms on 9,050 artificial instances. These are the instances we also use for our study. In summary, a label-correcting version with a label-selection strategy in a FIFO manner is concluded to be the fastest strategy on the instance classes provided. The authors do not investigate a node-selection strategy with the argument that it is harder to implement and is less efficient (cf. also [14]).

In an older study from 2001 [10], also label-selection and node-selection strategies are compared. The authors conclude that, in general, label-selection methods are faster than node-selection methods. However, the test set is rather small, consisting of only 8 artificial grid-graph instances ranging from 100 to 500 nodes and 2 to 4 objectives and 18 artificial random-graph instances ranging from 500 to 40,000 nodes and densities of 1.5 to 30 with 2 to 4 objectives.

In the work by Delling and Wagner [7], the authors solve a variant of the multiobjective shortest path problem where a preprocessing is allowed and we

---

[1] http://www.amprion.de/.

want to query the Pareto-front of paths between a pair of nodes as fast as possible. The authors use a variant of SHARC to solve this problem. Though being a different problem, this study is the first computational study where an implementation is tested on real-world road networks instead of artificial instances. The instances have sizes of 30,661, 71,619 and 892,392 nodes and 2 to 4 objectives. Though, the largest instance could only be solved using highly correlated objective functions resulting in Pareto-front sizes of only at most 2.5 points on average.

1.2 Our Contribution

We investigate the efficiency of label-correcting methods for the multiobjective shortest path problem. We focus on label-correcting methods, because the literature (cf. [13,15]) and our experience show that label-setting algorithms do not perform well on instances with more than two objectives. Hence, we investigate the question if label-selection or node-selection methods are more promising and test codes based on the recent literature.

We also perform the first computational study of these algorithms not only on artificial instances but also on real-world road networks based on the road network of Western Europe provided by the PTV AG for scientific use and instances emerging from the power grid optimization context. The road network sizes vary from 23,094 to 159,945 nodes and include three objective functions. The artificial instances are taken from the latest study on the MOSP problem [13].

Moreover, we propose a new pruning technique which performs very well on the road networks, achieving large speed-ups. We also show the limits of this technique and reason under which circumstances it works well.

1.3 Organization

In Sect. 2, we describe the basic techniques of labeling algorithms in multiobjective optimization. The tree-deletion pruning is the concern of Sect. 3. Because the implementation of the algorithms is crucial for this computational study, we give details in Sect. 4. The computational study and results then follow in Sect. 5.

2   Multiobjective Labeling Algorithms

In general, a labeling algorithm for enumerating the Pareto-front of paths in a graph maintains a set of labels L_u at each node $u \in V$. A label is a tuple which represents a path from s to a node $u \in V$ and it consists of the cost vector of the path, the associated node u and, for retrieving the actual path, a reference to the predecessor label. The algorithms are initialized by setting each label-set to $\emptyset$ and adding the label $(\mathbf{0}, s, \mathbf{nil})$ to L_s.

These algorithms can be divided into two groups, depending if they select either a label or a node in each iteration. These strategies are called *label* or *node selection strategies*, respectively. When we select a label $\ell = (\mathbf{v}, u, \hat{\ell})$ at a node u, this label is *pushed* along all out-arcs $a = (u, w)$ of u, meaning that a new label is created at the head of the arc with cost $\mathbf{v} + c(a)$, predecessor label ℓ and associated node w. This strategy is due to [16] for $d > 2$. If we follow a node-selection strategy, all labels in L_u will be pushed along the out-arcs of a selected node u. This strategy was first proposed in [4] for the biobjective case and has been generalized by many authors in subsequent works. Nodes or labels that are ready to be selected are called *open*.

After pushing a set of labels, the label sets at the head of each considered arc are *cleaned*, i.e. all *dominated* labels are removed from the modified label sets. We say a label $\ell = (\mathbf{v}, u, \hat{\ell})$ dominates a label $\ell' = (\mathbf{v}', u', \hat{\ell}',)$ if $\mathbf{v} \leq \mathbf{v}'$ and $\mathbf{v} \neq \mathbf{v}'$.

There are many ways in which an open label or node can be selected. A comparative study was conducted by Paixao and Santos [14]. For example, a pure FIFO strategy seems to work best in the aforementioned study. But also other strategies are possible: For example, we can sort the labels by their average cost, i.e., $\sum_{i \in \{1,...,d\}} \mathbf{v}_i / d$ and always select the smallest one. A less expensive variant is due to [2,13], where we decide depending on the top label $\ell = (\mathbf{v}, u, \hat{\ell})$ in a FIFO queue Q where to place a new label $\ell' = (\mathbf{v}', u', \hat{\ell}')$: If $\mathbf{v}'$ is lexicographically smaller than $\mathbf{v}$, then it is placed at the front of Q, otherwise it is placed at the back of Q.

Both available computational studies on labeling algorithms for the multiobjective shortest path problem with more than 2 criteria suggest that label-selection strategies are far superior compared to node-selection strategies [10,13].

2.1   Label-Setting vs. Label-Correcting Algorithms

In the paper by Martins [12], the author describes an algorithm using a label-selection strategy. The algorithm selects the next label by choosing the lexicographically smallest label among all labels in $\mathcal{L} := \bigcup_{u \in V} L_u$. In general, whenever we select a lexicographically smallest label $\ell = (\mathbf{v}, u, \hat{\ell})$ in $\mathcal{L}$, this label represents a nondominated path from s to u. Labeling algorithms having the property that whenever we select a label we know that the represented path is a Pareto-optimal path, are called *label-setting algorithms*. Labeling algorithms which do not have this property, and thus sometimes delete or correct a label, are called *label-correcting algorithms*. On the plus side, in label-setting algorithms, we never select a label which will be deleted in the process of the algorithm. But selecting these labels is not trivial. For example, selecting a lexicographically smallest label requires a priority queue data structure, whereas the simplest label-selection strategy requires only a simple FIFO queue.

There is a recent paper by Erb et al. [8], which suggests that in the biobjective case, a label-setting approach can outperform the label-correcting method they considered. But in many studies for 3 and more objectives, the label-correcting algorithms are superior to the label-setting variants. See for example [10,13,15]. It is not clear why this is the case. One possible reason for this is that the cost of the data structure can not make up for the advantage of not pushing too many unneeded labels. Concluding these considerations, we do not evaluate label-setting algorithms in our study.

3   Tree-Deletion Pruning

The main difficulty which is faced by label-correcting algorithms is that we can push labels which will later be dominated by a new label. To address this issue, let us take a label-selection algorithm into consideration which selects the next label in a FIFO manner. In Fig. 1, we see the situation where a label ℓ at node v, which encodes a path a from s to v, is dominated by a label ℓ', which encodes a different path b from s to v. Based on the label ℓ, we might already have built a tree of descendant labels c. If we proceed with the usual label-correcting algorithm, first the descendant labels of ℓ will be pushed and later be dominated by the descendants of ℓ'. To avoid the unnecessary pushes of descendants of ℓ, we can delete the whole tree c after the label ℓ is deleted. Instead of traversing the queue of open labels to delete the tree labels, we mark labels as being deleted and actually delete them when they are popped from the queue. This pruning method will be called tree-deletion pruning (TD). We can employ this method in label-correcting algorithms using both, the label-selection and node-selection strategies.

Fig. 1. Illustration of the tree-deletion pruning

Table 1. Overview on the artificial test instances

Name	n	d	Name	n	d
CompleteN-medium	10–200	3	CompleteN-large	10–140	6
CompleteK-medium	50	2–15	CompleteK-large	100	2–9
GridN-medium	441–1225	3	GridN-large	25–289	6
GridK-medium	81	2–15	GridK-large	100	2–9
RandomN-medium	500–10000	3	RandomN-large	1000–20000	6
RandomK-medium	2500	2–15	RandomK-large	5000	2–9

This strategy is similar to the parent-checking heuristic for the single-objective shortest path problem in the work by Cherkassky et al. [5]. But it is not so clear how the more lightweight methods can be implemented in the multiobjective setting.

4   Implementation Details

We implemented both label-correcting algorithms, a version of the FIFO label-selection (LS) and FIFO node-selection (NS) algorithms in C++11. The reason for choosing these variants is that the LS-algorithm is the fastest method in the latest comparative study [13]. Both alternatives are also implemented using tree-deletion pruning (LS-TD and NS-TD, respectively). Pseudocode can be found in Algorithms 1 and 2.

Algorithm 1. Abstract version of the LS algorithm

Require: Graph $G = (V, A)$, nodes $s, t \in V$ and objective function $c : A \to \mathbb{Q}^d$
Ensure: List R of pairs (p, y) for all $y \in c(\mathcal{P}_{s,t})$ and some $p \in \mathcal{P}_{s,t}$ such that $c(p) = y$
1: $L_u \leftarrow \emptyset$ for all $u \in V \backslash \{s\}$
2: $\ell \leftarrow (\mathbf{0}, s, \mathbf{nil})$
3: $L_s \leftarrow \{\ell\}$
4: Q.push(ℓ)
5: **while** not Q.empty **do**
6:     $\ell = (\mathbf{v}, u, \hat{\ell}) \leftarrow Q$.pop
7:     **for** $(u, w) \in A$ **do**
8:         Push ℓ along (u, w) and add the new label ℓ' to L_w
9:         Clean L_w ▷ Tree-deletion for every deleted label in L_w
10:        **if** ℓ' is nondominated in L_w **then**
11:            Q.push(ℓ')
12: Reconstruct paths for each label in L_t and output path/vector pairs

---

We use the OGDF[2] for the representation of graphs, because it is a well tested graph library. For cache efficiency, we try to use `std::vector` for collections of data wherever possible.

Node Selection. In the node-selection variants we use our own implementation of a ring buffer based on `std::vector` to implement the queue of open nodes. Also, only those labels of a node are pushed, which have not been pushed before.

Label Selection. In the label-selection variant we use a `std::deque` to implement the queue of open labels. A ring buffer cannot be used easily, because the size of the queue can vary a lot, i.e., it can be as small as 1 or larger than 2^n.

Tree-Deletion Pruning. The successors of each label are stored in an `std::list`. The reason for this is that the successor-lists are constructed empty and most of them remain empty for the whole process of the algorithm. Construction of empty `std::lists` is the cheapest operation among the creation of all other relevant data structures.

---

[2] http://ogdf.net/.

Algorithm 2. Abstract version of the NS algorithm

Require: Graph $G = (V, A)$, nodes $s, t \in V$ and objective function $c : A \to \mathbb{Q}^d$
Ensure: List R of pairs (p, y) for all $y \in c(\mathcal{P}_{s,t})$ and some $p \in \mathcal{P}_{s,t}$ such that $c(p) = y$
1: $L_u = \emptyset$ for all $u \in V \backslash \{s\}$
2: $L_s = \{(\mathbf{0}, s, \mathbf{nil})\}$
3: Q.push(s)
4: **while** not Q.empty **do**
5:    $u \leftarrow Q$.pop
6:    **for** $(u, w) \in A$ **do**
7:       **for** not yet pushed label ℓ in L_u **do**
8:          Push ℓ along (u, w) and add the new label to L_w
9:          Clean L_w ▷ Tree-deletion for every deleted label in L_w
10:          **if** at least one new label survived the cleaning process and $w \notin Q$ **then**
11:             Q.push(w)
12: Reconstruct paths for each label in L_t and output path/vector pairs

---

Cleaning Step. While there is a considerable literature on finding the subset of minimal vectors in a set of vectors, it is not clear which method to use in practice. In the node-selection algorithm, we could use the fact that we try to find the nondominated vectors of two sets of nondominated vectors, which has been successfully exploited for the biobjective problem in [4]. To make the comparison between label-selection and node-selection strategies more focused on the strategies themselves, we use a simple pairwise comparison between all pushed labels and all labels at the head of the arc. A more sophisticated method would make the node-selection strategy only faster. Moreover, the studies on multiobjective labeling algorithms also employ this method.

5   Computational Study

The experiments were performed on an Intel Core i7-3770, 3.4 GHz and 16 GB of memory running Ubuntu Linux 12.04. We compiled the code using LLVM 3.4 with compiler flag -O3.

In the computational study, we are concerned with the following questions: (1) Is label selection really faster than node selection? (2) In which circumstances is tree-deletion pruning useful? We will answer these questions in the following subsections.

5.1   Instances

We used three sets of instances. First, we took the instances of [13] and tested our implementations on them. The aforementioned study is the latest study which tested implementations of labelling algorithms for MOSP. The properties of these instances can be seen in Table 1.

The random graph instances are based on a Hamiltonian cycle where arcs are randomly added to the graph. In the complete graph instances, arcs are added between each pair of nodes in both directions. The grid graphs are all square. The arc costs were choosen uniformly random in $[1, 1000] \cap \mathbb{N}$. For each problem type, there are 50 randomly drawn instances. In summary they make up a set of 9,050 test problems. The instances are available online, see [13] for details.

The second set of instances is similar to the instances from [7]. They are based on the road network of Western Europe provided by PTV AG for scientific use. We conducted our experiments on the road network of the Czech Republic (CZE, 23,094 nodes, 53,412 edges), Luxembourg (LUX, 30,661 nodes, 71,619 edges), Ireland (IRL, 32,868 nodes, 71,655 edges) and Portugal (PRT, 159,945 nodes, 372,129 edges, only for the tree-deletion experiments). As metrics we used similar metrics as in [7]: travel distance, cost based on fuel consumption and travel time. The median Pareto-front sizes are 13.0, 30.5, 12.5 and 133.5, respectively and thus comparable or larger than those in [7]. For each of the instances we drew 50 pairs of source and target nodes uniformly at random.

We also evaluated TD on a set of instances coming from the finding of Pareto-optimal power grid lines. The test area is a small 6 km times 3 km square near Münster, North Rhine-Westfalia in Germany. The considered criteria for this first study are bird preservation areas (BCA), landscape preservation areas (LPA), existing power lines (EPL), freeways (FW), settlement areas (SA), and length (L). We superimpose an undirected grid graph (i.e., there exists an arc in both directions) with eight neighbors per inner node and raster pitch of 100 m on the area we investigate. The resulting graph has 1859 nodes and 12674 edges. For each criterion, we add one objective. In the case of the BCA, LPA, and SA criteria, an edge gets the cost of the length of traversing this area. The settlement area gets a buffer of 400 m which follows German and North Rhine-Westfalian law. For the EPL and FW criteria, the case is the other way around: Instead of trying to avoid these linear infrastructures, it is preferable to build the new line in a 200 m buffer around these. To represent this, we add an area to avoid everywhere outside the 200 m buffer around the infrastructure. Thus, an edge gets the cost of the length of not traversing the buffer. To get different instances, we try out different combinations of criteria.

5.2   Running Times Node Selection vs. Label Selection

In Fig. 2 we see a selection of the results on the `RandomN-large`, `RandomK-large`, `GridK-large` as well as the real-world road network instance sets.

To evaluate the results we decided to show box plots, because the deviation of the running times is very large and it is easier to recognize trends. The box plots give a direct overview on the quartiles (box dimension), median (horizontal line inside the box), deviation from the mean (*whiskers*: lines above and below the box) and outliers (points above and below the whiskers).

We see that the node-selection strategy performs better on all these instances than the label-selection strategy. The node-selection strategy is up to a factor of 3 faster on both test sets. This is also true for the other large and medium sized instances where the maximum factors range from 1.4 to 3.17. Detailed box plots can be found in the appendix.

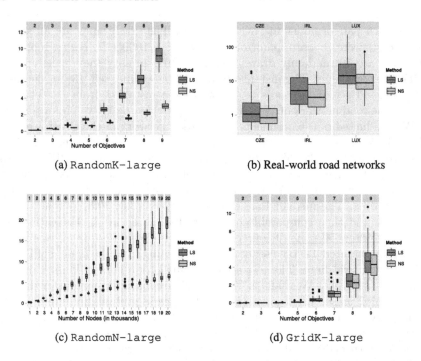

(a) RandomK-large

(b) **Real-world road networks**

(c) RandomN-large

(d) GridK-large

Fig. 2. Comparison of the running times (in seconds) of the label-selection (LS) and node-selection (NS) strategies

Also on the real-world road networks the results are positive. The node-selection strategy is up to factor of 3.16 faster than the label-selection strategy.

A partial explanation can be attributed to memory management: In the node-selection strategy a consecutive chunk of memory which contains the values of the labels pushed along an arc can be accessed in one cache access. While in the label-selection strategy only one label is picked in each iteration, producing potentially many cache misses when the next label—potentially at a very different location—is accessed.

5.3 Tree-Deletion Pruning

The results concerning TD are ambiguous. First, to see how well TD might work, we performed a set of experiments showing how many labels are touched by the label-correcting algorithms which could have been deleted when using tree-deletion pruning. That is, instead of deleting a label, we mark them deleted and every time we push a label which is marked as being deleted, we count this event. In Fig. 3, we see the results of this experiment on the CompleteN-large, CompleteK-large, GridK-large and RandomK-large test sets.

We observe that on these instances, especially the node-selection strategy tends to produce larger obsolete trees than the label-selection strategy. We also observe this behavior on the complete-graph instances of medium and large size.

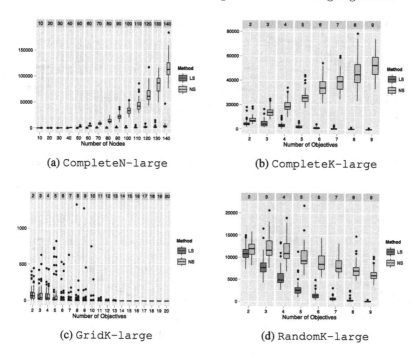

(a) CompleteN-large

(b) CompleteK-large

(c) GridK-large

(d) RandomK-large

Fig. 3. Measuring how many nodes have been touched which could have been deleted by tree-deletion pruning in the label-selection (LS) and node-selection (NS) strategies

The situation is different on the grid-graph instances, where both algorithms have a similar tendency to produce obsolete trees.

Another observation is that when increasing the number of objectives, the number of obsolete trees which could have been deleted decreases in the grid and random graph instances (see Fig. 3c and d). This happens because when looking at instances with a large number of objectives and totally random objective values, most labels remain nondominated in the cleaning step. The complete-graph instances are an exception here. Since there are so many path between a pair of nodes, a large number of labels is pushed and very many combinations of cost exist and so still many labels are dominated.

Hence, what we expect is that on instances where a large number of labels is dominated in the cleaning step the tree-deletion pruning is very useful. The results of the comparison of the running times can be seen in Fig. 4. It can be seen on the real-world road networks that the tree-deletion pruning works very well and we can achieve a speed-up of up to 3.5 in comparison to the pure node-selection strategy.

On the instances from the power grid line optimization set, we first see that we cannot optimize all objectives at once because of memory constraints which was 16 GB in these experiments. The combinations with at least three objectives which could be solved are shown in Table 2. We observe that the TD strategy

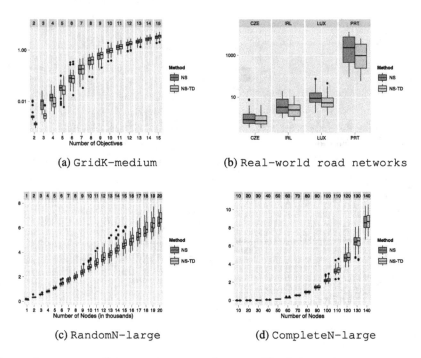

(a) GridK-medium

(b) Real-world road networks

(c) RandomN-large

(d) CompleteN-large

Fig. 4. Comparison of the running times (in seconds) of the node-selection strategy with (NS-TD) and without (NS) tree-deletion pruning

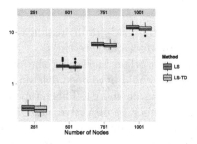

Fig. 5. Comparison of the running times (in seconds) of the node-selection strategy with and without TD on the correlated random networks

again improves the performance of the node-selection strategy on these instances with a speed-up of up to 2.54.

On the artificial benchmark instances however, the results are not so clear. TD works well on medium sized grid graphs with a small number of objectives and also on complete graphs of any size. On the other instances of the artificial benchmark set, TD performs slightly worse than the pure node-selection strategy, especially on the large instances.

This behavior can be explained by the large number of labels which are dominated in the road network instances. The size of the Pareto-fronts are small

Table 2. Running times and Pareto-front sizes on the power grid optimization instances. In the penultimate instance, the NS implementation exceeded the memory limit of 16 GB.

Objectives	Running time (s)		$|c(\mathcal{P}_{s,t})|$
	NS	NS-TD	
BPA, LPA, L	4.04	**3.72**	377
BPA, EPL, L	13.41	**6.37**	1170
BPA, FW, L	1.13	**1.10**	280
BPA, SA, L	172.23	**76.80**	693
LPA, EPL, L	0.30	**0.24**	109
LPA, FW, L	6.59	**4.95**	634
LPA, SA, L	11.35	**7.49**	640
EPL, FW, L	1.65	**1.13**	428
EPL, SA, L	134.21	**52.82**	3767
FW, SA, L	40.98	**20.97**	1301
BPA, LPA, EPL, L	1549.16	**941.01**	11902
BPA, EPL, FW, L	—	**508.33**	13449
LPA, EPL, FW, L	336.38	**222.30**	8106

compared to the instances of the artificial test set. So, we hypothesize that the pruning strategy is especially useful if many labels are dominated in the cleaning step and large obsolete trees can be deleted in this process. TD seems also to work better on denser networks.

To test this hypothesis, we created a new set of random graphs. To make the graphs more dense than in the previous instances, we drew 0.3 times the possible number of edges and to match the small sized Pareto-fronts of the instances from [7] with a high correlation of the objective functions, i.e., we used a Gauss copula distribution with a fixed correlation of 0.7. If the hypothesis is false, TD should run slower than the pure node-selection strategy on these instances.

But the results in Fig. 5 show that TD beats the pure node-selection strategy on these graphs. TD achieves a speed-up of up to 1.14. Using a wilcoxon signed rank test we can also see that the hypothesis that TD is slower than the pure node-selection strategy on these instances can be refused with a p-value of less than 10^{-3}.

6   Conclusion

To conclude, we showed in this paper that node-selection strategies in labeling algorithms for the MOSP problem can be advantageous, especially if implemented carefully. So node-selection strategies should not be neglected as an option for certain instance classes.

We showed that the tree-deletion pruning we introduced in the multiobjective labeling algorithms in this paper, works well on the real-world road networks and the instances coming from the power grid optimization context. On the artificial instances it does not seem to work too well, which we can explain by the very low densities and unrealistic objective functions used in these instances. To show that TD works well when having larger correlations as in the real-world road networks and higher densities, we also created instances which had the potential to refute this hypothesis. But the hypothesis passed the test.

References

1. Bachmann, D., Bökler, F., Dokter, M., Kopec, J., Schwarze, B., Weichert, F.: Transparente Identifizierung und Bewertung von Höchstspannungstrassen mittels mehrkriterieller Optimierung. Energiewirtschaftliche Tagesfragen (2015)
2. Bertsekas, D.P., Guerriero, F., Musmanno, R.: Parallel asynchronous label-correcting methods for shortest paths. J. Optim. Theory Appl. **88**(2), 297–320 (1996)
3. Bökler, F., Ehrgott, M., Morris, C., Mutzel, P.: Output-sensitive complexity of multiobjective combinatorial optimization. J. Multi-Criteria Decis. Anal. (2017, to appear)
4. Brumbaugh-Smith, J., Shier, D.: An empirical investigation of some bicriterion shortest path algorithms. Eur. J. Oper. Res. **43**(2), 216–224 (1989)
5. Cherkassky, B.V., Goldberg, A.V., Radzik, T.: Shortest paths algorithms: theory and experimental evaluation. Math. Program. **73**(2), 129–174 (1996)
6. Clímaco, J.C.N., Pascoal, M.M.B.: Multicriteria path and tree problems: discussion on exact algorithms and applications. Int. Trans. Oper. Res. **19**(1–2), 63–98 (2012)
7. Delling, D., Wagner, D.: Pareto paths with SHARC. In: Vahrenhold, J. (ed.) SEA 2009. LNCS, vol. 5526, pp. 125–136. Springer, Heidelberg (2009). doi:10.1007/978-3-642-02011-7_13
8. Erb, S., Kobitzsch, M., Sanders, P.: Parallel bi-objective shortest paths using weight-balanced B-trees with bulk updates. In: Gudmundsson, J., Katajainen, J. (eds.) SEA 2014. LNCS, vol. 8504, pp. 111–122. Springer, Heidelberg (2014). doi:10.1007/978-3-319-07959-2_10
9. Greco, S., Ehrgott, M., Figueira, J.R. (eds.): Multiple Criteria Decision Analysis: State of the Art Surveys. International Series in Operations Research & Management Science, 2nd edn. Springer, Berlin (2016)
10. Guerriero, F., Musmanno, R.: Label correcting methods to solve multicriteria shortest path problems. J. Optim. Theory Appl. **111**(3), 589–613 (2001)
11. Hansen, P.: Bicriterion path problems. In: Fandel, G., Gal, T. (eds.) Multiple Criteria Decision Making Theory and Application. LNEMS, vol. 177, pp. 109–127. Springer, Heidelberg (1979). doi:10.1007/978-3-642-48782-8_9
12. Martins, E.Q.V.: On a multicriteria shortest path problem. Eur. J. Oper. Res. **16**, 236–245 (1984)
13. Paixao, J., Santos, J.: Labelling methods for the general case of the multiobjective shortest path problem: a computational study. In: Madureira, A., Reis, C., Marques, V. (eds.) Computational Intelligence and Decision Making. ISCA, vol. 61, pp. 489–502. Springer, Dordrecht (2013). doi:10.1007/978-94-007-4722-7_46

14. Paixao, J.M., Santos, J.L.: Labelling methods for the general case of the multi-objective shortest path problem - a computational study. Technical report 07-42, Universidade de Coimbra (2007)
15. Raith, A., Ehrgott, M.: A comparison of solution strategies for biobjective shortest path problems. Comput. OR **36**(4), 1299–1331 (2009)
16. Tung, C.T., Chew, K.L.: A multicriteria Pareto-optimal path algorithm. Eur. J. Oper. Res. **62**, 203–209 (1992)

Minimum Weight Connectivity Augmentation for Planar Straight-Line Graphs

Hugo A. Akitaya[1], Rajasekhar Inkulu[2], Torrie L. Nichols[3],
Diane L. Souvaine[1], Csaba D. Tóth[1,3](✉), and Charles R. Winston[1]

[1] Tufts University, Medford, MA, USA
{hugo.alves_akitaya,diane.souvaine,charles.winston}@tufts.edu
[2] Indian Institute of Technology Guwahati, Guwahati, India
rinkulu@iitg.ernet.in
[3] California State University Northridge, Los Angeles, CA, USA
torrie.nichols.643@my.csun.edu, csaba.toth@csun.edu

Abstract. We consider edge insertion and deletion operations that increase the connectivity of a given planar straight-line graph (PSLG), while minimizing the total edge length of the output. We show that every connected PSLG $G = (V, E)$ in general position can be augmented to a 2-connected PSLG $(V, E \cup E^+)$ by adding new edges of total Euclidean length $\|E^+\| \leq 2\|E\|$, and this bound is the best possible. An optimal edge set E^+ can be computed in $O(|V|^4)$ time; however the problem becomes NP-hard when G is disconnected. Further, there is a sequence of edge insertions and deletions that transforms a connected PSLG $G = (V, E)$ into a plane cycle $G' = (V, E')$ such that $\|E'\| \leq 2\|\mathrm{MST}(V)\|$, and the graph remains connected with edge length below $\|E\| + \|\mathrm{MST}(V)\|$ at all stages. These bounds are the best possible.

1 Introduction

Connectivity augmentation is a classical problem in combinatorial optimization. A graph is τ-connected (resp., τ-edge-connected) if the subgraph obtained by deleting any $\tau - 1$ vertices (resp., edges) is connected. Given a graph $G = (V, E)$ and a parameter $\tau \in \mathbb{N}$, add a set of new edges E^+ of minimum cardinality or weight such that the augmented graph $G' = (V, E \cup E^+)$ is τ-connected (resp., τ-edge-connected). Efficient algorithms are known for both connectivity and edge-connectivity augmentation over abstract graphs and constant τ [9,20]. In this paper we consider weighted connectivity augmentation for planar straight-line graphs (PSLGs). The vertices are points in Euclidean plane, the edges are noncrossing line segments between the corresponding vertices, and the weight of an edge is its Euclidean length.

The edge- and node-connectivity of a planar graph is at most 5 by Euler's theorem. Further, not every PSLG can be augmented to a 3-connected

Research on this paper was supported in part by the NSF awards CCF-1422311 and CCF-1423615. Akitaya was supported by the Science Without Borders program.

S.-H. Poon et al. (Eds.): WALCOM 2017, LNCS 10167, pp. 204–216, 2017.
DOI: 10.1007/978-3-319-53925-6_16

(resp., 4-edge-connected) PSLG; see [13] for feasibility conditions. Finding the minimum *number* of edges to augment a given PSLG to τ-connectivity or τ-edge-connectivity is NP-complete [18] for $2 \leq \tau \leq 5$; the reduction requires the input graph G to be disconnected (the NP-hardness claim for connected input [18, Corollary 2] turned out to be flawed). Worst case bounds are known for the most important cases: Every PSLG G with n vertices can be augmented to 2-edge-connectivity with at most $\lfloor (4n-4)/3 \rfloor$ edges [2]; and at most $\lfloor (2n-2)/3 \rfloor$ new edges if G is already connected [19]. At most $b-1$ suffice for 2-connectivity, where b is the number of 2-blocks in G [1]. All these bounds are the best possible.

Our Results. We show that every connected PSLG $G = (V, E)$ with n vertices in general position can be augmented to a 2-connected PSLG $G' = (V, E \cup E^+)$ by adding new edges of total Euclidean length $\|E^+\| \leq 2\|E\|$ (Sect. 3). A set E^+ that minimizes $\|E^+\|$ can be computed in $O(|V|^4)$ time (Sect. 4); however the problem becomes NP-hard when G is disconnected (Sect. 5). Further, there is a sequence of edge insertions and deletions that transforms a connected PSLG G into a plane cycle $G' = (V, E')$ such that the graph at all stages remains connected with the sum of Euclidean edge lengths below $\|E\| + \|\mathrm{MST}(V)\|$, and, at termination, $\|E'\| \leq 2\|\mathrm{MST}(V)\|$, where $\mathrm{MST}(V)$ denotes the Euclidean minimum spanning tree of V; these bounds are the best possible (Sect. 6). Our proof is constructive, and yields a polynomial-time algorithm for computing such sequences. Complete proofs are provided in the full paper [3].

Related Previous Work. Biconnectivity augmentation over *planar* graphs (where no embedding of G is given) is also NP-complete [14]. Over planar graphs *with fixed combinatorial embedding*, biconnectivity augmentation remains NP-hard for disconnected graphs; but there is a near-linear time algorithm when the input is already connected [11]. Frederickson and Ja'Ja' [10] show that the *weighted* augmentation of a tree to be 2-connected or 2-edge-connected (without planarity constraint) is NP-complete even if the weights are restricted to $\{1, 2\}$. The problem is APX-hard, and breaking the approximation ratio of 2 is a major open problem [15]. In the *geometric* setting, we show (in Sect. 4) that the minimum-weight augmentation of a planar straight-line tree to a 2-connected (2-edge-connected) PSLGs can be computed efficiently.

The length of the edges in planar connectivity augmentation was studied only recently in the context of wireless networks. Given a PLSG $G = (V, E)$ where the vertices induce a 2-edge-connected unit disk graph, Dobrev et al. [7] compute a 2-edge-connected PSLG by adding edges of length at most 2. Kranakis et al. [16] studied the combined problem of adding the minimum *number* of edges of *bounded length*: A 2-edge-connected augmentation is possible such that $|E^+|$ is at most the number of bridges in G and $\max_{e' \in E^+} \|e'\| \leq 3 \max_{e \in E} \|e\|$. However, finding the minimum number of new edges of bounded length is NP-hard.

2   Preliminaries

Let $G = (V, E)$ be a planar straight-line graph (PSLG), where V is a set of n points in the plane, no three of which are collinear, and E is a set of open line segments between pairs of points in V. The length of an edge uv, denoted $\|uv\|$, is the Euclidean distance between u and v; the total length of the edges is $\|E\| = \sum_{e \in E} \|e\|$. Denote by F the set of faces of G. The *faces* of G are the connected components of the complement of all vertices and edges of G, that is, $\mathbb{R}^2 \setminus (V \cup \bigcup_{e \in E} e)$. If G is connected, then every bounded face is simply connected.

A *walk* is an alternating sequence of points (vertices) and line segments (edges) whose consecutive elements are incident, and hence it is uniquely described by the sequence of its vertices $w = (p_0, \ldots, p_t)$. A walk is *closed* if $p_0 = p_t$. A walk is called a *path* if no vertex appears more than once. Every face $f \in F$ defines a closed walk $(p_0, \ldots, p_t)$ that contain all edges on the boundary of F, called *facial walk*, where every edge $p_{i-1}p_i$ is incident to the face f, and consecutive edges in the path, $p_{i-1}p_i$ and p_ip_{i+1}, are also consecutive in the counterclockwise rotation of all edges incident to p_i (see Fig. 2(a)). Note that every edge $e \in E$ occurs twice in the facial walks of the faces of G: A *cut vertex* (resp., *bridge*) occurs twice in some facial walk.

A walk $p = (p_0, \ldots, p_t)$ is *convex* if $0 < \angle p_{i-1}p_ip_{i+1} < \pi$ for $i = 1, \ldots, t-1$ (for closed walks, $i = 1, \ldots, t$), where $\angle p_{i-1}p_ip_{i+1}$ is the measure of the minimum counterclockwise angle that rotates the ray $\overrightarrow{p_ip_{i-1}}$ into $\overrightarrow{p_ip_{i+1}}$. A vertex of G is called *convex* if $0 < \angle p_{i-1}p_ip_{i+1} < \pi$ for each pair $p_{i-1}p_i$ and p_ip_{i+1} of consecutive edges in the counterclockwise rotation of edges incident to p_i; and *reflex* otherwise. A convex walk $p = (p_0, \ldots, p_t)$ is *safe* if not all of its vertices are collinear, and the vertices lie on the boundary of the convex hull of $\{p_0, \ldots, p_t\}$ with the possible exception of the first or last vertex (Fig. 1(a) and (b)).

For a walk p, let $\gamma_p : [0, 1] \to \mathbb{R}^2$ be a piecewise linear arc from p_0 to p_t that traverses the edges of p in the given order. A *homotopy* between two walks, $p = (p_0, \ldots, p_t)$ and $q = (q_0, \ldots, q_{t'})$, is a continuous function $h : [0, 1] \times [0, 1] \to \mathbb{R}^2$ such that $h(0, \cdot) = \gamma_p(\cdot)$, $h(1, \cdot) = \gamma_q(\cdot)$, $h(\cdot, 0) = p_0 = q_0$, $h(\cdot, 1) = p_t = q_{t'}$, $h(a, b) \in \bigcup_{f \in F} f$ for all $(a, b) \in (0, 1) \times (0, 1)$. Intuitively, the walk p can be continuously deformed into q in the face f. The walks p and q are *homotopic* if such a homotopy exists. Let $p = (p_0, \ldots, p_t)$ be a walk contained in the boundary walk of a face $f \in F$. The shortest walk from p_0 to p_t homotopic to p, denoted geod(p), is called the *geodesic* between p_0 and p_t (dotted paths in Fig. 1).

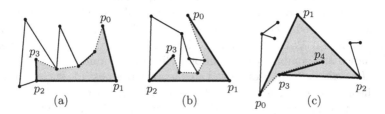

Fig. 1. (a) A convex path $p = (p_0, \ldots, p_3)$ with vertices in convex position. (b) A convex path $p = (p_0, \ldots, p_3)$, where $p_3 \in \text{conv}(p_0, p_1, p_2)$. (c) A convex path $p = (p_0, \ldots, p_4)$ with four edges, where geod(p) contains edge p_3p_4.

Given a walk $p = (p_0, \ldots, p_t)$ and polygonal environment $\bigcup_{f \in F} f$ with n vertices, the geodesic can be computed in $O(n \log n)$ time [4]. It is known [8,12] that all interior vertices of $\mathrm{geod}(p)$ are reflex vertices of G; and if p is a convex chain, then so is $\mathrm{geod}(p)$ (it may be a straight-line segment). Geodesics play a crucial role in our worst-case bounds, since $\|\mathrm{geod}(p)\| \leq \|p\|$ by definition, and the edges of $\mathrm{geod}(p)$ do not cross any existing edges of G. We show the following.

Lemma 1. *Let $G = (V, E)$ be a PSLG, and let $p = (p_0, \ldots, p_t)$ be a safe convex walk contained in some facial walk of G. Then $\mathrm{geod}(p)$ is a simple path that does not contain any vertices of p except for p_0 and p_t at its endpoints.*

Proof. Let H be the convex hull of $\{p_0, \ldots, p_t\}$ and ∂H its boundary. By assumption, the vertices of p lie on ∂H with the possible exception of p_0 and p_t, which may lie in the interior of H. By construction, $\mathrm{geod}(p)$ lies in $\mathrm{int}(H)$ with the possible exception of its endpoints p_0 and p_t. Hence, $\mathrm{geod}(p)$ does not contain any interior vertex of p. $\qquad\square$

Lemma 2. *Let $G = (V, E)$ be a PSLG, and let $p = (p_0, \ldots, p_t)$ be a convex walk contained in some facial walk of G. If $\mathrm{geod}(p_1, \ldots, p_t)$ is a simple path that contains none of the vertices $p_0, p_2, \ldots, p_{t-1}$, then $\mathrm{geod}(p)$ is also a simple path that does not contain any of the vertices $p_1, \ldots, p_{t-1}$.*

Proof. By Lemma 1, $C = (p_1, \ldots, p_t) \cup \mathrm{geod}(p_1, \ldots, p_t)$ is a simple cycle. Let $p_1 r$ be the first edge of $\mathrm{geod}(p_1, \ldots, p_t)$. Since $\mathrm{geod}(p_1, \ldots, p_t)$ is homotopic to $(p_1, \ldots, p_t)$, the edge $p_1 r$ lies in the angular domain $\angle p_0 p_1 p_2$, and $r \notin \{p_0, p_2\}$. Consequently, the edge $p_0 p_1$ lies in the exterior of the cycle C. Note that $\mathrm{geod}(p) = \mathrm{geod}(p_0, p_1, r) \cup \mathrm{geod}(r, p_1, \ldots, p_t)$.

Since $\mathrm{geod}(p_1, \ldots, p_t)$ is a convex chain, the interior of triangle $\Delta(p_0, p_1, r)$ lies in the exterior of C. On the other hand, $\mathrm{geod}(p_0, p_1, r)$ lies in $\Delta(p_0, p_1, r)$, and so it is disjoint from the vertices $p_2, \ldots, p_t$. We conclude that $\mathrm{geod}(p)$ is a simple path that does not contain any of the vertices $p_1, \ldots, p_{t-1}$. $\qquad\square$

3   Bounds on the Sum of Edge Lengths

Let G be a PSLG with no three collinear vertices. Denote by $\mathcal{P} = \mathcal{P}(G)$ the set of maximal convex walks contained in the facial walks of G (note that $\mathcal{P}$ can be computed with a graph traversal in $O(|E|)$ time). Partition $\mathcal{P}$ into three subsets: $\mathcal{P}_0$ contains the convex walks that consist of a single edge; $\mathcal{P}_1$ contains the closed convex walks $(p_0, \ldots, p_t)$, i.e., $p_0 = p_t$; and $\mathcal{P}_2$ contains all open convex walks of two or more edges. We define a *dual graph* D where the nodes correspond to the convex walks in $\mathcal{P}_1 \cup \mathcal{P}_2$, and two nodes are adjacent if and only if the corresponding convex chains share an edge in E.

Lemma 3. *Let $G = (V, E)$ be a connected PSLG with $|V| \geq 3$. Then*

(a) *every edge in E is part of a convex chain in $\mathcal{P}_1 \cup \mathcal{P}_2$,*
(b) *the dual graph D is connected.*

The proof of Lemma 3 is provided in [3]. When we modify a given PSLG with edge insertion operations, we prove the following.

Theorem 1. *Let $G = (V, E)$ be a connected PSLG with $|V| \geq 3$ and no three collinear vertices. Then G can be augmented to a 2-edge-connected PSLG $G' = (V, E \cup E^+)$ such that $\|E^+\| \leq 2\|E\|$, and this bound is the best possible.*

Proof. We prove the upper bound constructively, augmenting a connected PSLG $G = (V, E)$ incrementally into a 2-edge-connected PSLG $G' = (V, E \cup E^+)$. Then decompose every convex walk in $\mathcal{P}$ of two or more edges into edge-disjoint convex walks of two or three edges. Let $\mathcal{C}$ be the set of all resulting convex paths of two or three edges (that is, we discard convex walks that consist of a singe edge and convex cycles of 3 edges). See Fig. 2(b) for an illustration.

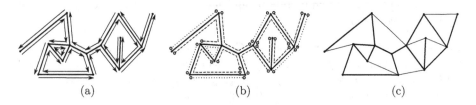

(a)   (b)   (c)

Fig. 2. (a) A PSLG $G = (V, E)$ with its boundary walks. (b) Dashed lines indicate convex paths of two or three edges in $\mathcal{C}$; dotted lines indicate maximal convex paths of a single edge, and triangles. (c) The 2-edge-connected PSLG $G' = (V, E \cup E^+)$ produced by our algorithm.

For each convex path $p \in \mathcal{C}$, augment G with the edges of geod(p) (refer to Fig. 2(c)), and denote by G' the resulting graph. Note that G' is 2-edge-connected since every edge in E is part of a cycle by Lemma 3(a): Each cycle is either a triangle in G, or a cycle $p \cup$ geod(p) for some $p \in \mathcal{C}$; and every edge in E^+ is part of a cycle by construction. By definition, edges in E^+ do not cross any edge in E. The cycles $p \cup$ geod(p) are interior disjoint, hence the defined geodesics cannot cross each other. Therefore G' is a PSLG.

Next, we derive an upper bound for $\|E^+\|$. Every $e \in E$ appears twice in the boundary walks of the faces of G, and so it appears in at most two convex paths in $\mathcal{C}$. By definition, $\|\text{geod}(p)\| \leq \|p\|$ for every $p \in \mathcal{C}$. Overall, we have

$$\|E^+\| = \sum_{p \in \mathcal{C}} \|\text{geod}(p)\| \leq \sum_{p \in \mathcal{C}} \|p\| \leq \sum_{p \in \mathcal{P}} \|p\| = 2\|E\|.$$

We now show a matching lower bound. For every $\varepsilon > 0$, let G_ε be defined on four vertices $p_1 = (0, 0)$, $p_2 = (0, \varepsilon)$, $p_3 = (1, 0)$, and $p_4 = (1, \varepsilon)$ with edge set $E = \{p_1 p_2, p_2 p_3, p_3 p_4\}$; refer to Fig. 3. Since p_1 and p_4 are leaves in G, and $p_1 p_4$ would cross $p_2 p_3$, both $p_1 p_3$ and $p_2 p_4$ have to be added. We have $\lim_{\varepsilon \to 0} \|E^+\|/\|E\| = 2$, and the ratio $\|E^+\|/\|E\| \leq 2$ is the best possible.   □

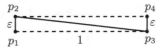

Fig. 3. A PSLG $G = (V, E)$ with three solid edges, where the augmentation to 2-edge-connectivity requires the addition of the dashed edges.

We strengthen Theorem 1 to vertex-connectivity.

Theorem 2. *Let $G = (V, E)$ be a connected PSLG with $|V| \geq 3$ and no three collinear vertices. Then G can be augmented to a 2-connected PSLG $G = (V, E \cup E^+)$ such that $\|E^+\| \leq 2\|E\|$, and this bound is the best possible.*

Proof. We prove the upper bound constructively. Consider the convex walk in $\mathcal{P}_1 \cup \mathcal{P}_2$, defined above. We augment G into $G' = (V, E \cup E^+)$ such that the vertex set of each convex walk in $\mathcal{P}_1 \cup \mathcal{P}_2$ induces a 2-connected subgraph in G'; and every new edge in E^+ is part of one of these subgraphs. Note that this implies that G' is 2-connected: By Lemma 3(a), every vertex is part of a 2-connected subgraph; if two 2-connected subgraphs share two vertices, then their union is 2-connected. By Lemma 3(b) the union of the subgraphs induced by the convex chains is 2-connected. In the remainder of the proof, we consider a single convex walks $p \in \mathcal{P}_1 \cup \mathcal{P}_2$.

Case 1: $p = (p_0, \ldots, p_t) \in \mathcal{P}_1$. If the vertices $p_0, \ldots, p_{t-1}$ are distinct, then p is a cycle, and all vertices of the walk are part of a 2-connected component. Otherwise $p_1 = p_{t-1}$. In this case, $(p_1, \ldots, p_{t-1})$ forms a convex polygon, whose interior contains $p_0 = p_t$ but no other vertices. Consequently, $t \geq 4$, and the only cut vertex along the walk is p_1. Add the edge $p_0 p_2$, where $\|p_0 p_2\| \leq \|p_0 p_1\| + \|p_1 p_2\| \leq \|p\|$ by the triangle inequality (Fig. 4(a)). As a result, the vertices of p induce a 2-connected subgraph.

Case 2: $p = (p_0, \ldots, p_t) \in \mathcal{P}_2$. We decompose p into edge-disjoint walks recursively as follows. If geod(p) does not contain any interior vertex of p, then we are done. Otherwise, let H be the convex hull of $\{p_0, \ldots, p_t\}$ and let $p' = (p_i, \ldots, p_j)$ be the subchain along ∂H. By Lemma 1, the set of interior vertices of geod(p') does not contain any vertex of p'. Starting from p', successively append the edges of p proceeding p_i or following p_j while the path p' maintains the property that p' contains no interior vertices of geod(p'). Then recurse on any prefix or suffix path in $p \setminus p'$. We obtain a decomposition of p into subpaths p' such that $p' \cup \text{geod}(p')$ is a simple cycle (Fig. 4(b)). By Lemma 2, any two such consecutive paths share two vertices. Consequently, the union of the cycles $p' \cup \text{geod}(p')$ is a 2-connected graph.

Analogously to the proof of Theorem 1, we have $\|E^+\| \leq 2\|E\|$, and the same lower bound construction shows that this bound is the best possible.   □

4   Algorithms for Connectivity Augmentation from 1 to 2

Let F be a face of a PSLG G, and let $W_F = (p_0, \ldots, p_n)$, $p_0 = p_n$ be the (closed) facial walk of F. We define the graph $G_F = (V_F, E_F)$, where V_F is

Fig. 4. Examples for (a) case 1 and (b) case 2.

the set of vertices in W_F and $E_F = \{\{p_i, p_{i+1}\}$ where $i \in \{1, \ldots, n\}\}$. Given a walk $W_{s,t} = (p_s, \ldots, p_t)$ contained in W_F, a vertex v_c is a *cut vertex relative to* $W_{s,t}$ if it appears more than once in $W_{s,t}$. For every pair $1 \leq i < j \leq n$, we introduce a weight function, corresponding to the feasibility of an edge $p_i p_j$. Let $f(i, j) = \|p_i p_j\|$ if the line segment $p_i p_j$ does not cross any edge in E_F, lies in the face F and in the wedges $\angle p_{i-1} p_i p_{i+1}$ and $\angle p_{j-1} p_j p_{j+1}$; and let $f(i, j) = \infty$ otherwise. Note that two feasible edges, $\{p_i, p_j\}$ and $\{p_{i'}, p_{j'}\}$, do not cross if their indices do not interleave (i.e., if $1 \leq i < j \leq i' < j' \leq n$).

We present a dynamic programming algorithm A that finds a set E^+ of edges of minimum total weight such that $(V_F, E_F \cup E^+)$ is a 2-connected PSLG. We call an optimal solution E_{OPT}.

Remark 1. Every edge in E_{OPT} lies in the face F. Indeed, suppose $(V_F, E_F \cup E^+)$ is a 2-connected PSLG and $\{p_i, p_j\} \in E^+$ is outside F. Then p_i and p_j are part of a simple cycle formed by some edges of G_F, and $(V_F, E_F \cup (E^+ \setminus \{p_i, p_j\}))$ is also 2-connected, showing that E^+ is not optimal.

By Remark 1, each face of the input can be treated independently. If we insert an edge in face F, then it decomposes F into two faces F_1 and F_2 that can be considered independently. However, defining subproblems in terms of faces might generate an exponential number of subproblems. Instead we define our subproblems in term of continuous intervals of the facial walk of F.

We characterize an optimal solution E_{OPT} for G_F in terms of local properties of the subproblems $W_{s,t}$. Let p_c be a cut vertex with respect to a walk $W_{s,t}$. The vertices between two consecutive occurrences of p_c in $W_{s,t}$ are called *descendants* of p_c. A *non-descendant* of p_c in $W_{i,j}$ is a vertex in W_F that is neither a descendant in $W_{i,j}$ nor equal to p_c. The *k-th group* of descendants is the set of vertices between the k-th and $(k + 1)$-st occurrence of p_c. A set E' of feasible edges *satisfies* a group if there is a cycle in the graph $G'_F = (V_F, E_F \cup E')$ that contains a descendant in that group and a non-descendant of p_c; and E' satisfies a cut vertex p_c if it satisfies all of its groups. If $(V_F, E_F \cup E^+)$ is a 2-connected PSLG, then E^+ satisfies all cut vertices in $W_{1,n}$. Indeed, suppose E^+ does not satisfy a vertex p_c, then the deletion of p_c would disconnect one of its groups of descendants from the rest of G_F, hence p_c would be a cut vertex in $(V_F, E_F \cup E^+)$.

Let $C[s, t]$, $s \leq t$, be the minimum weight of an edge set E' that satisfies all groups of all cut vertices relative to $W_{s,t}$ and such that $\{p_i, p_j\} \in E'$, $i, j \in$

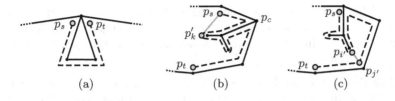

Fig. 5. Examples of cases (ii), (iii) and (iv) of the dynamic programming.

$\{s, \ldots, t\}$ and $(V_F, E_F \cup E')$ is a PSLG. Algorithm A uses the following recursive relation to compute subproblems $C[s, t]$ and returns $C[1, n]$.

(i) If $W_{s,t}$ does not contain any cut vertex relative to $W_{s,t}$, then $C[s, t] = 0$.
(ii) If $p_s = p_t$ and $s \neq t$, then $C[s, t] = \infty$.
(iii) If p_s is not a cut vertex relative to $W_{s,t}$, then $C[s, t] = \min\{C[s + 1, t], \min_{k \in \{s+2, \ldots, t-1\}}\{C[s, k] + C[k, t] + f(s, k)\}\}$.
(iv) If p_s is a cut vertex relative to $W_{s,t}$, let $X = \{$descendants of p_s in $W_{s,t}\} \times \{$non-descendants of p_s in $W_{s,t}\}$. Set $C[s, t] = \min_{(p_i, p_j) \in X}\{C[s, i] + C[i, j] + C[j, t] + f(i, j)\}$.

Correctness. We show that A correctly computes $C[s, t]$ and that $C[1, n]$ corresponds to an optimal edge set E'. The base case (i) is trivial.

In case (ii), $p_s = p_t$ is a cut vertex of W_F, and all other vertices in $W[s, t]$ are descendants of p_s. Therefore, there is no edge incident to a non-descendant of p_s in $W_{s,t}$, and p_s cannot be satisfied in $W_{s,t}$ (Fig. 5(a)).

In case (iii), since p_s does not have descendants, the set E' corresponding to $C[s, t]$ either has no edge incident to p_s or it has an edge between p_s and some descendant p_k of a cut vertex p_c in $W_{s,t}$. In the former case, we have $C[s, t] = C[s + 1, t]$ since E' satisfies all cut vertices relative to $W_{s+1,t}$. In the latter, the edge $\{p_s, p_k\}$ creates a cycle satisfying the group of descendants containing v_k for every cut vertex in $W_{c,k}$. It divides F into two faces: F_1 (resp., F_2) whose facial walk contains $W_{s,k}$ (resp., $W_{k,t}$). Every group that still needs to be satisfied is either in F_1 or F_2. If E_1 (resp., E_2) is the set with minimum weight between vertices in $W_{s,k}$ (resp., $W_{k,t}$) that satisfies the groups of descendants in F_1 (resp., F_2), then $E' = E_1 \cup E_2 \cup \{\{s, k\}\}$.

In case (iv), all non-descendants of p_s in $W_{s,t}$ are in the set $\{p_{k+1}, \ldots, p_t\}$, where p_k is the last occurrence of p_s in $W_{s,t}$. E' must contain at least one edge in X in order to satisfy p_s. Let $\{p_i, p_j\} \in X$ be an edge that minimizes i, breaking ties by maximizing j. Then, every edge $\{p_{i'}, p_{j'}\} \in E'$ such that $(p_{i'}, p_{j'}) \in X$ is between vertices in $W_{i,j}$. In particular, there exist no edge in E' between a vertex in $W_{s,i}$ and $W_{j,t}$. We can partition E' into E_1, E_2, and E_3 such that they each contain only edges between vertices in $W_{s,i'}$, $W_{i',j'}$, and $W_{j',t}$ respectively, each of them being the minimum-weight set that satisfies the groups of descendants in their respective subproblems. The edges in E_1, E_2, E_3 cannot cross since they correspond to edge-disjoint walks in W_F. Thus, we have $C[s, t] = C[s, i'] + C[i', j'] + C[j', t] + f(i', j')$.

Let E' be the edge set corresponding to $C[1,n]$. Since E' satisfies all cut vertices in $W_{1,n}$, which corresponds to the cut vertices of G_F, the graph $(V_F, E_F \cup E')$ is 2-connected. Since E' is a minimum-weight edge set that satisfies all groups of $W_{1,n}$ by definition, we have $\|E'\| = \|E_{OPT}\|$.

Running Time. The feasible edge weights $f(i,j)$ can be precomputed in $O(n^2)$ time [6,17] for all $i, j \in \{1, \dots, n\}$. There are $O(n^2)$ subproblems, and each can be computed in $O(n^2)$ time since it depends on $O(n^2)$ smaller subproblems. Hence algorithm A takes $O(n^4)$ time.

Theorem 3. *For a connected PSLG* (V, E)*, an edge set* E^+ *of minimum weight such that* $(V, E \cup E^+)$ *is 2-connected can be computed in* $O(|V|^4)$ *time.*

Proof. We identify the faces of (V, E) and run algorithm A in all faces. By Remark 1, the union of the optimal solutions for each face is the optimal solution for G. Since the size of the union of all facial walks is $O(|V|)$, algorithm A takes $O(|V|^4)$ time. $\qquad\square$

Theorem 4. *For a connected PSLG* (V, E)*, an edge set* E^+ *of minimum weight such that* $(V, E \cup E^+)$ *is 2-edge-connected can be computed in* $O(|V|^4)$ *time.*

Theorems 3 and 4 extend to any nonnegative weight function. In particular, a minimum cardinality edge set E^+ can also be computed in $O(|V|^4)$ time.

5 Hardness of Connectivity Augmentation from 0 to 2

Theorem 5. *Given a (disconnected) PSLG* $G = (V, E)$ *and a positive integer* k*, deciding whether there exists an edge set* E^+ *such that* $\|E^+\| \le k$ *and* $(V, E \cup E^+)$ *is a 2-edge-connected PSLG is NP-hard.*

Proof. We reduce from PLANAR-MONOTONE-3SAT, which is NP-complete [5]. An instance of this problem is given by a plane bipartite graph between n *variables* and m *clauses* such that the variables are embedded on the x-axis, no edge crosses the x-axis and every clause has degree 2 or 3. A clause is called *positive* if it is embedded on the upper half-plane and *negative* otherwise. PLANAR-MONOTONE-3SAT asks if there is an assignment from {true, false} to variables such that each positive (resp., negative) clause is adjacent to at least one true (resp., false) variable. Given such an instance we build a PSLG $G = (V, E)$ as follows. We divide the reduction into variable, wire and clause gadgets.

Variable Gadget. A gadget of a variable adjacent to two positive and one negative clause is shown in Fig. 6(a). The gray boxes and small disks represent 2-connected components and leaves respectively. Each leaf requires at least one edge for the augmentation and the closest node from each of them is 1 unit apart. A pair of leaves can possibly share an edge and the i-th gadget for each $i \in \{1, \dots, n\}$ has an even number of leaves t_i, hence the gadget requires at least $t_i/2$ length. There are exactly two possible ways to achieve this bound and they encode the true/false value of the variable. Figure 6(a) can be generalized to

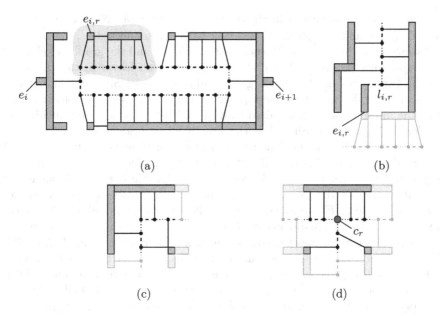

Fig. 6. Dotted and dashed lines represent **true** and **false** assignments respectively. Shaded rectangles represent cycles. (a) Variable gadget. (b) Connection between variable and wire gadgets. (c) Turn in a wire gadget. (d) Clause gadget.

other variables by repeating, omitting and changing the length and number of leaves of the component highlighted in the figure.

Wire Gadget. The wire gadget that connects the i-th variable to the r-th clause, denoted as the (i, r)-wire, share the edge $e_{i,r}$ with the i-th variable gadget and its first leaf is called $l_{i,r}$ (Fig. 6(b)). Turns, if needed, can be done as in Fig. 6(c).

Clause Gadget. The r-th clause gadget contains a special leaf called c_r shown in red in Fig. 6(d) that is located at an odd distance L_1 from at most three leaves, say $l_{i,r}, l_{j,r}, l_{k,r}$. If the clause is incident to two variables, use the turn Fig. 6(c) as a clause gadget, naming the upper left leaf c_r. If the r-th clause is positive, we place all $l_{i,r}$ at an even distance from each other, which makes the position of c_r always realizable. Use reflections through the x-axis for negative clauses.

Let $t_{i,r}$ be the number of leaves of the (i, r)-wire. We set $k = \sum_{i=1}^{n} t_i/2 + \sum_{\text{All } (i,r)\text{-wires}} (\lceil t_{i,r}/2 \rceil + 1)$.

Assume that the PLANAR-MONOTONE-3SAT instance has a satisfying assignment. We build E^+ as follows. For each variable assigned **false** (resp., **true**) add the edges shown as dashed (resp., dotted) lines of Fig. 6(a) to E^+. For each (i, r)-wire add to E^+ all dashed lines shown in Fig. 6(b), (c) and the closest dashed line in Fig. 6(d) from the (i, r)-wire if the i-th variable is assigned **false** and the r-th clause is positive; or if the i-th variable is assigned **true** and the r-th clause is negative. Add the dotted lines otherwise. We obtain E^+ such that $\|E^+\| = k$ and the graph $(V, E \cup E^+)$ is 2-edge-connected.

Assume that there exists E^+ such that $\|E^+\| = k$ and the graph $(V, E \cup E^+)$ is 2-edge-connected. We call (i, r)-*leaves* the set of all leaves in the (i, r)-wire and the two leaves in the clause gadget adjacent to it. Since the number of (i, r)-leaves is odd and the closest point from any such leaf is at least 1 unit away, the minimum length required to have at least an edge in E^+ incident to each leaf is $\lceil t_{i,r}/2 \rceil + 1$. Since k is the sum of all such lower bounds, then: (i) the subset of E^+ that is incident to a (i, r)-leaf must have exactly $\lceil t_{i,r}/2 \rceil + 1$ unit length edges; and (ii) the subset of E^+ that is incident to the i-th variable gadget must be either the set of dashed or dotted lines in Fig. 6(a). Assume that E^+ contains the dotted lines in Fig. 6(a) in the i-th variable gadget. Then, for a positive clause r, E^+ must contain an edge between $l_{i,r}$ and a point in the maximal 2-connected component of G that contains $e_{i,r}$, or else the bridge in G that connects such 2-connected component will remain a bridge in $(V, E \cup E^+)$. Therefore, all other (i, r)-leaves must be matched in order to satisfy (i). Since $(V, E \cup E^+)$ is 2-edge-connected, c_r is connected to some other leaf. Then, if it is connected to a (i, r)-leaf, the i-th variable gadget uses the dotted edges (`true`). By applying the symmetric argument for negative clauses, all such clauses must be incident to a variable gadget using dashed edges (`false`). Then we have also a satisfying assignment for the Planar-Monotone-3SAT instance.   □

Corollary 1. *Given a (disconnected) PSLG $G = (V, E)$ and $k > 0$, finding a set E^+ such that $\|E^+\| \leq k$ and $(V, E \cup E^+)$ is a 2-connected PSLG is NP-hard.*

Proof. The same reduction used in the proof of Theorem 5 also works for 2-connectivity. Notice that the length required by leaves is the same and any E^+, $\|E^+\| \leq k$, that augments G to 2-edge-connected also achieves 2-connectivity.   □

6   Dynamic Plane Graphs

Theorem 6. *Let $G = (V, E)$ be a connected PSLG with $|V| \geq 3$ and no three collinear vertices. Then there exists a sequence of edge insertion and deletion operations that transforms G into a plane cycle $G' = (V, E')$ such that $\|E'\| \leq 2\|MST(V)\|$ and that every intermediate graph is a connected planar straight-line graph of weight at most $\|E\| + \|MST(V)\|$. These bounds are the best possible.*

Proof sketch. The upper bound is proven with an algorithm while the lower bound is based on the graph shown in Fig. 3. The algorithm is divided into five phases. Phase 1 applies successive edge deletions until a spanning tree of V is obtained. Phase 2 applies a sequence of edge replacements (an insertion followed by a deletion) to transform the graph into a spanning tree contained in the Delaunay triangulation of V. Phase 3 applies edge replacements within the Delaunay triangulation to transform the graph into the MST of V. Phase 4 first created a cycle by inserting an edge of the convex hull, and then grows a weakly simple cycle, by replacing an edge of the component with a path that visits one additional vertex, until it spans V. Finally, phase 5 transforms a weakly simple cycle into a simple cycle by connecting the two neighbors of a repeated vertex with a geodesic. See the full paper [3] for a complete proof.

References

1. Abellanas, M., García, A., Hurtado, F., Tejel, J., Urrutia, J.: Augmenting the connectivity of geometric graphs. Comput. Geom. **40**, 220–230 (2008)
2. Akitaya, H.A., Castello, J., Lahoda, Y., Rounds, A., Tóth, C.D.: Augmenting planar straight line graphs to 2-edge-connectivity. In: Di Giacomo, E., Lubiw, A. (eds.) GD 2015. LNCS, vol. 9411, pp. 563–564. Springer, Heidelberg (2015). doi:10. 1007/978-3-319-27261-0_52
3. Akitaya, H.A., Inkulu, T., Nichols, T.L., Souvaine, D.L., Tóth, C.D., Winston, C.R.: Minimum weight connectivity augmentation for planar straight-line graphs. Preprint, arXiv:1612.04780 (2016)
4. Bespamyatnikh, B.: Computing homotopic shortest paths in the plane. J. Algorithms **49**, 284–303 (2003)
5. de Berg, M., Khosravi, A.: Optimal binary space partitions for segments in the plane. Int. J. Comput. Geom. Appl. **22**, 187–205 (2012)
6. Chen, D.Z., Wang, H.: A new algorithm for computing visibility graphs of polygonal obstacles in the plane. J. Comput. Geom. **6**, 316–345 (2015)
7. Dobrev, S., Kranakis, E., Krizanc, D., Morales-Ponce, O., Stacho, L.: Approximating the edge length of 2-edge connected planar geometric graphs on a set of points. In: Fernández-Baca, D. (ed.) LATIN 2012. LNCS, vol. 7256, pp. 255–266. Springer, Heidelberg (2012). doi:10.1007/978-3-642-29344-3_22
8. Efrat, A., Kobourov, S.G., Lubiw, A.: Computing homotopic shortest paths efficiently. Comput. Geom. **35**, 162–172 (2006)
9. Frank, A.: Connections in Combinatorial Optimization. Oxford Lecture Series in Mathematics and Its Applications, vol. 12. Oxford University Press, Oxford (2011)
10. Frederickson, G.N., Ja'Ja', J.: Approximation algorithms for several graph augmentation problems. SIAM J. Comput. **10**, 270–283 (1981)
11. Gutwenger, C., Mutzel, P., Zey, B.: Planar biconnectivity augmentation with fixed embedding. In: Fiala, J., Kratochvíl, J., Miller, M. (eds.) IWOCA 2009. LNCS, vol. 5874, pp. 289–300. Springer, Heidelberg (2009). doi:10.1007/978-3-642-10217-2_29
12. Hershberger, J., Snoeyink, J.: Computing minimum length paths of a given homotopy class. Comput. Geom. **4**, 63–98 (1994)
13. Hurtado, F., Tóth, C.D.: Plane geometric graph augmentation: a generic perspective. In: Pach, J. (ed.) Thirty Essays on Geometric Graph Theory, pp. 327–354. Springer, Heidelberg (2013)
14. Kant, G., Bodlaender, H.L.: Planar graph augmentation problems. In: Dehne, F., Sack, J.-R., Santoro, N. (eds.) WADS 1991. LNCS, vol. 519, pp. 286–298. Springer, Heidelberg (1991). doi:10.1007/BFb0028270
15. Kortsarz, G., Nutov, Z.: A simplified 1.5-approximation algorithm for augmenting edge-connectivity of a graph from 1 to 2. ACM Trans. Algorithms **12**, Article no. 23 (2016)
16. Kranakis, E., Krizanc, D., Ponce, O.M., Stacho, L.: Bounded length 2-edge augmentation of geometric planar graphs. Discret. Math. Algorithms Appl. **4**(3) (2012). doi:10.1142/S179383091250036X

17. Overmars, M.H., Welzl, E.: New methods for computing visibility graphs. In: Proceedings of 14th Symposium on Computational Geometry, pp. 164–171. ACM Press, New York (1988)
18. Rutter, I., Wolff, A.: Augmenting the connectivity of planar and geometric graphs. J. Graph Algorithms Appl. **16**, 599–628 (2012)
19. Tóth, D.C.: Connectivity augmentation in planar straight line graphs. Eur. J. Combin. **33**, 408–425 (2012)
20. Végh, L.A.: Augmenting undirected node-connectivity by one. SIAM J. Discret. Math. **25**, 695–718 (2011)

A Fast Deterministic Detection of Small Pattern Graphs in Graphs Without Large Cliques

Mirosław Kowaluk[1] and Andrzej Lingas[2]([⊠])

[1] Institute of Informatics, University of Warsaw, Warsaw, Poland
kowaluk@mimuw.edu.pl
[2] Department of Computer Science, Lund University,
Lund, Sweden
Andrzej.Lingas@cs.lth.se

Abstract. We show that for several pattern graphs on four vertices (e.g., C_4), their induced copies in host graphs with n vertices and no clique on $k + 1$ vertices can be deterministically detected in time $\tilde{O}(n^\omega k^\mu + n^2 k^2)$, where $\tilde{O}(f)$ stands for $O(f(\log f)^c)$ for some constant c, and $\mu \approx 0.46530$. The aforementioned pattern graphs have a pair of non-adjacent vertices whose neighborhoods are equal. By considering dual graphs, in the same asymptotic time, we can also detect four vertex pattern graphs, that have an adjacent pair of vertices with the same neighbors among the remaining vertices (e.g., K_4), in host graphs with n vertices and no independent set on $k + 1$ vertices.

By using the concept of Ramsey numbers, we can extend our method for induced subgraph isomorphism to include larger pattern graphs having a set of independent vertices with the same neighborhood and n-vertex host graphs without cliques on $k+1$ vertices (as well as the pattern graphs and host graphs dual to the aforementioned ones, respectively).

Keywords: Induced subgraph isomorphism · Matrix multiplication · Witnesses for Boolean matrix product · Time complexity

1   Introduction

The problems of detecting subgraphs or induced subgraphs of a host graph that are isomorphic to a pattern graph are basic in graph algorithms. They are generally termed as *subgraph isomorphism* and *induced subgraph isomorphism* problems, respectively. Such well-known NP-hard problems as the independent set, clique, Hamiltonian cycle or Hamiltonian path can be regarded as their special cases.

Recent examples of applications of some variants of subgraph isomorphism include bio-molecular networks [1], social networks [19], and automatic design of processor systems [20]. In the aforementioned applications, the pattern graphs are typically of fixed size which allows for polynomial-time solutions.

For a pattern graph on k vertices and a host graph on n vertices, the fastest known general algorithms for subgraph isomorphism and induced subgraph

© Springer International Publishing AG 2017
S.-H. Poon et al. (Eds.): WALCOM 2017, LNCS 10167, pp. 217–227, 2017.
DOI: 10.1007/978-3-319-53925-6_17

isomorphism run in time $O(n^{\omega(\lfloor k/3 \rfloor, \lceil (k-1)/3 \rceil, \lceil k/3 \rceil)})$ [4,12,17], where $\omega(p, q, r)$ denotes the exponent of fast matrix multiplication for rectangular matrices of size $n^p \times n^q$ and $n^q \times n^r$, respectively [15]. For $k \geq 6$, they also run in time $O(m^{\omega(\lfloor k/3 \rfloor, \lceil (k-1)/3 \rceil, \lceil k/3 \rceil)/2})$ [4,12], where m denotes the number of edges in a (connected) host graph. Further, we shall denote $\omega(1, 1, 1)$ by just ω. It is known that $\omega \leq 2.373$ [16,23] and for example $\omega(1, 2, 1) \leq 3.257$ [15].

There are several known examples of pattern graphs of fixed size k for which one succeeded to design algorithms for subgraph isomorphism or/and induced subgraph isomorphism yielding asymptotic time upper bounds in terms of n lower than those offered by the aforementioned general method. For instance, for each pattern graph on k vertices having an independent set of size s, an isomorphic subgraph of an n-vertex graph can be (deterministically) detected in time $O(n^{\omega(\lceil (k-s)/2 \rceil, 1, \lfloor (k-s)/2 \rfloor)}) \leq O(n^{k-s+1})$, assuming $k = O(1)$ [13]. Also, an induced subgraph isomorphic to the generalized diamond $K_k - e$, i.e., K_k with a single edge removed, as well as an induced subgraph isomorphic to the path on k vertices, P_k, can be detected in $O(n^{k-1})$ time [10,21] which improves the general bound from [4] for $k \leq 5$.

More recent examples yields the randomized algorithm of Vassilevska Williams et al. for detecting an induced subgraph isomorphic to a pattern graph on four vertices different from K_4 and the four isolated vertices ($4K_1$) [22], subsuming similar randomized approach from [6]. Their algorithm runs in the same asymptotic time as that based on matrix multiplication for detecting triangles (i.e., K_3) from [11], i.e., in $O(n^\omega)$ time. The authors of [22] succeeded to obtain a deterministic version of their algorithm also running in the triangle asymptotic time for the diamond which is K_4 with one removed edge, denoted by $K_4 - e$. In fact there are few known earlier examples of pattern graphs on four vertices, different from the diamond, for which isomorphic induced subgraphs can be deterministically detected in an n vertex host graph substantially faster than by the general method. The earliest example is P_4, a path on four vertices, which can be detected in $O(n + m)$ time [2], where m is the number of edges in the host graph. The other example is a paw which is a triangle connected to the fourth vertex by an edge, denoted by $K_3 + e$. It can be detected in $O(n^\omega)$ time [18]. The third example is a claw which is a star with three leaves, it can be also detected in $O(n^\omega)$ time [4]. (Analogous upper bounds hold for the pattern graphs that are the complement to one of the aforementioned pattern graphs.)

Summarizing, there are twelve pairwise non-isomorphic pattern graphs on four vertices. For P_4 and its complement one can detect deterministically an isomorphic induced subgraph in an n vertex host graph in $O(n^2)$ time, while for the diamond, paw and claw, it takes $O(n^\omega)$ time. Thus, only for the cycle on four vertices, C_4, and its complement, and K_4 and its complement, there are no known deterministic algorithms for the induced subgraph isomorphism that are asymptotically faster than the general method yielding the $O(n^{\omega(1,2,1)})$-time bound. See also Table 1.

In this paper, we show in particular that if an n-vertex host graph does not contain a clique on $k + 1$ vertices then for the pattern graph C_4 the induced

Table 1. Known upper bounds on the deterministic time complexity of induced subgraph isomorphism for pattern graphs on four vertices.

4-vertex pattern graph	Deterministic time complexity
P_4	$O(n^2)$ [2]
Claw	$O(n^{2.373})$ [4]
Paw $K_3 + e$	$O(n^{2.373})$ [18]
Diamond $K_4 - e$	$O(n^{2.373})$ [22]
C_4	$O(n^{3.257})$ [4]
K_4	$O(n^{3.257})$ [4]

subgraph isomorphism can be solved in time $\tilde{O}(n^\omega k^\mu + n^2 k^2)$, where $\tilde{O}(f)$ stands for $O(f(\log f)^c)$ for some constant c, and $\mu \approx 0.46530$. We also show that if the host graph does not contain a clique on $k + 1$ vertices then one can detect an independent set on four vertices in the host graph in time $\tilde{O}(n^\omega k^\mu + n^2 k^2)$. Note that our upper time-bounds subsume the general $O(n^{\omega(1,2,1)}) \approx O(n^{3.257})$ bound for pattern graphs on four vertices for $k = O(n^{0.628})$. Our method works for all pattern graphs on four vertices that have a pair of non-adjacent vertices with the same neighborhood and all n-vertex host graphs without cliques on $k + 1$ vertices in time $\tilde{O}(n^\omega k^\mu + n^2 k^2)$. By considering the dual graphs, we obtain the same asymptotic upper bound for the detection of an induced subgraph isomorphic to a given pattern graph with four vertices and a pair of adjacent vertices with the same neighbors among the remaining vertices in an n-vertex host graph without independent sets on $k + 1$ vertices. (We denote the class of aforementioned 4-vertex pattern graphs and the class of pattern dual to them by $F_s^-(4)$ and $F_s^+(4)$, respectively.)

By using the concept of Ramsey numbers, we can extend our method for induced subgraph isomorphism to include larger pattern graphs having a set of independent vertices with the same neighborhood and n-vertex host graphs without cliques on $k + 1$ vertices (as well as the pattern graphs and host graphs dual to the aforementioned ones, respectively). In particular, we obtain an $\tilde{O}(n^\omega(\frac{k(k+3)}{2})^\mu + n^2(\frac{k(k+3)}{2})^3)$ bound on the time complexity of induced subgraph isomorphism for pattern graphs with five vertices among which three are independent and have the same neighbors and n-vertex host graphs without cliques on $k + 1$ vertices. (We denote the class of these 5-vertex pattern graphs by $F_s^-(5)$ and the class of graphs dual to them by $F_s^+(5)$, respectively.) See Table 2 for the summary of our results for pattern graphs on four and five vertices.

Our paper is structured as follows. In the next section, we provide basic definitions and facts. In Sect. 3, we present our algorithms for the detection of induced subgraphs isomorphic to pattern graphs on four vertices, in graphs without large cliques or large independent sets. In Sect. 4, we extend our algorithms for induced subgraph isomorphism to include larger pattern graphs. We conclude with Final Remarks.

Table 2. Upper bounds on the deterministic time complexity of induced subgraph isomorphism for pattern graphs on four and five vertices and restricted host graphs presented in this paper ($\mu \approx 0.46530$). $K_{2,3}$ denotes the complete bipartite graph with two vertices on one side and three vertices on the other side. $(2,3) - fan$ can be obtained from $K_{2,3}$ by connecting the two vertex side by an edge.

Pattern graph class	Deterministic time complexity	Host graph with
$F_s^-(4)$, e.g., C_4 and $4K_1$	$\tilde{O}(n^\omega k^\mu + n^2 k^2)$	no $k+1$ clique
$F_s^+(4)$, e.g., K_4 and $2K_2$	$\tilde{O}(n^\omega k^\mu + n^2 k^2)$	no $k+1$ ind. set
$F_s^-(5)$, e.g., $K_{2,3}$, $(2,3) - fan$	$\tilde{O}(n^\omega (\frac{k(k+3)}{2})^\mu + n^2 (\frac{k(k+3)}{2})^3)$	no $k+1$ clique
$F_s^+(5)$, e.g., K_5 and $K_5 - e$	$\tilde{O}(n^\omega (\frac{k(k+3)}{2})^\mu + n^2 (\frac{k(k+3)}{2})^3)$	no $k+1$ ind. set

2   Preliminaries

A *subgraph* of the graph $G = (V, E)$ is a graph $H = (V_H, E_H)$ such that $V_H \subseteq V$ and $E_H \subseteq E$.

An *induced subgraph* of the graph $G = (V, E)$ is a graph $H = (V_H, E_H)$ such that $V_H \subseteq V$ and $E_H = E \cap (V_H \times V_H)$.

The *neighborhood* of a vertex v in a graph G is the set of all vertices in G adjacent to v.

For $q \geq 4$, we shall distinguish the family $F_s^-(q)$ of pattern graphs H on q vertices $v_1, v_2, ..., v_q$ such that $v_1, ..., v_{q-2}$ form an independent set and have the same neighbors among the remaining two vertices v_{q-1}, v_q. We shall also denote the family of pattern graphs dual to those in $F_s^-(q)$ by $F_s^+(q)$. Note that the latter family consists of pattern graphs H on q vertices $v_1, v_2, ..., v_q$ such that $v_1, ..., v_{q-2}$ form a clique and have the same neighbors among the remaining two vertices v_{q-1}, v_q.

The *adjacency matrix* A of a graph $G = (V, E)$ is the $0 - 1$ $n \times n$ matrix where for $1 \leq i, j \leq n$, $A[i, j] = 1$ iff $\{i, j\} \in E$.

A *witness* for an entry $B[i, j]$ of the Boolean matrix product B of two Boolean matrices A_1 and A_2 is any index k such that $A_1[i, k]$ and $A_2[k, j]$ are equal to 1 [9].

Fact 1. *The fast matrix multiplication algorithm runs in $O(n^\omega)$ time, where ω is not greater than 2.3728639 [16] (cf. [23]).*

Fact 2. *The k-witness algorithm from [9] takes as input an integer k and two $n \times n$ Boolean matrices, and returns a list of q witnesses for each positive entry of the Boolean matrix product of those matrices, where q is the minimum of k and the total number of witnesses for this entry. It runs in $\tilde{O}(n^\omega k^{(3-\omega-\alpha)/(1-\alpha)} + n^2 k)$ time, where $\alpha \approx 0.30298$ (see [15]). One can rewrite the upper time bound as $\tilde{O}(n^\omega k^\mu + n^2 k)$, where $\mu \approx 0.46530$ [9].*

For two natural numbers p, s, the *Ramsey number* $R(p, s)$ is the minimum number ℓ such that any graph on at least ℓ vertices contains a complete subgraph (clique) on p vertices or an independent set on s vertices.

Fact 3. *For p, $s \geq 2$, $R(p, s) \leq \binom{p+s-2}{s-1}$ holds [3].*

3   The Algorithms

The idea of our algorithm for detecting induced subgraphs isomorphic to C_4 depicted in Fig. 1 is simple. First, we check if there is a pair of non-adjacent vertices connected by more than k different paths of length two in the host graph, by computing the arithmetic square of the adjacency matrix of the host graph. If so, there must be a pair of non-adjacent middle vertices of the aforementioned paths of length two, since all the middle vertices cannot induce a clique on more than k vertices. We conclude that the host graph contains a subgraph isomorphic to C_4. Otherwise, for each pair of non-adjacent vertices i, j, we examine the set of all middle vertices of paths of length two connecting i with j for the containment of a pair of non-adjacent vertices. Since the aforementioned middle vertices correspond to witnesses for the (i, j) entry of the Boolean square of the adjacency matrix of the host graph, we can use the k-witness algorithm from [9] described in Fact 2 to compute the middle vertices.

Lemma 1. *Algorithm 1 is correct.*

Proof. Let B be the Boolean square of the adjacency matrix A. Witnesses of $B[i, j]$ are just middle vertices of different paths of length two connecting the vertices i and j. The answer YES is returned if $A[i, j] = 0$ and the set of witnesses contains at least two vertices and it does not form a clique. Then, i and j are not adjacent, and there are two paths of length two connecting i and j whose middle vertices are not adjacent. Thus, there is an induced C_4 in G.

Input: a graph G on n vertices given by its adjacency matrix A and an integer parameter $k \in [2, n]$ such that G does not contain a $(k + 1)$-clique.
Output: if G contains an induced subgraph isomorphic to C_4 then YES else NO
1: $C \leftarrow A \times A$ (arithmetic product)
2: **for** $i = 1$ **to** n **do**
3:     **for** $j = 1$ **to** n **do**
4:         **if** $i \neq j \land A[i, j] = 0 \land C[i, j] > k$ **then**
5:         **return** YES and stop
6:     **end for**
7: **end for**
8: **for** $i = 1$ **to** n **do**
9:     **for** $j = 1$ **to** n **do**
10:        **if** $i \neq j \land A[i, j] = 0$ **then** $W[i, j] \leftarrow$ the set of witnesses for the $B[i, j]$ entry of the Boolean product B of A with A
11:        **if** $W[i, j]$ contains at least two vertices and it does not induce a clique in G **then**
12:        **return** YES and stop
13:    **end for**
14: **end for**
15: **return** NO

Fig. 1. An algorithm for detecting an induced subgraph isomorphic to C_4 in a graph with no $(k + 1)$-clique.

To prove the correctness of the answer NO suppose that there is an induced C_4 in G. We may assume w.l.o.g. that it is induced by the vertices q, r, i, j, where $A[i,j] = 0$ and $A[q,r] = 0$. Note that q and r belong to the set $W[i,j]$ of witnesses of $B[i,j]$. Since $A[q,r] = 0$, $W[i,j]$ does not induce a clique. Consequently, Algorithm 1 would return YES. □

Lemma 2. *Algorithm 1 runs in time* $O(n^\omega + T(n,k) + n^2k^2)$, *where* $T(n,k)$ *stands for the time necessary to solve the k-witness problem for two $n \times n$ Boolean matrices.*

Proof. The computation of C takes $O(n^\omega)$ time. The first double loop takes $O(n^2)$ time. To implement the second double loop we need to solve the k-witness problem. It takes $T(n,k)$ time. After that, the second double loop takes $O(n^2k^2)$ time since the considered sets $W[i,j]$ are of size not exceeding k, and we can test if $l \leq k$ vertices induce a clique in $O(k^2)$ time. □

By combining Lemmata 1, 2 with Fact 2, we obtain our first main result.

Theorem 1. *Let G be a graph on n vertices with no clique on $k + 1$ vertices. We can decide if G contains an induced subgraph isomorphic to C_4 in time* $\tilde{O}(n^\omega k^\mu + n^2k^2)$, *where* $\mu \approx 0.46530$.

Recall that $F_s^-(4)$ is the family of graphs H on four vertices v_1, v_2, v_3, v_4 such that v_1, v_2 are not adjacent and have the same neighbors among the remaining two vertices v_3, v_4. Clearly, C_4 belongs to $F_s^-(4)$. Also, $4K_1$ (i.e., an independent set on four vertices), $K_4 - e$, $P_3 + K_1$ and the paw belong to $F_s^-(4)$.

We can immediately generalize Algorithm 1 and Theorem 1 to include the detection of a given member $H \in F_s^-(4)$ in the graph G satisfying the requirements of Theorem 1.

Consider the distinguished pair of non-adjacent vertices v_1, v_2 in H. Let v_3, v_4 be the two remaining vertices. Finally, let $\bar{H}$ stand for the graph dual to H, and $\bar{A}$ for the adjacency matrix of $\bar{H}$. In the generalized Algorithm 1, we match v_3, v_4 with i, j, respectively, and v_1, v_2, with witnesses for the entry corresponding to (i,j) of an appropriated Boolean matrix product. In case of C_4, both v_3 and v_4 are neighbors of v_1 and v_2, so we use the Boolean product of A with A. Generally, for $H \in F_s^-(4)$, we have to replace the Boolean product $A \times A$ with that of two matrices A_1, A_2 defined as follows:

- if $\{v_3, v_1\}$, $\{v_3, v_2\}$ are edges of H then $A_1 = A$,
- otherwise, $\{v_3, v_1\}$, $\{v_3, v_2\}$ are edges of $\bar{H}$ and $A_1 = \bar{A}$,
- if $\{v_1, v_4\}$, $\{v_2, v_4\}$ are edges of H then $A_2 = A$,
- otherwise, $\{v_1, v_4\}$, $\{v_2, v_4\}$ are edges of $\bar{H}$ and $A_2 = \bar{A}$.

Also, to count the number of witnesses for respective entries of the Boolean product of A_1 and A_2, in the first step of the generalized algorithm, we compute the arithmetic product of A_1 and A_2 treated as arithmetic matrices, instead of the arithmetic product of A with A. Finally, if $\{v_3, v_4\}$ is an edge of H then we have to replace the condition $A[i,j] = 0$ with $A[i,j] = 1$.

By arguing analogously as in the proof of Theorem 1, we obtain the following generalization.

Input: a graph G on n vertices given by its adjacency matrix A, a pattern graph H on
vertices v_1, v_2, v_3, v_4 satisfying the requirements of the membership in $F_s^-(4)$, and
an integer parameter $k \in [2, n]$ such that G does not contain a $(k+1)$-clique.
Output: if G contains an induced subgraph isomorphic to H then YES else NO
1: $\bar{A} \leftarrow$ the adjacency matrix of the complement of G
2: **if** $\{v_3, v_1\}$ is an edge of H **then** $A_1 \leftarrow A$ **else** $A_1 \leftarrow \bar{A}$
3: **if** $\{v_1, v_4\}$ is an edge of H **then** $A_2 \leftarrow A$ **else** $A_2 \leftarrow \bar{A}$
4: $C \leftarrow A_1 \times A_2$ (arithmetic product)
5: **if** $\{v_3, v_4\}$ is an edge of H **then** $a \leftarrow 1$ **else** $a \leftarrow 0$
6: **for** $i = 1$ **to** n **do**
7:     **for** $j = 1$ **to** n **do**
8:         **if** $i \neq j \wedge A[i,j] = a \wedge C[i,j] > k$ **then**
9:             **return** YES and stop
10:     **end for**
11: **end for**
12: **for** $i = 1$ **to** n **do**
13:     **for** $j = 1$ **to** n **do**
14:         **if** $i \neq j \wedge A[i,j] = a$ **then** $W[i,j] \leftarrow$ the set of witnesses for the $B[i,j]$ entry
        of the Boolean product B of A_1 with A_2
15:         **if** $W[i,j]$ contains at least two vertices and it does not induce a clique in G
        then
16:             **return** YES and stop
17:     **end for**
18: **end for**
19: **return** NO

Fig. 2. An algorithm for detecting an induced subgraph isomorphic to $H \in F_s^-(4)$ in
a graph with no $(k+1)$-clique.

Theorem 2. *Let $H \in F_s^-(4)$, and let G be a graph on n vertices with no clique
on $k+1$ vertices. We can decide if G contains an induced subgraph isomorphic
to H in time $\tilde{O}(n^\omega k^\mu + n^2 k^2)$, where $\mu \approx 0.46530$.*

Recall that $F_s^+(4)$ is the family of graphs on four vertices dual to those in
$F_s^-(4)$, i.e., graphs that have a pair of adjacent vertices with the same neighborhood. Note that in particular K_4, $K_3 + K_1$, $2K_2$, and again $K_4 - e$ belong to
$F_s^+(4)$.

By considering dual graphs, we obtain the following corollary from Theorem 2.

Corollary 1. *Let $H \in F_s^+(4)$, and let G be a graph with n vertices and no independent set on $k+1$ vertices. We can decide if G contains an induced subgraph
isomorphic to H in time $\tilde{O}(n^\omega k^\mu + n^2 k^2)$, where $\mu \approx 0.46530$.*

Corollary 2. *Let $H \in F_s^-(4) \cap F_s^+(4)$, and let G be a graph on n vertices
which does not contain a clique on $k+1$ vertices or an independent set on $k+1$
vertices. We can decide if G contains an induced subgraph isomorphic to H in
time $\tilde{O}(n^\omega k^\mu + n^2 k^2)$, where $\mu \approx 0.46530$.*

Note that $F_s^-(4) \cap F_s^+(4) = \{K_4 - e, \ K_2 + 2K_1\}$.

4   Extensions to Larger Pattern Graphs

Recall that for $q \geq 4$, $F_s^-(q)$ stands for the family of graphs H on q vertices $v_1, v_2, ..., v_q$ such that $v_1, ..., v_{q-2}$ form an independent set and have the same neighbors among the remaining two vertices v_{q-1}, v_q.

We can easily generalize Algorithm 2 and Theorem 2 to include the detection of a given member $H \in F_s^-(q)$, where $q \geq 4$, in the graph G satisfying the requirements of Theorem 2.

In Algorithm 2, it is sufficient for a pair of vertices i, j satisfying $A[i,j] = a$ to deduce or verify that the set of witnesses for the (i,j) entry of the Boolean matrix product contains a pair of non-adjacent vertices in order to detect an induced subgraph isomorphic to the pattern graph H. Now, we have to deduce or verify that the aforementioned set of witnesses contains an independent set on $q - 2$ vertices in G instead. If the set of witnesses contains no less than the Ramsey number $R(k + 1, q - 2)$ of vertices then it fulfills this requirement since we assume that the host graph G does not contain a clique on $k + 1$ vertices. Otherwise, we have to go through all subsets of $q - 2$ witnesses to check if any of them forms an independent set. If we do not know the exact Ramsey number $R(k + 1, q - 2)$ but only an upper bound t on $R(k + 1, q - 2)$, we have to use t instead. Thus the threshold on the number of witness becomes now $t - 1$ instead of k, and we have to use Fact 2 to find up to $t - 1$ witnesses. The generalized algorithm is depicted in Fig. 3.

Lemma 3. *Algorithm 3 runs in time* $\tilde{O}(n^\omega (t - 1)^\mu + n^2 (t - 1)^{q-2})$, *where* $\mu \approx 0.46530$.

Proof. As a straightforward generalization of Algorithm 2, Algorithm 3 runs in time $O(n^\omega + T(n,t) + n^2 t^{q-2})$, where $T(n,t)$ stands for the time necessary to solve the $(t - 1)$-witness problem for two $n \times n$ Boolean matrices. By Fact 2, we obtain the lemma. □

Theorem 3. *Let* $q \geq 4$, $H \in F_s^-(q)$, *and let* G *be a graph on* n *vertices with no clique on* $k + 1$ *vertices. Next, let* t *be a known upper bound on* $R(k + 1, q - 2)$. *We can decide if* G *contains an induced subgraph isomorphic to* H *in time* $\tilde{O}(n^\omega (t - 1)^\mu + n^2 (t - 1)^{q-2})$, *where* $\mu \approx 0.46530$.

Proof. We use Algorithm 3. Its correctness follows from the discussion preceding its pseudocode. Lemma 3 yields the time bound. □

By Fact 3, we can easily conclude that $R(k + 1, 3) \leq \frac{(k+2)(k+1)}{2}$. Hence, we obtain the following corollary.

Corollary 3. *Let* $H \in F_s^-(5)$, *and let* G *be a graph on* n *vertices with no clique on* $k + 1$ *vertices. We can decide if* G *contains an induced subgraph isomorphic to* H *in time* $\tilde{O}(n^\omega (\frac{k(k+3)}{2})^\mu + n^2 (\frac{k(k+3)}{2})^3)$.

Note that $H \in F_s^-(5)$ includes $K_{2,3}$, $(2,3) - fan$ (see Table 2), $K_2 + 3K_1$ and $5K_1$ among other things. We leave to the reader stating the results implied by Theorem 3 and Corollary 3 for the dual pattern and host graphs.

Input: a graph G on n vertices given by its adjacency matrix A, a pattern graph H on $q \geq 4$ vertices $v_1, ..., v_q$ satisfying the requirements of the membership in $F_s^-(q)$, an integer parameter $k \in [2, n]$ such that G does not contain a $(k+1)$-clique, and an upper bound t on the Ramsey number $R(k+1, q-2)$.

Output: if G contains an induced subgraph isomorphic to H then YES else NO

1: $\bar{A} \leftarrow$ the adjacency matrix of the complement of G
2: **if** $\{v_q, v_1\}$ is an edge of H **then** $A_1 \leftarrow A$ **else** $A_1 \leftarrow \bar{A}$
3: **if** $\{v_1, v_{q-1}\}$ is an edge of H **then** $A_2 \leftarrow A$ **else** $A_2 \leftarrow \bar{A}$
4: $C \leftarrow A_1 \times A_2$ (arithmetic product)
5: **if** $\{v_{q-1}, v_q\}$ is an edge of H **then** $a \leftarrow 1$ **else** $a \leftarrow 0$
6: **for** $i = 1$ to n **do**
7:     **for** $j = 1$ to n **do**
8:         **if** $i \neq j \wedge A[i, j] = a \wedge C[i, j] \geq t$ **then**
9:             **return** YES and stop
10:        **end for**
11: **end for**
12: **for** $i = 1$ to n **do**
13:     **for** $j = 1$ to n **do**
14:         **if** $i \neq j \wedge A[i, j] = a$ **then** $W[i, j] \leftarrow$ the set of witnesses for the $B[i, j]$ entry of the Boolean product of A_1 with A_2
15:             **if** $W[i, j]$ contains an independent set on $q - 2$ vertices in G **then**
16:                 **return** YES and stop
17:        **end for**
18: **end for**
19: **return** NO

Fig. 3. An algorithm for detecting an induced subgraph isomorphic to $H \in F_s^-(q)$, $q \geq 4$, in a graph with no $(k+1)$-clique.

5   Final Remarks

The authors of [12] have shown that if one knows the number of induced subgraphs of an n vertex host graph that are isomorphic to a given 4-vertex pattern then one can compute the analogous number for each of the twelve 4-vertex pattern graphs in $O(n^\omega)$ time. This result has been generalized to include pattern graphs on more than four vertices in [13]. Thus, in particular, if we knew that the host graph is free from K_4 then we could compute for each pattern graph H on four vertices the number of induced subgraphs isomorphic to H in $O(n^\omega)$ time. Generally, graphs with some forbidden subgraphs, induced subgraphs, or minors (e.g., planar graphs) are widely studied in algorithmics. Typically, the forbidden subgraphs are of small fixed size. In our approach the forbidden $(k+1)$ clique (or, an $(k+1)$ independent set, respectively) can be very large, e.g., even larger than $\sqrt{n}$ and we can still obtain an upper time-bound on detecting for instance induced subgraphs isomorphic to C_4 better than the known $O(n^{\omega(1,2,1)})$ one.

One of the reviewers posed an interesting question of whether or not our pattern detection algorithms can be extended to include pattern finding algorithms without substantially increasing their running times. For instance, consider our algorithm for C_4 detection. Suppose that i, j is a pair of non-adjacent vertices in the input graph. If C_4 is detected by finding a pair of non-adjacent witnesses among at most k witnesses of $B[i, j]$, then we can easily locate an induced subgraph isomorphic to C_4. However, if C_4 is detected by checking that $B[i, j]$ has more than k witnesses then we need to know at least $k + 1$ witnesses of $B[i, j]$ in order to locate a pair of non-adjacent ones. Thus, we can extend our algorithm for C_4 and the other ones to the finding variant by increasing the number of witnesses to compute for each positive entry of B by one. This increases the running times of our algorithms solely marginally.

Acknowledgments. The research has been supported in part by the grant of polish National Science Center 2014/13/B/ST6/00770 and Swedish Research Council grant 621-2011-6179, respectively.

References

1. Alon, N., Dao, P., Hajirasouliha, I., Hormozdiari, F., Sahinalp, S.C.: Biomolecular network motif counting and discovery by color coding. Bioinformatics (ISMB 2008) **24**(13), 241–249 (2008)
2. Corneil, D.G., Perl, Y., Stewart, L.K.: A linear recognition algorithm for cographs. SIAM J. Comput. **14**(4), 926–934 (1985)
3. Chung, F.R.K., Grinstead, C.M.: A survey of bounds for classical ramsey numbers. J. Graph Theory **7**, 25–37 (1983)
4. Eisenbrand, F., Grandoni, F.: On the complexity of fixed parameter clique and dominating set. Theoret. Comput. Sci. **326**, 57–67 (2004)
5. Eschen, E.M., Hoàng, C.T., Spinrad, J., Sritharan, R.: On graphs without a C4 or a diamond. Discret. Appl. Math. **159**(7), 581–587 (2011)
6. Floderus, P., Kowaluk, M., Lingas, A., Lundell, E.-M.: Detecting and counting small pattern graphs. SIAM J. Discret. Math. **29**(3), 1322–1339 (2015)
7. Floderus, P., Kowaluk, M., Lingas, A., Lundell, E.-M.: Induced subgraph isomorphism: are some patterns substantially easier than others? Theoret. Comput. Sci. **605**, 119–128 (2015)
8. Garey, M.R., Johnson, D.S.: Computers and Intractability - A Guide to the Theory of NP-Completeness. Bell Laboratories, Murray Hill (1979)
9. Gąsieniec, L., Kowaluk, M., Lingas, A.: Faster multi-witnesses for Boolean matrix product. Inf. Process. Lett. **109**, 242–247 (2009)
10. Hoàng, C.T., Kaminski, M., Sawada, J., Sritharan, R.: Finding and listing induced paths and cycles. Discret. Appl. Math. **161**(4–5), 633–641 (2013)
11. Itai, A., Rodeh, M.: Finding a minimum circuit in a graph. SIAM J. Comput. **7**, 413–423 (1978)
12. Kloks, T., Kratsch, D., Müller, H.: Finding and counting small induced subgraphs efficiently. Inf. Process. Lett. **74**(3–4), 115–121 (2000)
13. Kowaluk, M., Lingas, A., Lundell, E.-M.: Counting and detecting small subgraphs via equations and matrix multiplication. SIAM J. Discret. Math. **27**(2), 892–909 (2013)
14. Kuramochi, M., Karypis, G.: Finding frequent patterns in a large sparse graph. Data Min. Knowl. Disc. **11**, 243–271 (2005)

15. Le Gall, F.: Faster algorithms for rectangular matrix multiplication. In: Proceedings of 53rd Symposium on Foundations of Computer Science (FOCS), pp. 514–523 (2012)
16. Le Gall, F.: Powers of tensors and fast matrix multiplication. In: Proceedings of 39th International Symposium on Symbolic and Algebraic Computation, pp. 296–303 (2014)
17. Nešetřil, J., Poljak, S.: On the complexity of the subgraph problem. Commentationes Math. Univ. Carol. **26**(2), 415–419 (1985)
18. Olariu, S.: Paw-free graphs. Inf. Process. Lett. **28**, 53–54 (1988)
19. Schank, T., Wagner, D.: Finding, counting and listing all triangles in large graphs, an experimental study. In: Nikoletseas, S.E. (ed.) WEA 2005. LNCS, vol. 3503, pp. 606–609. Springer, Heidelberg (2005). doi:10.1007/11427186_54
20. Wolinski, C., Kuchcinski, K., Raffin, E.: Automatic design of application-specific reconfigurable processor extensions with UPaK synthesis kernel. ACM Trans. Des. Autom. Electron. Syst. **15**(1), 1–36 (2009)
21. Vassilevska, V.: Efficient algorithms for path problems in weighted graphs. Ph.D. thesis, CMU, CMU-CS-08-147 (2008)
22. Williams, V.V., Wang, J.R., Williams, R., Yu, H.: Finding four-node subgraphs in triangle time. In: Proceedings of SODA, pp. 1671–1680 (2015)
23. Williams, V.V.: Multiplying matrices faster than Coppersmith-Winograd. In: Proceedings of 44th Annual ACM Symposium on Theory of Computing (STOC), pp. 887–898 (2012)

Approximation Algorithm for the Distance-3 Independent Set Problem on Cubic Graphs

Hiroshi Eto[1], Takehiro Ito[2,3], Zhilong Liu[4(✉)], and Eiji Miyano[4]

[1] Kyushu University, Fukuoka 812-8581, Japan
h-eto@econ.kyushu-u.ac.jp
[2] Tohoku University, Sendai 980-8579, Japan
takehiro@ecei.tohoku.ac.jp
[3] CREST, JST, 4-1-8 Honcho, Kawaguchi, Saitama 332-0012, Japan
[4] Kyushu Institute of Technology, Fukuoka 820-8502, Japan
liu@theory.ces.kyutech.ac.jp, miyano@ces.kyutech.ac.jp

Abstract. For an integer $d \geq 2$, a distance-d independent set of an unweighted graph $G = (V, E)$ is a subset $S \subseteq V$ of vertices such that for any pair of vertices $u, v \in S$, the number of edges in any path between u and v is at least d in G. Given an unweighted graph G, the goal of MAXIMUM DISTANCE-d INDEPENDENT SET problem (MaxDdIS) is to find a maximum-cardinality distance-d independent set of G. In this paper we focus on MaxD3IS on cubic (3-regular) graphs. For every fixed integer $d \geq 3$, MaxDdIS is NP-hard even for planar bipartite graphs of maximum degree three. Furthermore, when $d = 3$, it is known that there exists no σ-approximation algorithm for MaxD3IS oncubic graphs for constant $\sigma < 1.00105$. On the other hand, the previously best approximation ratio known for MaxD3IS on cubic graphs is 2. In this paper, we improve the approximation ratio into 1.875 for MaxD3IS on cubic graphs.

1 Introduction

Let G be an unweighted graph; we denote by $V(G)$ and $E(G)$ the sets of vertices and edges, respectively, and let $n = |V(G)|$. An *independent set* (or *stable set*) of G is a subset $S \subseteq V(G)$ of vertices such that $\{u, v\} \notin E$ holds for all $u, v \in S$. In theoretical computer science and combinatorial optimization, one of the most important and most investigated computational problems is the MAXIMUM INDEPENDENT SET problem (MaxIS for short): Given a graph G, the goal of MaxIS is to find an independent set S of maximum cardinality in G.

In this paper, we consider a generalization of MaxIS, named the MAXIMUM DISTANCE-d INDEPENDENT SET problem (MaxDdIS for short). For an integer $d \geq 2$, a *distance-d independent set* of an unweighted graph G is a subset $S \subseteq V(G)$ of vertices such that for any pair of vertices $u, v \in S$, the distance (i.e., the number of edges) of any path between u and v is at least d in G. For an integer

---

This work is partially supported by JSPS KAKENHI Grant Numbers JP15J05484, JP15H00849, JP16K00004, and JP26330017.

S.-H. Poon et al. (Eds.): WALCOM 2017, LNCS 10167, pp. 228–240, 2017.
DOI: 10.1007/978-3-319-53925-6_18

$d \geq 2$, MaxDdIS is formulated as the following class of problems [1,5]: Given an unweighted graph G, the goal of MaxDdIS is to find a maximum-cardinality distance-d independent set of G.

When $d = 2$, MaxDdIS (i.e., MaxD2IS) is equivalent to the original MaxIS. Zuckerman [13] proved that MaxD2IS cannot be approximated in polynomial time, unless P = NP, within a factor of $n^{1-\varepsilon}$ for any $\varepsilon > 0$. Moreover, MaxD2IS remains NP-hard even if the input graph is a cubic planar graph, a triangle-free graph, or a graph with large girth. Fortunately, however, it is well known that MaxD2IS can be solved in polynomial time when restricted to, for example, bipartite graphs [10], chordal graphs [7], circular-arc graphs [8], comparability graphs [9], and many other classes [3,11,12].

For every fixed integer $d \geq 3$, Eto et al. [5] proved that MaxDdIS is NP-hard even for planar bipartite graphs of maximum degree three. Furthermore, they showed that it is NP-hard to approximate MaxDdIS on bipartite graphs and chordal graphs within a factor of $n^{1/2-\varepsilon}$ ($\varepsilon > 0$) for every fixed integer $d \geq 3$ and every fixed *odd* integer $d \geq 3$, respectively. On the other hand, interestingly, they showed that MaxDdIS on chordal graphs is solvable in polynomial time for every fixed *even* integer $d \geq 2$. As the other positive results, Agnarsson et al. [1] showed the tractability of MaxDdIS on interval graphs, trapezoid graphs, and circular-arc graphs.

In this paper, we focus only on cubic (i.e., 3-regular) graphs as input. For $d = 2$ i.e., MaxD2IS (MaxIS), it is known that it is NP-hard even for cubic planar graphs. Furthermore, Chlebík and Chlebíková [4] proved the 1.0107-inapproximability for MaxD2IS on cubic graphs. On the other hand, we can obtain polynomial-time 1.2-approximation algorithms for MaxD2IS on cubic graphs by applying the $\frac{\Delta+3}{5}$-approximation algorithm proposed by Berman and Fujito [2] for the problem on general graphs of maximum degree $\Delta \leq 613$.

When $d = 3$, it is known that there exists no σ-approximation algorithm for MaxD3IS on cubic graphs for constant $\sigma < 1.00105$. As for approximability, by using the above $\frac{\Delta+3}{5}$-approximation algorithm for MaxD2IS as a subroutine, we can obtain a 2.4-approximation algorithm for MaxD3IS on cubic graphs. The previously best approximation ratio known for MaxD3IS on cubic graphs is 2 [6]. In this paper, we refine the 2-approximation algorithm and design a 1.875-approximation algorithm for MaxD3IS on cubic graphs. Due to the page limitation, however, we omit some details and proofs from this extended abstract.

2   Preliminaries

Let $G = (V, E)$ be an unweighted graph, where V and E denote the set of vertices and the set of edges, respectively. $V(G)$ and $E(G)$ also denote the vertex set and the edge set of G, respectively. We denote an edge with endpoints u and v by $\{u, v\}$. A path P of length ℓ from a vertex v_0 to a vertex v_ℓ is represented as a sequence of vertices such that $P = \langle v_0, v_1, \cdots, v_\ell \rangle$. A cycle C of length ℓ is similarly written as $C = \langle v_0, v_1, \cdots, v_{\ell-1}, v_0 \rangle$. For a pair of vertices u and v, the length of a shortest path from u to v, i.e., the distance between u and v is denoted by $dist_G(u, v)$.

For a graph G and its vertex v, we denote the (open) neighborhood of v in G by $D_1(v) = \{u \in V(G) \mid \{v, u\} \in E(G)\}$, i.e., for any $u \in D_1(v)$, $dist_G(v, u) = 1$ holds. More generally, for $d \geq 1$, let $D_d(v) = \{w \in V(G) \mid dist_G(v, w) = d\}$ be the subset of vertices that are distance-d away from v. Similarly, let $D_1(S)$ be the open neighborhood of a subset S of vertices, $D_2(S)$ be the open neighborhood of $D_1(S) \cup S$, and so on. That is, $D_k(S) = D_1 \left(\bigcup_{i=1}^{k-1} D_i(S) \cup S \right)$. The degree of v is denoted by $deg(v) = |D_1(v)|$. A graph is r-regular if the degree $deg(v)$ of every vertex v is exactly $r \geq 0$, and a 3-regular graph is often called cubic graph. A graph G_S is a subgraph of a graph G if $V(G_S) \subseteq V(G)$ and $E(G_S) \subseteq E(G)$. For a subset of vertices $U \subseteq V$, let $G[U]$ be the subgraph induced by U. For a positive integer $d \geq 1$ and a graph G, the dth power of G, denoted by $G^d = (V(G), E^d)$, is the graph formed from $V(G)$, where all pairs of vertices $u, v \in G$ such that $dist_G(u, v) \leq d$ are connected by edges $\{u, v\}$'s.

Let $OPT(G)$ be an optimal distance-3 independent set on input G. We say that an algorithm ALG is a σ-approximation algorithm for MaxDdIS or that ALG's approximation ratio is at most σ if $|OPT(G)| \leq \sigma \cdot |ALG(G)|$ holds for any input G, where $ALG(G)$ is a distance-d independent set returned by ALG. Then, we can obtain the following proposition on the upper bound of $|OPT(G)|$ [6]:

Proposition 1 [6]. *Consider a cubic graph $G = (V, E)$ with $|V| = n$ vertices. Then, the size $|OPT(G)|$ of every optimal solution of MaxD3IS is at most $\frac{n}{4}$.*

Let ALG$_2$ be a $\frac{\Delta+3}{5}$-approximation algorithm for MaxD2IS on graphs with the maximum degree Δ proposed in [1]. Then, we obtain the following simple algorithm: First, construct the second power G^2 of an input cubic graph G, and then obtain a distance-2 independent set of G^2 by using ALG$_2$. Note that any distance-2 independent set of G^2 is a distance-3 independent set of G and the maximum degree of G^2 of the cubic graph G is nine. Hence the above algorithm achieves a 2.4-approximation ratio for MaxD3IS on cubic graphs [6].

The previously best approximation ratio for MaxD3IS on cubic graphs is 2 [6]:

Proposition 2 [6]. *There exists a polynomial-time 2-approximation algorithm for MaxD3IS on cubic graphs.*

The basic strategies of the above algorithm are quite straightforward; the algorithm iteratively picks a vertex v into the distance-3 independent set, say, $D3IS(G)$, and eliminates all the vertices in $\{v\} \cup D_1(v) \cup D_2(v)$ from candidates of the solution: Given a graph G, (i) in the first iteration of the algorithm, the first vertex, say, s_1, is selected into $D3IS(G)$, then $B_1 = \{s_1\} \cup D_1(s_1) \cup D_2(s_1)$ is removed from $V(G)$, and set $V = V(G) \backslash B_1$ since any vertex in B_1 cannot be a candidate of the solution. One can see that $|B_1| \leq 10$. (ii) In the second iteration, the second vertex, say, s_2, is selected from neighbor vertices in $D_1(B_1)$ of B_1 into $D3IS(G)$, and then $B_2 = \{s_2\} \cup D_1(s_2) \cup D_2(s_2)$ is removed from V. Note that $|B_1 \cap B_2| \geq 2$ holds since there must exist at least two vertices between s_1 and s_2 from the fact $dist_G(s_1, s_2) \geq 3$. That is, $|B_2 \backslash B_1| \leq 8$ and thus at most eight vertices are removed from V in the second iteration. Similarly, when s_i for $3 \le i \le \ell$ are selected into $D3IS(G)$, at most eight vertices are removed from V. Therefore, roughly speaking, the algorithm can find one vertex among eight ones, i.e., a distance-3 independent set of $n/8$ vertices among n vertices. Since $|OPT(G)| \le n/4$, the approximation ratio is 2 $(= (n/4)/(n/8) \ge |OPT(G)|/|ALG(G)|)$.

3 Approximation Algorithm for MaxD3IS on Cubic Graphs

Algorithm. In this section, we improve the approximation ratio from the previous 2 to 1.875 for MaxD3IS on cubic graphs. Now we make a simple observation; see Fig. 1(a). In the previous algorithm in [6], if s_{i-1} is selected in the $(i-1)$th iteration and black vertices are removed from the solution candidates, then we select, for example, v_1 into a solution $D3IS(G)$ in the ith iteration since $dist_G(s_{i-1}, v_1) = 3$, and remove eight "gray" vertices, v_1 through v_8, from the solution candidates. In other words, we can select one vertex v_1 into the solution among (at most) eight candidates in $\{v_1\} \cup D_1(v_1) \cup D_2(v_1) \setminus B$, where B is a set of "non-candidate vertices." For the case in Fig. 1, however, if we select a *neighbor* v_2 of v_1 into $D3IS(G)$, then at most seven vertices in $\{v_2\} \cup D_1(v_2) \cup D_2(v_2) \setminus B$ $(= \{v_1, v_2, v_3, v_4, v_5, v_6, v_9\})$ are removed; now we could select one among seven candidates. As a desirable example, if we can averagely select one vertex into $D3IS(G)$ among seven vertices in an iteration, then we can find a solution of size $n/7$, i.e., we achieve the 7/4-approximation ratio. Hence, it is our goal to find a vertex s such that $|\{s\} \cup D_1(s) \cup D_2(s) \setminus B|$ is as small as possible in each iteration. As another desirable example, if v_1 has two neighbors in B as shown in Fig. 1(b), then $|\{v_1\} \cup D_1(v_1) \cup D_2(v_1) \setminus B| \le 4$. In the following, we show that we can averagely select one vertex among "15/2" vertices, which implies the approximation ratio of $(n/4)/(2n/15) = 15/8 = 1.875$.

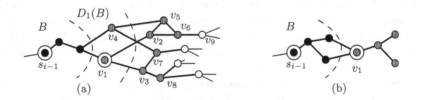

Fig. 1. Observations (a) and (b)

Our new algorithm ALG basically selects (i) the *first candidate* vertex v_f from $D_1(B)$ if $|\{v_f\} \cup D_1(v_f) \cup D_2(v_f) \setminus B| \le 7$, but (ii) a neighbor u of v_f if $|\{v_f\} \cup D_1(v_f) \cup D_2(v_f) \setminus B| \ge 8$. Unfortunately, however, there are *special* subgraphs such that for any neighbor $u \in D_1(v_f)$ of the first candidate v_f, $|\{u\} \cup D_1(u) \cup D_2(u) \setminus B| \ge 8$ must hold. Therefore, ALG initially finds such special subgraphs and gives some special treatments to them as preprocessing.

There are five special subgraphs, SG_1, SG_2, SG_3, SG_4 and SG_5 illustrated in Figs. 2(a), (b), (c), (d), and (e), respectively. The special subgraph SG_1 consists of nine "gray" vertices. On the other hand, each of SG_1 through SG_4 has eight vertices, including the first candidate v_f and its two neighbors in $D_1(v_f)$. Note that the black vertex v in $D_1(v_f)$ may be not in B.

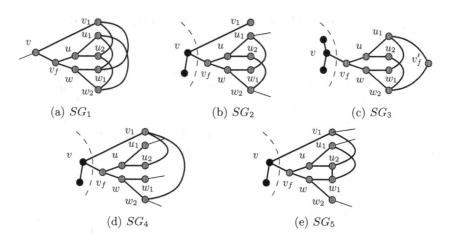

(a) SG_1 (b) SG_2 (c) SG_3

(d) SG_4 (e) SG_5

Fig. 2. Special subgraphs (a) SG_1, (b) SG_2, (c) SG_3, (d)SG_4, (e) SG_5

(a) See Fig. 2(a). The first special subgraph SG_1 has nine gray vertices, the first candidate v_f, its three neighbor vertices v, u and w, two neighbors u_1 and u_2 of u, two neighbors w_1 and w_2 of w, and the top vertex v_1, where $dist_G(v_f, v_1) = 2$. The vertex v_1 is connected to either of u_1 and u_2 and either of w_1 and w_2. As shown in Fig. 2(a), assume that the graph has two edges $\{v_1, u_2\}$ and $\{v_1, w_1\}$. Furthermore, there are three edges, $\{u_1, w_1\}$, $\{u_1, w_2\}$, and $\{u_2, w_2\}$. For SG_1, our algorithm ALG selects u_1 which is not connected to v_1, and v into $D3IS(G)$, and eliminates nine vertices in $V(SG_1)$ and three vertices in $(D_1(v) \cup D_2(v)) \setminus V(SG_1)$, i.e., (at most) 12 vertices in $\{\{v, u_1\} \cup D_1(\{v, u_1\}) \cup D_2(\{v, u_1\})\}$ from the solution candidates. That is, we can averagely select one vertex among six ones.

(b) See Fig. 2(b). The second special subgraph SG_2 has eight gray vertices, $V(SG_2) = \{v_f, u, w, v_1, u_1, u_2, w_1, w_2\}$, where $dist_G(v_f, v_1) = 2$. Furthermore, (b1) neither of u_1 and u_2 (w_1 and w_2, resp.) is connected to w (u, resp.), and (b2) u_1 is connected to either w_1 or w_2, and u_2 is connected to the other. Without loss of generality, assume that u_1 (u_2, resp.) is connected to w_1 (w_2, resp.) as shown in Fig. 2(b). (b3) Either of $(dist_G(u_1, w_2), dist_G(u_2, w_1)) = (1, 3)$, $(1, 1)$, and $(3, 3)$ holds. Note that the case $(dist_G(u_1, w_2), dist_G(u_2, w_1)) = (3, 1)$ is essentially the same as the case $(dist_G(u_1, w_2), dist_G(u_2, w_1)) = (1, 3)$. If $dist_G(u_1, w_2) = dist_G(u_2, w_1) = 3$, then ALG selects u_2 and w_1 into $D3IS(G)$. If $(dist_G(u_1, w_2), dist_G(u_2, w_1)) = (1, 3)$, then ALG selects u_2 and w_1 into $D3IS(G)$. If $dist_G(u_1, w_2) = dist_G(u_2, w_1) = 1$, then ALG selects one arbitrary vertex in

$\{u_1, u_2, w_1, w_2\}$ into $D3IS(G)$. One can see that the case where $dist_G(u_1, w_2) = dist_G(u_2, v_1) = dist_G(w_1, v_1) = 1$ is essentially equivalent to SG_1.

(c) See Fig. 2(c). The third special subgraph SG_3 has eight gray vertices, v_f, u, w, u_1, u_2, w_1, w_2, and v'_f, where $dist_G(v'_f, v_f) \geq 3$. The conditions (c1) and (c2) are the same as (b1) and (b2), respectively. (c3) The conditions on $dist_G(u_1, w_2)$ and $dist_G(u_2, w_1)$ are different from the above: $dist_G(u_1, w_2) = 2$ or $dist_G(u_2, w_1) = 2$ holds. That is, there is exactly one vertex between u_1 and w_2, or exactly one vertex between u_2 and w_1. For SG_3 with $dist_G(u_1, w_2) = 2$ in Fig. 2(c), ALG selects v_f and v'_f into $D3IS(G)$.

(d) See Fig. 2(d). The fourth special subgraph SG_4 consists of eight gray vertices, v_f, u, w, u_1, u_2, w_1, w_2, and v_1, where $dist_G(v_f, v_1) = 2$. (d1) is the same as (b1). (d2) The vertex v_1 is connected to one of u_1 and u_2, and one of w_1 and w_2. Now, without loss of generality, we assume that there are two edges $\{v_1, u_2\}$ and $\{v_1, w_2\}$ as shown in Fig. 2(d). (d3) There is no edge $\{u_2, w_2\}$. (d4) There exists neither edge $\{u_1, u_2\}$ nor $\{w_1, w_2\}$. Therefore, possibly, there are further three edges, $\{u_1, w_1\}$, $\{u_1, w_2\}$, and $\{u_2, w_1\}$. If $dist_G(u_1, w_1) = 1$ and $dist_G(u_1, w_2) \geq 2$ (i.e., SG_4 does not have the edge $\{u_1, w_2\}$), then ALG selects two vertices w_2 and u into $D3IS(G)$. If $dist_G(u_1, w_2) \geq 2$ and $dist_G(u_2, w_1) = 1$, then ALG selects two vertices w_2 and u into $D3IS(G)$. If $dist_G(u_1, w_1) \geq 2$ and $dist_G(u_2, w_1) = 1$, then ALG selects two vertices w and u_1 into $D3IS(G)$.

(e) See Fig. 2(e). The fifth special subgraph SG_5 consists of eight gray vertices, v_f, u, w, u_1, u_2, w_1, w_2, and v_1. (e1) is the same as (b1). (e2) There are three edges, $\{v_1, u_2\}$, $\{u_2, w_1\}$, and $\{u_1, w_1\}$. One can verify that if the graph has an edge $\{v_1, w_2\}$, it can be regarded as SG_4, and if there is an edge $\{u_1, w_2\}$, it can be regarded as SG_1 or SG_2. Therefore, all the three vertices v_1, u_1 and w_2 have neighbors which are not in SG_5. If the black vertex v is not in B, then ALG selects v and w_1 into $D3IS(G)$, and $|\{v, w_1\} \cup D_1(\{v, w_1\}) \cup D_2(\{v, w_1\})| \leq 13$. If v is in B, then ALG selects w and v_1 into $D3IS(G)$.

Recall that our algorithm ALG first finds every special subgraph and determines a (part of) solution in the special subgraphs as the preprocessing phase. After that, ALG iteratively executes the general phase, that is, it selects (i) the first candidate vertex v_f from $D_1(B)$ if $|\{v_f\} \cup D_1(v_f) \cup D_2(v_f) \setminus B| \leq 7$, but (ii) a neighbor u of v_f if $|\{v_f\} \cup D_1(v_f) \cup D_2(v_f) \setminus B| \geq 8$ into the distance-3 independent set. The following is the detailed description of ALG. In the preprocessing phase (**Phase 1**), the first candidate vertex v_f is selected and removed from a set F; the subgraph induced by $\{v_f\} \cup D_1(v_f) \cup D_2(v_f)$ is repeatedly checked whether it is identical to SG_1; after all SG_1's have been processed, the subgraph induced by $\{v_f\} \cup D_1(v_f) \cup D_2(v_f)$ is checked whether it is one of the four special subgraphs SG_2, SG_3, SG_4, and SG_5; and v_f is stored into a set C of "already checked" vertices. The vertex s_i in the distance-3 independent set is stored in $D3IS(G)$; its (closed) neighbors in $\{s_i\} \cup D_1(s_i) \cup D_2(s_i)$ are eliminated from V and stored into B.

Algorithm ALG

Input: Cubic graph $G = (V, E)$.

Output: Distance-3 independent set $D3IS(G)$ of G.

Initialization: Set $C = \emptyset$, $B = \emptyset$, $D3IS(G) = \emptyset$, and $F = \emptyset$.

Phase 1. Find all special subgraphs and determine a partial solution in them.
/* The vertices in all the special subgraphs SG_1, SG_2, SG_3, SG_4, and SG_5 are labeled as shown in Figures 2(a), (b), (c), (d), and (e), respectively. */

Step 0. Select arbitrarily one vertex v from V and set $F = F \cup \{v\}$.

Step 1. (i) If $B \cup C \neq V$ and thus $F \neq \emptyset$, then select arbitrarily one vertex $v_f \in F$, and set $C = C \cup \{v_f\}$. If the induced subgraph $G[\{v_f\} \cup D_1(v_f) \cup D_2(v_f) \setminus B]$ includes SG_1, then set $D3IS(G) = D3IS(G) \cup \{v, u_1\}$, $B = B \cup \{v, u_2\} \cup D_1(\{v, u_2\}) \cup D_2(\{v, u_2\})$, $F = D_1(B \cup C)$. Repeat **Step 1**.
(ii) If $B \cup C = V$, then set $C = \emptyset$ and $F = D_1(B) \setminus B$, and goto **Step 2**.

Step 2. (i) If $B \cup C \neq V$, then select $v_f \in F$ and set $C = C \cup \{v_f\}$. If the induced subgraph $G[\{v_f\} \cup D_1(v_f) \cup D_2(v_f) \setminus B]$ does not include any of the special subgraphs SG_2, SG_3, SG_4, and SG_5, then set $F = D_1(B \cup C)$ and repeat **Step 2**. If $G[\{v_f\} \cup D_1(v_f) \cup D_2(v_f) \setminus B]$ includes SG_2, SG_3, SG_4 and SG_5, then execute **Case 2-1**, **Case 2-2**, **Case 2-3**, and **Case 2-4**, respectively. (ii) If $B \cup C = V$, then goto **Phase 2**.

Case 2-1: (i) If $dist_G(u_1, w_2) = dist_G(u_2, w_1) = 3$, then set $D3IS(G) = D3IS(G) \cup \{u_2, w_1\}$ and $B = B \cup \{u_2, w_1\} \cup D_1(\{u_2, w_2\}) \cup D_2(\{u_2, w_1\})$. (ii) If $dist_G(u_1, w_2) = 1$ and $dist_G(u_2, w_1) = 3$, then set $D3IS(G) = D3IS(G) \cup \{u_2, w_1\}$ and $B = B \cup \{u_2, w_1\} \cup D_1(\{u_2, w_1\}) \cup D_2(\{u_2, w_1\})$. (iii) If $dist_G(u_1, w_2) = dist_G(u_2, w_1) = 1$, then set $D3IS(G) = D3IS(G) \cup \{u_1\}$ and $B = B \cup \{u_1\} \cup D_1(\{u_1\}) \cup D_2(\{u_1\})$. Set $F = D_1(B \cup C)$ and goto **Step 2**.

Case 2-2: Set $D3IS(G) = D3IS(G) \cup \{v_f, v_f'\}$ and $B = B \cup \{v_f, v_f'\} \cup D_1(\{v_f, v_f'\}) \cup D_2(\{v_f, v_f'\}))$. Set $F = D_1(B \cup C)$ and goto **Step 2**.

Case 2-3: (i) If $dist_G(u_1, w_2) \geq 2$ and $dist_G(u_1, w_1) = 1$, then set $D3IS(G) = D3IS(G) \cup \{u, w_2\}$ and $B = B \cup \{u, w_2\} \cup D_1(\{u, w_2\}) \cup D_2(\{u, w_2\})$. (ii) If $dist_G(u_1, w_2) \geq 2$ and $dist_G(u_2, w_1) = 1$, then $D3IS(G) = D3IS(G) \cup \{u, w_2\}$ and $B = B \cup \{u, w_2\} \cup D_1(\{u, w_2\}) \cup D_2(\{u, w_2\})$. (iii) If $dist_G(u_1, w_1) \geq 2$ and $dist_G(u_2, w_1) = 1$, then $D3IS(G) = D3IS(G) \cup \{w, u_1\}$ and $B = B \cup \{w, u_1\} \cup D_1(\{w, u_1\}) \cup D_2(\{w, u_1\})$. Set $F = D_1(B \cup C)$ and goto **Step 2**.

Case 2-4: (i) If $v \notin B$ (i.e., $v \in C$), then set $D3IS(G) = D3IS(G) \cup \{v, w_1\}$, $C = C \setminus \{v\}$, and $B = B \cup \{v, w_1\} \cup D_1(\{v, w_1\}) \cup D_2(\{v, w_1\})$. If $v \in B$, then $D3IS(G) = D3IS(G) \cup \{w, v_1\}$ and $B = B \cup \{w, v_1\} \cup D_1(\{w, v_1\}) \cup D_2(\{w, v_1\})$. Set $F = D_1(B \cup C)$ and goto **Step 2**.

Phase 2. If $B \neq V$, then repeat the following **Step 3**. Otherwise, goto **Termination**.

Step 3. Select one vertex v_f from $D_1(B) \setminus B$ such that $|\{v_f\} \cup D_1(v_f) \cup D_2(v_f) \setminus B|$ is minimum among all vertices in F. **(Case 3-1)** If $|(\{v_f\} \cup D_1(v_f) \cup D_2(v_f) \setminus B| \leq 7$, then $D3IS(G) = D3IS(G) \cup \{v_f\}$ and $B \doteq B \cup$

$\{v_f\} \cup D_1(\{v_f\}) \cup D_2(\{v_f\})$. **(Case 3-2)** If $|(\{v_f\} \cup D_1(v_f) \cup D_2(v_f) \setminus B| \geq 8$ and at most one vertex in $D_2(v_f) \setminus B$ is adjacent to vertices in $B \cup D_2(v_f)$, then set $D3IS(G) = D3IS(G) \cup \{v_f\}$ and $B = B \cup \{v_f\} \cup D_1(\{v_f\}) \cup D_2(\{v_f\})$. **(Case 3-3)** If $|\{v_f\} \cup D_1(v_f) \cup D_2(v_f) \setminus B| \geq 8$ and at least two vertices in $D_2(v_f) \setminus B$ are adjacent to vertices in $B \cup D_2(v_f)$, then select one, say, u, of two vertices in $D_1(v_f)$ such that $|\{u\} \cup D_1(u) \cup D_2(u) \setminus B|$ is minimum. **(Case 3-4)** If $|\{v_f\} \cup D_1(v_f) \cup D_2(v_f) \setminus B| \geq 8$ and $|\{u\} \cup D_1(u) \cup D_2(u) \setminus B| = |\{w\} \cup D_1(w) \cup D_2(w) \setminus B| = 7$ for $u, w \in D_1(v_f)$ and u is in a cycle $\langle u, u_1, u_2 \rangle$, then Set $D3IS(G) = D3IS(G) \cup \{u\}$ and $B = B \cup \{u\} \cup D_1(\{u\}) \cup D_2(\{u\})$. Goto **Step 3**.
Termination. Terminate and output $D3IS(G)$ as a solution.   [End of ALG]

Approximation Ratio. The algorithm ALG always outputs a feasible solution since ALG eliminates all vertices in $\{s\} \cup D_1(s) \cup D_2(s)$ from the solution candidates if s is in the solution. In this section, we will investigate the approximation ratio of ALG. We first give notation used in the following. Suppose that given a graph G, ALG outputs $ALG(G) = D3IS(G) = \{s_1, s_2, \cdots, s_\ell\}$. Also, without loss of generality, suppose that ALG selects those ℓ vertices into $D3IS(G)$, one by one in the order, i.e., first s_1, next s_2, and so on. Let v_i denote the first candidate vertex when the ith vertex s_i is selected into $D3IS(G)$, and it is called the *ith first candidate*. Also, we call s_i the ith *solution vertex*.

For a vertex v, let $B(v) = \{v\} \cup D_1(v) \cup D_2(v)$ be a set of vertices such that $dist_G(u, v) \leq 2$ for any $u \in B(v)$. Especially, for the ith solution vertex s_i in $ALG(G)$ ($i = 1, \cdots, \ell$), we call $B(s_i)$ the ith *solution block*. Let $B^-(s_i) = B(s_i) \cap (\bigcup_{j=1}^{i-1} B(s_j))$ and $B^+(s_i) = B(s_i) \setminus (\bigcup_{j=1}^{i-1} B(s_j))$, and we call $B^-(s_i)$ and $B^+(s_i)$ the ith *old solution block* and the ith *new solution block*, respectively.

Consider the time when the ith solution s_i is selected and $\bigcup_{j=1}^{i} B(s_j)$ are removed from V. Then, we define the *separate vertices* in $B(s_i)$ by $SV(s_i) = D_1(V \setminus (\bigcup_{j=1}^{i} B(s_j))) \cap B^+(s_i)$ for each i ($1 \leq i \leq \ell - 1$). Let $SV(ALG) = \bigcup_{i=1}^{\ell-1} SV(s_i)$. Also, we define the *near separate vertices* from s_i by $SV_{near}(s_i) = (D_1(s_i) \cup D_2(s_i)) \cap (\bigcup_{j=1}^{i-1} SV(s_j))$. Note that $SV_{near}(s_i)$ is *not in* $B^+(s_i)$. Let $B^*(s_i) = B^+(s_i) \cup SV_{near}(s_i)$. Moreover, let $SV_{near}(ALG) = \bigcup_{i=1}^{\ell-1} SV_{near}(s_i)$ and $SV_{far} = SV(ALG) \setminus SV_{near}(ALG)$. Now consider ℓ integers, δ_1 through δ_ℓ, which are associated with ℓ new solution blocks, $B^+(s_1)$ through $B^+(s_\ell)$, and initially set $\delta_1 = \cdots = \delta_\ell = 0$. Recall that each separate vertex sv in SV_{far} must be connected to one or two vertices not in $B(s_i)$. Now suppose that sv in SV_{far} is connected to a vertex in $B^+(s_j)$. Then, we set $\delta_j = +1$. Suppose that sv is connected to two vertices, one in $B^+(s_j)$ and one in $B^+(s_k)$ for $j \neq k$. Then, if $j > k$, then we set $\delta_j = +1$; otherwise, $\delta_k = +1$. Therefore, $\sum_{i=1}^{\ell} \delta_i = |SV_{far}|$ holds.

Lemma 1. *Suppose that $|B(v_i) \setminus \bigcup_{j=1}^{i-1} B(s_j)| = 8$. Also, suppose that the ith solution vertex s_i is selected in **Phase 2** of ALG, and s_i is not the first candidate v_i. Then, $|B^+(s_i)| \leq 7$ holds, and furthermore, if $|B^+(s_i)| = 7$, then s_i must be in a cycle of length at most three. (The proof is omitted.)*

Lemma 2. *Suppose that s_i ($2 \leq i \leq \ell$) is selected into $D3IS(G)$ in **Phase 2** of ALG. Then, $|B^*(s_i)| \leq 9$ holds. (The proof is omitted.)*

Lemma 3. *Suppose that given a graph $G = (V(G), E(G))$, only **Phase 1** is executed in ALG. Then, $|V(G)|/|ALG(G)| \leq 7.5$ is satisfied. (The proof is omitted.)*

From Lemma 1, if $|B^+(v_i)| = 8$ for $2 \leq i \leq \ell$ holds, then we can assume that s_i is always identical to v_i. Then, we can obtain the following lemma:

Lemma 4. *Suppose that $|B^+(v_i)| = 8$ for $2 \leq i \leq \ell$ and v_i is selected into $D3IS(G)$ in **Phase 2** of ALG, i.e., $s_i = v_i$ Then, $|SV(s_i)| - \delta_i \geq 4$ is satisfied.*

Proof. Since $|B^+(s_i)| = 8$ holds and $s_i = v_i$, then **(Case 3-2)** in **Phase 2** must be executed. Therefore, we can assume that at most one vertex in $D_2(s_i) \setminus \bigcup_{j=1}^{i-1} B(s_j)$ is adjacent to vertices in $\bigcup_{j=1}^{i-1} B(s_j) \cup D_2(v_i)$, which means that at least four vertices in $D_2(s_i) \setminus \bigcup_{j=1}^{i-1} B(s_j)$ are not connected to any vertex in $\bigcup_{j=1}^{i-1} B(s_j) \cup D_2(v_i)$. Therefore, we obtain $|SV(s_i)| \geq 4$. Suppose that $\delta_i \geq 1$, i.e., there must exist one vertex, say, sv, which is connected to a vertex, say, u in $D_2(s_i) \cap B^+(s_i)$, and the other neighbors of sv are in $\bigcup_{j=1}^{i-1} B(s_j)$. Hence $|B^+(u)| \leq 7$ holds, which implies that for the first candidate v_i in $B^+(s_i)$, $|B^+(s_i)| \leq |B^+(v_i)| \leq 7$, which is a contradiction. Hence $\delta_i = 0$. As a result, $|SV(s_i)| - \delta_i \geq 4$ is satisfied. This completes the proof of this lemma. $\square$

If $\delta_i > |SV(s_i)|$ is satisfied, then we call the iteration when s_i is selected *negative iteration*; otherwise, *positive iteration*.

Lemma 5. *Suppose that $s_i \in D3IS(G)$ is selected in **Phase 2** of ALG. Then, if the iteration is negative, then $|B^+(s_i)| \leq 6$ is satisfied.*

Proof. Since the current iteration is *negative*, there must exist a *separate* vertex, say, sv, in $\bigcup_{j=1}^{i-1} SV(s_j)$ such that sv is connected to some vertices in $D_2(s_i) \cap B^+(s_i)$ but not connected to any vertex in $\sum_{j=i+1}^{\ell} B^+(s_j)$, and $\delta_i \geq 1$. If each vertex, say, v', in $D_2(s_i) \cap B^+(s_i)$ is connected to two separate vertices, then $|B^+(s_i)| \leq |B^+(v')| \leq 4$, which is a contradiction. If v' is connected to one separate vertex and another in $\sum_{j=1}^{i-1} B(s_j)$, then $|B^+(s_i)| \leq |B^+(v')| \leq 5$, again contradiction. Also, if a vertex, say, v'', in $D_2(s_i) \cap B^+(s_i)$ is connected to two vertices in $\sum_{j=1}^{i-1} B(s_j)$, then $|B^+(s_i)| \leq |B^+(v'')| \leq 6$. Therefore, there must be at least one separate vertex, say, sv, which is connected to a non-separate one, say, $\overline{sv}$. Now, $|B^+(v_i)| \leq 7$ and thus $s_i = v_i$. In the following, we assume that $|B^+(v_i)| = 7$. Here there are only two cases as illustrated in Fig. 3.

(Case 1) See Fig. 3(a). Suppose that $\delta_i = 1$ and thus $|SV(s_i)| = 0$. Then, there is at least one separate vertex sv which is connected to u_1, u_2 or w_1 but not connected to any in $\sum_{j=i+1}^{\ell} B^+(s_j)$. If $dist_G(sv, w_1) = 1$, then $|B^+(s_i)| = |B^+(v_i)| \leq |B^+(w_1)| \leq 5$. If $dist_G(sv, u_1) = 1$ (or equivalently, $dist_G(sv, u_2) = 1$), then u_1 must be connected to one of $\{u_2, u_3, v\}$. If $dist_G(u_1, u_2) = 1$, then

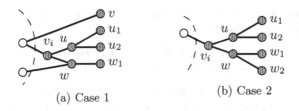

(a) Case 1   (b) Case 2

Fig. 3. Two cases in the proof of Lemma 5.

$|B^+(s_i)| = |B^+(v_i)| \leq |B^+(u_1)| \leq 5$. Moreover, if $dist_G(u_1, u_3) = 1$ and $dist_G(u_2, w_1) = 1$, then $|B^+(s_i)| = |B^+(v_i)| \leq |B^+(u_1)| \leq 6$. If $dist_G(u_1, u_3) = 1$ and $dist_G(u_2, v) = 1$, then $dist_G(v, w_1) = 1$ must hold and thus the graph is identical to SG_4. As a result, this lemma holds. Now, suppose that $\delta_i \geq 2$, and there are at least two separate vertices, say, sv_1 and sv_2, which are connected to different two vertices in $\{u_1, u_2, w_1\}$. Suppose that $dist_G(u_1, sv_1) = dist_G(u_2, v_2) = 1$. If v is connected to both u_1 and u_2, then $|B^+(s_i)| = |B^+(v_i)| \leq |B^+(u_1)| \leq 5$. If v is connected to u_1 (or u_2) and w_1, then $|B^+(s_i)| = |B^+(v_i)| \leq |B^+(u_1)| \leq 6$. Therefore, this lemma holds.

(Case 2) See Fig. 3(b). First, without loss of generality, let u_1 be the non-separate vertex which is connected to vertices in $D_2(s_i) \setminus \bigcup_{j=1}^{i-1} B(s_j)$. If $dist_G(u_1, u_2) = 1$, then $|B^+(u_1)| \leq 5 < 6$, and thus this lemma holds. Next, consider the case $dist_G(u_1, w_1) = 1$. Note that $dist_G(u_1, w_2) = 1$ is essentially equivalent to $dist_G(u_1, w_1) = 1$. There are two cases: (2-1) $\delta_i = 1$ and (2-2) $\delta_i \geq 2$. (2-1) If $\delta_i = 1$, then $|SV(s_i)| = 0$. One can see that w_2 cannot be connected to u_2 since such a graph is SG_1 or SG_2. Now suppose that w_1 is not connected to any in $B^+(v_i) \cap D_2(v_i)$. Then, w_1 must have two neighbors in $\bigcup_{j=1}^{i-1} B(s_j)$ and hence $|B^+(w_1)| \leq 6$. Therefore, w_1 must be connected w_2 and another vertex in $\bigcup_{j=1}^{i-1} B(s_j)$, which implies that $|B^+(w_1)| \leq 6$. (2-2) The condition $\delta \geq 2$ means that there are two far separate vertices for $B^+(s_i)$. If only one vertex in $B^+(s_i)$ is connected to those far separate vertices, then $|B^+(u)| \leq 4$ and thus $|B^+(v_i)| \leq 4$, which is a contradiction. Therefore, we assume that two far separate vertices, sv_1 and sv_2, are connected two vertices in $D_2(v_i) \cap B^+(v_i)$, i.e., one to one. Note that two vertices in $D_2(v_i) \cap B^+(v_i)$ cannot be connected to vertices in $\bigcup_{j=1}^{i-1} B(s_j)$. The essentially different cases we have to consider are: (i) $dist_G(sv_1, u_1) = dist_G(sv_2, u_2) = 1$, and (ii) $dist_G(sv_1, u_2) = dist_G(sv_2, w_1) = 1$. (i) $dist_G(sv_1, u_1) = dist_G(sv_2, u_2) = 1$. If $dist_G(u_1, u_2)$, then $|B^+(u_1)| \leq 4$. If $dist_G(u_1, w_1) = 1$, then u_2 must be connected to w_1 since the graph with $dist_G(u_2, w_2) = 1$ is identical to SG_2 or SG_3. Therefore, $|B^+(u_2)| \leq 5$, contradiction. (ii) $dist_G(sv_1, u_2) = dist_G(sv_2, w_1) = 1$. The similar arguments to the above ones can be applied and details are omitted here. □

Corollary 1. *Suppose that $s_i \in D3IS(G)$ is selected in* **Phase 2** *of ALG. Then, if $|B^+(s_i)| = 7$, then $|SV(s_i)| \geq \delta_i$ is satisfied.*

Proof. From Lemma 5, the iteration when s_i is selected must be positive, i.e., $|SV(s_i)| \geq \delta_i$ hold from the definition. □

Lemma 6. *Suppose that $s_i \in D3IS(G)$ is selected in* **Phase 2** *of* ALG*. If $|B^+(s_i)| \geq 5$, then $\delta_i \leq \beta_i$, where $\beta_i = |D_2(s_i) \cap B^+(s_i)|$.*

Proof. Suppose for contradiction that $\delta_i > \beta_i$. Since δ_i edges are incident with β_i vertices, there must exist one vertex, say, $u \in D_2(v_i) \setminus \bigcup_{j=1}^{i-1} B(s_j)$ which is the endpoint of at least two of δ_i edges, i.e., the vertex u is connected to at least two far separate vertices. Furthermore, they cannot be connected to any vertex in $B^+(s_j)$ for $i < j \leq \ell$. One can verify that $|B^+(u)| \leq 4$ holds for the vertex u. Since ALG selects v_i such that $|B^+(v_i)|$ is minimum as the first candidate, $|B^+(v_i)| \leq |B^+(u)| \leq 4$ must holds for v_i, which is a contraction to the assumption that $|B^+(s_i)| \geq 5$. As a result, $\delta_i \leq \beta_i$. □

Lemma 7. *Suppose that $s_i \in D3IS(G)$ is selected in* **Phase 2** *of* ALG*. Then, (1) if $|B^+(s_i)| = 6$ or (2) $|B^+(s_i)| = 5$, then $|SV(s_i)| - \delta_i \geq -2$. (3) If $|B^+(s_i)| = 4$, then $|SV(s_i)| - \delta_i \geq -4$. (The proof is omitted.)*

In the following, we assume that ALG selects ℓ_1 vertices, s_1 through s_{ℓ_1}, and ℓ_2 vertices, s_{ℓ_1+1} through $s_{\ell_1+\ell_2}$, into $D3IS(G)$ in **Phase 1** and **Phase 2**, respectively. That is, $\ell = \ell_1 + \ell_2$. Let i_k denote the number of the solution vertices s_i such that $|B^+(s_i)| = k$ for $5 \leq k \leq 8$. Also, let $i_{\leq 4}$ denote the number of the solution vertices s_i such that $|B^+(s_i)| \leq 4$. Let $SV'(ALG) = \bigcup_{i=\ell_1+1}^{\ell} SV(s_i)$ and $SV'_{near}(ALG) = \bigcup_{i=\ell_1+1}^{\ell} SV_{near}(s_i)$. Then, if **Phase 1** is executed (i.e., at least one special subgraph is included in the input graph G), then let p be the number of vertices which are put into B in **Phase 1** and connected to vertices in $\bigcup_{i=\ell_1+1}^{\ell} B^+(s_i)$; otherwise, i.e., if no special subgraphs are not included in G and thus **Phase 1** is not executed, then let p be equal to $|SV(s_1)|$.

Lemma 8. *(1) If* **Phase 1** *of* ALG *is not executed, then $|SV_{near}(ALG)| \geq p + 4i_8 - 2i_6 - 3i_5 - 4i_{\leq 4}$ is satisfied. (2) Suppose that* **Phase 1** *is executed and $s_i \in D3IS(G)$ is selected in* **Phase 2** *for $\ell_1 + 1 \leq i \leq \ell$. Then $|SV'_{near}(ALG)| \geq p + 4i_8 - 2i_6 - 3i_5 - 4i_{\leq 4}$ is satisfied.*

Proof. (1) We first assume that **Phase 1** is *not* executed. Since $|SV_{near}(ALG)| = |SV(ALG)| - |SV_{far}|$, it satisfies $|SV_{near}(ALG)| = |SV(ALG)| - |SV_{far}| \geq \sum_{i=1}^{\ell} |SV(s_i)| - \sum_{i=1}^{\ell} \delta_i = \sum_{i=1}^{\ell}(|SV(s_i)| - \delta_i)$. By Lemmas 4 and 7, and Corollary 1, we can know $\sum_{i=1}^{\ell}(|SV(s_i)| - \delta_i) \geq (|SV(s_1)| - 0) + \sum_{i=2}^{\ell}(|SV(s_i)| - \delta_i) \geq p + 4i_8 - 2i_6 - 3i_5 - 4i_{\leq 4}$. (2) Suppose that **Phase 1** is executed and $s_i \in D3IS(G)$ is selected in **Phase 2** for $\ell_1 + 1 \leq i \leq \ell$. Then, $|SV'_{near}(ALG)| = (p + |SV'(ALG)|) - |SV_{far}| \geq p + \sum_{i=\ell_1+1}^{\ell}(|SV(s_i)| - \delta_i)$. By Lemmas 4 and 7, and Corollary 1, $|SV'_{near}(ALG)| \geq p + 4i_8 - 2i_6 - 3i_5 - 4i_{\leq 4}$. This completes the proof of this lemma. □

Corollary 2. *(1) If* **Phase 1** *of* ALG *is not executed, then it satisfies $4i_8 \leq 9\ell + 1 + 2i_6 + 3i_5 + 4i_{\leq 4} - n - p$. (2) Suppose that* **Phase 1** *is executed and $s_i \in D3IS(G)$ is selected in* **Phase 2** *for $\ell_1 + 1 \leq i \leq \ell$. Let $n_2 = |\bigcup_{i=\ell_1+1}^{\ell} B^+(s_i)|$. Then, $4i_8 \leq 9\ell_2 + 2i_6 + 3i_5 + 4i_{\leq 4} - n_2 - p$ is satisfied.*

Proof. (1) Suppose that **Phase 1** is *not* executed. From Lemma 8, $\sum_{i=1}^{\ell}(|B^*(s_i)| - |B^+(s_i)|) \geq |SV_{near}(ALG)| \geq p + 4i_8 - 2i_6 - 3i_5 - 4i_{\leq 4}$. Since $|B^*(s_i)| \leq 9$ holds for $i \geq 2$ from Lemma 2, $10 + 9(\ell-1) \geq |B^+(s_1)| + 9(\ell-1) - n \geq p + 4i_8 - 2i_6 - 3i_5 - 4i_{\leq 4}$ and we can obtain the inequality $4i_8 \leq 9\ell + 1 + 2i_6 + 3i_5 + 4i_{\leq 4} - n - p$. (2) Suppose that **Phase 1** is executed. From Lemma 8, we know $\sum_{i=\ell_1+1}^{\overline{\ell}}(|B^*(s_i)| - |B^+(s_i)|) \geq |SV'_{near}(ALG)| \geq p + 4i_8 - 2i_6 - 3i_5 - 4i_{\leq 4}$. Furthermore, since $|B^*(s_i)| \leq 9$ holds for $i \geq 2$ from Lemma 2, the following inequality holds: $9\ell_2 - n_2 \geq \sum_{i=\ell_1+1}^{\ell}(|B^*(s_i)| - |B^+(s_i)|) \geq p + 4i_8 - 2i_6 - 3i_5 - 4i_{\leq 4}$. Hence, we get $4i_8 \leq 9\ell_2 + 2i_6 + 3i_5 + 4i_{\leq 4} - n_2 - p$. □

Theorem 1. *ALG achieves an approximation ratio of* $1.875 + O(\frac{1}{n})$.

Proof. We need to investigate the following three situations: (1) $1 \leq \ell_1 < \ell$, i.e., both **Phase 1** and **Phase 2** are executed, (2) $\ell_1 = 0$, i.e., **Phase 1** is not executed, and (3) $\ell_1 = \ell$, i.e., **Phase 2** is not executed. (1) One can see that $7.5\ell_1 + 8i_8 + 7i_7 + 6i_6 + 5i_5 + 4i_{\leq 4} \geq n$ holds. From $\ell = \ell_1 + i_8 + i_7 + i_6 + i_5 + i_{\leq 4}$, we obtain $4\ell + i_5 + 2i_6 + 3i_7 + 4i_8 + 3.5\ell' \geq n$. Furthermore, since $i_7 = \ell - \ell_1 - i_8 - i_6 - i_5 - i_{\leq 4}$ holds, we get $4\ell + i_5 + 2i_6 + 3(\ell - \ell_1 - i_8 - i_6 - i_5 - i_{\leq 4}) + 4i_8 + 3.5\ell_1 \geq n$. That is, $7\ell - 2i_5 - i_6 - 3i_{\leq 4} + i_8 + 0.5\ell_1 \geq n$ holds. Recall that $4i_8 \leq 9\ell_2 + 2i_6 + 3i_5 + 4i_{\leq 4} - n_2 - p$ as shown in Corollary 2. Since $\ell_2 = \ell - \ell_1$ and $n_2 \geq n - 7.5\ell_1$, we get $4i_8 \leq 9\ell + 2i_6 + 3i_5 + 4i_{\leq 4} - n - 1.5\ell_1 - p$. Since $\ell_1 \leq \ell - 1$, we obtain $\ell \geq (5n + 1.5)/37.5 > n/7.5$. (2) $\ell_2 = \ell$ and $n_2 = n$. Obviously, $p \geq 1$. From $|B^+(s_1)| \leq 10$ and the definitions on i_k, $10 + 8i_8 + 7i_7 + 6i_6 + 5i_5 + 4i_{\leq 4} \geq |B^+(s_1)| + 8i_8 + 7i_7 + 6i_6 + 5i_5 + 4i_{\leq 4} \geq n$ holds. Note that $1 + i_8 + i_7 + i_6 + i_5 + i_{\leq 4} = \ell$. Hence, we obtain $7\ell + i_8 - 2i_5 - i_6 - 3i_{\leq 4} + 3 \geq n$. From Corollary 2, $7\ell + (9\ell + 2i_6 + 3i_5 + 4i_{\leq 4} - n)/4 - 2i_5 - i_6 - 3i_{\leq 4} + 3 \geq n$ holds. Therefore, we obtain $\ell \geq (5n - 12)/37 > (5n - 12)/37 \geq n/7.5 - 12/37$. (3) From Lemma 3, $\ell \geq n/7.5$. Since $|OPT(G)| \leq \frac{n}{4}$ holds from Proposition 1, ALG achieves the approximation ratio of $1.875 + O(1/n)$. □

References

1. Agnarsson, G., Damaschke, P., Halldórsson, M.H.: Powers of geometric intersection graphs and dispersion algorithms. Discret. Appl. Math. **132**, 3–16 (2004)
2. Berman, P., Fujito, T.: On approximation properties of the independent set problem for low degree graphs. Theory Comput. Syst. **32**(2), 115–132 (1999)
3. Brandstädt, A., Giakoumakis, V.: Maximum weight independent sets in hole- and co-chair-free graphs. Inf. Process. Lett. **112**, 67–71 (2012)
4. Chlebík, M., Chlebíková, J.: Complexity of approximating bounded variants of optimization problems. Theoret. Comput. Sci. **354**, 320–338 (2006)
5. Eto, H., Guo, F., Miyano, E.: Distance-d independent set problems for bipartite and chordal graphs. J. Comb. Optim. **27**(1), 88–99 (2014)
6. Eto, H., Ito, T., Liu, Z., Miyano, E.: Approximability of the distance independent set problem on regular graphs and planar graphs. In: Chan, T.-H.H., Li, M., Wang, L. (eds.) COCOA 2016. LNCS, vol. 10043, pp. 270–284. Springer, Heidelberg (2016)

7. Gavril, F.: Algorithms for minimum coloring, maximum clique, minimum covering by cliques, and maximum independent set of chordal graph. SIAM J. Comput. **1**, 180–187 (1972)
8. Gavril, F.: Algorithms on circular-arc graphs. Networks **4**, 357–369 (1974)
9. Golumbic, M.C.: The complexity of comparability graph recognition and coloring. Computing **18**, 199–208 (1977)
10. Harary, F.: Graph Theory. Addison-Wesley, Reading (1969)
11. Lozin, V.V., Milanič, M.: A polynomial algorithm to find an independent set of maximum weight in a fork-free graph. J. Discret. Algorithm **6**, 595–604 (2008)
12. Minty, G.J.: On maximal independent sets of vertices in claw-free graphs. J. Comb. Theory Ser. B **28**, 284–304 (1980)
13. Zuckerman, D.: Linear degree extractors and the inapproximability of max clique and chromatic number. Theory Comput. **3**(1), 103–128 (2007)

Computational Geometry II

Online Inserting Points Uniformly on the Sphere

Chun Chen[1], Francis C.M. Lau[2], Sheung-Hung Poon[3], Yong Zhang[1,2], and Rong Zhou[1(✉)]

[1] Shenzhen Institutes of Advanced Technology, Chinese Academy of Sciences, Shenzhen, People's Republic of China
{chun.chen,zhangyong,rong.zhou}@siat.ac.cn
[2] Department of Computer Science, The University of Hong Kong, Hong Kong, People's Republic of China
fcmlau@cs.hku.hk
[3] School of Computing and Informatics, Universiti Teknologi Brunei, Mukim Gadong A, Brunei Darussalam
sheung.hung.poon@gmail.com

Abstract. In many scientific and engineering applications, there are occasions where points need to be inserted uniformly onto a sphere. Previous works on uniform point insertion mainly focus on the offline version, i.e., to compute N positions on the sphere for a given interger N with the objective to distribute these points as uniformly as possible. An example application is the Thomson problem where the task is to find the minimum electrostatic potential energy configuration of N electrons constrained on the surface of a sphere. In this paper, we study the online version of uniformly inserting points on the sphere. The number of inserted points is not known in advance, which means the points are inserted one at a time and the insertion algorithm does not know when to stop. As before, the objective is achieve a distribution of the points that is as uniform as possible at each step. The uniformity is measured by the gap ratio, the ratio between the maximal gap and the minimal gap of any pair of inserted points. We give a two-phase algorithm by using the structure of the regular dodecahedron, of which the gap ratio is upper bounded by 5.99. This is the first result for online uniform point insertion on the sphere.

1 Introduction

In this paper, we consider the problem of inserting points onto the sphere such that the inserted points are as uniformly spaced as possible. There are many applications, e.g., the *Thomson problem* [13] which was introduced by the physicist Sir Joseph John Thomson in 1904; the objective is to determine the configuration of N electrons on the surface of a unit sphere that minimizes the electrostatic potential energy, which translates directly into the problem of placing N points on the surface of the sphere as uniformly as possible. The minimum energy configuration of the Thomson problem and other configurations with

© Springer International Publishing AG 2017
S.-H. Poon et al. (Eds.): WALCOM 2017, LNCS 10167, pp. 243–253, 2017.
DOI: 10.1007/978-3-319-53925-6_19

uniform point distribution on the unit sphere play important roles in many scientific and engineering applications [3,6,8,9], e.g., 3D projection reconstruction of Computed Tomography (CT) or Magnetic Resonance Images (MRI).

From the perspective of computer science, the traditional Thomson problem is offline, i.e., the number of points is known in advance and the objective is to place these points on the sphere as uniformly as possible. Inserting points in an online fashion is also an interesting problem, which might have its application in real life. An example is to assign an unknown number of volunteers as they show up to an area where a rescue mission is in progress. In the online version, the points are inserted over a time span, and the strategy has no idea about the number of points to be inserted. The position of a point cannot be changed after it has been inserted.

Any solution to the online problem is to be measured by the uniformity of the distribution of the points. There are several ways to define the uniformity of a set of points. Some studies define the uniformity according to the minimal pairwise distance [7,11]. In discrepancy theory [5,10], uniformity is defined as the ratio between the maximal and minimal number of points in a fixed shape within the area. In this paper, uniformity is defined to be the gap ratio, which is the ratio between the maximal gap and the minimal gap between any pair of points.

Formally, let S be the surface of a 3-dimensional unit sphere. The task is to insert a sequence of points onto S. Let p_i be the i-th point to be inserted and $S_i = \{p_1, \ldots, p_i\}$ be the configuration in S after inserting the i-th point. In configuration S_i, let the maximal gap be $G_i = max_{p \in S} min_{q \in S_i} 2 \cdot \widehat{d}\ (p, q)$, the minimal gap be $g_i = min_{p,q \in S_i, p \neq q} \widehat{d}\ (p, q)$, where $\widehat{d}\ (p, q)$ is the spherical distance between points p and q, i.e., the shortest distance along the surface of the sphere from p to q. In other words, the maximal gap is the spherical diameter of the largest empty circle while the minimal gap is the minimal spherical distance between two inserted points. We call $r_i = G_i/g_i$ the i-th gap ratio. The objective is to insert points onto S as uniformly as possible so that the maximal gap ratio (min max$_i\,r_i$) during the insertion of the whole set of points is minimized.

The problem of uniformly inserting points in a given area has been studied before. Teramoto et al. [12] and Asano and Teramoto [2] showed that the Voronoi insertion is a good strategy on the plane; moreover, the gap ratio of the Voronoi insertion is proved to be at most 2. They also studied insertion onto a one-dimensional line; if the algorithm knows the number n of the points to be inserted, an insertion strategy with maximal gap ratio $2^{\lfloor n/2 \rfloor/(\lfloor n/2 \rfloor +1)}$ can be derived. If the points must be inserted at the fixed grid points, Asano [1] gave an insertion strategy with uniformity 2 for the one dimensional case. For insertion on two-dimensional grid, Zhang et al. [14] proved the lower bound to be at least 2.5 and gave an algorithm with the maximal gap ratio 2.828. Recently, Bishnu et al. [4] considered some variants and measurements of the insertion on the Euclidean space.

In the remainder of this paper, we present a strategy for online inserting points uniformly onto the surface of a sphere with a maximal gap ratio of no more than 5.99. This is the first result for the problem of online insertion of points on sphere.

2   Point Insertion Strategy

A simple intuitive idea is to greedily insert the incoming point at the "center" of the largest empty spherical surface area. The early steps of such a greedy approach are simple; however, when many points have been inserted, the shapes of different local structures may vary significantly and the configuration may become very complicated. As a result, the computational cost of finding the largest empty spherical surface area and then computing its center may become prohibitive.

Observe that once some points have been inserted, the sphere is partitioned into local structures and the next point insertion within the area of some local structure will only affect the local configuration, i.e., the spherical distances (including the max gap and min gap) outside this area do not change. Based on this observation, a two-phase strategy can be devised. In the first phase, we use a polyhedron to approximate the sphere and points are inserted at the vertices of the polyhedron. After all vertices of the polyhedron are occupied, the second phase starts. In the second phase, we recursively compute the point positions in all faces of the polyhedron, and the inserted points on the sphere are the projections of these positions onto the sphere.

As mentioned before, the computation cost of the simple greedy approach is large due to the complicated local structures on the sphere when the a large number of points have been inserted. In this paper, a regular dodecahedron is used to simulate the shape of the sphere. A regular dodecahedron has twelve identical regular pentagonal faces and twenty vertices. In the first phase, handling the insertion of twenty points on the sphere is quite straightforward and the gap ratio is not large. The main advantage of the regular dodecahedron lies in the processing of the second phase. Since all faces of the decahedron are identical, we only need to consider how to insert points onto the sphere with respect to a regular pentagon. When the number of points inserted increases, the refinement of the regular pentagon contains local structures which can be categorized as three types (see Sect. 2.2). Then according to these three types of local shapes, we can impose a recursive procedure to compute the next point insertion positions.

Since changing the radius of the sphere does not affect the gap ratio, for the convenience of computation, we assume that the radius of the sphere is $\sqrt{3}$. Thus, the length of each edge of the corresponding regular dodecahedron is $\frac{4}{\sqrt{5}+1}$. In the following, we give the details of the two phases of our algorithm.

2.1   The First Phase

In our strategy, the sphere can be divided into 12 sections by projecting the 20 vertices and all edges of the regular dodecahedron onto the sphere, as shown in Fig. 1. On the dodecahedron, eight orange vertices with coordinates

(± 1, ± 1, ± 1) form a cube (dotted lines). Let O be the center of the cube and let $\phi = (1 + \sqrt{5})/2 \approx 1.618$ be the golden ratio. Four green vertices lying at $(0, \pm 1/\phi, \pm\phi)$ form a rectangle on the yz-plane. Four blue vertices lying at $(\pm 1/\phi, \pm\phi, 0)$ form a rectangle on the xy-plane. Four pink vertices lying at $(\pm\phi, 0, \pm 1/\phi)$ form a rectangle on the xz-plane.

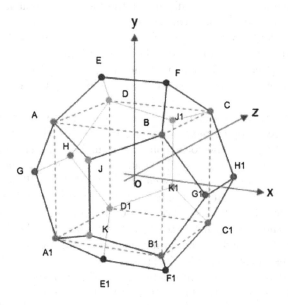

Fig. 1. Vertex distribution of the regular dodecahedron. (Color figure online)

The insertion strategy of the twenty points is as follows.

1. First insert eight points at orange vertices with coordinates (± 1, ± 1, ± 1), i.e., the vertices of the cube ($A, B, C, D, A_1, B_1, C_1, D_1$). The order of the inserted points is A, C_1, followed by an arbitrary order of the remaining 6 points.
2. Then insert the remaining twelve points in any arbitrary order.

Lemma 1. *During the insertion at the first eight vertex points of the dodecahedron, the gap ratio is no more than 2.55.*

Proof. After the insertion of two points at A and C_1, the maximal gap and the minimal gap are both the spherical distance between these two points, i.e.,

$$G_2 = g_2 = \widehat{d}\,(A, C_1).$$

In this case, the gap ratio $r_2 = 1$.

Note that the radius of the sphere is $\sqrt{3}$. Arbitrarily choose any one of the remaining six points, w.l.o.g., say B. After the insertion at B, the maximal gap is still $G_3 = \widehat{d}\,(A, C_1) = \sqrt{3}\pi$ while the minimal gap decreases to $g_3 = \widehat{d}\,(A, B)$, which is the value of $\sqrt{3}\angle AOB$. Since $|OA| = |OB| = \sqrt{3}$ and $|AB| = 2$, $\angle AOB = 2 \cdot \arcsin(\frac{\sqrt{3}}{3}) = 1.231$ and $g_3 = 2.132$. Thus, the gap ratio $r_3 = G_3/g_3 = 2.55$.

After inserting any other points in this sub-phase, the value of the minimal gap does not change while the value of the maximal gap may decrease. After all the eight points have been inserted, the maximal gap

$$G_8 = \sqrt{3} \cdot \angle AOB_1 = 3.309$$

while the minimal gap $g_8 = g_3$. Thus at this stage, the gap ratio is $G_8/g_8 = 1.55$.

Hence, the maximal gap ratio for inserting the first eight points is 2.55. □

Now we analyze the gap ratio for inserting the remaining twelve points at the vertices of the dodecahedron.

Lemma 2. *During the process of inserting the remaining twelve vertex points of the dodecahedron, the maximal gap ratio for the sphere is at most 2.615.*

Proof. W.l.o.g., assume that the first point inserted in this sub-phase is E, and thus the minimal gap $g_9 = \overset{\frown}{d}(A, E)$. At this stage, since all eight points on the cube have been inserted, the maximal gap $G_9 = \overset{\frown}{d}(A, B_1) = 3.309$.

$$g_9 = \sqrt{3} \cdot 2 \cdot \arcsin \frac{|AE|/2}{R} = \sqrt{3} \cdot 2 \cdot \arcsin \frac{2}{(\sqrt{5}+1)\sqrt{3}} = 1.264.$$

Thus, the gap ratio

$$r_9 = \frac{G_9}{g_9} = 2.618.$$

For the remaining eleven points in this sub-phase, the minimal gap will not decrease while the maximal gap will not increase. Hence, the maximal gap in this sub-phase is at most 2.615. □

Lemma 3. *The maximal gap ratio in the first phase is 2.618.*

2.2   The Second Phase

After all the vertices on the dodecahedron have been inserted, the second phase begins. As mentioned before, the regular dodecahedron has some good property and consequently, further point insertions can be done recursively on the sphere with respect to the corresponding structures of the faces of the dodecahedron after some points have been inserted onto them. Moreover, in our strategy, we first compute positions on the faces of the regular dodecahedron, and the true insertions will be done at the projected positions of these computed points on the sphere.

By such an implementation, point insertion on the sphere is reduced to the point insertion on the plane faces of the dodecahedron, which is much easier to handle. However, the gap ratio on the plane (pentagon) is smaller than that on the sphere. In the remaining part of this subsection, we first show that the difference of the gap ratios on sphere and on plane is quite small, and then we give the strategy of how to insert points on a pentagon.

The Difference of the Gap Ratio. W.l.o.g., we consider point insertion on the pentagon $AJBFE$. First, we consider the situation where two inserted points are both on the sphere and on the pentagon. Since only five vertices of the pentagon satisfy such condition, there are two cases to be examined. Let the edge of the pentagon be ℓ, i.e., $|AJ| = \ell$. Let O' be the center of the pentagon $AJBFE$. Since $|AJ|^2 = |AO'|^2 + |JO'|^2 - 2|AO'| \cdot |JO'| \cos(2\pi/5)$, we have $|AO'| = |JO'| = 0.85\ell$, and $|OO'| = \sqrt{R^2 - |AO'|^2} = 1.11\ell$.

- First, we consider the subsituation that the two inserted points are not adjacent vertices of the pentagon. W.l.o.g., let A and B denote these two inserted points. Since the angle $\angle AJB = 3\pi/5$, we can see that $|AB| = 1.618\ell$. By the property of the regular dodecahedron, the radius of the sphere $R = 1.4\ell$. We then have

$$\widehat{d}\,(A,B) = R \cdot \angle AOB = R \cdot 2 \cdot \arcsin \frac{|AB|}{2R} = 1.231R.$$

Thus,

$$\frac{\widehat{d}\,(A,B)}{|AB|} = 1.066.$$

- Then we consider the subsituation that the two inserted points are adjacent vertices of the pentagon. W.l.o.g., let A and J denote these two inserted points. By a similar analysis, we have

$$\widehat{d}\,(A,J) = R \cdot \angle AOJ = R \cdot 2 \cdot \arcsin \frac{|AJ|}{2R} = 0.73R.$$

Thus,

$$\frac{\widehat{d}\,(A,J)}{|AJ|} = 1.023.$$

By the above analysis, we can see that for two positions that are on the sphere, the ratio between the spherical distance and the direct distance is monotonically increasing with respect to the corresponding subtended angle.

Next we consider the situation that both of the two inserted points lie inside the pentagon.

As shown in Fig. 2, C and D are two points lying inside the pentagon, and C' and D' are their projections on the sphere respectively. Let $|C'D'|$ be the direct distance between C' and D'; thus,

$$\frac{|C'D'|}{|CD|} \leq \frac{R}{|OO'|} = 1.26.$$

According to the previous description, $\widehat{d}\,(C',D')/|C'D'| \leq \widehat{d}\,(A,B)/|AB|$ since with respect to the sphere, the subtended angle is maximized for A and B. Thus,

$$\frac{\widehat{d}\,(C',D')}{|CD|} = \frac{\widehat{d}\,(C',D')}{|C'D'|} \cdot \frac{|C'D'|}{|CD|} \leq 1.34.$$

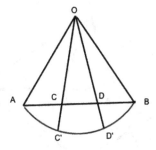

Fig. 2. Points on the plane and the corresponding projections on the sphere.

For any two spherical distances $\widehat{d}\,(A,B)$ and $\widehat{d}\,(C,D)$, its ratio is upper bounded by

$$\frac{\widehat{d}\,(A,B)}{\widehat{d}\,(C,D)} \le 1.34 \cdot \frac{|AB|}{|CD|}.$$

Therefore, the comparison between two spherical distances can be reduced to the comparison between two direct distances and the ratio would not change much.

Insertion on the Pentagon. In this part, we describe how to compute the point insertion positions on the pentagon.

The pentagon can be recursively partitioned into smaller polygons of one of three shapes, as shown in Fig. 3. For the pentagon $AJBFE$, by connecting non-adjacent vertices, we can see that the pentagon is partitioned into eleven parts, one smaller pentagon $ajbfe$, five isosceles triangles with vertex angle $\frac{\pi}{5}$ (Aae, Jja, Bbj, Ffb and Eef), and five isosceles triangles with vertex angle $\frac{3\pi}{5}$ (AJa, JBj, BFb, FEf and EAe).

The isosceles triangles can be further partitioned into smaller isosceles triangles of the above two shapes, as shown in Fig. 4. For example, in isosceles triangle AJj, by adding the point p_1, two isosceles triangles, say Jjp_1 and AJp_1 emerge, which are still of the above two shapes. For isosceles triangle AeE, by adding a point h_1, AeE is partitioned into two isosceles triangles Aeh_1 and Ah_1E, both of which are again of the above two shapes.

In the above description, we can see that the pentagon can be recursively partitioned into three types of polygons. Such property can be used in the insertion strategy in order to reduce the complexity of the computation.

Insertion Strategy. In the insertion strategy, three queues Q_1, Q_2 and Q_3 are used to store these three types of shapes; for each type, the sizes of the objects in the corresponding queue are non-increasing, where the size of a polygonal shape is defined to be the length of its longest edge.

When a new point comes in, the following procedure applies.

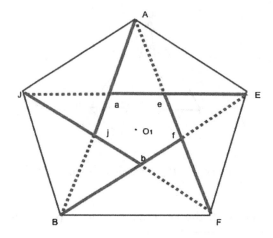

Fig. 3. The insertion in the pentagon.

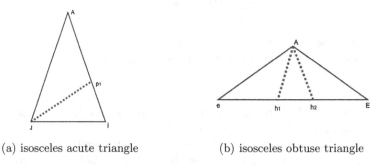

(a) isosceles acute triangle   (b) isosceles obtuse triangle

Fig. 4. The insertion in isosceles triangle.

- Compare the insertions on the heads of these three queues and select the one with the largest minimal gap if the point is inserted at an appropriate position. As shown in Fig. 3, the inserted points in pentagon $AJBFE$ can be a, j, b, f or e, respectively; as shown in Fig. 4(a), the inserted point in the isosceles acute triangle AJj is p_1; as shown in Fig. 4(b), the inserted point in the isosceles obtuse triangle AeE is h_1 or $h2$. Assume that the selected polygon is P.
- Determine the point insertion position x on the corresponding polygon P.
- Note that P is partitioned into some smaller polygons. Remove P from the head of the queue and then add these smaller polygons at the tail of the corresponding queues.
- Find the point insertion position X which is the projection of x onto the sphere.

Note that when we are processing a triangle, the inserted point is on an edge of the triangle, which is also on the edge of another triangle with the same size and of the same type or the other type. In this case, both of these two triangles will be partitioned and removed from the queue, and the newly created smaller triangles will be added at the tails of the corresponding queues.

Lemma 4. *After a point is inserted in the above operation, in each of the three queues, the sizes of the objects are still in non-increasing order.*

Proof. This lemma can be proved as follows.

– First, we consider the pentagon. Initially, there are twelve pentagons with the same size. When each of them is processed, a smaller one will appear and the larger one will be removed from the queue. Since the queue is ordered in the initial stage, the order property will hold at any time.
– Then we consider the isosceles triangle. Initially, the queue is empty and thus it is ordered. After the partition of a pentagon or an isosceles triangle, if the order does not hold, i.e., the size of the newly created polygon is larger than the size of the tail polygon in the same queue. This means that the selection criteria is violated. Contradiction!

Hence, this lemma follows.   □

Gap Ratio Analysis. In this part, we analyze the gap ratio of the above strategy. Since there are three different shapes, we will study all these three cases one by one.

– First, we consider point insertion in a pentagon; the inserted points are shown in Fig. 3. Let O_1 be the center of the pentagon $AJBFE$, as shown in Fig. 3. In this case, the maximal gap is twice of $|O_1A|$, which is the radius of the circumcircle of the pentagon. Thus, at this stage, the maximal gap

$$G = 2 \cdot |O_1A| = 2 \cdot 0.85\ell = 1.7\ell$$

where ℓ is the length of the pentagon.
During the point insertion operation, the minimal gap is lower bounded by the length of the smaller pentagon. Thus, the minimal gap g is at least

$$|BE| - |Ba| - |eE| = |BE| - 2 \cdot |Ba|.$$

Note that the length of an edge x of a triangle XYZ can be computed by $x = \sqrt{y^2 + z^2 - 2yz \cos \angle X}$. After computation, we have $|BE| = 1.617\ell$, $|Ba| = 0.618\ell$. Thus, $g \geq 0.38\ell$. Therefore, the gap ratio just after point insertion on the pentagon is at most

$$\frac{G}{g} \cdot 1.34 = 5.99.$$

– Then we consider point insertion in an isosceles acute triangle. Since all such isosceles acute triangles are of the same shape, after insertion, the gap ratio will be the same too. This case can be analyzed as shown in Fig. 4(a), i.e., by considering the insertion in the acute triangle AJj. Due to the selection criteria, the maximal gap is the spherical diameter. Thus,

$$G = \frac{2 \cdot |AJ| \cdot |Jj| \cdot |jA|}{\sqrt{(|AJ| + |Jj| + |jA|)(-|AJ| + |Jj| + |jA|)(|AJ| - |Jj| + |jA|)(|AJ| + |Jj| - |jA|)}}.$$

Since $|AJ| = |jA| = \ell$ and $|Jj| = \sqrt{|AJ|^2 + |jA|^2 - 2 \cdot |AJ| \cdot |jA| \cos \angle JAj} = 0.618\ell$, we have $G = 1.05\ell$.

After insertion, the minimal gap is the distance between p_1 and j, Thus,

$$g = \sqrt{|Jj|^2 + |Jp_1|^2 - 2 \cdot |Jj| \cdot |Jp_1| \cos \angle jJp_1}.$$

Note that the triangle Jjp_1 is still an isosceles acute triangle with the angle $\angle jJp_1 = \pi/5$, and we have $g = 0.382\ell$. Therefore, the gap ratio is at most

$$\frac{G}{g} \cdot 1.34 = 3.68.$$

– Lastly, we consider point insertion in an isosceles obtuse triangle. Similar to the previous case, we only need to consider the insertion on the triangle AeE, which is shown in Fig. 4(b). Since it is an isosceles obtuse triangle, the maximal gap is at most twice the distance $|eh_1|$. After insertion, the triangle is partitioned into an isosceles obtuse triangle Aeh_1 and an isosceles acute triangle Ah_1E. In this case, the minimal gap is the distance between e and h_1, i.e., $|eh_1|$. Thus, the gap ratio at this stage is at most

$$\frac{G}{g} \cdot 1.34 \leq 2.68.$$

Combining all the above cases, we have the following concluding theorem.

Theorem 1. *The maximal gap ratio of the insertion strategy is at most 5.99.*

3   Conclusion and Discussion

Uniform insertion of points is an interesting problem in computer science. With the help of the dodecahedron and the pentagon, we give a two-phase insertion strategy with gap ratio of no more than 5.99 in the paper.

Is it possible to further reduce the gap ratio by using other structures? How about some regular simpler structure, e.g., isocahedron? If we split the isocahedron into four congruent sub-triangles regularly, the gap ratio will be larger since the newly inserted points are on the side of the isocahedron. From the definition, the maximal gap is the spherical diameter of the largest empty circle while the minimal gap is the minimal spherical distance between two inserted points. So, if points are inserted on the side of some configuration, the ratio might be not good.

Acknowledgements. This research is supported by National Key Research and Development Program of China under Grant 2016YFB0201401, National High Technology Research and Development Program of China under Grant Nos. 2014AA01A302, 2015AA050201, China's NSFC grants (Nos. 61402461, 61433012, U1435215), and Shenzhen basic research grant JCYJ20160229195940462.

References

1. Asano, T.: Online uniformity of integer points on a line. Inf. Process. Lett. **109**, 57–60 (2008)
2. Asano, T., Teramoto, S.: On-line uniformity of points. In: Book of Abstracts for 8th Hellenic-European Conference on Computer Mathematics and its Applications, Athens, Greece, September, pp. 21–22 (2007)
3. Badanidiyuru, A., Kleinberg, R., Singer, Y.: Analytically exact spiral scheme for generating uniformly distributed points on the unit sphere. J. Comput. Sci. **2**(1), 88–91 (2011)
4. Bishnu, A., Desai, S., Ghosh, A., Goswami, M., Paul, S.: Uniformity of point samples in metric spaces using gap ratio. In: Jain, R., Jain, S., Stephan, F. (eds.) TAMC 2015. LNCS, vol. 9076, pp. 347–358. Springer, Heidelberg (2015). doi:10.1007/978-3-319-17142-5_30
5. Chazelle, B.: The Discrepancy Method: Randomness and Complexity. Cambridge University Press, Cambridge (2000)
6. Cook, J.M.: An efficient method for generating uniformly distributed points on the surface on an n-dimensional sphere (remarks). Commun. ACM **2**(10), 26 (1959)
7. Collins, C.R., Stephenson, K.: A circle packing algorithm. Comput. Geom.: Theory Appl. **25**(3), 233–256 (2003)
8. Hicks, J.S., Wheeling, R.F.: An efficient method for generating uniformly distributed points on the surface of an n-dimensional sphere. Commun. ACM **2**(4), 17–19 (1959)
9. Koay, C.G.: Distributing points uniformly on the unit sphere under a mirror reflection symmetry constraint. J. Comput. Sci. **5**(5), 696–700 (2014)
10. Matoušek, J.: Geometric Discrepancy. Springer, Heidelberg (1991)
11. Nurmela, K.J., Östergård, P.R.J.: More optimal packings of equal circles in a square. Discrete Comput. Geom. **22**(3), 439–457 (1999)
12. Teramoto, S., Asano, T., Katoh, N., Doerr, B.: Inserting points uniformly at every instance. IEICE Trans. Inf. Syst. **E89–D**(8), 2348–2356 (2006)
13. Thomson, J.J.: On the structure of the atom: an investigation of the stability and periods of oscillation of a number of corpuscles arranged at equal intervals around the circumference of a circle; with application of the results to the theory of atomic structure. Philos. Mag. Ser. 6 **7**(39), 237–265 (1904)
14. Zhang, Y., Chang, Z., Chin, F.Y.L., Ting, H.-F., Tsin, Y.H.: Uniformly inserting points on square grid. Inf. Process. Lett. **111**, 773–779 (2011)

Computing the Center Region and Its Variants

Eunjin Oh[✉] and Hee-Kap Ahn

Department of Computer Science and Engineering,
POSTECH, Pohang, South Korea
{jin9082,heekap}@postech.ac.kr

Abstract. We present an $O(n^2 \log^4 n)$-time algorithm for computing the center region of a set of n points in the three-dimensional Euclidean space. This improves the previously best known algorithm by Agarwal, Sharir and Welzl, which takes $O(n^{2+\epsilon})$ time for any $\epsilon > 0$. It is known that the complexity of the center region is $\Omega(n^2)$, thus our algorithm is almost tight.

The second problem we consider is computing a colored version of the center region in the two-dimensional Euclidean space. We present an $O(n \log^4 n)$-time algorithm for this problem.

1 Introduction

Let P be a set of n points in $\mathbb{R}^d$. The *(Tukey) depth* of a point x in $\mathbb{R}^d$ with respect to P is defined to be the minimum number of points in P contained in a closed halfspace containing x. A point in $\mathbb{R}^d$ of largest depth is called a *Tukey median*. The Tukey median is a generalization of the standard median in the one-dimensional space to a higher dimensional space.

Helly's theorem implies that the depth of a Tukey median is at least $\lceil n/(d+1)\rceil$. In other words, there always exists a point in $\mathbb{R}^d$ of depth at least $\lceil n/(d+1)\rceil$. Such a point is called a *centerpoint* of P. We call the set of all points in $\mathbb{R}^d$ of depth at least $\lceil n/(d+1)\rceil$ the *center region* of P.

The Tukey median and centerpoint are considered as alternative concepts of the center of points. They are robust against outliers and do not rely on distances. Moreover, they are invariant under affine transformations [4].

In this paper, we consider two problems. The first problem is computing the center region of points in $\mathbb{R}^3$. Previously, Agarwal, Sharir and Welzl presented an algorithm for this problem which takes $O(n^{2+\epsilon})$ time for any $\epsilon > 0$ [3]. The constant hidden in the big-O notation is proportional to ϵ. Moreover, as ϵ goes to 0, the constant goes to infinity. We present an algorithm for this problem which takes $O(n^2 \log^4 n)$ time, improving the algorithm by Agarwal, Sharir and Welzl. It answers an open problem given in this paper.

Then we consider a *colorful center region*, in which each point has exactly one color among $k \in \mathbb{N}$ colors. We represent each color as an integer between

This work was supported by the NRF grant 2011-0030044 (SRC-GAIA) funded by the government of Korea.

S.-H. Poon et al. (Eds.): WALCOM 2017, LNCS 10167, pp. 254–265, 2017.
DOI: 10.1007/978-3-319-53925-6_20

1 and k. Then the *colorful (Tukey) depth* of x in $\mathbb{R}^d$ is naturally defined to be the minimum number of different colors of points contained in a closed halfspace containing x. A colorful Tukey median is a point in $\mathbb{R}^d$ of largest colorful depth.

The colorful depth and the colorful Tukey median have some properties similar to the (standard) depth and Tukey median. We prove that the colorful depth of a colorful Tukey median is at least $\lceil k/(d+1) \rceil$. Then the colorful centerpoint and colorful center region are defined naturally. We call a point in $\mathbb{R}^d$ with colorful depth at least $\lceil k/(d+1) \rceil$ a *colorful centerpoint*. The set of points of all colorful centerpoints is called the *colorful center region*.

Previous Work. In $\mathbb{R}^2$, the first nontrivial algorithm for computing a Tukey median is given by Matoušek [8]. Their algorithm computes the set of all points of Tukey depth at least a given value as well as a Tukey median. The algorithm takes $O(n \log^5 n)$ time for computing a Tukey median and $O(n \log^4 n)$ time for computing the region of Tukey depth at least a given value.

Although it is the best known algorithm for computing the region of Tukey depth at least a given value, a Tukey median can be computed faster. Langerman and Steiger [7] present an algorithm to compute a Tukey median of points in $\mathbb{R}^2$ in $O(n \log^3 n)$ deterministic time. Later, Chan [4] gave an algorithm to compute a Tukey median of points in $\mathbb{R}^d$ in $O(n \log n + n^{d-1})$ expected time.

A centerpoint of points in $\mathbb{R}^2$ can be computed in linear time [5]. On the other hand, it is not known whether a centerpoint of points in $\mathbb{R}^d$ for $d > 2$ can be computed faster than a Tukey median. Note that a Tukey median is a centerpoint.

The center region of points in $\mathbb{R}^2$ can be computed using the algorithm by Matoušek [8]. For $\mathbb{R}^3$, Agarwal, Sharir and Welzl present an $O(n^{2+\epsilon})$-time algorithm for any $\epsilon > 0$ [3]. However, it is not known whether the center region of points in $\mathbb{R}^d$ can be computed efficiently for $d > 3$.

All these results consider every input point to be identical, except for its position. Now, suppose that there are k different types of facilities and we have n facilities of these types. Then the standard definitions of the center of n points including centerpoints, center regions, and Tukey medians do not give a good representative of the n facilities of these types. Motivated from this, the center of colored points and its variants have been studied in the literature [1,2,6].

Another motivation of colored points comes from discrete imprecise data. Suppose that we have k imprecise points and each imprecise point has a candidate set of points. We know that each imprecise point lies in exactly one point in its candidate set, but we do not know the exact position. We can consider points of the same color to be points in some candidate set.

However, the centers of colored points defined in most previous results [1,2,6] are sensitive to distances, which are not adequate to handle imprecise data. Therefore, a more robust definition of a center of colored points is required. We believe that the colorful center region and colorful Tukey median can be alternative definitions of the center of colored points.

Our Result. We present an algorithm to compute the center region of n points in $\mathbb{R}^3$ in $O(n^2 \log^4 n)$ time. This is faster than the previously best known algorithm [3]. Moreover, it is almost tight as the complexity of the center region is $\Omega(n^2)$ [3].

We also present an algorithm to compute the colorful center region of n points in $\mathbb{R}^2$ in $O(n \log^4 n)$ time. We obtain this algorithm by modifying the algorithm for computing the (standard) center region of points in $\mathbb{R}^2$ in [5].

We would like to mention that a colorful Tukey median can be computed by modifying the algorithms for the standard version of a Tukey median without increasing the running times, which take $O(n \log n + n^{d-1})$ expected time in $\mathbb{R}^d$ [4] and $O(n \log^3 n)$ deterministic time in $\mathbb{R}^2$ [7].

All missing proofs can be found in the full version of this paper.

2   Preliminaries

In this paper, we use a duality transform that maps a set of input points to a set of hyperplanes. Then we transform each problem into an equivalent problem in the dual space and handle the problem using the arrangement of the hyperplanes. The Tukey depth is closely related to the level of an arrangement. This is a standard way to deal with the Tukey depth [3,4,7,8]. Thus, in this section, we introduce a duality transform and some definitions for an arrangement.

Duality Transform. A standard duality transform maps a point $x \in \mathbb{R}^d$ to the hyperplane $x^* = \{z \in \mathbb{R}^d : \langle x, z \rangle = 1\}$ and vice versa, where $\langle x, z \rangle$ is the scalar product of x and z for any two points $x, z \in \mathbb{R}^d$. Then x lies below a hyperplane s if and only if the point s^* lies below the hyperplane x^*.

Level of an Arrangement. Let H be a set of hyperplanes in $\mathbb{R}^d$. A point $x \in \mathbb{R}^d$ has *level* i if exactly i hyperplanes lie below x (or pass through x.) Note that any point in the same cell in the arrangement of H has the same level. For an integer $\ell > 0$, the *level* ℓ in the arrangement of H is the set of all points of level at most ℓ. We define the level of an arrangement of a set of convex polygonal curves in a similar way.

3   Computing the Center Region in $\mathbb{R}^3$

Let S be a set of n points in $\mathbb{R}^3$. In this section, we present an $O(n^2 \log^4 n)$-time algorithm for computing the set of points of Tukey depth at least ℓ with respect to S for a given value ℓ. We achieve our algorithm by modifying the algorithm by Agarwal et al. [3].

3.1   The Algorithm by Agarwal, Sharir and Welzl

We first describe how the algorithm by Agarwal, Sharir and Welzl works. Using the standard duality transform, they map the set S of points to a set S^* of n planes in $\mathbb{R}^3$. Then the problem reduces to computing the convex hull of points of the level ℓ in the arrangement of the planes in S^*.

Let Λ_ℓ be the level ℓ in the arrangement of the planes in S^*. They compute a convex polygon K_h for each plane $h \in S^*$ with the property that $\mathsf{CH}(\Lambda_\ell \cap h) \subset K_h \subset \mathsf{CH}(\Lambda_\ell) \cap h$. By definition, the convex hull of K_h's over all planes h in S^* is the convex hull of Λ_ℓ. Thus, once we have such a convex polygon K_h for every plane h, we can compute the set of points of Tukey depth at least ℓ with respect to S^*.

To this end, they sort the planes in S^* in the following order. Let h^+ be the closed halfspace bounded from below by a plane h, and h^- be the closed halfspace bounded from above by a plane h. We use $\langle h_1, \ldots, h_n \rangle$ to denote the sequence of the planes in S^* sorted in this order. This order satisfies the following property: for any index i, the level of a point $x \in h_i$ in the arrangement of S^* is the number of halfplanes containing x among all halfplanes $h_j^+ \cap h_i$ for $j \le i$ and all halfplanes $h_{j'}^- \cap h_i$ for all $j' > i$.

In the following, we consider each plane in S^* one by one in this order and show how to define and compute K_h for each plane h. Let $K_i = K_{h_i}$ for any i.

For h_1, the convex hull of the level ℓ in the arrangement of all lines in $\{h_1 \cap h_j : 1 < j \le n\}$ satisfies the property for K_1. So, let K_1 be the convex hull of such points. It can be computed in $O(n \log^4 n)$ time by the algorithm in [8].

Now, suppose that we have handled all planes $h_1, \ldots, h_{j-1}$ and we have $K_1, \ldots, K_{j-1}$. Let $\Gamma_j = \{K_i \cap h_j : 1 \le i < j\}$. Let R_j be the convex hull of all line segments (rays, or lines) in Γ_j, which is clearly contained in $\mathsf{CH}(\Lambda_\ell)$. Then Agarwal, Sharir and Welzl define $K_j = \mathsf{CH}(R_j \cup (\Lambda_\ell \cap h_j))$. Moreover, they show that this set consists of at most two connected components. Due to this property, they can give a procedure to compute the intersection of K_j with a given line segment. More precisely, they give the following lemma.

Lemma 1 (Lemma 2.11. [3]). *Given a triangle $\triangle \subset h_j$, the set Z_j of edges of the convex hull of K_j that intersect the boundary of $\triangle$, a segment $e \subset \triangle$, the subset $G \subset H_j$ of the m planes that cross $\triangle$, and an integer $u < m$, such that the level u of the arrangement of G coincides with Λ_ℓ within $\triangle$, the edge of K_j intersecting e can be computed in $O(m \log^3(m + s))$ time*[1].

We denote this procedure by INTERSECTION$(e, Z_j, \triangle, G, u)$. By applying this procedure with inputs satisfying the assumption in the lemma, we can obtain the edge of K_j intersecting e.

To compute the boundary of K_j, they recursively subdivide the plane h_j into a number of triangles using $1/r$-nets.

In the following, we show how to compute K_j within a given triangle $\triangle$ in $O(m^{1+\epsilon})$ time, where m is the number of lines in $H = \{h_i \cap h_j : i \ne j\}$ intersecting $\triangle$. We assume that we are given a triangle $\triangle$, a set of lines in H intersecting $\triangle$, a set Z_j of edges of the convex hull of K_j intersecting the boundary of $\triangle$, and an integer u such that the level u of the arrangement of G coincides with Λ_ℓ within $\triangle$.

---

[1] The authors in [3] roughly analyze this procedure and mention that this procedure takes $O(m \, \mathrm{polylog}(m + s))$ time. We analyze the running time of this procedure to give a more tight bound.

Initially, we have a (degenerate) triangle $\triangle = h_j$, and a set $G = H = \{h_i \cap h_j : i \neq j\}$ of lines, an empty set Z_j, and an integer $u = \ell$.

Consider the set system $(G, \{\{h \in G : h \cap \tau \neq \phi\} : \tau$ is a triangle$\})$. Let $r \in \mathbb{R}$ be a sufficiently large number. We compute a $1/r$-net $N \subset G$ of size $O(r \log r)$ and triangulate every cell in the arrangement of N restricted to $\triangle$.

For each side e of the triangles, we apply INTERSECTION$(e, Z_j, \triangle, G, u)$. Then we can obtain partial information of K_j. Note that $\triangle'$ does not intersect the boundary of K_j for a triangle $\triangle'$ none of whose edge crosses the boundary of K_j. Therefore, it is sufficient to consider only the triangles some of whose edges cross the boundary of K_j. There are $O(|N|\alpha(|N|))$ such triangles.

Moreover, for each such triangle $\triangle'$, it is sufficient to consider the lines crossing $\triangle'$ in G. So, let G' be the set of all lines intersecting $\triangle'$. A line lying above $\triangle'$ does not affect the level of a point in $\triangle'$, so we do not need to consider it. For a line lying below $\triangle'$, the level of a point in $\triangle'$ in the arrangement of G coincides with the level of the point in the arrangement of G minus one. Thus, let u' be u minus the number of lines in G lying below $\triangle'$. The following lemma summarizes this argument.

Lemma 2. *Consider a triangle $\triangle$ and a set G of lines such that the level u of the arrangement of G coincides with Λ_ℓ within $\triangle$. For a triangle $\triangle' \subset \triangle$, the level u' of the arrangement of G' coincides with Λ_ℓ within $\triangle$, where u' is u minus the number of lines in G lying below $\triangle'$ and G' is the set of lines in G intersecting $\triangle'$.*

By recursively applying this procedure, we can obtain $K_j \cap \triangle$. This means that we can obtain K_j because $\triangle$ is initially set to h_j.

Now, we analyze the running time of this procedure. Let $T(m, \mu)$ be the running time of the subproblem within $\triangle$, where m is the number of lines interesting $\triangle$ in G and μ is the number of vertices of K_j lying inside $\triangle$. Then we have the following recurrence inequality.

$$T(m, \mu) \leq \sum_{\triangle'} T(\frac{m}{r}, \mu') + O(m \log^3(m + s) + \mu)$$

for $m \geq Ar \log r$, where A is some constant independent of r. This inequality holds because the number of lines in G intersecting a triangle $\triangle'$ is $O(m/r)$ by the property of $1/r$-nets.

Solving this inequality, we have $T(m, \mu) = O(m^{1+\epsilon})$ for any constant $\epsilon > 0$. Initially, we have n lines in H, thus we can compute K_j in $O(n^{1+\epsilon})$ time. Therefore, we can compute K_j for all $1 \leq j \leq n$ in $O(n^{2+\epsilon})$ time, and compute Λ_ℓ in the same time.

Theorem 1 (Theorem 2.10. [3]). *Given a set S of n points in $\mathbb{R}^3$ and an integer $\ell \geq 0$, the set of points of depth at least ℓ can be computed in $O(n^{2+\epsilon})$ time for any ϵ.*

3.2 Our Algorithm

In this subsection, we show how to compute K_j for an integer $1 \leq j \leq n$ in $O(n \log^4 n)$ time. This leads to the total running time of $O(n^2 \log^4 n)$ by replacing the corresponding procedure in the algorithm by [3]. Recall that the previous algorithm by Agarwal, Sharir, Welzl considers the triangles in the triangulation of the arrangement of an $1/r$-net. Instead, we consider finer triangles.

Again, consider a triangle $\triangle$, which is initially set to the plane h_j. We have a set G of line segments (lines, or rays), which is initially set to $H = \{h_j \cap h_i : i \neq j\}$, and an integer u, which is initially set to ℓ. We compute a $1/r$-net N of the set system defined on the lines in H intersecting $\triangle$ as the previous algorithm does. Then we triangulate the cells in the arrangement of N.

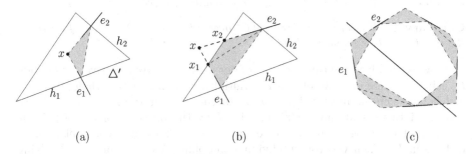

(a)   (b)   (c)

Fig. 1. (a) The gray triangle is a triangle we obtained from e_1 and e_2. (b) If x lies outside of $\triangle$, we obtain two triangles. (c) The gray region is the convex hull of all triangles we obtained. Any line segment intersects at most four triangles in E'.

For each side e of the triangles, we compute the edge of K_j intersecting e by applying INTERSECTION$(e, Z_j, \triangle, G, u)$. Let E be the set of triangles at least one of whose sides intersect K_j, and K be the set of edges of K_j intersecting triangles in E. The previous algorithm applies this procedure again for the triangles in E. But, our algorithm subdivides the triangles in E further.

We sort the edges in K in clockwise order along K_j. (We can do this although we do not know K_j.) For two consecutive edges e_1 and e_2 in K, let x be the intersection of two lines containing e_1 and containing e_2. See Fig. 1(a). Note that both e_1 and e_2 intersect a common triangle $\triangle'$ in E. Let h_1 and h_2 be two sides of $\triangle'$ intersecting e_1 and e_2, respectively.

If x is contained in $\triangle'$, then we consider the triangle with three corners x, $e_1 \cap h_1$, and $e_2 \cap h_2$. See Fig. 1(a). If x is not contained in $\triangle'$, let x_1 be the intersection of the line containing e_1 with the side of $\triangle'$ other than h_1 and h_2. See Fig. 2(b). Similarly, let x_2 be the intersection of the line containing e_2 with the side of $\triangle'$ other than h_1 and h_2. In this case, we consider two triangles; the triangle with corners $e_1 \cap h_1$, x_1, $e_2 \cap h_2$ and the triangle with corners x_1, x_2, $e_2 \cap h_2$.

Now, we have one or two triangles for each two consecutive edges in K. Let E' be the set of such triangles. By construction, the union of all triangles in E' contains all vertices of K_j. In other words, once we have the intersection of the boundary of K_j with each triangles in E', we can compute the intersection of the boundary of K_j with $\triangle$. Thus, it is sufficient to consider the triangles in E'.

For each triangle in E', we compute the intersection of the boundary of K_j with the triangle recursively as the previous algorithm does. For each triangle $\triangle' \in E'$, we define G' to be the set of lines in G intersecting $\triangle'$. And we define u' to be u minus the number of line segments lying below $\triangle'$. By Lemma 2, the level u' of the arrangement of G' coincides with Λ_ℓ within $\triangle$.

The following lemma and corollary allow us to obtain a faster algorithm. For an illustration, see Fig. 1(c).

Lemma 3. *A line intersects at most four triangles in E'.*

Corollary 1. *The total complexity of G' over all triangles $\triangle' \in E'$ is four times the number of line segments in G.*

Now, we analyze the running time of our algorithm. We iteratively subdivide h_j using $1/r$-nets until we obtain K_j. Initially, we consider the whole plane h_j, which intersects at most n lines in H. In the ith iteration, each triangle we consider intersects at most n/r^i lines in H by the property of $1/r$-nets. This means that in $O(\log_r n)$ iterations, every triangle intersects a constant number of lines in H. Then we stop subdividing the plane. We can compute K_j lying inside each triangle in the final iteration in constant time.

Consider the running time for each iteration. For each triangle, we first compute a $1/r$-net in time linear to the number of lines crossing the triangle. Then we apply the procedure in Lemma 1 for each edge in the arrangement of the $1/r$-net. This takes $O(m/r^2 \log^3 m)$ time, where m is the number of lines in H crossing the triangle.

In each iteration, we have $O(n)$ triangles, because every triangle contains at least one vertex of K_j. Moreover, the sum of the numbers of lines intersecting the triangles is $O(n)$ by Corollary 1. This concludes that the running time for each iteration is $O(n \log^3 n)$.

Since we have $O(\log_r n)$ iterations, we can compute K_j in $O(n \log^4 n)$ time. Recall that the convex hull of the level ℓ of the arrangement of the n planes is the convex hull of K_j's for all indices $1 \leq j \leq n$. Therefore, we can compute the level ℓ in $O(n^2 \log^4 n)$ time, and compute the set of points of depth at least ℓ in the same time.

Theorem 2. *Given a set of n points in $\mathbb{R}^3$ and an integer $\ell \geq 0$, the set of points of depth at least ℓ can be computed in $O(n^2 \log^4 n)$ time.*

4   Computing the Colorful Center Region in $\mathbb{R}^2$

Now, we consider the colored version of the Tukey depth in $\mathbb{R}^2$. Let $\ell > 0$ be an integer at most n. Let P be a set of n points in $\mathbb{R}^2$ each of which has exactly one color from 1 to k. For each color i, we assume that there exists a point in P which has color i.

In this section, we present an algorithm to compute the set of all points of colorful depth at least ℓ with respect to P. By setting $\ell = \lceil k/(d+1) \rceil$, we can compute the center region using this algorithm.

4.1   Properties of the Colorful Tukey Depth

Before describing the algorithm, we show some properties of the colorful Tukey depth, which are analogous to properties of the (standard) Tukey depth. The proofs in this section are similar to proofs for the standard version.

The following lemma gives a lower bound of the colorful depth of a colorful Tukey median.

Lemma 4. *The colorful depth of a colorful Tukey median is at least $\lceil k/(d+1) \rceil$.*

By definition, Lemma 4 implies the following corollary. However, a colorful centerpoint is not unique.

Corollary 2. *A colorful centerpoint always exists.*

4.2   A Duality Transform

Let S be a set of n points in $\mathbb{R}^2$. Now, we present an algorithm to compute the set of all points of colorful depth at least a given value ℓ with respect to S. Our algorithm follows the approach of the algorithm in [5]. As their algorithm does, we use a duality of points and lines. Then our problem reduces to computing the convex hull of a level of the arrangement of convex polygonal curves, not lines.

The standard duality transform maps a point s to a line s^*, and a line h to a point h^*. Let $S^* = \{s^* : s \in S\}$. Each line s^* in S^* has the same color as s. Now, we consider the colorful depth of a point in the dual space. We define a *colorful level* of a point $x \in \mathbb{R}^2$ with respect to S^* to be the number of different colors of lines lying below x or containing x. Then a point $x \in \mathbb{R}^2$ with respect to S has colorful depth at least ℓ if and only if all points in the line x^* have colorful level at least ℓ and at most $k - \ell$.

Note that a line in S^* with color i lies below a point x if and only if x lies above the lower envelope of lines in S^* of color i. With this property, we can give an alternative definition of the colorful level. For each color i, we consider the lower envelope C_i of lines in S^* which have color i. The colorful level of a point $x \in \mathbb{R}^2$ is the number of lower envelopes C_i lying below x or containing x. In other words, the colorful level of a point with respect to P^* is the level of the point with respect to the set of the lower envelopes C_i for $i = 1, \ldots, k$.

Thus, in the following, we consider the arrangement of the lower envelopes C_i for $i = 1, \ldots, k$. A cell in this arrangement is not necessarily convex. Moreover, this arrangement does not satisfy the property in Lemma 4.1 of [8]. Thus, the algorithm in [8] does not handle the colored version directly.

Let L_ℓ be the set of points in $\mathbb{R}^2$ of colorful level at most ℓ. Similarly, let U_ℓ be the set of points in $\mathbb{R}^2$ of colorful level at least ℓ. By definition, a point of colorful depth at most ℓ belongs neither to L_ℓ nor to $U_{k-\ell}$. Moreover, such a point lies outside of both the convex hull of L_ℓ and the convex hull of $U_{k-\ell}$.

Thus, once we have the convex hull of L_ℓ and the convex hull of $U_{k-\ell}$, we can compute the set of all points of colorful depth at most ℓ in linear time. In the following, we show how to compute the convex hull of L_ℓ. The convex hull of $U_{k-\ell}$ can be computed analogously.

4.3   Computing the Intersection of a Line and a Convex Hull

Let h be a vertical line in $\mathbb{R}^2$. In this subsection, we give a procedure to compute the intersection of h and the convex hull of L_ℓ. This procedure is used as a subprocedure of the algorithm in Sect. 4.4. We slightly modify the procedure by Matoušek [8], which deals with the standard (noncolored) version of the problem. Let $\mathsf{CH}(L_\ell)$ be the convex hull of L_ℓ.

We apply the two-level parametric search by Megiddo [9]. In the first level, we check whether a point x in h lies above $\mathsf{CH}(L_\ell)$ or not. This procedure is used as a subprocedure in the second level. In the second level, we compute the intersection of h and $\mathsf{CH}(L_\ell)$.

Lemma 5. *It can be checked in $O(n \log n + k \log^2 n)$ time whether a given point x lies above the convex hull of L_ℓ. In addition, we can compute the lines tangent to $\mathsf{CH}(L_\ell)$ passing through x in the same time.*

Lemma 6. *Given a vertical line h, the intersection of h and $\mathsf{CH}(L_\ell)$ can be computed in $O(\min\{n \log^3 n, n \log^2 n + k \log^4 n\})$ time.*

4.4   Computing the Convex Hull of L_ℓ

We are given a set of k polygonal curves (lower envelopes) of total complexity $O(n)$ and an integer ℓ. Let $\mathcal{C}$ be the set of the line segments which are the edges of the k polygonal curves. Recall that L_ℓ is the set of points of colorful level at most ℓ. That is, L_ℓ is the set of points lying above (or contained in) at most ℓ polygonal curves. In this subsection, we give an algorithm to compute the convex hull of L_ℓ.

Basically, we subdivide the plane into $O(n)$ vertical slabs such that the interior of each vertical slab does not contain any vertex of the k polygonal curves. We say a vertical slab is *elementary* if its interior contains no vertex of the k polygonal curves. Let A be an elementary vertical slab and $\mathcal{Q}_A$ be the set of the intersection of A with the line segments in $\mathcal{C}$ intersecting A. That is, $\mathcal{Q}_A = \{e \cap A : e \in \mathcal{C} \text{ and } e \cap A \neq \phi\}$.

Consider the arrangement of the line segments in $\mathcal{Q}_A$ restricted to A. See Fig. 2(a). Let $\mathsf{CH}(L_{\ell,A})$ denote the convex hull of points of level at most ℓ in this arrangement. Note that $\mathsf{CH}(L_{\ell,A})$ is contained in $\mathsf{CH}(L_\ell) \cap A$, but it does not coincide with $\mathsf{CH}(L_\ell) \cap A$.

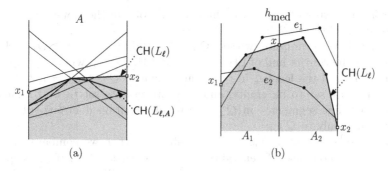

Fig. 2. (a) $\mathsf{CH}(L_\ell) \cap A$ coincides with the convex hull of x_1, x_2, and $\mathsf{CH}(L_{\ell,A})$. (b) We put e_1 only to $\mathcal{Q}'_{A_2}$ and e_2 to both sets.

By the following observation, we can compute $\mathsf{CH}(L_\ell) \cap A$ once we have $\mathsf{CH}(L_{\ell,A})$.

Observation 1. *The intersection of the boundary of $\mathsf{CH}(L_\ell)$ with A coincides with the convex hull of x_1, x_2, and $\mathsf{CH}(L_{\ell,A})$, where x_1 and x_2 are the intersections of $\mathsf{CH}(L_\ell)$ with the vertical lines bounding A.*

The subdivision of $\mathbb{R}^2$ with desired property is trivial. We consider all vertical lines passing through endpoints of the line segments in $\mathcal{C}$. However, the total complexity of $\mathcal{Q}_A$ over all slabs A is $\Omega(n^2)$. To obtain a near-linear time algorithm, we have to avoid considering all line segments in $\mathcal{Q}_A$. We will choose a subset $\mathcal{Q}'_A$ of $\mathcal{Q}_A$ and a value ℓ' such that $\mathsf{CH}(L_\ell) \cap A$ coincides with the convex hull of the level ℓ' with respect to $\mathcal{Q}'_A$, and the total complexity of $\mathcal{Q}'_A$ is linear. In the following, we show how to choose $\mathcal{Q}'_A$ for every elementary slab.

Subdividing the Region into Two Vertical Slabs. Initially, the subdivision of $\mathbb{R}^2$ is the plane itself. We subdivide each region further. After $O(\log n)$ iterations, every region in the subdivision is elementary. While we subdivide a region, we choose a set $\mathcal{Q}'_A$ for every region A. In the final subdivision, such a set satisfies the desired property.

Now, we consider a vertical slab A. Assume that we already have $\mathcal{Q}'_A$ and a level ℓ' for A. We assume further that we already have the intersection points x_1 and x_2 of $\mathsf{CH}(L_\ell)$ with the vertical line bounding A. We will compute the convex hull $\mathsf{CH}(L_{\ell',A})$ of the level ℓ' of the arrangement of $\mathcal{Q}'_S$ recursively.

We find the vertical line h_{med} passing through the median of the endpoints of line segments in $\mathcal{C}$ with respect to their x-coordinates in $O(n)$ time. The vertical line subdivides the slab A into two subslabs. Let A_1 be the subslab lying left to the vertical line, and A_2 be the other subslab. We compute $\mathsf{CH}(L_{\ell',A}) \cap h_{\text{med}}$ by applying the algorithm in Lemma 6 in $O(\min\{N \log^3 N, N \log^2 N + k \log^4 N\}) = O(N \log^3 N)$ time, where $N = |\mathcal{Q}'_A|$. We denote the intersection point by x. While computing x, we can obtain the slope τ of the edge of $\mathsf{CH}(L_\ell)$ containing

x. (If x is the vertex of the convex hull, we can obtain the slope τ of the edge lying left to x.) See Fig. 2(b).

We show how to compute two sets $\mathcal{Q}'_{A_1}, \mathcal{Q}'_{A_2}$ and two integers ℓ'_1, ℓ'_2 such that $\mathsf{CH}(L_{\ell',A})$ is the convex hull of $\mathsf{CH}(L_{\ell'_1,A_1})$ and $\mathsf{CH}(L_{\ell'_2,A_2})$, where $\mathsf{CH}(L_{\ell'_t,A_t})$ is the convex hull of the level ℓ'_t in the arrangement of $\mathcal{Q}'_{A_t}$ for $t = 1, 2$. The two sets are initially set to be empty, and two integers are set to be ℓ'. Then we consider each line segment s in $\mathcal{Q}'_A$. If s is fully contained in one subslab A_t, then we put s only to $\mathcal{Q}'_{A_t}$.

Otherwise, s intersects h_{med}. If s lies above x, then we compare the slope of s and τ. Without loss of generality, we assume that $\tau \geq 0$. If the slope of s is larger than τ, $\mathsf{CH}(L_{\ell',A}) \cap S_2$ does not intersect s. Thus, a point of level at most ℓ' in the arrangement of $\mathcal{Q}'_A$ restricted to A_2 has level at most ℓ' in the arrangement of $\mathcal{Q}'_A \setminus \{s\}$ restricted to A_2. This means that we do not need to put s to $\mathcal{Q}'_{A_2}$. We put s only to $\mathcal{Q}'_{A_1}$. The case that the slope of s is at most τ is analogous.

Now, consider the case that s lies below x. If both endpoints are contained in the interior of S, we put s to both $\mathcal{Q}'_{A_1}$ and $\mathcal{Q}'_{A_2}$. Otherwise, s crosses one subslab, say A_1. In this case, we put s to $\mathcal{Q}'_{A_2}$. For A_1, we check whether s lies below the line segment connecting x_1 and x. If so, we set ℓ'_1 to $\ell'_1 - 1$ and do not put s to the set for A_1. This is because $\mathsf{CH}(L_{\ell',A})$ contains s. Otherwise, we put s to the set for A_1.

After considering every line segment in $\mathcal{Q}'_A$, we have two sets $\mathcal{Q}'_{A_t}$ and two integers ℓ_t for $t = 1, 2$. The line segments in $\mathcal{Q}'_{A_t}$ with the same color form a convex polygonal curve. However, the endpoints of each convex polygonal curve do not necessarily lie on the lines bounding A_t. We remove the part of the line segments lying outside of A_t. For a convex polygonal curve some of whose endpoints lie in the interior of A_t, we extend such an endpoint a in the direction opposite to the edge of the curve incident to a until it hits the boundary of A_t. Then the endpoints of each convex polygonal curves obtained from the updated set A_t lie on the lines bounding A_t.

Moreover, these sets and integers satisfy the following. $\mathsf{CH}(L_{\ell',A})$ is the convex hull of $\mathsf{CH}(L_{\ell'_1,A_1})$ and $\mathsf{CH}(L_{\ell'_2,A_2})$, where $\mathsf{CH}(L_{\ell'_t,A_t})$ is the convex hull of the level ℓ'_t in the arrangement of $\mathcal{Q}'_{A_t}$ for $t = 1, 2$. Thus, we can recursively compute $\mathsf{CH}(L_{\ell'_1,A_1})$ and $\mathsf{CH}(L_{\ell'_2,A_2})$ and merge them to obtain $\mathsf{CH}(L_{\ell',A})$.

We analyze the running time of the procedure. In the ith iteration, each vertical slab in the subdivision contains at most $n/2^i$ endpoints of the line segments in $\mathcal{C}$. Thus, we can complete the subdivision in $O(\log n)$ iterations.

Each iteration takes $O(\sum_j n_j \log^3 n_j)$ time, where n_j is the complexity of $\mathcal{Q}'_{A_j}$ for the jth slab A_j. By construction, each line segment in $\mathcal{C}$ is contained in at most two sets $\mathcal{Q}'_A$ and $\mathcal{Q}'_{A'}$ for two vertical slabs A and A' in the same iteration. Therefore, each iteration takes $O(\sum_j n_j \log^3 n_j) = O(n \log^3 n)$ time, except for the final iteration.

Computing the Convex Hull Inside an Elementary Vertical Slab. In the final iteration, we have $O(n)$ elementary vertical slabs. Each elementary vertical slab has a set of line segments whose total complexity is $O(n)$. In addition, each elementary vertical slab has an integer. For each elementary vertical slab with integer ℓ', we have to compute the convex hull of the level ℓ' in the arrangement of its line segments.

Matoušek [8] gave an $O(n \log^4 n)$-time algorithm to compute the convex hull of the level ℓ in the arrangement of lines. In our problem, we want to compute the convex hull of the level ℓ in the arrangement of lines restricted to a vertical slab. The algorithm in [8] works also for our problem (with modification). This modification is straightforward, so we omit this procedure.

Lemma 7. *The convex hull of L_ℓ can be computed in $O(n \log^4 n)$ time.*

Theorem 3. *Given a set P of n colored points in $\mathbb{R}^2$ and an integer ℓ, the set of points of colorful depth at most ℓ with respect to P can be computed in $O(n \log^4 n)$ time.*

Corollary 3. *Given a set P of n colored points in $\mathbb{R}^2$, the colorful center region of P can be computed in $O(n \log^4 n)$ time.*

References

1. Abellanas, M., Hurtado, F., Icking, C., Klein, R., Langetepe, E., Ma, L., Palop, B., Sacristán, V.: Smallest color-spanning objects. In: Heide, F.M. (ed.) ESA 2001. LNCS, vol. 2161, pp. 278–289. Springer, Heidelberg (2001). doi:10.1007/3-540-44676-1_23
2. Abellanas, M., Hurtado, F., Icking, C., Klein, R., Langetepe, E., Ma, L., Palop, B., Sacristán, V.: The farthest color Voronoi diagram and related problems. Technical report, University of Bonn (2006)
3. Agarwal, P.K., Sharir, M., Welzl, E.: Algorithms for center and tverberg points. ACM Trans. Algorithms **5**(1), 1–20 (2008)
4. Chan, T.M.: An optimal randomized algorithm for maximum Tukey depth. In: Proceedings of the Fifteenth Annual ACM-SIAM Symposium on Discrete Algorithms (SODA 2004), pp. 430–436 (2004)
5. Jadhav, S., Mukhopadhyay, A.: Computing a centerpoint of a finite planar set of points in linear time. Discrete Comput. Geom. **12**(3), 291–312 (1994)
6. Khanteimouri, P., Mohades, A., Abam, M.A., Kazemi, M.R.: Computing the smallest color-spanning axis-parallel square. In: Cai, L., Cheng, S.-W., Lam, T.-W. (eds.) ISAAC 2013. LNCS, vol. 8283, pp. 634–643. Springer, Heidelberg (2013). doi:10.1007/978-3-642-45030-3_59
7. Langerman, S., Steiger, W.: Optimization in arrangements. In: Alt, H., Habib, M. (eds.) STACS 2003. LNCS, vol. 2607, pp. 50–61. Springer, Heidelberg (2003). doi:10.1007/3-540-36494-3_6
8. Matousek, J.: Computing the center of a planar point set. In: Discrete and Computational Geometry: Papers from the DIMACS Special Year. American Mathematical Society (1991)
9. Megiddo, N.: Applying parallel computation algorithms in the design of serial algorithms. J. ACM **30**(4), 852–865 (1983)

Fault-Tolerant Spanners in Networks with Symmetric Directional Antennas

Mohammad Ali Abam[1], Fatemeh Baharifard[2], Mohammad Sadegh Borouny[1], and Hamid Zarrabi-Zadeh[1(✉)]

[1] Department of Computer Engineering, Sharif University of Technology, Tehran, Iran
zarrabi@sharif.edu
[2] Institute for Research in Fundamental Sciences (IPM), Tehran, Iran

Abstract. Let P be a set of points in the plane, each equipped with a directional antenna that can cover a sector of angle α and range r. In the symmetric model of communication, two antennas u and v can communicate to each other, if and only if v lies in u's coverage area and vice versa. In this paper, we introduce the concept of *fault-tolerant spanners* for directional antennas, which enables us to construct communication networks that retain their connectivity and spanning ratio even if a subset of antennas are removed from the network. We show how to orient the antennas with angle α and range r to obtain a k-fault-tolerant spanner for any positive integer k. For $\alpha \geq \pi$, we show that the range 13 for the antennas is sufficient to obtain a k-fault-tolerant 3-spanner. For $\pi/2 < \alpha < \pi$, we show that using range $6\delta + 19$ for $\delta = \lceil 4/|\cos\alpha| \rceil$, one can direct antennas so that the induced communication graph is a k-fault-tolerant 7-spanner.

1 Introduction

Omni-directional antennas, whose coverage area are often modelled by a disk, have been traditionally employed in wireless networks. However, in many recent applications, omni-directional antennas have been replaced by directional antennas, whose coverage region can be modelled as a sector with an angle α and a radius r (also called transmission range), where the orientation of antennas can vary among the nodes of the network. The point is that by a proper orientation of directional antennas, one can generate a network with lower radio wave overlapping and higher security than the traditional networks with omni-directional antennas [4].

There are two main models of communication in networks with directional antennas. In the *asymmetric* model, each antenna has a directed link to any node that lies in its coverage area. In the *symmetric* model, there exists a link between two antennas u and v, if and only if u lies in the coverage area of v, and v lies in the coverage area of u. The symmetric model of communication is more practical, especially in networks where two nodes must handshake to each other before transmitting data [6].

© Springer International Publishing AG 2017
S.-H. Poon et al. (Eds.): WALCOM 2017, LNCS 10167, pp. 266–278, 2017.
DOI: 10.1007/978-3-319-53925-6_21

In this paper, we consider the symmetric model for communication in directional antennas, and study two properties of the communication graphs: *k-connectivity* and *spanning ratio*. A network is *k-connected* if it remains connected after removing or destroying any $k - 1$ of its nodes. Furthermore, if after some failure of nodes, it still has some desirable properties, we say that the network is *fault-tolerant*. Therefore, the fault-tolerance property is more general than the connectivity. A network is called a *spanner*, if there is a short path between any pairs of nodes, within a guaranteed ratio to the shortest paths between those nodes in an underlying base graph. This ratio is called the *stretch factor*. A *fault-tolerant spanner* has the property that when a small number of nodes fail, the remaining network still contains short paths between any pair of nodes. (See [14] for an overview of the properties of geometric spanner networks.)

Related Work. The problem of orienting directional antennas to obtain a strongly connected network was first studied by Caragiannis et al. [5] in the asymmetric model. They showed that the problem is NP-hard for $\alpha < 2\pi/3$, and presented a polynomial time algorithm for $\alpha \geq 8\pi/5$ with optimal radius. The problem was later studied for other values of α, and approximation algorithms were provided to minimize the transmission range of connected networks [1,7]. However, the communication graphs obtained from these algorithm could have a very large stretch factor, such as $O(n)$, compared to the original unit disk graph (i.e., the omni-directional graph of radius 1). Therefore, subsequent research was shifted towards finding a proper orientation such that the resulting graph becomes a *t*-hop spanner [4,11]. In a *t-hop spanner*, the number of hops (i.e., links) in a shortest link path between any pair of nodes is at most t times the number of hops in the shortest link path between those two nodes in the base graph, which happens to be a unit disk graph in this case.

The connectivity of communication graphs in the symmetric model was first studied by Ben-Moshe et al. [3] in a limited setting where the orientation of antennas were chosen from a fixed set of directions. Carmi et al. [6] later considered the general case, and proved that for $\alpha \geq \pi/3$, it is always possible to orient antennas so that the induced graph is connected. In their presented algorithm, the radius of the antennas were related to the diameter of the nodes. Subsequent work considered the stretch factor of the communication graph. Aschner et al. [2] studied the problem for $\alpha = \pi/2$ and obtained a symmetric connected network with radius $14\sqrt{2}$ and a stretch factor of 8, assuming that the unit disk graph of the nodes is connected. Recently, Dobrev et al. [8] proved that for $\alpha < \pi/3$ and radius one, the problem of connectivity in the symmetric model is also NP-hard. They also showed how to construct spanners for various values of $\alpha \geq \pi/2$. A summary of the current records for the radius and the stretch factor of the communication graphs in the symmetric model is presented in Table 1.

The problem of *k*-connectivity in wireless networks has been also studied in the literature, mostly for omni-directional networks [12,13], where the objective is to assign transmission range such that the network can sustain fault nodes and remain connected. The stretch factor of the constructed network is also

Table 1. Summary of the previous results for netwroks with symmetric directional antennas. In all these results, the unit disk graph of the nodes (antennas) is assumed to be connected. Here, $\delta = \sqrt{3 - 2\cos\alpha(1 + 2\sin\frac{\alpha}{2})}$.

Angle of antenna	Stretch factor	Radius	Ref.
$\pi/2$	8	$14\sqrt{2}$	[2]
$\pi/2$	7	33	[9]
	5	718	
$\pi/2 \leq \alpha < 2\pi/3$	9	10	[8]
$2\pi/3 \leq \alpha < \pi$	–	5	
$2\pi/3 \leq \alpha < \pi$	6	6	
$\alpha \geq \pi$	–	$\max(2, 2\sin\frac{\alpha}{2} + 1)$	
$\alpha \geq \pi$	3	$\max(2, 2\sin\frac{\alpha}{2} + \delta)$	

studied in some limited settings. In [10], a setting is studied where antennas are on a unit segment or a unit square, and a sufficient condition is obtained on the angle of directional antennas so that the energy consumption of the k-connected networks is lower when using directed rather than omni-directed antennas. In [15], a tree structure is built on directed antennas, and a fault-tolerance property is maintained by adding additional links to tolerate failure in limited cases, namely, when only a node or a pair of adjacent nodes fail.

Our Results. In this paper, we study the problem of finding fault-tolerant spanners in networks with symmetric directional antennas. The problem is formally defined as follows. Given a set P of n points in the plane, place antennas with angle α and radius r on P, so that the resulting communication graph is a k-fault-tolerant t-spanner. A graph G on the vertex set P is a k-fault-tolerant t-spanner, if after removing any subset $S \subseteq P$ of nodes with $|S| < k$, the resulting graph $G \setminus S$ is a t-spanner of the unit disk graph of P. In the rest of the paper, we assume that the unit distance is sufficiently large to ensure that the unit disk graph of P is k-connected. To the best of our knowledge, this is the first time that fault-tolerance is studied in networks with symmetric directional antennas.

We show that for any $\alpha \geq \pi$, we can place antennas with angle α and radius 9, such that the resulting communication graph is k-connected. Moreover, we show that by increasing the radius to 13, we can guarantee that the resulting graph is a k-fault-tolerant 3-spanner. When $\pi/2 < \alpha < \pi$, we consider two cases depending on whether the distribution of antennas is sparse or dense. We prove that for sparse distribution, we can place antennas with angle α and radius $6\delta + 19$, where $\delta = \lceil 4/|\cos\alpha| \rceil$, such that the resulting communication graph is a k-fault-tolerant 7-spanner. Moreover, for dense distribution, we prove that our algorithm yields a k-fault-tolerant 4-spanner using radius δ. Our results are summarized in Table 2.

We recall that the k-connectivity of the unit disk graph is assumed in the rest of the paper. In other words, we compared the radius and stretch factor of our k-connected directional network to those of a k-connected omni-directional network. While this assumption is reasonable, it is possible to relax it, and only assume the connectivity of the unit disk graph, which is the minimum requirement assumed in the related (non-fault-tolerant) work. If we replace the k-connectivity assumption with 1-connectivity, the radius and stretch factor of our constructed network is increased by a factor of k, as explained in Sect. 5.

Table 2. Summary of our results for netwroks with symmetric directional antennas. In these results, the unit disk graph of the nodes is assumed to be k-connected. Here, $\delta = \lceil 4/|\cos\alpha| \rceil$.

Angle of antenna	Stretch factor	Radius	Ref.
$\alpha \geq \pi$	–	9	Theorem 1
$\alpha \geq \pi$	3	13	Theorem 2
$\pi/2 < \alpha < \pi$ (sparse)	7	$6\delta + 19$	Theorem 3
$\pi/2 < \alpha < \pi$ (dense)	4	δ	Theorem 3

2   Preliminaries

Let P be a set of points in the plane, and G be a graph on the vertex set P. For two points $p, q \in P$, we denote by $\delta_G(p, q)$ the shortest hop (link) distance between p and q in G. If the graph G is clear from the context, we simply write $\delta(p, q)$ instead of $\delta_G(p, q)$. Throughout this paper, the *length* of a path in a graph refers to the number of edges on that path. For two points p and q in the plane, the Euclidean distance between p and q is denoted by $\|pq\|$.

Let $\mathcal{B}(c, r)$ denote a (closed) disk of radius r centered at c. We define $\mathcal{A}(c, r) \equiv \mathcal{B}(c, r) - \mathcal{B}(c, r-1)$ to be an *annulus* of width 1 enclosed by two concentric circles of radii $r - 1$ and r, centered at c. Note that by our definition, $\mathcal{A}(c, r)$ is open from its inner circle, and is closed from the outer circle.

A graph G is k-connected, if removing any set of at most $k - 1$ vertices leaves G connected. Given a point set P, we denote by $\mathrm{UDG}(P)$ the unit disk graph defined by the set of disks $\mathcal{B}(p, 1)$ for all $p \in P$. We say that P is k-connected, if $\mathrm{UDG}(P)$ is k-connected. Let $G = \mathrm{UDG}(P)$. A graph H on the vertex set P is a t-spanner of G, if for any two vertices u and v in G, we have $\delta_H(u, v) \leq t \cdot \delta_G(u, v)$. We say that the subgraph $H \subseteq G$ is a k-fault-tolerant t-spanner of G, if for all sets $S \subseteq P$ with $|S| < k$, the graph $H \setminus S$ is a t-spanner of $G \setminus S$.

Fact 1. *Let G and H be two k-connected graphs, and E be a set of edges between the vertices of G and H. If E contains a matching of size k, then the graph $G \cup H \cup E$ is k-connected.*

Fact 2. *Let G be a k-connected graph, and v be a new vertex adjacent to at least k vertices of G. Then $G + v$ is k-connected.*

Lemma 1. *Let P be a k-connected point set, and r be a positive integer. If $|P| \geq rk$, then for any point $p \in P$, $\mathcal{B}(p, r)$ contains at least rk points of P.*

Proof. Fix a point p, and let q be the furthest point from p in P. If $\|pq\| \leq r$, then the disk $P \subseteq \mathcal{B}(p, r)$, and we are done. Otherwise, consider the annuli $A_i = \mathcal{A}(p, i)$ for $1 \leq i \leq r + 1$, and let $A_0 = \{p\}$. Each A_i must be non-empty, because otherwise, p is disconnected from q in $\mathrm{UDG}(P)$. Now, we claim that each A_i, for $1 \leq i \leq r$, contains at least k points. Otherwise, if $|A_i| < k$ for some $1 \leq i \leq r$, then removing the points of A_i disconnects A_{i-1} from A_{i+1}, contradicting the fact that P is k-connected. □

3   Antennas with $\alpha \geq \pi$

In this section, we present our algorithm for orienting antennas with angle at least π. The main ingredient of our method is a partitioning algorithm which we describe below.

Partitioning Algorithm. The following algorithm builds a graph H on the input point set P. The graph will induce a partitioning on the input set, as described in Lemma 2. In the following algorithm, p is an arbitrary point of P, and r is a positive integer.

Algorithm 1. PARTITION(P, p, r)

1: add vertex p to graph H
2: $P = P \setminus \mathcal{B}(p, 2r)$
3: **while** $\exists q \in P \cap \mathcal{B}(p, 2r + 1)$ **do**
4:     PARTITION(P, q, r)
5:     add edge (p, q) to graph H

Lemma 2. *Let P be a k-connected point set, p be an arbitrary point in P, and $|P| \geq kr$ for a positive integer r. Let $H = (V, E)$ be the graph obtained from PARTITION(P, p, r). For each $v \in V$, we define $Q_v = P \cap \mathcal{B}(v, r)$. Moreover, we define F_v to be the set of all points in $P \setminus \cup_{u \in V} Q_u$ closer to v than any other point in V (ties broken arbitrarily). Then the followings hold:*

(a) H is connected, and for each edge $(u, v) \in E$, $2r < \|uv\| \leq 2r + 1$.
(b) P is partitioned into disjoint sets Q_v and F_v.
(c) Q_v has at least kr points, for all $v \in V$.
(d) F_v is contained in $\mathcal{B}(v, 2r)$, for all $v \in V$.

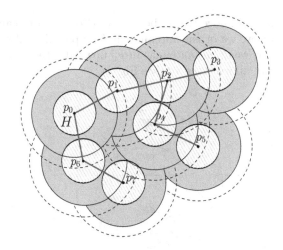

Fig. 1. A partitioning obtained by Algorithm 1. The induced graph H is shown by dark edges.

Proof.

(a) The graph H computed by the algorithm is obviously connected, as each new vertex created by calling PARTITION in line 4 is connected in line 5 to a previous vertex of H. Moreover, lines 2 and 3 of the algorithm enforce that any two adjacent vertices in H have distance between $2r$ and $2r + 1$.

(b) The sets F_v are disjoint by their definition. The sets Q_v are also disjoint, because any two vertices in H have distance more than $2r$ by line 2 of the algorithm.

(c) This is a corollary of Lemma 1.

(d) This is clear from lines 2 and 3 of the algorithm.   □

We call each set Q_v a *group*, and the points in F_v the *free points* associated to the group Q_v. We call v the *center* of Q_v. Two groups Q_u and Q_v are called *adjacent groups*, if there is an edge (u, v) in the graph H.

Orienting Antennas. Here, we show how to place antennas with angle at least π on a point set P, so that the resulting communication graph becomes k-connected, with a guaranteed stretch factor. In the rest of this section, we describe our method for $\alpha = \pi$. However, the method is clearly valid for any larger angle.

Theorem 1. *Given a k-connected point set P with at least $2k$ points in the plane, we can place antennas with angle π and radius 9 on P, such that the resulting communication network is k-connected.*

Proof. We run Algorithm 1 with $r = 2$ on the point set P to obtain the graph $H = (V, E)$. For each $v \in V$, let Q_v and F_v be the sets defined in Lemma 2. Since $r = 2$, each set Q_v has at least $2k$ points. We partition Q_v by a horizontal line ℓ_v into two equal-size subsets U_v and D_v, each of size at least k, where points in

U_v (resp., in D_v) are all above (resp., below) ℓ_v. (Points on ℓ_v can be placed in either U_v or D_v.) Now, we orient antennas in D_v upward, and antennas in U_v downward. Moreover, we orient antennas in F_v upward if they are below ℓ_v, and downward if they are above or on ℓ_v (see Fig. 2).

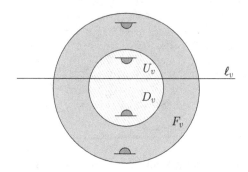

Fig. 2. The orientation of antennas with angle π in $Q_v \cup F_v$.

Let G_π be the communication graph obtained by the above orientation, where the radius of each antenna is set to $4r+1 = 9$. Since each node in D_v has distance at most $2r$ to any node in U_v, Q_v forms a complete bipartite graph, with each part having size at least k, and hence, it is k-connected. Now, we show that the graph on $Q = \cup Q_v$ is k-connected. Note that the distance between the centers of any two adjacent groups Q_u and Q_v is at most $2r+1$, and the farthest points in the groups have distance at most $4r+1$. By setting the radius of antennas to $4r+1$, either all members of D_u connect to all members of U_v, or all members of U_u connect to all members of D_v. So there is a matching of size k between any two adjacent groups, and hence, Q is k-connected by Fact 1. Since F_v is contained in $\mathcal{B}(v, 2r)$, the farthest points in $Q_v \cup F_v$ are at distance $4r$, and hence, each node in F_v connects to at least k nodes in Q_v. Therefore, the whole communication graph is k-connected by Fact 2.   □

Theorem 2. *Given a k-connected point set P with at least $2k$ points in the plane, we can place antennas with angle π and radius 13 on P, such that the resulting communication network is a k-fault-tolerant 3-spanner.*

Proof. We use the same orientation described in the proof of Theorem 1. Now, we show that by setting radius of antennas to $6r+1 = 13$, the resulting graph G_π is a k-fault-tolerant 3-spanner. Fix a set $S \subseteq P$ with $|S| < k$. We show that for any edge $(p, q) \in \text{UDG}(P) \setminus S$, there is a path between p and q in $G_\pi \setminus S$ of length at most 3. Let $T_v = Q_v \cup F_v$. Suppose $p \in T_u$ and $q \in T_v$. Assume w.l.o.g. that ℓ_u is below or equal to ℓ_v. Since $\|pq\| \leq 1$, the centers of Q_u and Q_v are at most $4r+1$ apart. Therefore, by setting the radius to $6r+1$, we have a matching of size k between D_u and U_v in G_π. We distinguish the following four cases based on the order of points and lines on the y-axis:

- $p \le \ell_u$ and $q \le \ell_v$. Since $|S| < k$, there is a vertex $w \in U_v \setminus S$ such that p and q are both connected to w. Therefore, $\delta_G(p, q) = 2$ in this case.
- $p \le \ell_u$ and $q > \ell_v$. Since $|S| < k$, there is an edge $(w, x) \in (D_u \setminus S, U_v \setminus S)$. Now, the path $\langle p, x, w, q \rangle$ is a path of length 3 in G.
- $p > \ell_u$ and $q > \ell_v$. Since $|S| < k$, there is a vertex $w \in D_u \setminus S$ such that p and q are both connected to w. Therefore, $\delta_G(p, q) = 2$ in this case.
- $p > \ell_u$ and $q < \ell_v$. This case is analogous to the second case.   □

4 Antennas with $\pi/2 < \alpha < \pi$

We now pay our attention to a more challenging case where the goal is to orient the antennas with angle $\pi/2 < \alpha < \pi$ on a point set P, so that the resulting communication graph becomes k-connected. Let $\delta = \lceil 4/|\cos \alpha| \rceil$. We distinguish two cases based on the distribution of P on the plane. P is called α-*sparse* if the diameter of P (i.e. the distance of the farthest pair of points in P) is at least δ. Otherwise, P is called α-*dense*.

Lemma 3. *If P is α-sparse, then the diameter of $P \cap \mathcal{B}(p, \delta + 3)$ is at least δ, for any $p \in P$.*

Proof. Let (q, q') be the farthest pair of points in P. If both q and q' are contained in $\mathcal{B}(p, \delta + 3)$, we are done. Otherwise, at least one of q and q' (say q) is outside $\mathcal{B}(p, \delta + 3)$. Since $\mathrm{UDG}(P)$ is connected, $\mathcal{A}(p, \delta + 1)$ must contain some point t of P. Since t is inside $\mathcal{B}(p, \delta + 1)$ and $\|tp\| > \delta$, the diameter of $P \cap \mathcal{B}(p, \delta + 3)$ is at least δ.   □

Algorithm Sketch. We first sketch the whole algorithm, and then go into details of each part. The algorithm is almost similar to the one given in the previous section for $\alpha = \pi$. We run Algorithm 1 with $r = \delta + 3$ on the point set P to obtain the graph $H = (V, E)$. We then add edges to H to make any two vertices of H whose distance is at most $4r + 1$ adjacent. For each $v \in V$, let Q_v and F_v be the sets defined in Lemma 2. We orient antennas in $Q_v \cup F_v$ such that the resulting graph is k-connected. We then make the radius of the antennas large enough, so that for any two adjacent groups Q_u and Q_v, their union (and consequently $Q = \cup Q_v$) becomes k-connected.

Observation 1. *If P is α-dense, $H = (V, E)$ is a single vertex.*

We start explaining how to make each Q_v k-connected. We define α-*cone* to be a cone with angle $2\alpha - \pi$. Let $\sigma(c)$ be an α-cone with apex c and let $\bar{\sigma}(c)$ be the reflection of $\sigma(c)$ about c. Our algorithm relies on the following lemma.

Lemma 4.

- *If the diameter of Q_v is at least δ, then there is an α-cone $\sigma(c)$ for some point c on the plane such that both $\sigma(c)$ and $\bar{\sigma}(c)$ contain at least $2k$ points of Q_v.*

– If the diameter of Q_v is less than δ but Q_v contains at least $8k \cdot \pi/(2\alpha - \pi)$ points, then there is an α-cone $\sigma(c)$ for some point c on the plane such that both $\sigma(c)$ and $\bar{\sigma}(c)$ contain at least $2k$ points of Q_v.

The proof of this lemma is omitted in this version due to lack of space.

We recall that if P is α-sparse, the diameter of each set Q_v is at least δ. If P is α-dense, we only have one set Q_v and then we only need the extra assumption that P contains at least $8k \cdot \pi/(2\alpha - \pi)$ points, in order to use the lemma in our algorithm.

Orienting $Q_v \cup F_v$. Let $\sigma(c)$ be the α-cone obtained in Lemma 4. Let ℓ_1 and ℓ_2 be the lines passing through the sides of $\sigma(c)$ (and $\bar{\sigma}(c)$ as well), and let ℓ be the bisector of the angle $2\pi - 2\alpha$ whose sides are ℓ_1 and ℓ_2 (see Fig. 3 to get more intuition). We define and depict four types of orienting antennas with angle α in Fig. 3 naming O_1, O_2, O_3 and O_4. In each type, each side is parallel to one of the lines $\ell_1, \ell_2,$ and ℓ.

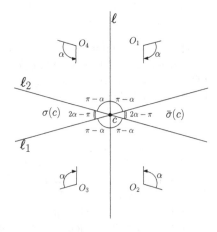

Fig. 3. Cones $\sigma(c)$ and $\bar{\sigma}(c)$, and four orientations with angle α.

Backbone Antennas. We select $2k$ point of $Q_v \cap \sigma(c)$ and arbitrarily partition them into two sets D_v and U_v of size k. Similarly, we select $2k$ point of $Q_v \cap \bar{\sigma}(c)$ and arbitrarily partition them into two sets $\bar{D}_v$ and $\bar{U}_v$ of size k. We use types $O_1, O_2, O_3,$ and O_4 for orienting antennas in $D_v, U_v, \bar{D}_v,$ and $\bar{U}_v$, respectively. We call each of these four sets a backbone set. Regardless of the antennas radii, this orientation holds the following properties:

- Each antenna in $D_v \cup U_v$ covers each antenna in $\bar{D}_v \cup \bar{U}_v$ and vice versa.
- Each point in the plane is covered by all antennas in one of the backbone sets.

To orient antenna p in $Q_v \cup F_v$ other than backbone antennas, we detect which backbone set covers p (i.e. p is visible from all antennas in the backbone set). Let O_i be the orientation type used to orient the backbone set. We orient p with type $\bar{O}_i$ where $\bar{O}_i$ is the reflection of O_i about its apex. Figure 4 depicts how to orient antennas depending on their subdivisions induced by ℓ_1, ℓ_2, and ℓ.

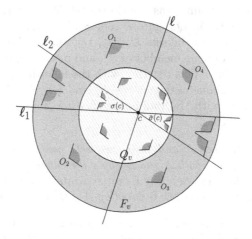

Fig. 4. The orientation of antennas with angle $\pi/2 < \alpha < \pi$ in $Q_v \cup F_v$

Radius. If P is α-dense, we set the radius to be δ as the distance of any two antennas is at most δ. For the α-sparse set P, we need that any two visible backbone antennas u' and v' from two adjacent groups Q_u and Q_v cover each other. Since their distance is at most $\|u'u\| + \|uv\| + \|vv'\| \leq r + 4r + 1 + r \leq 6(\delta + 3) + 1$, we set the radius to be $6\delta + 19$.

k-Connectivity. For any $v \in V$, the induced graph over $(D_v \cup U_v, \bar{D}_v \cup \bar{U}_v)$ is a bipartite complete graph. Moreover, any antenna in $Q_v \cup F_v$ other than the backbone antennas has a direct connection with at least k backbone antennas. All these simply imply that the induced graph over $Q_v \cup F_v$ is k-connected.

Lemma 5. *Suppose $p, q \in Q_v \cup F_v$ and q is a backbone antenna. p and q are in connection with each other via at most three links, even if at most $k-1$ antennas are destroyed.*

Proof. Assume w.l.o.g. that $q \in D_v$. We know p is visible from all members of one backbone set. This backbone set can be either $D_v, U_v, \bar{D}_v$, or $\bar{U}_v$. If this backbone set is either $D_v, \bar{D}_v$ or $\bar{U}_v$, we reach q from p with at most two links.

Otherwise, with 3 links we can get q from p. Since each backbone set has k members and any member of $D_v \cup U_v$ is visible to $\bar{D}_v \cup \bar{U}_v$ and vice versa, the proof works even if at most $k-1$ antennas are destroyed.   □

Graph $H = (V, E)$ has only one vertex if P is α-dense. Therefore, using Lemma 5 we can simply show any two points are in connection with other via at most 4 links even if $k-1$ antennas are destroyed. Note that any antenna is either a backbone antenna or directly connected to a backbone antenna. Next we assume P is α-sparse.

Here, we need to show the connection of two adjacent groups Q_v and Q_u remain safe even if $k-1$ antennas are destroyed. We partition the backbone antennas in Q_v (similarly in Q_u) into k sets S_v^i ($i = 1, \ldots, k$) of size 4, each containing one antenna from the sets D_v, U_v, $\bar{D}_v$, and $\bar{U}_v$. We know each point in the plane is visible from one member of S_v^i, and moreover, two sets S_v^i and S_u^i can be separated by a line. This together with the following proposition implies that there are two backbone antennas $p \in S_v^i$ and $q \in S_u^i$ which are visible to each other, and hence, with the radius specified for antennas they are in the coverage area of each other.

Proposition 1 ([2]). *Let A and B be two sets containing 4 antennas with angle at least $\pi/2$. Suppose both A and B cover the entire plane regardless of the antennas radius. If there exists a line ℓ that separates A and B, then by setting the radius unbounded, the network induced by $A \cup B$ is connected.*

The above discussion shows that there are at least k distinct links between the backbone antennas of two adjacent groups Q_v and Q_u. Therefore, even if $k-1$ antennas are destroyed, the connection between Q_v and Q_u remains safe. This together with Lemma 5 implies that for any two antennas $p \in Q_v \cup F_v$ and $q \in Q_u \cup F_u$, there is a connection via at most 7 links.

Stretch Factor. Let p and q be two arbitrary points in P, and let $x_0 = p, x_1, \ldots, x_t = q$ be the shortest link distance between p and q in $\mathrm{UDG}(P) \setminus S$, where S is the fault set with size at most $k-1$. Since $\|x_i x_{i+1}\| \leq 1$, either there exists $v \in V$ such that $x_i, x_{i+1} \in Q_v \cup F_v$, or there exist two adjacent $u, v \in V$ such that $x_i \in Q_v \cup F_v$ and $x_{i+1} \in Q_u \cup F_u$. This shows that in the communication graph obtained by our algorithm, each link (x_i, x_{i+1}) either exist or is replaced by a path of length at most 4 in the α-dense set P, and a path of length at most 7 in the α-sparse set P. Therefore, our resulting graph is a 4-spanner and a 7-spanner for the α-dense set P and the α-sparse set P, respectively.

Putting all these together, we get the main theorem of this section.

Theorem 3. *Suppose P is a k-connected point set in the plane, and α is a given angle in the range $(\pi/2, \pi)$. Let $\delta = \lceil 4/|\cos\alpha| \rceil$. Then, the followings hold:*

  - *If P is α-sparse, we can place antennas with angle α and radius $6\delta + 19$ on P, such that the resulting communication network is a k-fault-tolerant 7-spanner.*

- If P is α-dense and contains at least $8k \cdot \pi/(2\alpha-\pi)$ points, we can place antennas with angle α and radius δ on P, such that the resulting communication network is a k-fault-tolerant 4-spanner.

5 Concluding Remarks

In this paper, we studied the problem of constructing fault-tolerant spanners in networks with symmetric directional antennas, and presented the first algorithms for placing antennas with angles $\alpha > \pi/2$, so that the resulting communication graph is a k-fault-tolerant t-spanner, for small stretch factors $t \leq 7$.

Throughout this paper, we assumed that $\mathrm{UDG}(P)$ is k-connected. This assumption can be relaxed to the connectivity of $\mathrm{UDG}(P)$ at the expense of increasing the radius and stretch factor. If we replace the k-connectivity with a 1-connectivity assumption, the radius of antennas implied by Lemma 1 is multiplied by k, and hence, the radius and stretch factor of our constructed network is increased by a factor of k. For example, on a point set whose UDG is connected, our algorithm constructs a k-fault-tolerant spanner with radius $13k$ and stretch factor $3k$. A natural open problem is to find fault-tolerant spanners with smaller radius and/or stretch factors. The case $\pi/3 \leq \alpha \leq \pi/2$ is also open for further investigation.

References

1. Aloupis, G., Damian, M., Flatland, R.Y., Korman, M., Zkan, O., Rappaport, D., Wuhrer, S.: Establishing strong connectivity using optimal radius half-disk antennas. Comput. Geom. Theory Appl. **46**(3), 328–339 (2013)
2. Aschner, R., Katz, M., Morgenstern, G.: Symmetric connectivity with directional antennas. Comput. Geom. Theory Appl. **46**(9), 1017–1026 (2013)
3. Ben-Moshe, B., Carmi, P., Chaitman, L., Katz, M., Morgenstern, G., Stein, Y.: Direction assignment in wireless networks. In: Proceedings of 22nd Canadian Conference on Computational Geometry, pp. 39–42 (2010)
4. Bose, P., Carmi, P., Damian, M., Flatland, R., Katz, M., Maheshwari, A.: Switching to directional antennas with constant increase in radius and hop distance. Algorithmica **69**(2), 397–409 (2014)
5. Caragiannis, I., Kaklamanis, C., Kranakis, E., Krizanc, D., Wiese, A.: Communication in wireless networks with directional antennas. In: Proceedings of 20th ACM Symposium Parallel Algorithms Architecture, pp. 344–351 (2008)
6. Carmi, P., Katz, M., Lotker, Z., Rosen, A.: Connectivity guarantees for wireless networks with directional antennas. Comput. Geom. Theory Appl. **44**(9), 477–485 (2011)
7. Damian, M., Flatland, R.: Spanning properties of graphs induced by directional antennas. In: Proceedings of 20th Annual Fall Workshop Computational Geometry (2010)
8. Dobrev, S., Eftekhari, M., MacQuarrie, F., Manuch, J., Ponce, O.M., Narayanan, L., Opatrny, J., Stacho, L.: Connectivity with directional antennas in the symmetric communication model. Comput. Geom. Theory Appl. **55**, 1–25 (2016)

9. Dobrev, S., Plžík, M.: Improved spanners in networks with symmetric directional antennas. In: Gao, J., Efrat, A., Fekete, S.P., Zhang, Y. (eds.) ALGOSENSORS 2014. LNCS, vol. 8847, pp. 103–121. Springer, Heidelberg (2015). doi:10.1007/978-3-662-46018-4_7

10. Kranakis, E., Krizanc, D., Williams, E.: Directional versus omnidirectional antennas for energy consumption and k-connectivity of networks of sensors. In: Higashino, T. (ed.) OPODIS 2004. LNCS, vol. 3544, pp. 357–368. Springer, Heidelberg (2005). doi:10.1007/11516798_26

11. Kranakis, E., MacQuarrie, F., Morales-Ponce, O.: Connectivity and stretch factor trade-offs in wireless sensor networks with directional antennae. Theoret. Comput. Sci. **590**, 55–72 (2015)

12. Li, N., Hou, J.C.: FLSS: a fault-tolerant topology control algorithm for wireless networks. In: Proceedings of 10th Annual International Conference on Mobile Computing and Networking, pp. 275–286 (2004)

13. Li, X.Y., Wan, P.J., Wang, Y., Yi, C.W.: Fault tolerant deployment and topology control in wireless networks. In: Proceedings of the 4th ACM International Symposium on Mobile Ad Hoc Networking and Computing, pp. 117–128 (2003)

14. Narasimhan, G., Smid, M.: Geometric Spanner Networks. Cambridge University Press, Cambridge (2007)

15. Shirazipourazad, S., Sena, A., Bandyopadhyay, S.: Fault-tolerant design of wireless sensor networks with directional antennas. Pervasive Mob. Comput. **13**, 258–271 (2014)

Gathering Asynchronous Robots in the Presence of Obstacles

Subhash Bhagat[✉] and Krishnendu Mukhopadhyaya

ACM Unit, Indian Statistical Institute, Kolkata, India
{sbhagat_r,krishnendu}@isical.ac.in

Abstract. This work addresses the problem of *Gathering* a swarm of point robots when the plane of deployment has non-intersecting transparent convex polygonal obstacles. While multiplicity detection is enough for gathering three or more asynchronous robots without obstacles, it is shown that in the presence of obstacles, gathering may not be possible even in the FSYNC model with all of multiplicity detection, memory, chirality and direction-only axis agreement. Initial configurations for which gathering is impossible are characterized. For other configurations, a distributed algorithm for the gathering problem is proposed without any extra assumption on the capabilities of the robots. The algorithm works even if the configuration contains points of multiplicities.

Keywords: Gathering · Asynchronous · Oblivious · Polygonal obstacle · Swarm robots

1 Introduction

A *Robot Swarm* is a distributed system of autonomous, memoryless, homogeneous mobile robots which can move freely on the infinite two-dimensional plane. The robots are anonymous, i.e., they cannot be distinguished by a unique identity or by their appearances. They do not communicate with each other directly. The communication is done implicitly by sensing the positions of other robots. The robots are memoryless or oblivious in the sense that they do not remember any information from the previous computational cycles. Each robot has its own local coordinate system. The robots follow the same execution cycle Look-Compute-Move [10].

We consider the *ASYNC* [8] model, where robots are activated asynchronously and independently from other robots. The time taken to complete an action is unpredictable but finite. Chirality or orientation of axes may be different for different robots. The primary objective of research in this field is to find strategies for cooperation, control and interaction to solve fundamental classes of problems like geometric pattern formation, gathering, convergence, flocking etc. Researchers have also identified sets of conditions under which certain problems are unsolvable.

The *Gathering* problem (also known as Homing/Rendezvous/Point Formation) is defined as bringing multiple autonomous mobile robots into a point which

© Springer International Publishing AG 2017
S.-H. Poon et al. (Eds.): WALCOM 2017, LNCS 10167, pp. 279–291, 2017.
DOI: 10.1007/978-3-319-53925-6_22

is not fixed in advance. The existing works have considered the plane of motion to be free of obstacles. We consider a model in which the region of deployment has a finite number of non-interesting convex polygonal obstacles. The robots can see through the obstacles but they can not pass through them. An instance of this model is the case when the obstacles are holes in the plane which hinder movements but cause no visual obstruction.

1.1 Related Works

The problem of gathering has received extensive attention in the field of swarm robotics [10]. In *FSYNC* model, the gathering problem is solvable without any extra assumption [10]. Suzuki and Yamashita proved that in the *SSYNC* model, gathering problem is not solvable for two robots without any agreement on the local coordinate systems even with strong multiplicity detection [14]. Prencipe proved that for $n > 2$ robots, there does not exists any deterministic algorithm for the gathering problem in absence of multiplicity detection and any form of agreement on the local coordinate systems [13]. Flocchini et al. solved the gathering problem in the *ASYNC* model with oblivious robots having limited visibility and knowledge of a common direction [9]. Cieliebak et al. proposed an algorithm for gathering in the *ASYNC* model with multiplicity detection for $n \geq 5$ robots [4]. All of these works have considered robots to be dimensionless i.e., point robots. The gathering problem for *fat robots* (represented as unit discs) have also been investigated by the researchers [1,5,11]. The gathering problem under different fault models have been addressed by many researchers [2,3,6].

To the best of our knowledge, this paper is the first attempt to study the problem of gathering in the *ASYNC* model with the assumptions that the plane of motion has non-intersecting convex polygonal obstacles.

1.2 Our Contribution

This paper proposes a distributed algorithm for gathering if the initial configuration, consisting of the obstacles as well as the robots, does not have any rotational symmetry. The algorithm does not assume any extra capability for the robots. The obstacles provide some fixed reference points. However, the gathering problem is not solvable in general, if the configuration of the obstacles and the robots has rotational symmetry, even if robots have multiplicity detection capability, memory, chirality and direction-only axis agreement [7] and robots are fully synchronous.

2   Robot Model and Terminology

This work adopts the basic the *ASYNC* model (CORDA model). The robots are represented as points in the two dimensional Euclidean plane. The visibility range of a robot is assumed to be unlimited. We consider non-rigid motion of the robots i.e., a robot may stop before reaching its destination. However, to ensure finite time reachability to the destination point, there exists a fixed value $\delta > 0$ such that in each movement, where the robot does not reach its destination, it moves a distance not less than δ towards its destination [10]. The value of δ is not known to the robots.

Initially the robots are stationary and in arbitrary positions. Multiple robots may share same position on the plane. However, the robots do not have multiplicity detection capabilities i.e., they can not identify multiple robots at a point. The total number of robots in the system is not known to the robots. The region of deployment of the robots has finite non-zero number of non-intersecting convex polygonal obstacles. The robots have full visibility through the polygons but they can not pass through them. No robot can lie inside or on the boundary of a polygon.

Let $\mathcal{R} = \{r_1, r_2, \ldots, r_n\}$ be the set of n homogeneous robots. The position occupied by a robot $r_i \in \mathcal{R}$ at time t is denoted by $r_i(t)$. Let $\mathcal{R}(t)$ be the collection of all such positions occupied by the robots in $\mathcal{R}$ at time t. Let $\mathcal{P} = \{P_1, P_2, \ldots, P_m\}$ denote the set of mutually non-intersecting convex polygonal obstacles where $m \geq 1$. For a polygon $P_i \in \mathcal{P}$, let P_{iv} denote the set of vertices of P_i. Let $\mathcal{P}_v = \{P_{1v}, P_{2v}, \ldots, P_{mv}\}$ be the collection of all such sets of vertices for the polygons in $\mathcal{P}$. Note that in our model, $P_{iv} \cap P_{jv} = \emptyset$ for $i \neq j$. The set of all vertices of all the polygons in $\mathcal{P}$ is denoted by $\hat{\mathcal{P}}_v$. The center of gravity of a point set A is denoted by $CoG(A)$. We use $\mathcal{O}$ to denote $CoG(\hat{\mathcal{P}}_v)$. If $\mathcal{O}$ lies inside a polygon, the polygon containing $\mathcal{O}$ is called the *central polygon* and is denoted by P_c. By a configuration $\mathcal{C}(t)$ at time t, we mean the set $\mathcal{P} \cup \mathcal{R}(t)$.

The distance of point x from a polygon P_i is the minimum euclidean distance between x and a point y in P_i and it is denoted by $dist(x, P_i)$. The distance between two polygons $P_i, P_j \in \mathcal{P}$, is the minimum distance between two points in P_i and P_j and it is denoted by $dist(P_i, P_j)$. The minimum of all the distances of the robot positions in $\mathcal{R}(t)$ from a polygon $P_i \in \mathcal{P}$ at time t is denoted by $\sigma_i(t)$ i.e., $\sigma_i(t) = min\{dist(r_j(t), P_i) : \forall r_j(t) \in \mathcal{R}(t)\}$. Let $\Sigma(t) = min\{\sigma_i(t) : \forall P_i \in \mathcal{P}\}$.

Let

$$D_p = \begin{cases} min\{dist(P_i, P_j), \forall P_i, P_j \in \mathcal{P}\} & \text{if } |\mathcal{P}| > 1 \\ l & \text{if } |\mathcal{P}| = 1 \end{cases}$$

where l is the length of the smallest side of the polygon in $\mathcal{P}$,

$$\Delta(t) = \begin{cases} min\{\sigma_c(t), D_p/4\} & \text{if } \mathcal{O} \text{ lies inside } P_c \\ D_p/4 & \text{otherwise} \end{cases}$$

and

$$\zeta(t) = \begin{cases} min\{\Sigma(t), D_p/4\} & \text{if } \mathcal{O} \text{ lies inside } P_c \\ D_p/4 & \text{otherwise} \end{cases}$$

For every polygon, we define an extended version of it by expanding the boundaries. A polygon in $\mathcal{P} \setminus \{P_c\}$ is extended by an amount $\zeta(t)$. The polygon P_c is extended by an amount $\Delta(t)$. Note that these extended polygons are also convex and non-overlapping. Let P_{ai} be the extended version of P_i. The polygon P_{ai} is called the auxiliary polygon of P_i. The auxiliary polygons are used in the path computations for the robots. Let $\overline{ab}$ denote the line segment joining two points a and b (including the end points a and b).

3 Preliminaries

This section describes some basic results which are used to develop the gathering algorithm.

Definition 1. *A straight line $\mathcal{L}$ is called a line of symmetry for $\mathcal{P}$, if (i) for every polygon $P_i \in \mathcal{P}$, such that $\mathcal{L}$ passes through the interior of P_i, $\mathcal{L}$ is a line of symmetry for P_i (ii) for every polygon $P_i \in \mathcal{P}$, such that $\mathcal{L}$ does not pass through the interior of P_i, there exists a $P_j \in \mathcal{P}$, such that P_j is the mirror image of P_i about $\mathcal{L}$.*

Definition 2. *A point O is called the center of rotational symmetry for $\mathcal{P}$ with an angle of symmetry $0 < \theta \leq \pi$, if (i) for every polygon $P_i \in \mathcal{P}$, such that O lies in the interior of P_i, O is the center of rotational symmetry for P_i with θ as an angle of symmetry (ii) for every polygon $P_i \in \mathcal{P}$, such that O does not lie in the interior of P_i, there exists a $P_j \in \mathcal{P}$, such that P_j can be obtained by rotating P_i by an angle θ about O.*

Definition 3. *A straight line $\mathcal{L}$ is a line of symmetry for $\mathcal{P} \cup \mathcal{R}(t)$, if $\mathcal{L}$ is a line of symmetry for $\mathcal{P}$ as well as for $\mathcal{R}(t)$. Similarly, a point O is the center of rotational symmetry for $\mathcal{P} \cup \mathcal{R}(t)$, if O is the center of rotational symmetry for $\mathcal{P}$ as well as for $\mathcal{R}(t)$ with the same angle of symmetry.*

Note that if a set of points $\mathcal{A}$ (polygons $\mathcal{P}$) has more than one line of symmetry, then $\mathcal{A}$ ($\mathcal{P}$) has rotational symmetry with center of gravity of $\mathcal{A}$ ($\mathcal{P}$) as the center of rotational symmetry. A set of points $\mathcal{A}$ (polygons $\mathcal{P}$) is asymmetric if it has neither a line of symmetry nor rotational symmetry. We use the notion of ordering as defined in [11]. A set of points $\mathcal{A}$ (polygons $\mathcal{P}$) is orderable, if a deterministic algorithm can produce a unique sequencing of the points in $\mathcal{A}$ (polygons in $\mathcal{P}$) such that the sequencing does not depend on the local coordinate systems.

Result 1 [11]. *If a set of points $\mathcal{A}$ (polygons $\mathcal{P}$) does not have any line of symmetry or rotational symmetry, then the points in $\mathcal{A}$ (polygons in $\mathcal{P}$) are orderable.*

Observation 1. *If a set of points $\mathcal{A}$ has only one line of symmetry $\mathcal{L}$, then positive and negative directions can be defined along $\mathcal{L}$.*

Observation 2. *If a set of points $\mathcal{A}$ has both a line of symmetry and rotational symmetry, then A has multiple lines of symmetry.*

4   Algorithm GatheringObstacles()

Our objective is to compute a unique point which remains intact under the motion of robots and all robots can agree on that point. If such a point exists in the initial configuration $\mathcal{P} \cup \mathcal{R}(t_0)$, then we are done. Otherwise, we convert the initial configuration into one where such a point can be defined. The point $\mathcal{O}$ is fixed and invariant under orthogonal coordinate transformations. If $\mathcal{O}$ lies outside the obstacles, then $\mathcal{O}$ serves our purpose. If $\mathcal{O}$ lies inside an obstacle, we adopt different strategies to have such a point. If the set $\mathcal{P}$ does not have rotational symmetry, a gathering point is defined in terms of the points in $\hat{\mathcal{P}}_v$. Since the obstacles are stationary, this point is also fixed. When $\mathcal{P}$ has rotational symmetry, the positions of the robots along with the polygons are considered to compute a unique gathering point. If $\mathcal{P} \cup \mathcal{R}(t)$ has no rotational symmetry, a unique robot position is created on the boundary of P_{ac} which serves the purpose. During the creation of such unique point, the movements of the robots are coordinated in such a way that they do not create any rotational symmetry of $\mathcal{P} \cup \mathcal{R}(t)$. If the initial configuration $\mathcal{P} \cup \mathcal{R}(t_0)$ has rotational symmetry, then gathering is not possible. When $\mathcal{O}$ lies inside or on the boundary of the obstacle P_c and $\mathcal{P} \cup \mathcal{R}(t)$ has no rotational symmetry, there are following possible scenarios:

(A) $\mathcal{P}$ has no rotational symmetry: In this case, $\mathcal{P}$ has at most one line of symmetry. If $\mathcal{P}$ is asymmetric, we fix an ordering of the vertices of the polygons in $\mathcal{P}$. Let p be the first vertex of P_c in that ordering such that $p \neq \mathcal{O}$. Draw the straight line $\mathcal{L}$ through p and $\mathcal{O}$. The direction from $\mathcal{O}$ along $\mathcal{L}$ on which p lies is defined as the positive direction of $\mathcal{L}$ and denoted by $\mathcal{L}^+$. This definition remains invariant since it involves only fixed points. If $\mathcal{P}$ has exactly one line of symmetry $\mathcal{L}$, by Observation 1, we can define the positive direction $\mathcal{L}^+$ along $\mathcal{L}$. In both the cases, we consider the point r on $\mathcal{L}^+$ which is $D_p/4$ distance apart from P_c. Since this point is defined in terms of polygons in $\mathcal{P}$, it is a fixed point and we take this point as our desired gathering point. Robots move towards r.

(B) $\mathcal{P}$ has rotational symmetry: The Observation 2 implies that either (i) $\mathcal{P}$ has more than one line of symmetry or (ii) $\mathcal{P}$ has no line of symmetry. In scenario (ii), $\mathcal{P} \cup \mathcal{R}(t)$ is asymmetric. When $\mathcal{P}$ has multiple lines of symmetry, these lines of symmetry define different wedges w.r.t. $\mathcal{O}$. When $\mathcal{P}$ has rotational symmetry but no line of symmetry, the wedges are defined in the following way: consider the robot positions having minimum Euclidean distance from $\mathcal{O}$. Let $S(t)$ be the set of these robot positions. Since $\mathcal{P} \cup \mathcal{R}(t)$ is orderable, consider the robot position $r_i(t) \in S(t)$, which has highest ordering among the robot positions in $S(t)$. Let $\mathcal{L}_i(t)$ be the line joining $r_i(t)$ and $\mathcal{O}$. The line $\mathcal{L}_i(t)$ divides the plane into two non-overlapping wedges.

The set $\mathcal{P} \cup \mathcal{R}(t)$ has at most one line of symmetry. Our approach focuses on the number of robot positions on P_{ac} and performs actions accordingly.

(B.1) P_{ac} contains no robot position on its boundary: If in the initial configuration, there is no robot on P_{ac}, we create at least one robot position on P_{ac}. In the following discussion, we shall be talking about paths from robot positions to the auxiliary polygons. Any path computation will serve our purpose as long as: (i) the computation is done with auxiliary polygons (ii) if the robot stops in the middle, the newly computed path would again be the remaining portion of the original path and (iii) each path is completely contained inside one of the wedges as defined above. Since polygons are non-interesting and no polygon disconnects a wedge, it is possible to find such paths. In particular one may consider the piece-wise linear voronoi diagram for convex sites [12] with the extended polygons. The path may be obtained by projecting a perpendicular on a voronoi edge and then following voronoi edges to the destination, with the final segment being a perpendicular from the destination point. When $\mathcal{P}$ has multiple lines of symmetry, for each wedge, we compute the intersection point between its bisector and P_{ac}. On the other hand, when $\mathcal{P}$ has rotational symmetry but no line of symmetry, the intersection point between $\mathcal{L}_i(t)$ and P_{ac} is considered. These intersection points are called *wedge points*. The shortest paths are computed from each robot position to the wedge point of the wedge it belongs to. If a robot position lies on a line of symmetry, both the adjacent wedge points are equidistant. Any one of them may be considered for path computation.

(a) $\mathcal{P}$ **has no line of symmetry:** If the line segment $\overline{r_i(t)\mathcal{O}}$ intersects P_{ac} only, then the robots at $r_i(t)$ move towards the intersection point of $\overline{r_i(t)\mathcal{O}}$ and P_{ac} along the line segment $\overline{r_i(t)\mathcal{O}}$. Other robots in the system waits until P_{ac} has one robot position on its boundary. Now suppose, the line segment $\overline{r_i(t)\mathcal{O}}$ intersects at least one obstacle other than P_c. Let P_m be the nearest obstacle to $r_i(t)$ among such obstacles. The robots at $r_i(t)$ computes a point $\hat{r}$ on the line segment $\overline{r_i(t)\mathcal{O}}$ such that distance of $\hat{r}$ from P_m is $\frac{1}{2}\Sigma(t)$ (any distance less than $\Sigma(t)$ will work). The robots at $r_i(t)$ move towards $\hat{r}$ along the line segment $\overline{r_i(t)\mathcal{O}}$. Other robots wait until, at least one of these robots reaches a point on $\overline{r_i(t)\mathcal{O}}$, which realizes $\Sigma(t')$ for some $t' > t$. Let w be the unique point on $\overline{r_i(t)\mathcal{O}}$ which realizes $\Sigma(t')$. The movements of the robots at $r_i(t)$ towards $\hat{r}$ do not change the asymmetry of $\mathcal{P} \cup \mathcal{R}(t)$. Once such a point w is created, the robots compute the set $F(t')$ of robot positions over $\mathcal{R}(t')\backslash\{w\}$, having shortest paths to the wedge point (the shortest paths should completely lie within one of the wedges defined by $\mathcal{L}_i(t)$). The robot position in $F(t')$, which appears first in an ordering of $\mathcal{P} \cup \mathcal{R}(t')$, is selected. The robots at this point move towards the wedge point along the corresponding shortest path(s). Rest of the robots wait until at least one robot reaches P_{ac}. During this whole process, $\mathcal{P} \cup \mathcal{R}(t)$ remains asymmetric.

(b) $\mathcal{P}$ **has multiple lines of symmetry:** Let $T(t)$ be the set of all robot positions, which have the shortest paths to the corresponding wedge points. The following are the different possibilities:

b.1. $|T(t)| = 1$: Let $r_i(t)$ be the corresponding robot position. If the point $r_i(t)$ does not lie on any wedge boundary, then the robots at this point move to the corresponding wedge point along the shortest path(s). Otherwise, the robots at $r_i(t)$ choose any of one of the wedge points corresponding to the neighbouring wedges of $r_i(t)$ and move towards it. Since multiple robots may be located at $r_i(t)$, it may happen that two robots at $r_i(t)$ choose two different wedge points as their destinations. The other robots do not move till at least one robot reaches P_{ac}.

b.2. $|T(t)| > 1$: Since $\mathcal{P} \cup \mathcal{R}(t)$ has at most one line of symmetry, we have following cases only:

b.2.1. $\mathcal{P} \cup \mathcal{R}(t)$ is asymmetric: The set $\mathcal{P} \cup \mathcal{R}(t)$ is orderable [11]. We fix any ordering. Let $r_i(t)$ be the point in $T(t)$, which appears first in this ordering. If the point $r_i(t)$ does not lie on the boundary of the wedges, the robots at this point move to the corresponding wedge point along the shortest path. Otherwise, the robots at $r_i(t)$ choose any one of the wedge points in the neighbouring wedges and move towards it. If $r_i(t)$ contains multiple robots, the robots at this points may move to both of the wedge points.

b.2.2. $\mathcal{P} \cup \mathcal{R}(t)$ has exactly one line of symmetry: Let $\mathcal{L}$ be the line of symmetry of $\mathcal{P} \cup \mathcal{R}(t)$. We define positive and negative directions along $\mathcal{L}$ as stated in Observation 1. Let $\mathcal{L}^+$ and $\mathcal{L}^-$ denote the positive and negative sides of $\mathcal{L}$ respectively, with $\mathcal{O}$ as the origin.

(i) **$\mathcal{L}$ passes through at least one member of $T(t)$:** If $\mathcal{L}$ passes through robot positions in $T(t)$, consider the robot position in $T(t)$ which lies on $\mathcal{L}^+$ and has smallest Euclidean distance from $\mathcal{O}$. Let $r_i(t)$ be this robot position. The robots at $r_i(t)$ arbitrarily choose any one of the wedge points in the neighbouring wedges and move towards wedge point(s) along the shortest path(s).

(ii) **$\mathcal{L}$ does not pass through any member of $T(t)$:** In this case, we move the robots in such a way that even if symmetry occurs, $\mathcal{L}$ remains as the unique line of symmetry for the whole configuration. We consider the pair of points in $T(t)$ which are closest to $\mathcal{L}^+$. If there is more than one such pair, we choose the one having smallest euclidean distance from $\mathcal{O}$. Let these points be $r_l(t)$ and $r_k(t)$. If these two points do not lie on the wedge boundaries, the active robots at these two points move towards the corresponding wedge points. Otherwise, the robots at each point has two options. For each of the points, the robots select the wedge point which makes smaller angle with $\mathcal{L}^+$.

(B.2) P_{ac} contains exactly one robot position on its boundary: The robots which observe this point on P_{ac}, move towards it along paths not crossing P_{ac}.

(B.3) P_{ac} contains more than one robot position on its boundary: If at any time there are more than one robot position on P_{ac}, we create a unique fixed robot position on P_{ac}.

If possible, we divide the perimeter of P_{ac} into two polygonal chains such that (i) one of them contains all the robot positions (the robot positions at the joint belong to both the chains) and (ii) the chain containing the robots is not longer than the other chain.

Let both the chains have same length and v_i and v_j be the two joints of the chains. The robots do one of the followings to change the lengths of the chains: (1) if $\mathcal{P} \cup \mathcal{R}(t)$ is asymmetric, the robot positions are orderable. Without loss of generality, suppose that v_i has higher order than v_j. The robots at v_i move along the boundary of P_{ac} towards the next nearest vertex of P_{ac} on the chain satisfying any one of the followings: (a) if P_{ac} contains more than two robot positions, the chain contains the other robot positions (b) if P_{ac} contains exactly two robot positions, the chain contains the vertex of P_{ac}, having highest order among the vertices of P_{ac} (2) if $\mathcal{P} \cup \mathcal{R}(t)$ has one line of symmetry $\mathcal{L}$, the positive and the negative directions are defined along $\mathcal{L}$. The robots at v_i and v_j move towards the corresponding next nearest vertices of P_{ac} on the chain intersected by $\mathcal{L}^+$. The robots move along the boundary of P_{ac}.

Once we have a polygonal chain having length less than half of the perimeter of P_{ac} and containing all the robots, we move the robots towards the middle vertex of the corresponding chain. If the middle vertex is not unique, the robots move to the nearest one. In each movement, a robot tries to reach the next vertex on its path. This process continues till all robots lie on a single edge or reach at a single vertex of P_{ac}. If the robots lie on a single edge, they gather at the mid point of the edge.

If such a division is not possible, we update P_{ac} with a newer version which is obtained by expanding the boundaries of P_c by an amount $\Delta(t)/2$. It may be noted that the expansion can be made by any amount less than $\Delta(t)$. Now we have a new P_{ac} with no robot position on its boundary. We follow the same strategies as described in (B.1). According to the strategies in (B.1), the new P_{ac} would contain at most two robots on its boundary. Since robots on old P_{ac} move to the boundary of new P_{ac} along the straight lines and these straight lines completely lie in the corresponding wedges in which robots lie, the updated configuration $\mathcal{P} \cup \mathcal{R}(t')$ can have at most one line of symmetry. Suppose, there are two robot positions on the updated P_{ac}. If these two robot positions define two polygonal chains of the updated P_{ac} with unequal lengths, then we are done. Otherwise, the two polygonal chains would have same length and the robots follow the same strategy as described above to reduce the length of one of the chains. This implies that the process of creating new polygonal chains with unequal lengths will terminate in finite time. Note that none of our strategies, described above, creates this scenario. This scenario is possible only in one of the initial configurations. Also this is the only case in which P_{ac}, initially defined, is changed. In all other cases P_{ac} remains intact.

4.1   Correctness of GatheringObstacles()

In this section we prove that if the initial configuration does not have rotational symmetry, the robots gather in finite time by executing *GatheringObstacles()*. To prove the correctness of our approach, we show that in possible cases, our approach provides, within finite time, a unique gathering point which remains intact under the motion of the robots.

Lemma 1. *Suppose $\mathcal{P} \cup \mathcal{R}(t_0)$ has no rotational symmetry. During the whole execution of algorithm GatheringObstacles(), $\mathcal{P} \cup \mathcal{R}(t)$ does not become rotationally symmetric.*

Proof. The obstacles are static. If $\mathcal{P}$ has no rotational symmetry, then the lemma is true. If $\mathcal{P}$ has rotational symmetry, during the execution of algorithm *GatheringObstacles()*, following are the possible scenarios:

- P_{ac} **contains exactly one robot position on its boundary:** The robots on P_{ac} do not move. Whenever other robots move to this point, they move along the shortest paths having no intersections with P_{ac}. Hence, the lemma is true.

- P_{ac} **contains no robot position on its boundary:** In this scenario, $\mathcal{P}$ has either multiple lines of symmetry or no line of symmetry. First consider the case when $\mathcal{P}$ has multiple lines of symmetry. The set $\mathcal{P} \cup \mathcal{R}(t)$ has at most one line of symmetry. Robots having positions in $T(t)$ move towards at most two wedge points. All the shortest paths to these wedge points lie in the corresponding wedges in which the wedge points lie. If the robots move towards a single wedge point, the lemma follows immediately. If the robots move towards two wedge points, the angle made by these two wedge points at $\mathcal{O}$ is less than π. This implies the lemma.

  Next, consider the case when $\mathcal{P}$ has no line of symmetry. In this case, $\mathcal{P} \cup \mathcal{R}(t)$ is asymmetric. A robot position $r_i(t)$ is selected. The robots at $r_i(t)$ move straight either to the boundary of P_{ac} or move to a point which would be a unique robot position in the system realizing the quantity $min\{\Sigma(t'), \frac{D_p}{4}\}$, $t' > t$. The robots move along the straight line to the corresponding destination point. Until, at least one robot reaches the destination point, other robots do not move. Thus, these movements of the robots do not make $\mathcal{P} \cup \mathcal{R}(t)$ rotationally symmetric.

- P_{ac} **contains more than one robot position on its boundary:** Suppose, it is possible to define two polygonal chains, one with all robot positions. Since the robots move along the chain which contains them, to create a unique robot position on the chain and the length of this chain is at most half of the perimeter of P_{ac}, the movements of these robots do not make $\mathcal{P} \cup \mathcal{R}(t)$ rotationally symmetric. If it is not possible to define such polygonal chains, the robots move to the updated version of P_{ac}. At most two robot positions are selected and the robots at these positions are asked to move to the boundary of the new P_{ac}. The robots follow same strategies as in the case of (B.1) when initial P_{ac} contains no robot positions. Thus, in this case, lemma follows from the same arguments as in the case when P_{ac} contains no robot positions.   □

Lemma 2. *Let $\mathcal{P} \cup \mathcal{R}(t_0)$ have no rotational symmetry. If $\mathcal{P}$ has rotational symmetry and initially P_{ac} does not contain any robot position on its boundary, during the execution of algorithm GatheringObstacles(), P_{ac} will have at least one robot position on its boundary in finite time.*

Proof. Since P_{ac} does not contain any robot position, by definition, the polygon P_{ac} is defined by the obstacles. This implies that P_{ac} is independent of the movements of the robots. By Lemma 1, $\mathcal{P} \cup \mathcal{R}(t)$ does not become rotationally symmetric during the whole execution of the algorithm. This implies that at least one robot reaches the boundary of P_{ac} in finite time.   □

Lemma 3. *Let $\mathcal{P}$ have rotational symmetry and $\mathcal{P} \cup \mathcal{R}(t_0)$ have no rotational symmetry. Suppose, P_{ac} contains more than one robot position on its boundary and it cannot be decomposed into two polygonal chains as described in (B.3) of Sect. 4. During the execution of algorithm GatheringObstacles(), there exits a time $t > t_0$ such that $\Delta(t')$ does not change, $\forall t' \geq t$. Furthermore, P_{ac} at time t can be decomposed into two desired polygonal chains.*

Proof. According to algorithm *GatheringObstacles*(), at most two robot positions are selected and the robots at these positions are asked to move towards at most two points on the boundary of the updated P_{ac} which is defined by the quantity $\frac{\Delta(t_0)}{2}$. The robots move along straight lines towards their respective destination points. Once at least a robot starts moving towards the updated P_{ac}, the value of $\Delta()$ is dependent solely on the robots positions. There are two possibilities: (i) at least one robot reaches the boundary of the computed P_{ac} at time t or (ii) at least one robot position is created by some static robot (in between old and the computed P_{ac}) such that this robot position realizes the value $\Delta(t)$ and all other robots agree on this value. These imply that $\Delta(t')$ does not change for $t' \geq t$. The polygon P_{ac} at time t' depends only on the robot positions and it contains at most two robot positions. By Lemma 1, $\mathcal{P}\, cup\mathcal{R}(t)$ has at most one line of symmetry. Hence, P_{ac} can be decomposed into two desired polygonal chains.   □

During the whole execution of algorithm *GatheringObstacles*(), none of the strategies creates a configuration in which P_{ac} contains more than one robot position such that it cannot be decomposed into two polygonal chains. Only an initial configuration can have such scenario. The above lemma implies that the process of updating P_{ac} in (B.3) of Sect. 4, when P_{ac} contains more than one robot position on its boundary and it cannot be decomposed into two polygonal chains, terminates in finite time.

Lemma 4. *Let $\mathcal{P} \cup \mathcal{R}(t_0)$ have no rotational symmetry. Suppose $\mathcal{P}$ has rotational symmetry and P_{ac} contains more than one robot position on its boundary. During the execution of algorithm GatheringObstacles(), there is a time $t \geq t_0$ such that P_{ac} does not change after time t and P_{ac} has exactly one robot position on its boundary, $\forall t' \geq t$.*

Proof. First consider the initial configuration $\mathcal{P} \cup \mathcal{R}(t_0)$. Suppose the circumference of P_{ac} can be decomposed into two polygonal chains, one containing all the robots. Our approach asks the robots to move along the corresponding polygonal chain of P_{ac} to merge the robot positions on P_{ac} into one point. Since $\mathcal{P} \cup \mathcal{R}(t)$ has at most one line of symmetry, the robots can agree on the corresponding polygonal chain to move along. The polygonal chain is of finite length. In each movement, robots move towards the nearest vertex of the chain in the direction of its destination. Hence after a finite time, we are left with either one robot position or two robot positions, sharing same edge, on the polygonal chain. In the later case robots move to the mid point of the corresponding edge. This creates exactly one robot position on P_{ac} within finite time. In this process, P_{ac}

remains same. If the circumference of P_{ac} cannot be decomposed into two desired polygonal chains, then by Lemma 3, there exits t such that P_{ac} becomes static and it can be decomposed into two desired polygonal chains. Hence, the lemma follows from the same foregoing arguments. □

Lemma 5. *Algorithm GatheringObstacles() deterministically solves the gathering problem within finite time provided the initial configuration $\mathcal{P} \cup \mathcal{R}(t_0)$ does not have rotational symmetry,*

Proof: When $CoG(\mathcal{P})$ i.e., $\mathcal{O}$ lies outside the obstacles, then we are done. The point $\mathcal{O}$ is static and serves as the point of gathering. Otherwise, according to our algorithm:

- When $\mathcal{P}$ is asymmetric or has one line of symmetry, our algorithm chooses a point r based on the vertices of the polygons. Thus, r is a fixed point and is not affected by the movements of the robots.
- When $\mathcal{P}$ has more than one line of symmetry, our approach involves the robot positions in $\mathcal{R}(t)$. If $\mathcal{P} \cup \mathcal{R}(t)$ does not have rotational symmetry, according to our algorithm we have the following: (i) if P_{ac} does not change and contains exactly one robot position on its boundary, this unique robot position on P_{ac} serves as the gathering point. Since the robots move towards this point along paths not crossing P_{ac} (such paths exists because polygons are non-overlapping), the robot position on P_{ac} remains fixed under the movements of the robots (ii) otherwise, by Lemmas 2 and 4, P_{ac} becomes static and contains exactly one robot position in finite time, which implies the corresponding of the lemma.

Therefore, algorithm GatheringObstacles() deterministically solves the gathering problem in finite time. □

Combining the above results, we have the following theorem.

Theorem 1. *The gathering problem is solvable in finite time, without any extra assumption on the capabilities of the robots, in the presence of non-intersecting convex polygonal obstacles, when the initial configuration of the polygons and the robot positions does not have rotational symmetry.*

Theorem 2. *If $\mathcal{P} \cup \mathcal{R}(t)$ has rotational symmetry around $\mathcal{O}$ which lies inside an obstacle in $\mathcal{P}$, there does not exist any deterministic algorithm that solves the gathering problem starting from $\mathcal{P} \cup \mathcal{R}(t)$. The problem remains unsolvable even if we arm the robots strong multiplicity detection, memory, chirality and direction-only axis agreement and have them move in FSYNC model.*

5   Conclusion

This paper studies the gathering problem in the presence of transparent polygonal obstacles. Obstacles are fixed objects. They help us by providing fixed reference points. One may identify certain fixed points (i.e., independent of axes and robot positions) based on these reference points. However if all these points fall within the obstacles and hence are unreachable, the advantage is lost. On the other hand obstacles also create problems in finding paths. A distributed algorithm has been proposed which solves the gathering problem without any extra assumption on the capabilities of the robots if the initial configuration does not have any rotational symmetry. The gathering is impossible if the initial configuration has rotational symmetry. The problem remains unsolvable even if robots have multiplicity detection, memory, chirality and direction-only axis agreement in the *FSYNC* model. While it would be interesting to consider opaque obstacles, it is not sure whether much can be expected from that model.

Acknowledgement. The authors would like to express their gratitude to Peter Widmayer and Matus Mihalak for their valuable inputs.

References

1. Agathangelou, C., Georgiou, C., Mavronicolas, M.: A distributed algorithm for gathering many fat mobile robots in the plane. In: ACM Symposium on Principles of Distributed Computing (PODC 2013), pp. 250–259. ACM, New York (2013)
2. Agmon, N., Peleg, D.: Fault-tolerant gathering algorithms for autonomous mobile robots. SIAM J. Comput. **36**(1), 56–82 (2006)
3. Bhagat, S., Chaudhuri, S.G., Mukhopadhyaya, K.: Fault-tolerant gathering of asynchronous oblivious mobile robots under one-axis agreement. J. Discrete Algorithms **36**, 50–62 (2016)
4. Cieliebak, M., Flocchini, P., Prencipe, G., Santoro, N.: Distributed computing by mobile robots: gathering. SIAM J. Comput. **41**(4), 829–879 (2012)
5. Czyzowicz, J., Gasieniec, L., Pelc, A.: Gathering few fat mobile robots in the plane. Theor. Comput. Sci. **410**(6–7), 481–499 (2009)
6. Défago, X., Gradinariu, M., Messika, S., Raipin-Parvédy, P.: Fault-tolerant and self-stabilizing mobile robots gathering. In: Dolev, S. (ed.) DISC 2006. LNCS, vol. 4167, pp. 46–60. Springer, Heidelberg (2006). doi:10.1007/11864219_4
7. Efrima, A., Peleg, D.: Distributed algorithms for partitioning a swarm of autonomous mobile robots. Theor. Comput. Sci. **410**(14), 1355–1368 (2009)
8. Flocchini, P., Prencipe, G., Santro, N., Widmayer, P.: Hard tasks for weak robots: the role of common knowledge in pattern formation by autonomous mobile robots. In: Aggarwal, A., Pandu Rangan, C. (eds.) ISAAC 1999. LNCS, vol. 1741, pp. 93–102. Springer, Heidelberg (1999). doi:10.1007/3-540-46632-0_10
9. Flocchini, P., Prencipe, G., Santoro, N., Widmayer, P.: Gathering of asynchronous robots with limited visibility. Theor. Comput. Sci. **337**(1), 147–168 (2005)
10. Flocchini, P., Prencipe, G., Santoro, N.: Distributed Computing by Oblivious Mobile Robots-Synthesis Lectures on Distributed Computing Theory. Morgan & Claypool Publishers, San Rafael (2012)
11. Chaudhuri, S.G., Mukhopadhyaya, K.: Leader election and gathering for asynchronous fat robots without common chirality. J. Discrete Algorithms **33**, 171–192 (2015)
12. McAllister, M., Kirkpatrick, D., Snoeyink, J.: A compact piecewise-linear voronoi diagram for convex sites in the plane. J. Discrete Comput. Geom. **15**(1), 73–105 (1996)

13. Prencipe, G.: Impossibility of gathering by a set of autonomous mobile robots. Theor. Comput. Sci. **384**(2–3), 222–231 (2007)
14. Suzuki, I., Yamashita, M.: Distributed anonymous mobile robots: formation of geometric patterns. SIAM J. Comput. **28**(4), 1347–1363 (1999)

Space-Efficient Algorithms

A Space-Efficient Algorithm for the Dynamic DFS Problem in Undirected Graphs

Kengo Nakamura$^{(\boxtimes)}$ and Kunihiko Sadakane

Graduate School of Information Science and Technology,
The University of Tokyo, Tokyo, Japan
{kengo_nakamura,sada}@mist.i.u-tokyo.ac.jp

Abstract. Depth-first search (DFS) is a well-known graph traversal algorithm and can be performed in $O(n + m)$ time for a graph with n vertices and m edges. We consider dynamic DFS problem, that is, to maintain a DFS of an undirected graph G when edges and vertices are gradually inserted into or deleted from G. We present an algorithm for this problem which takes worst case $O(\sqrt{mn} \cdot \text{polylog}(n))$ time per update and requires only $(3m + o(m)) \log n + O(m)$ bits of space. This is the first sublinear worst case update time algorithm for this problem which requires only $O(m \log n)$ bits. Moreover, the time complexity of this algorithm is close to, or under particular condition better than, the state-of-the-art algorithm of Chen et al. [5], which requires $O(m \log^2 n)$ bits of space.

1 Introduction

Depth-first search (DFS) is a fundamental algorithm for searching graphs. As a result of performing DFS, a rooted tree which spans all vertices reachable from the root is constructed. This rooted tree is called *DFS tree*, which is used as a tool for many graph algorithms such as finding strongly connected components of digraphs and detecting articulation vertices or bridges of undirected graphs. Generally, for a graph with n vertices and m edges, DFS can be performed in $O(n + m)$ time, and DFS tree can be constructed in the same time.

The graph structure that appears in the real world often changes gradually with time. Therefore the following problem is considered: when a graph G and its DFS tree T are given, for an online sequence of updates on G, we try to rebuild a DFS tree for G after each update. This problem is called *dynamic DFS problem*. Here single update on the graph is one of the following four operations: insertion or deletion of one edge, or those of one vertex (and accompanying edges).

Until recently, there are few papers for the dynamic DFS problem, despite of the simplicity of DFS in static setting. For directed acyclic graphs, Franciosa et al. [6] proposed an incremental (i.e. supporting only insertion of edges) algorithm and later Baswana and Choudhary [2] proposed a decremental (i.e. supporting only deletion of edges) algorithm. For undirected graphs, Baswana and Khan [3] proposed an incremental algorithm. However, these algorithms support only either of insertion or deletion. Moreover, none of these algorithms achieve

© Springer International Publishing AG 2017
S.-H. Poon et al. (Eds.): WALCOM 2017, LNCS 10167, pp. 295–307, 2017.
DOI: 10.1007/978-3-319-53925-6_23

Table 1. Comparison of required space and worst case update time for dynamic DFS.

	[1]	[5]	Ours
Space (bit)	$O(m \log^2 n)$	$O(m \log^2 n)$	$(3m + o(m)) \log n + O(m)$
Update time			
Under (I)	$O(\sqrt{mn} \log^{2.5} n)$	$O(\sqrt{mn} \log^{1.5} n)$	$O(\sqrt{mn} \log^{1.75} n / \sqrt{\log \log n})$
Under (II)	$O(\sqrt{mn} \log^{2.5} n)$	$O(\sqrt{mn} \log^{1.5} n)$	$O(\sqrt{mn} \log^{1.25} n)$
Under (III)	$O(n \log^3 n)$	$O(n)$	$O(n \log n)$

the worst case time complexity of $o(m)$ per single update though the amortized computational time is better than the static DFS algorithm. This means in the worst case the computational time becomes the same as the static algorithm.

Recently, Baswana et al. [1] proposed a dynamic DFS algorithm for undirected graphs which overcomes these two problems. This algorithm supports all four types of graph updates and achieves the worst case $O(\sqrt{mn} \cdot \text{polylog}(n))$ time per update. Thus it is expected that various graph problems in dynamic setting can be solved with it.

A drawback of their works is that these algorithms require $O(m \log^2 n)$ bits for the auxiliary data structures, which is $O(\log n)$ times larger space than the adjacency list of original graph. This is unsuitable for the large graph. Therefore we want to compress the required space of these algorithms.

Related Works and Our Results. Very recently, Chen et al. [5] improved the algorithms of Baswana et al. [1] and reduced the time complexity[1]. However, their algorithms still use $O(m \log^2 n)$ bits. We use a part of their ideas.

First, we improve the way to solve a query that is frequently used in the algorithm of [1]. Second, we compress the data structures used in [1,5] using wavelet tree [7]. Third, we propose a space-efficient algorithm for dynamic DFS problem in undirected graphs. Here we consider the following three conditions on the updates on the graph: (I) all four kinds of updates can appear, (II) three kinds of updates other than edge deletion can appear, and (III) only edge insertion appear. Our algorithm requires only $(3m+o(m)) \log n+O(m)$ bits under every condition, and the worst case update time is better than [1]. Moreover, the update time is, under (II) better than and under (I) close to, the results of [5]. These results are summarized in Table 1. If amortized update time is permitted, our algorithm requires only $(2m + o(m)) \log n + O(m)$ bits.

2   Preliminaries

Throughout this paper, n denotes the number of vertices and m denotes the number of edges. We assume that a graph is always simple, i.e. has no self-loops

---

[1] However, in their paper there is a bit too strong assumption on queries used in their algorithms, and we succeed to remove this assumption. details are in Sect. 4.

or parallel edges, since they make no sense in constructing DFS tree. We also assume that $n = o(m)$, or more specifically, $m = \omega(n \log^{0.5} n)$.

Bit Vectors. Let $B[1..l]$ be a 0,1-sequence of length l, and consider two queries on B: $\text{rank}_c(i, B)$ returns the number of c in $B[1..i]$ and $\text{select}_c(i, B)$ returns the position of the i-th occurrence of c in B ($c = 0, 1$). Then there exists a data structure such that rank and select queries for $c = 0, 1$ can be answered in all $O(1)$ time and the required space is $l + O(l \log \log l / \log l) = l + o(l)$ bits. Moreover, the space can be reduced to $lH_0(B) + o(l)$ bits while keeping $O(1)$ query time [12], where $H_0(B) \leq 1$ is the zeroth-order empirical entropy of B. When 1 occurs v times in B, $lH_0(B) = v \log \frac{l}{v} + (l - v) \log \frac{l}{l-v} \leq v \log \frac{el}{v}$.

The bit vector described above is static, i.e. the 0,1-sequence B never changes once the data structure is built. Instead, we can consider a "dynamic" bit vector that allows insertion or deletion (indels) of bits. Then there exists a data structure for a binary sequence of length l such that rank, select queries and indels of one bit can be done in all $O(\log l / \log \log l)$ time and the required space is $l + o(l)$ bits [11]. The space can also be reduced to $lH_0(B) + o(l)$ bits while keeping the same query time [11].

Wavelet Trees. Let $S[1..l]$ be an integer sequence of symbols $[0, \sigma - 1]$. A wavelet tree [7] for S is a complete binary tree with σ leaves and $\sigma - 1$ internal nodes, each internal node of which has a bit vector. Each internal node v corresponds to an interval of symbols $[l_v, r_v] \subseteq [0, \sigma - 1]$; the root corresponds to $[0, \sigma - 1]$ and its left (right) child to $[0, \lfloor \sigma/2 \rfloor]$ ($[\lfloor \sigma/2 \rfloor + 1, \sigma - 1]$), and these intervals are recursively divided until leaves, each of which corresponds to one symbol. The bit vector $B_v[1..l_v]$ corresponding to the internal node v is defined as follows: let $S_v[1..l_v]$ be the subsequence of S which consists of elements with symbols $[l_v, r_v]$, and if the symbol $S_v[i]$ corresponds to the left child of v then $B_v[i] = 0$, otherwise $B_v[i] = 1$. The wavelet tree requires $(l + o(l)) \log \sigma$ bits of space, and can be built in $O(l \log \sigma / \sqrt{\log l})$ time [9].

3 Overview of the Algorithms of Baswana et al.

In this section, we give an overview of the algorithms proposed by Baswana et al. [1], and describe some lemmas used in this paper.

3.1 Fault Tolerant DFS Algorithm

We first refer to the algorithm for *fault tolerant DFS problem*. This problem is described as follows: when an undirected graph G and its DFS tree T are given, we try to rebuild a DFS tree for the new graph obtained by deleting $k (\leq n)$ edges or vertices from G. In this part U denotes a set of vertices and edges we want to delete from G, and $G - U$ denotes the new graph obtained by deleting vertices and edges in U from G.

The algorithm for this problem proposed in [1] is, roughly speaking, as follows: first the forest obtained by deleting vertices and edges in U from T is divided into paths and trees (they refer to this partition as *disjoint tree partition*), and then the new DFS tree T' for $G - U$ is gradually built while the partition is also gradually updated. The key point of reducing computational complexity is that taking advantage of partitioning, the number of edges accessed by this algorithm can be decreased from m. At this time, it must be ensured that the edges not accessed by this algorithm are not needed to construct the new DFS tree T'. In order to achieve this, they utilize a *reduced adjacency list* L and two queries Q and Q'. Here Q and Q' are defined as follows.

Problem 1 [1]. A connected undirected graph G and its DFS tree T, which is a directed rooted spanning tree, are given, and later a set U of vertices and edges is given. Then for three vertices w, x, y in $G - U$, the following query is considered. Among all edges which directly connect a subtree $T(w)$ of T rooted at w (a vertex w) and a path $path(x, y)$ in T connecting x and y, the query $Q(T(w), x, y)$ $(Q'(w, x, y))$ returns an edge whose endpoint on $path(x, y)$ is the nearest to x. If there are no edges between $T(w)$ (or w) and $path(x, y)$, this query should return $\emptyset$. In this query we can assume that $T(w)$ $(\{w\})$ and $path(x, y)$ have no common vertices and contain no vertices or edges in U. We can also assume that x is an ancestor of y in T, y is an ancestor of x in T, or $x = y$.

During the construction of T', the edges added to L are chosen carefully by Q and Q', and instead of the whole adjacency list of G, only L is accessed.

In fact, this fault tolerant DFS algorithm can be easily extended to handle insertion of vertices/edges other than deletion updates [1]. From now we consider each of the three conditions (I), (II) and (III) described in Sect. 1. Then the time complexity of algorithm can be summarized in the following lemma.

Lemma 1 [1]. *Suppose that the query Q (Q') can be solved in $O(f)$ $(O(f'))$ time if U has no edges, or $O(g)$ $(O(g'))$ time if U has an edge. Then for any $k(\leq n)$ updates on graph, a new DFS tree can be built in $O(k \cdot Gn)$ time under the condition (I), $O(k \cdot Fn)$ time under (II), and $O(fn)$ time under (III), where $F = f \log n + f'$ and $G = g \log n + g'$.*

3.2 Dynamic DFS Algorithm

Next we refer to the algorithm for dynamic DFS. Baswana et al. [1] proposed an algorithm for this problem by using the fault tolerant DFS algorithm as a subroutine. The heart of their result can be summarized in the following lemma.

Lemma 2 [1]. *Suppose that for any $k(\leq n)$ updates on the original undirected graph G, a new DFS tree can be built in $O(k \cdot g + h)$ time with a data structure $\mathcal{D}$ constructed in $O(f)$ time. Then for any online sequence of updates on the graph, a new DFS tree after each update can be built in amortized/worst case $O(\sqrt{fg} + h)$ time per update, if $\sqrt{f/g} \leq n$ holds.*

First we refer to the amortized (not worst case) update time algorithm. Their idea is to rebuild the data structure $\mathcal{D}$ which solves Q and Q' after every $c = \sqrt{f/g}$ updates. To explain this idea in detail, let G_j be the graph obtained by applying first jc updates on G, and T_j be the DFS tree for G_j reported by this algorithm. For the first c updates, use the data structure $\mathcal{D}_0$ (i.e. perform fault tolerant DFS with $\mathcal{D}_0$ for each arrival of update) constructed from the original graph G and DFS tree T. After c updates are processed, build the data structure $\mathcal{D}_1$ from G_1 and T_1, and use $\mathcal{D}_1$ for next c updates. Similarly, after jc updates are processed, the data structure $\mathcal{D}_j$ is built from G_j and T_j, and $\mathcal{D}_j$ is used for next c updates. In this way the construction time of $\mathcal{D}$ is amortized over c updates and achieves the amortized time complexity in Lemma 2.

Next we proceed to the worst case update time algorithm. The idea to achieve the efficient "worst case" update time described in [1] is to divide the building process of data structure over c updates. For the first $c = \sqrt{f/g}$ updates, use the data structure $\mathcal{D}_0$ built from the original graph G and DFS tree T. For the next c updates, use again $\mathcal{D}_0$ and build $\mathcal{D}_1$ gradually from G_1 and T_1. Similarly, from $(jc+1)$-st to $((j+1)c)$-th updates on the graph, use $\mathcal{D}_{j-1}$ for fault tolerant DFS and build $\mathcal{D}_j$ gradually from G_j and T_j. We call this moment phase j of the algorithm. In this way the construction time of data structures is divided, and the efficient worst case update time in Lemma 2 is achieved.

4 Query Reduction to Orthogonal Range Search Problem

In this section, it is shown that Q and Q' can be answered by solving *orthogonal range successor (predecessor) query*. This idea is first proposed by Chen et al. [5] and we use a part of their work, but our idea differs from theirs. Here the orthogonal range successor (predecessor) query is defined as follows.

Problem 2. On grid points in a 2-dimensional plane, k points are given. Then for any rectangular regions $R = [x_1, x_2] \times [y_1, y_2]$, we want to answer the point whose y-coordinate is the smallest (largest) among those within R. This query is called *orthogonal range successor (predecessor) query*, and we abbreviate it as ORS (ORP) query. If there are no points within R, the query should return $\emptyset$.

First we show how a set of points is constructed from G. The first step is the same as Baswana et al. [1]: a heavy-light (HL) decomposition [13] of T is calculated, and then the order $\mathcal{L}$ of vertices is decided according to the pre-order traversal of T, such that for the first time a vertex v is visited, the next vertex to visit is one that is directly connected with a heavy edge derived from the HL decomposition. Then the vertices of G are numbered from 0 to $n-1$ according to $\mathcal{L}$; the vertex id of v is denoted by $\text{num}(v)$. The next step is the same as Chen et al. [5]: we consider a grid $\mathcal{G}$ and, for each edge (i, j) of G, put two points on the coordinates $(\text{num}(i), \text{num}(j))$ and $(\text{num}(j), \text{num}(i))$ in $\mathcal{G}$. This is equivalent to consider the adjacency matrix of G, thus $2m$ points are placed.

Next we show that the query Q can be converted to single ORS/ORP query on $\mathcal{G}$. This idea is first proposed by Chen et al. [5], but their argument is a bit incomplete, as described below. Here we define some symbols: for two vertices a and b in G, $a \prec b$ means a is an ancestor of b in T, $a \preceq b$ means $a \prec b$ or $a = b$, and $a \parallel b$ means neither $a \preceq b$ nor $b \preceq a$ holds, i.e. a and b have no ancestor-descendant relation. Let p be the parent of w (which appears in the query $Q(T(w), x, y)$) in T. Then there are five patterns on the configuration of x, y, p as drawn in Fig. 1, assuming $x \preceq y$: (i) $p \prec x \preceq y$, (ii) $x \preceq p \preceq y$, (iii) $x \preceq y \prec p$, (iv) $x \parallel p$ and $y \parallel p$, and (v) $x \prec p$ and $y \parallel p$. However, Chen et al. [5] argued only the pattern (ii), which makes trouble when the online sequence of updates involves deletion of edges or vertices.

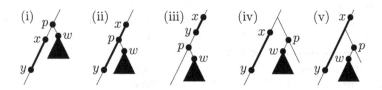

Fig. 1. The configurations of $path(x, y)$ and $T(w)$ in T that can appear in $Q(T(w), x, y)$.

Now we show a way to convert Q to single ORS/ORP query on $\mathcal{G}$ for any patterns. Let $[q, r]$ be an interval in $\mathcal{L}$ occupied by the vertices of $T(w)$ (the vertices of $T(w)$ occupy single interval in the vertex numbering since $\mathcal{L}$ is a pre-order traversal of T). We employ the fact that T is a DFS tree for an undirected graph G iff all non-tree edges in G are back edges, i.e. if $x \parallel y$ then x and y are not directly connected. For convenience, we call this fact DFS property. Note that Chen et al. [5] also employ this fact. Consider the case $x \preceq y$. Then the query Q is converted to single ORS query. The case $y \preceq x$ is almost the same as the case $x \preceq y$, except that Q is converted to single ORP query.

In fact, we prove that the ORS/ORP query to which Q is converted can be decided by the vertex id of x, y and p. First, if (a) $num(p) < num(x) \leq num(y)$ then the answer for $Q(T(w), x, y)$ is $\emptyset$: (a) can appear in pattern (i) or (iv), and in these patterns there are no edges between $T(w)$ and $path(x, y)$ due to DFS property. Second, if (b) $num(x) \leq num(p) \leq num(y)$ then the rectangle is $R = [q, r] \times [num(x), num(LCA(y, w))]$ where $LCA(y, w)$ is the lowest common ancestor of y and w in T: (b) can appear in (ii) or (v), and in these patterns all the edges between $T(w)$ and $path(x, y)$ are indeed between $T(w)$ and $path(x, LCA(y, w))$ again due to DFS property (in (ii), $LCA(y, w) = p$). Note that it does not matter if there are some branches forked from $path(x, LCA(y, w))$, since the vertices of these branches have no ancestor-descendant relations with those of $T(w)$. The LCA query can be solved in $O(1)$ time with a data structure of $O(n \log n)$ bits constructed in $O(n)$ time [4]. Finally, if (c) $num(x) \leq num(y) < num(p)$ then the rectangle is $R = [q, r] \times [num(x), num(y)]$. The inequality (c) can appear in (iii), (iv) or (v). In (iii) it does not matter if there are some branches forked from $path(x, y)$ because of DFS property. In (iv) there are no edges between $T(w)$ and $path(x, y)$, and the ORS query also returns $\emptyset$. In (v) the

ORS query reports correctly the edge connecting $T(w)$ and $path(x, LCA(y, w))$, since $\text{num}(x) \leq \text{num}(y) < \text{num}(p)$.

Finally we consider the query Q'. Using the method of Baswana et al. [1], Q' can be converted to $O(\log n)$ ORS/ORP queries on $\mathcal{G}$ thanks to the property of HL decomposition (note that the method of Baswana et al. [1] can be interpreted as solving $O(\log n)$ ORS/ORP queries). From Lemma 1, it does not matter if Q' is solved $O(\log n)$ times slower than Q is.

At this time it is noted that when U includes some edges (other than vertices) elimination of points from $\mathcal{G}$ must be supported, since deletion of edges from G means that. These observations are summarized in the following theorem.

Theorem 1. *Suppose that there exists a data structure $\mathcal{D}$ which solves single ORS/ORP query on an $l \times l$ grid $\mathcal{G}$ with k points in $O(f(l, k))$ time and supports deletion of one point from $\mathcal{G}$ in $O(g(l, k))$ time. Then Q (Q') can be answered in $O(f(n, 2m))$ ($O(f(n, 2m) \log n)$) time and deletion of single edge can be done in $O(g(n, 2m))$ time with $\mathcal{D}$. If U does not include edges, it is not necessary for $\mathcal{D}$ to support deletion of points.*

5   Compression of Data Structures

In this section, we show a way to solve Q and Q' space-efficiently.

The ORS/ORP query on an $l \times l$ grid with k points can be solved efficiently with wavelet tree [10]. The method is to build an integer array $S[1..k]$ which contains the y-coordinates of the points sorted by the corresponding x-coordinates. Additionally, a bit vector $B = 0^{m[0]} 1 0^{m[1]} 1 \cdots 0^{m[l-1]} 1$ is constructed (where $m[i]$ is the number of points whose x-coordinate is i), which enables us to inter-convert between the position in S and the x-coordinate with $O(1)$ `rank` and `select` queries. Then the ORS/ORP query is converted to the *range next (previous) value query* [10] (we abbreviate it as RN (RP) query) on S, that is, for any interval $[a, b]$ and integer p, to answer the smallest (largest) element in $S[a..b]$ which is not less than (not more than) p. The RN/RP query can be solved with $O(\log l)$ `rank` and `select` queries on the wavelet tree $\mathcal{W}$ for S [10]. Here $\mathcal{W}$ takes $(2m + o(m)) \log n$ bits and B takes $(n + 2m) H_0(B) + o(m)$ bits.

However, now $\mathcal{W}$ has information of both directions for each edge of G. This is redundant since G is an undirected graph, thus we want to hold information of only one direction for each edge. In fact, the placement of points on $\mathcal{G}$ is symmetric since the adjacency matrix of G is also symmetric. Therefore even if only half of these points are stored, i.e. for each two points (i, j) and (j, i) either one is stored, it is likely that we can solve ORS/ORP query by the following way: first solve query on halved points, second transpose R and again solve query on halved points, and finally combine these two results. However, at this time a problem arises: when R is transposed, a switch between x- and y-coordinate occurs, and we need to consider a symmetric variant of ORS/ORP query, that is, to answer the point whose "x-coordinate" is the smallest (largest) among those within R. Although it is sufficient to build data structures additionally for the symmetric variant, this doubles the required space, which makes no sense.

Recall that we already build an integer array $S[1..k]$. In fact, the symmetric variant of ORS/ORP query corresponds to the following query on S.

Problem 3. An integer sequence $S[1..l]$ is given. Then for any interval $[a, b] \subseteq [1, l]$ and two integers p, q, we want to answer the leftmost (rightmost) element in $S[a..b]$ which is not less than p and not more than q. We call this problem *range leftmost (rightmost) value query*, and abbreviate it as RL (RR) query. If there is no such element, the query should return 0.

These queries are generalization of *prevLess query* [8], which is the RR query with $a = 1$ and $p = 0$. It is already known that the prevLess query can be efficiently solved with the wavelet tree $\mathcal{W}$ for S [8].

Now we show that the RL/RR query can also be efficiently solved with $\mathcal{W}$. We show only the solution of RL query, since RR query can be solved by almost the same strategy. Recall that the interval of symbols $[p, q]$ can be covered with $O(1)$ internal nodes or leaves per level, since $\mathcal{W}$ is a complete binary tree. Let $V_1, \ldots, V_r$ be such nodes or leaves. First, we begin with the interval $[a, b]$ of position in the bit vector of the root of $\mathcal{W}$, and traverse these intervals of position while descending $\mathcal{W}$, until reaching one of $V_1, \ldots, V_r$. Traversing intervals can be done with $O(1)$ `rank` queries per node. Then the leftmost position of the traversing interval of each of $V_1, \ldots, V_r$ is the candidate for answer. Second, we traverse the positions of candidates using `select` queries while ascending to the root. If both children of a node v have a candidate, v has two candidates for answer during this process. Here only the left one of the two can be a candidate, thus it can be done with $O(1)$ `select` queries per node. We use $O(\log \sigma)$ `rank` and `select` queries overall. Hence we obtain the following lemma.

Lemma 3. *Suppose that $S[i] \in [0, \sigma - 1]$ for all $i = 1, \ldots, l$. Then using the wavelet tree $\mathcal{W}$ for S, both RL and RR queries can be answered with $O(\log \sigma)$ `rank` and `select` queries on $\mathcal{W}$.*

The pseudocode for solving RL query by the wavelet tree for S is given in Algorithm 1. Calling $\text{RL}(root, [p, q], [a, b], [0, \sigma - 1])$ yields the position i of answer for the query if the answer exists, otherwise returns 0. Note that if i is given, we can obtain $S[i]$ with $O(\log \sigma)$ `rank` and `select` queries on $\mathcal{W}$.

Now we can solve Q and Q' with information of only one direction for each edge. First, we concatenate the "halved" adjacency list of G into single integer array S (of length m). Then for a query rectangle $R = [x_1, x_2] \times [y_1, y_2]$ of ORS/ORP query, we first solve the ORS/ORP query with halved points by the corresponding RN/RP query on S. Second, we transpose the query rectangle R and solve the symmetric ORS/ORP query with halved points by the corresponding RL/RR query on S. Combining these two results yields the answer for the original ORS/ORP query. Here we should again construct the bit vector B as above, but it takes only $(n+m)H_0(B) + o(m) \leq n \log \frac{e(n+m)}{n} + o(m) = o(m) \log n$ bits. Note that if U has some edges, deletion of elements of S and B must be supported. That can be achieved by substituting B and the bit vectors in $\mathcal{W}$ for dynamic bit vectors. Therefore we can obtain the following theorem.

Algorithm 1. Range Leftmost Value by Wavelet Tree.

1: **function** RL($v, [p, q], [a, b], [\alpha, \omega]$)
2:    **if** $a > b$ **or** $[\alpha, \omega] \cap [p, q] = \emptyset$ **then return** 0
3:    **if** $[\alpha, \omega] \subseteq [p, q]$ **then return** a
4:    $\gamma \leftarrow \lfloor (\alpha + \omega)/2 \rfloor$
5:    $i \leftarrow$ RL(*left*(v), $[p, q]$, $[\text{rank}_0(B_v, a - 1) + 1, \text{rank}_0(B_v, b)]$, $[\alpha, \gamma]$)
6:    $j \leftarrow$ RL(*right*(v), $[p, q]$, $[\text{rank}_1(B_v, a - 1) + 1, \text{rank}_1(B_v, b)]$, $[\gamma + 1, \omega]$)
7:    **if** $i \neq 0$ **and** $j \neq 0$ **then return** $\max(\text{select}_0(B_v, i), \text{select}_1(B_v, j))$
8:    **else if** $i \neq 0$ **then return** $\text{select}_0(B_v, i)$
9:    **else if** $j \neq 0$ **then return** $\text{select}_1(B_v, j)$
10:    **else return** 0
11: **end function**

Theorem 2. *With a data structure which takes $(m + o(m)) \log n$ bits and can be built in $O(m\sqrt{\log n})$ time, Q (Q') can be solved in $O(\log n)$ ($O(\log^2 n)$) time if U has no edges, or Q (Q') can be solved in $O(\log^2 n / \log \log n)$ ($O(\log^3 n / \log \log n)$) time and deletion of one edge can be done in $O(\log^2 n / \log \log n)$ time otherwise.*

6 Space-Efficient Dynamic DFS

In this section, we show a way to solve dynamic DFS problem space-efficiently. following the algorithms of Baswana et al. [1].

6.1 Fault Tolerant DFS

We begin with the case of fault tolerant DFS problem. Following the original algorithm described in Sect. 3.1, the important point is that once a data structure for answering Q and Q' is built, the whole adjacency list of the original graph is no longer needed. Moreover, information used in the algorithm other than the data structure and the reduced adjacency list takes only $O(n \log n) = o(m) \log n$ bits: there are $O(n)$ words of information for the original DFS tree T (including the data representing the disjoint tree partition), $O(1)$ words of information attached to each vertex and each edge in T, a stack which have at most $O(n)$ elements, and a partially constructed DFS tree, but these sum up to only $O(n)$ words. Since we have already shown that the data structure takes $(m + o(m)) \log n$ bits, we have only to consider the size of the reduced adjacency list L. From [1], it can be concluded that for any $k (\leq n)$ graph updates, the number of edges in L is at most $O(nk \log n)$ under (I)/(II), and $O(n)$ under (III). The time complexity of this algorithm can be calculated from Lemma 1 and Theorem 2.

Theorem 3. *Let G be an undirected graph and suppose that the DFS tree T for G is given. Then for any $k (\leq n)$ updates on G, a new DFS tree can be built in $O(nk \log^3 n / \log \log n)$ time and $(m + o(m)) \log n + O(nk \log^2 n)$ bits of space under (I), $O(nk \log^2 n)$ time and $(m + o(m)) \log n + O(nk \log^2 n)$ bits of space under (II), and $O(n \log n)$ time and $(m + o(m)) \log n$ bits of space under (III), with a data structure constructed in $O(m\sqrt{\log n})$ time.*

6.2 Amortized Update Time Dynamic DFS

Next, we focus on the amortized update time dynamic DFS algorithm. During the "amortized time" algorithm described in Sect. 3.2, we should do both the reconstruction of $\mathcal{D}$ and the fault tolerant DFS, and store information of up to last c updates. Therefore we have to consider (a) how many edges the reduced adjacency list L may have, (b) how much space is required to store information of updates, and (c) how to rebuild the data structure space-efficiently.

First we consider (a). Since we rebuild $\mathcal{D}$ after every c updates, we solve the fault tolerant DFS problem with at most c updates. Therefore we can obtain an upper bound on the number of edges in L. Under the condition (I), we can say $f = m\sqrt{\log n}$, $g = n\log^3 n/\log\log n$ and $h = 0$ for Lemma 2, thus the size of L is at most $O(n\log n \cdot \sqrt{f/g}) = O(\sqrt{mn\log\log n}/\log^{0.25} n) = o(m)$. Under (II), we can say $f = m\sqrt{\log n}$, $g = n\log^2 n$ and $h = 0$, thus the upper bound is $O(\sqrt{mn}\log^{0.25} n)$ which is $o(m)$ under the assumption $m = \omega(n\log^{0.5} n)$. Under (III), we can say $f = m\sqrt{\log n}$, $g = \log n$ and $h = n\log n$ since $O(k \cdot g + h) = O(n\log n)$ holds for $k \leq n$, and the upper bound is $O(n)$, as described in Sect. 6.1. Hence we can conclude that L takes only $o(m)\log n$ bits in any conditions.

Next we consider (b), but this is almost the same as (a). Under (I)/(II), the number of edges inserted or deleted during c updates is up to $n\sqrt{f/g}$, since indels of one vertex involves indels of accompanying edges, which can be at most n arcs. This is smaller than the maximum size of L. Under (III), the number of edges inserted during c updates is up to $\sqrt{f/g} = \sqrt{m}/\log^{0.25} n$ which is $o(m)$.

Finally we consider (c). Let $\mathcal{L}_j$ be the order of vertices in G_j defined by the pre-order traversal of T_j. As described in Sect. 5, the data structure $\mathcal{D}_j$ is indeed a pair of bit vector B_j and wavelet tree $\mathcal{W}_j$. These data structures can be built from two arrays $M_j[0..n]$ and $S_j[1..m]$ meeting the following conditions.

1. $0 = M_j[0] \leq M_j[1] \leq \cdots \leq M_j[n] = m$.
2. The elements of S_j have one-to-one correspondence with the edges in G_j; $S_j[k]$ corresponds to the edge $(i, S_j[k])$ for $k = M_j[i] + 1, \ldots, M_j[i+1]$ $(i = 0, \ldots, n-1)$. Here the numbering of vertices from 0 to $n-1$ is decided according to $\mathcal{L}_j$. We call $S_j[M_j[i] + 1..M_j[i+1]]$ the block k of S_j.
3. The elements of S_j in each block are sorted in ascending order.

Note that S_j takes $m\log n$ bits and M_j takes $O(n\log m) = o(m)\log n$ bits.

Now we describe the way to rebuild the data structures space-efficiently. During use of B_{j-1} and $\mathcal{W}_{j-1}$, we also retain M_{j-1} and S_{j-1}. After jc updates are processed, i.e. T_j is reported, we rebuild the data structures as follows.

1. Destroy B_{j-1} and $\mathcal{W}_{j-1}$.
2. Sort the inserted/deleted edges during the last c updates in the same order as S_{j-1}, and create a new array S'_j by merging these information with S_{j-1}. Simultaneously M_{j-1} is updated to M'_j to meet S'_j.
3. Decide the order $\mathcal{L}_j$ of vertices in G_j by the pre-order traversal of T_j.
4. Create M_j and S_j from M'_j and S'_j (details are described below), and destroy M'_j and S'_j. Then build B_j and $\mathcal{W}_j$ from M_j and S_j.

Note that the number of edges inserted/deleted during c updates is up to $o(m)$, and sorting can be done in linear time and $o(m) \log n$ bits of working space by radix sort. Merging two sorted arrays can also be done in linear time. Therefore we have only to consider the process of creating M_j and S_j.

This process is divided into three steps as an example shown in Fig. 2: ReNumber, ReArrange and Reverse. In ReNumber, the elements of S'_j, which have an old numbering that came from $\mathcal{L}_{j-1}$, is replaced by the new numbering that came from $\mathcal{L}_j$. Note that the old-to-new and the new-to-old correspondence tables of these numberings can be both stored in only $O(n \log n)$ bits of space. Next, in ReArrange, a new array S''_j is created by sorting the blocks of S'_j in the ascending order of $\mathcal{L}_j$ using M'_j, and later M'_j and S'_j are destroyed. At the same time an array M''_j is created in the same manner as M'_j. The final step Reverse is almost the same as a counting sort for S''_j, but each element of a new array S_j is not the element of S''_j itself but the vertex id of the other end of edge derived from M''_j, and later M''_j and S''_j are destroyed. Note that the counter used in the counting sort takes only $O(n \log m)$ bits, and at the end of Reverse M_j can be easily built from this counter. This step reverses the direction of each edge retained in the array, but it makes no problem. Now M_j and S_j meet all conditions.

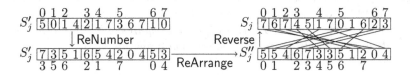

Fig. 2. An example of the process of creating M_j and S_j from M'_j and S'_j.

The whole process to rebuild data structures takes $O(m\sqrt{\log n})$ time, since building single wavelet tree takes $O(m\sqrt{\log n})$ time and the others take only $O(m)$ time. In the whole process, data structures or arrays which take $(m + o(m)) \log n$ bits are $\mathcal{W}_{j-1}$, S_{j-1}, S'_j, S''_j, S_j and $\mathcal{W}_j$, and at any time this algorithm retains at most two of them. Since all other data take only $o(m) \log n$ bits, the space required by the algorithm is $(2m + o(m)) \log n + O(m)$ bits where the $O(m)$ term is the working space of building wavelet tree. Combining these observations with Lemma 2 yields the following theorem.

Theorem 4. *Suppose that an original undirected graph G and its DFS tree T is given. Then for any online sequence of updates on G, a new DFS tree after each update can be built in amortized $O(\sqrt{mn} \log^{1.75} n/\sqrt{\log \log n})$ time under (I), $O(\sqrt{mn} \log^{1.25} n)$ time under (II), and $O(n \log n)$ time under (III). This algorithm requires only $(2m + o(m)) \log n + O(m)$ bits once data structures for the original graph are built.*

6.3 Worst Case Update Time Dynamic DFS

Finally we consider the worst case update time algorithm for the dynamic DFS, following the "worst case time" algorithm described in Sect. 3.2. To implement this space-efficiently, again we must consider (a), (b) and (c) described in Sect. 6.2, but two of them are almost the same argument. In the worst case time algorithm, we should solve the fault tolerant DFS problem with at most $2c$ updates and store information of last $2c$ updates. Thus both the maximum size of reduced adjacency list and the required space for the information of updates are doubled, but these doublings are absorbed in the big O notation.

Therefore we have only to consider (c). Here $\mathcal{D}_j$ denotes the pair of the bit vector B_j and the wavelet tree $\mathcal{W}_j$. Then during phase 0, $\mathcal{D}_0$ is used to perform fault tolerant DFS and rebuilding of data structures is not needed. During phase $j(\geq 1)$, $\mathcal{D}_{j-1}$ is used and the following processes are done gradually: first destroy $\mathcal{D}_{j-2}$ (this is not needed for phase 1), and then build M_j, S_j and $\mathcal{D}_j$ from M_{j-1}, S_{j-1} in the same way as Sect. 6.2. At the end of phase j there exist $\mathcal{D}_{j-1}$, M_j, S_j and $\mathcal{D}_j$, and we can continue to the next phase $j + 1$.

Finally we consider how much space is needed to implement this algorithm. In phase j, $\mathcal{D}_{j-1}$ takes $(m + o(m)) \log n$ bits, and rebuilding the data structures requires at most $(2m+o(m)) \log n + O(m)$ bits as described in Sect. 6.2. Therefore the total required space is $(3m + o(m)) \log n + O(m)$ bits.

Theorem 5. *Suppose the same assumption as Theorem 4. Then for any online sequence of updates, a new DFS tree after each update can be built in worst case $O(\sqrt{mn} \log^{1.75} n / \sqrt{\log \log n})$ time under (I), $O(\sqrt{mn} \log^{1.25} n)$ time under (II), and $O(n \log n)$ time under (III). This algorithm requires only $(3m+o(m)) \log n + O(m)$ bits once data structures for the original graph are built.*

References

1. Baswana, S., Chaudhury, S.R., Choudhary, K., Khan, S.: Dynamic DFS in undirected graphs: breaking the $O(m)$ barrier. In: Proceedings of SODA, pp. 730–739 (2016)
2. Baswana, S., Choudhary, K.: On dynamic DFS tree in directed graphs. In: Italiano, G.F., Pighizzini, G., Sannella, D.T. (eds.) MFCS 2015. LNCS, vol. 9235, pp. 102–114. Springer, Heidelberg (2015). doi:10.1007/978-3-662-48054-0_9
3. Baswana, S., Khan, S.: Incremental algorithm for maintaining DFS tree for undirected graphs. In: Esparza, J., Fraigniaud, P., Husfeldt, T., Koutsoupias, E. (eds.) ICALP 2014. LNCS, vol. 8572, pp. 138–149. Springer, Heidelberg (2014). doi:10.1007/978-3-662-43948-7_12
4. Bender, M.A., Farach-Colton, M.: The LCA problem revisited. In: Gonnet, G.H., Viola, A. (eds.) LATIN 2000. LNCS, vol. 1776, pp. 88–94. Springer, Heidelberg (2000). doi:10.1007/10719839_9
5. Chen, L., Duan, R., Wang, R., Zhang, H.: Improved algorithms for maintaining DFS tree in undirected graphs (2016). arXiv:1607.04913v2
6. Franciosa, P.G., Gambosi, G., Nanni, U.: The incremental maintenance of a Depth-First-Search tree in directed acyclic graphs. Inf. Process. Lett. **61**, 113–120 (1997)
7. Grossi, R., Gupta, A., Vitter, J.S.: High-order entropy-compressed text indexes. In: Proceedings of SODA, pp. 841–850 (2003)
8. Kreft, S., Navarro, G.: Self-indexing based on LZ77. In: Giancarlo, R., Manzini, G. (eds.) CPM 2011. LNCS, vol. 6661, pp. 41–54. Springer, Heidelberg (2011). doi:10.1007/978-3-642-21458-5_6

9. Munro, J.I., Nekrich, Y., Vitter, J.S.: Fast construction of wavelet trees. Theor. Comput. Sci. **638**, 91–97 (2016)
10. Navarro, G.: Wavelet trees for all. J. Discrete Algorithms **25**, 2–20 (2014)
11. Navarro, G., Sadakane, K.: Fully-functional static and dynamic succinct trees. ACM TALG **10**(3), 16 (2014)
12. Raman, R., Raman, V., Rao, S.S.: Succinct indexable dictionaries with applications to encoding k-ary trees, prefix sums and multisets. ACM TALG **3**(4), 43 (2007)
13. Sleator, D.D., Tarjan, R.E.: A data structure for dynamic trees. J. Comput. Syst. Sci. **26**, 362–391 (1983)

Time-Space Trade-Off for Finding the k-Visibility Region of a Point in a Polygon

Yeganeh Bahoo[1], Bahareh Banyassady[2]([✉]), Prosenjit Bose[3],
Stephane Durocher[1], and Wolfgang Mulzer[2]

[1] Department of Computer Science, University of Manitoba, Winnipeg, Canada
{bahoo,durocher}@cs.umanitoba.ca
[2] Institut für Informatik, Freie Universität Berlin, Berlin, Germany
{bahareh,mulzer}@inf.fu-berlin.de
[3] School of Computer Science, Carleton University, Ottawa, Canada
jit@scs.carleton.ca

Abstract. We study the problem of computing the k-visibility region in the memory-constrained model. In this model, the input resides in a randomly accessible read-only memory of $O(n)$ words, with $O(\log n)$ bits each. An algorithm can read and write $O(s)$ additional words of workspace during its execution, and it writes its output to write-only memory. In a given polygon P and for a given point $q \in P$, we say that a point p is inside the k-visibility region of q, if and only if the line segment pq intersects the boundary of P at most k times. Given a simple n-vertex polygon P stored in a read-only input array and a point $q \in P$, we give a time-space trade-off algorithm which reports the k-visibility region of q in P in $O(cn/s + n \log s + \min\{\lceil k/s \rceil n, n \log \log_s n\})$ expected time using $O(s)$ words of workspace. Here $c \leq n$ is the number of critical vertices for q, i.e., the vertices of P where the visibility region may change. We also show how to generalize this result for polygons with holes and for sets of non-crossing line segments.

Keywords: Memory-constrained model · k-visibility region · Time-space trade-off

1 Introduction

Memory constraints on mobile and distributed devices have led to an increasing concern among researchers to design algorithms that use memory efficiently. One common model to capture this notion is the *memory-constrained model* [2]. In this model, the input is provided in a randomly accessible read-only array of $O(n)$ words, with $O(\log n)$ bits each. There is an additional read/write memory consisting of $O(s)$ words of $O(\log n)$ bits each, which is called the *workspace* of the algorithm. Here, $s \in \{1, \ldots, n\}$ is a parameter of the model. The output is written to a write-only array.

This work was partially supported by DFG project MU/3501-2 and by the Natural Sciences and Engineering Research Council of Canada (NSERC).

S.-H. Poon et al. (Eds.): WALCOM 2017, LNCS 10167, pp. 308–319, 2017.
DOI: 10.1007/978-3-319-53925-6_24

Suppose we are given a polygon P and a query point $q \in P$. We say that the point $p \in P$ is k-*visible* from q if and only if the line segment pq properly intersects the boundary of P at most k times (p and q are not counted toward k). The set of k-visible points of P from q is called the k-*visibility region* of q within P, and it is denoted by $V_k(P, q)$; see Fig. 1. Visibility problems have played and continue to play a major role in computational geometry since the dawn of the field, leading to a rich history; see [16] for an overview. The concept of visibility through a single edge first appeared in [11]. Recently, k-visibility for $k > 1$ has been introduced and applied to model coverage of wireless devices whose radio signals can penetrate a given number k of walls [1,13]. There are some other results in this context; for example see [4,10,12,14,15,17,20]. While the 0-visibility region consists of one connected component, the k-visibility region may be disconnected in general. A previous work [3] presents an algorithm for a slightly different variant of this problem, which computes the set of points in the plane which are k-visible from q in presence of a polygon P in $O(n^2)$ time using $O(n)$ space. In this case the k-visibility region is a single connected component.

The optimal classic algorithm for computing the 0-visibility region runs in $O(n)$ time using $O(n)$ space [18]. In the memory-constrained model using $O(1)$ workspace, there is an algorithm which reports the 0-visibility region of a point $q \in P$ in $O(n\bar{r})$ time, where $\bar{r}$ denotes number of reflex vertices of P in output [6]. This algorithm scans the boundary of P in counterclockwise order and it reports the maximal chains of adjacent vertices of P which are 0-visible from q. More precisely, it starts from a visible vertex v_{start}, and it finds v_{vis}, the next visible reflex vertex with respect to q, in $O(n)$ time. The first intersection of the ray qv_{vis} with the boundary of P is called the shadow of v_{vis}. Depending on the type of v_{vis}, the end vertex of the maximal visible chain starting at v_{start}, is either v_{vis} or its shadow, and in each case the other one is the start vertex of the next maximal visible chain. Thus, in each iteration the algorithm reports a maximal visible chain and it repeats this procedure $\bar{r}$ times, where $\bar{r}$ is the number of visible reflex vertices of P. This takes $O(n\bar{r})$ time using $O(1)$ workspace. When the workspace is increased to $O(s)$, such that $s \in O(\log r)$ and r is the number of reflex vertices of P with respect to q, they present $O(nr/2^s + n \log^2 r)$ time or $O(nr/2^s + n \log r)$ randomized expected time algorithm. The method is based on a divide-and-conquer approach which uses the previous algorithm as base algorithm and in each step of the recursion, it splits a chain into two subchains with roughly half of the visible reflex vertices of the chain. Due to differences between the properties of the 0-visibility region and the k-visibility region, there seems to be no straightforward way to generalize this approach. In [5] a general method for transforming *stack-based* algorithms into the memory-constrained model is provided, which can be used as an alternative method to obtain a time-space trade-off to compute the 0-visibility region.

Here, we look at the more general problem of computing the k-visibility region of a simple polygon P from $q \in P$ using a small workspace, and we establish a trade-off between running time and workspace. Unless stated otherwise, all polygons will be understood to be simple.

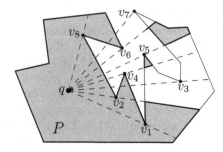

Fig. 1. The gray region is $V_2(P, q)$. The vertices $v_1, \ldots, v_8$ are critical for q. Here v_1, v_2, v_3 and v_6 are start vertices, while v_4, v_5, v_7 and v_8 are end vertices. The boundary of P is partitioned into 8 disjoint chains, i.e., the counterclockwise chain $v_3 v_5$.

2   Preliminaries and Definitions

We assume that our simple polygon P is given in a read-only array as a list of n vertices in counterclockwise (CCW) order along the boundary. This input array also contains a query point $q \in P$. The aim is to report $V_k(P, q)$, using $O(s)$ words of workspace. We assume that the vertices of P are in *weak general position*, i.e., the query point q does not lie on the line determined by any two vertices of P. Without loss of generality, assume that k is even and that $k < n$. If k is odd, we compute $V_{k-1}(P, q)$, which is by definition equal to $V_k(P, q)$, and if $k \geq n$ then P is completely k-visible. The boundary of $V_k(P, q)$ consists of part of the boundary of P and some chords that cross the interior of P to join two points on its boundary. We denote the boundary of a planar set U by ∂U.

Let $\theta \in [0, 2\pi)$, and let r_θ be the ray from q that forms a CCW angle θ with the positive-horizontal axis. An edge of P that intersects r_θ is called an *intersecting edge* of r_θ. The *edge list* of r_θ is defined as the sorted list of intersecting edges of r_θ, according to their intersection with r_θ (from q). The jth member of the edge list of r_θ is denoted $e_\theta(j)$. When rotating r_θ around q in CCW order, the edge list of r_θ does not change unless r_θ stabs a vertex v of P. If r_θ stabs v, then the edge lists of $r_{\theta-\varepsilon}$ and of $r_{\theta+\varepsilon}$ differ, for any small $\varepsilon > 0$. The difference is caused only by edges incident to v. If these edges lie on opposite sides of r_θ, then the edge list of $r_{\theta+\varepsilon}$ is obtained from the edge list of $r_{\theta-\varepsilon}$ by exchanging the incident edge of v, which is in the edge list of $r_{\theta-\varepsilon}$, with the other incident edge of v. If both incident edges of v lie on the same side of r_θ, we call v a *critical vertex*; see Fig. 1. If both incident edges of v lie on the right/left side of r_θ, then the edge list of $r_{\theta+\varepsilon}$ is obtained by removing/adding the two incident edges of v from/to the edge list of $r_{\theta-\varepsilon}$. For simplicity, if r_θ stabs a vertex v, we define the edge list of r_θ equal to the edge list of $r_{\theta+\varepsilon}$, for a small $\varepsilon > 0$. The number of critical vertices in P is denoted by c. A *chain* is defined as a maximal sequence of edges of P which does not contain a critical vertex, except at the beginning and at the end. The critical vertex v is called an *end vertex*/a *start vertex* if both incident edges to v lie on the right/left side of r_θ. The name is due to the fact that an end/start vertex shows the end/start of two chains in the edge list; see Fig. 1. The *angle* of a vertex v which lies on the ray r_θ refers to θ.

Observation 2.1. *Suppose we are given an edge e of a chain C of P, and a ray r_θ. We can find the edge of C which intersects r_θ (if it exists) by scanning the chain C of P in $O(|C|)$ time using $O(1)$ words of workspace.*

The above observation implies that, any edge of a chain may be used as a proper representative of the chain and its other edges. Thus, in the edge list, each edge refers to its containing chain. Obviously, in direction θ, only the first $k+1$ members of the edge list of r_θ are k-visible from q, which leads us to focus on chains and their order in the edge list. As we explained before, when rotating r_θ around q, the structure of the edge list of r_θ (i.e., the chains and their order) changes only when r_θ stabs a critical vertex v. We will see that in this case a segment on r_θ may belong to $\partial V_k(P, q)$. Obviously, v is k-visible if its position on r_θ is not after $e_\theta(k+1)$.

Lemma 2.2. *If r_θ stabs a k-visible end (or start) vertex v, then the segment on r_θ between $e_\theta(k)$ and $e_\theta(k+1)$ (or $e_\theta(k+2)$ and $e_\theta(k+3)$), if these two edges exist, is an edge of $V_k(P, q)$.*

Proof. If v is an end vertex, then for small enough $\varepsilon > 0$, the edges $e_\theta(k)$ and $e_\theta(k+1)$ are respectively $e_{\theta-\varepsilon}(k+2)$ and $e_{\theta-\varepsilon}(k+3)$, so they are not k-visible in direction $\theta - \varepsilon$. These edges are also $e_{\theta+\varepsilon}(k)$ and $e_{\theta+\varepsilon}(k+1)$, so they are k-visible in direction $\theta + \varepsilon$. Hence, the segment on r_θ between $e_\theta(k)$ and $e_\theta(k+1)$ belongs to $\partial V_k(P, q)$. Similarly, if v is a start vertex, the segment between $e_\theta(k+2)$ and $e_\theta(k+3)$ belongs to $\partial V_k(P, q)$; see Fig. 2.   □

Lemma 2.2 leads to the following definition: for a ray r_θ that stabs a k-visible end (or start) vertex v, the segment between $e_\theta(k)$ and $e_\theta(k+1)$ (or $e_\theta(k+2)$ and $e_\theta(k+3)$), if they exist, is called the *window* of r_θ; see Fig. 2.

Observation 2.3. *The boundary of $V_k(P, q)$ has $O(n)$ vertices.*

Proof. $\partial V_k(P, q)$ consists of part of ∂P and windows; thus, a vertex of $V_k(P, q)$ is either a vertex of P or an endpoint of a window. Since each critical vertex causes at most one window, the number of vertices of $V_k(P, q)$ is $O(n)$.   □

Obviously, if P has no critical vertex, then no window exists, and $\partial V_k(P, q) = \partial P$. Thus, we assume that P has at least one critical vertex. From now on, $e_i(j)$ denotes the jth intersecting edge of the ray qv_i, where v_i is a vertex of P. However, instead of $e_i(j)$, it suffices to find an arbitrary edge of the chain containing $e_i(j)$ and then apply Observation 2.1 to find $e_i(j)$. Therefore, we refer to any edge of the chain containing $e_i(j)$ by $e_i(j)$. In the following algorithms, for any critical vertex v_i, we determine $e_i(k+1)$ which helps to find the window of qv_i (if it exists), and also the part of ∂P which is in $\partial V_k(P, q)$. However, depending on how much workspace is available, we have different approaches for finding all $e_i(k+1)$. Details follow in the next sections.

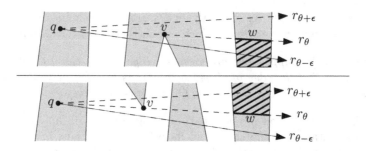

Fig. 2. The ray r_θ in the top/bottom figure stabs the end/start vertex v. The segment w is a window of 4-visible region. The tiled regions are not 4-visible for q.

3   A Constant-Memory Algorithm

In this section, we assume that only $O(1)$ words of workspace is available. Suppose v_0 and v_1 are the critical vertices with respectively first and second smallest (polar) angles. We start from qv_0 and we find $e_0(k+1)$ in $O(kn)$ time using $O(1)$ words of workspace. Basically, we perform a simple *selection* subroutine as follows: pass over the input $k+1$ times, and in each pass, find the next intersecting edge of qv_0 until the $(k+1)^{\text{th}}$ one, $e_0(k+1)$. If v_0 does not lie after $e_0(k+1)$ on qv_0, i.e., if v_0 is k-visible, we report the window of qv_0 (if it exists). Since the window is defined by $e_0(k)$ and $e_0(k+1)$ or by $e_0(k+2)$ and $e_0(k+3)$, it can be found in at most two passes over the input. Then we report the part of $\partial V_k(P,q)$ lying between qv_0 and qv_1 while scanning ∂P. In fact, for each edge $e \in P$ which is in the edge list of qv_0 and lies before $e_0(k+1)$ on qv_0, we report the segment of e which is between qv_0 and qv_1. We repeat the above procedure for v_1 except for determining $e_1(k+1)$ which is done in $O(n)$ time using $e_0(k+1)$ as follows: for $1 \le i$, if v_i is an end or a start vertex the incident edges to v_i are respectively in the edge list of qv_{i-1} or qv_i and not in the other one; see Fig. 3. Except for edges incident to v_i, all the other intersecting edges of qv_{i-1} intersect qv_i in the same order, and vice versa. Hence, if $e_{i-1}(k+1)$ lies before v_i on qv_i, then it defines $e_i(k+1)$. Otherwise, if v_i is an end/a start vertex, then the second right/left neighbour of $e_{i-1}(k+1)$ in the edge list of qv_{i-1}/qv_i defines $e_i(k+1)$. However, in all cases the chain of $e_i(k+1)$ is found by at most two passes over the input; applying Observation 2.1, the edge $e_i(k+1)$ is obtained. Notice that here and in the following algorithms, if there are less than $k+1$ intersecting edges on qv_{i-1}, we store the last intersecting edge of qv_{i-1}, and the number of intersecting edges of qv_{i-1}. We this edge instead of $e_{i-1}(k+1)$, in the same procedure as above, to find $e_i(k+1)$ or the last intersecting edge of qv_i and the number of intersecting edges of qv_i. The algorithm repeats the above procedure until all critical vertices have been processed. The number of critical vertices is c, and processing each of them takes $O(n)$ time, except for the selection subroutine during processing v_0, which takes $O(kn)$ time. Thus, the running time of the algorithm is $O(kn+cn)$, using $O(1)$ words of workspace. This leads to the following theorem:

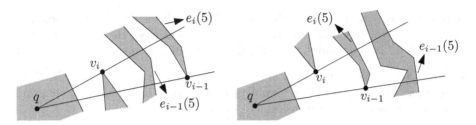

Fig. 3. Left: v_i is an end vertex and $e_i(5)$ is the second intersecting chain to the right of $e_{i-1}(5)$. Right: v_i is a start vertex and $e_i(5)$ is the second intersecting chain to the left of $e_{i-1}(5)$.

Theorem 3.1. *Given a simple polygon P with n vertices in a read-only array, a point $q \in P$, and a parameter $k \in \mathbb{N}$, there is an algorithm which reports $\partial V_k(P,q)$ in $O(kn + cn)$ time using $O(1)$ words of workspace, where c is the number of critical vertices of P.*

4   Memory-Constrained Algorithms

In this section, we assume that word of $O(s)$ workspace is available, and we show how to exploit the additional workspace to compute the k-visibility region faster. We provide two algorithms, the first one is presented only for better understanding of the second algorithm, which provides a better running time. In the first algorithm we process all the vertices in contiguous batches of size s in angular order. In each iteration we find the next batch of s vertices, and using the edge list of the last processed vertex, we construct a data structure which is used to find the windows of the batch. Using the windows we report $\partial V_k(P,q)$ in the interval of the batch. In the second algorithm we improve the running time by skipping the non-critical vertices. Specifically, in each iteration we find the next batch of s adjacent critical vertices, and similarly as the first algorithm, we construct a data structure for finding the windows, which requires a more involved approach to be updated. We first state Lemma 4.1, which is implicitly mentioned in [8] (see the second paragraph in the proof of Theorem 2.1).

Lemma 4.1. *Given a read-only array A of size n and an element $x \in A$, there is an algorithm that finds the s smallest elements in A, among those elements which are larger than x, in $O(n)$ time using $O(s)$ additional words of workspace.*

Proof. In the first step, insert the first $2s$ elements of A which are larger than x into workspace memory (without sorting them). Select the median of the $2s$ elements in memory in $O(s)$ time and remove the elements which are larger than the median. In the next step, insert the next batch of s elements of A which are larger than x into memory and again find the median of the $2s$ elements in memory and remove the elements which are larger than the median. Repeat the latter step until all the elements of A are processed. Clearly, at the end of each step, the s smallest elements of the ones which have been already processed, are in memory. Since the number of batches or steps is $O(n/s)$, the running time of the algorithm is $O(n)$ and it uses only $O(s)$ word of workspace.   □

Lemma 4.2. *Given a read-only array A of size n and a parameter $k \in \mathbb{N}$, there is an algorithm that finds the kth smallest element in A in $O(\lceil k/s \rceil n)$ time using $O(s)$ additional word of workspace.*

Proof. In the first step, apply Lemma 4.1 to find the first batch of s smallest elements in A and insert them into workspace memory in $O(n)$ time. If $k < s$ select the kth smallest element in memory in $O(s)$ time; otherwise, find the largest element in memory, which plays the role of x in Lemma 4.1. In step i, apply Lemma 4.1 to find the ith batch of s smallest elements in A and insert them into memory. If $k < i \cdot s$ select the $(k - (i-1)s)^{th}$ smallest element in memory in $O(s)$ time and stop; otherwise, find the largest element in memory and repeat. The element being sought is in the $\lceil k/s \rceil^{th}$ batch of s smallest elements of A; therefore, we can find it in $O(\lceil k/s \rceil n)$ time using $O(s)$ word of workspace.   □

In addition to our algorithm in Lemma 4.2 there are several other results on the selection problem in the read-only model; see Table 1 of [9]. There are $O(n \log \log_s n)$ expected time randomized algorithms for selection problem using $O(s)$ words of workspace in the read-only model [7,19]. Depending on k, s and n we choose one of the latter algorithms or the algorithm of Lemma 4.2. In conclusion, the running time of selection in the read-only model using $O(s)$ words of workspace, which is denoted by $T_{\text{selection}}$, is $O(\min\{\lceil k/s \rceil n, n \log \log_s n\})$ expected time. Next we describe how to apply Lemmas 4.1 and 4.2.

4.1   Algo 1: Processing All the Vertices

First we find the critical vertex v_0 with smallest angle. We apply Lemma 4.1 to find the batch of s vertices with smallest angles after v_0, and we sort them in workspace memory in $O(s \log s)$ time. For qv_0 we use the selection subroutine (with $O(s)$ word of workspace) to find $e_0(k + 1)$, and if v_0 is a k-visible vertex we report its window (if it exists).

Then, we apply Lemma 4.1 to find the two batches of $2s$ adjacent intersecting edges to the right and to the left of $e_0(k+1)$ on qv_0, we insert them in a balanced search tree T. In other words, in T we store all $e_0(j)$, for $k+1-2s \leq j \leq k+1+2s$, sorted according to their intersection with qv_0. These edges are candidates for the $(k+1)^{th}$ intersecting edge of the next s rays in angular order or $e_i(k+1)$, for $1 \leq i \leq s$. This is because, as we explained in Sect. 3, if $e_i(k+1)$ belongs to the edge list of qv_{i-1}, there is at most one edge between $e_{i-1}(k+1)$ and $e_i(k+1)$ in the edge list. Therefore, $e_i(k+1)$ is either an in the edge list of qv_0, and in this case there are at most $2i - 1$ edges between $e_0(k+1)$ and $e_i(k+1)$, or $e_i(k+1)$ is an edge which is inserted in T later; see Fig. 4. More specifically, after creating T, we start from v_1, the next vertex with smallest angle after v_0, and according to the type of v_1, we update T: if v_1 is a non-critical vertex we change the incident edge to v_1 which is already in T with the other incident edge to v_1;

if v_1 is an end (start) critical vertex, we remove (insert) the two edges which are incident to v_1. In all cases we update T only if the incident edges to v_1 are in the interval of the $2s$ intersecting edges of qv_0 in T, this takes $O(\log s)$ time. By updating T we can find $e_1(k+1)$ and the window of qv_1 (if it exists) using the position of $e_0(k+1)$ or its neighbours in T in $O(1)$ time.

In the same procedure for $1 \le i \le s$, using T and $e_{i-1}(k+1)$, we determine $e_i(k+1)$ and the window of qv_i, which take $O(s \log s)$ total time. Whenever we find and report a window, we insert its endpoints into a balanced search tree W in $O(\log s)$ time. In W the endpoints of windows are sorted according to the indices of the edges of P on which they lie. For reporting the part of $\partial V_k(P, q)$ between qv_0 and qv_s, we use W (as a sorted array) and also E which is the set of edges $e_i(k+1)$, $1 \le i \le s$. We know that, if there is no endpoint of a window on a segment, then the visibility of the segment is consistent on the entire segment. Using this, we walk on ∂P and for each edge e of P, we check if its endpoints, restricted to the interval of the batch, are k-visible or not (in $O(1)$ time using E). We also check if there is any endpoint of windows on e (in $O(|w_e|)$ time, where $|w_e|$ is the number of windows' endpoints on e). By having this information we report the k-visible segments of e restricted to the interval of the batch. Since the endpoints of windows are sorted according to their positions on ∂P, we do not check any member of W more than one time. It follows that the procedure of reporting the k-visible part of ∂P takes $O(n)$ time in each batch.

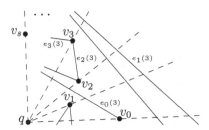

Fig. 4. The vertices $v_0, v_1, \ldots, v_s$ are the first batch of s vertices in angular order. The edge $e_1(3)$ is the second right neighbour of $e_0(3)$ because v_1 is an end vertex. The edge $e_2(3)$ is the second left neighbour of $e_1(3)$ which is inserted in T while processing v_2. The edge $e_3(3)$ is on the same chain as $e_2(3)$ because v_3 is a non-critical vertex.

After processing the first batch of vertices, we apply Lemma 4.1 to find the next batch of s vertices with smallest angle, and we sort them in memory in $O(s \log s)$ time. The last updated T is not usable anymore, because it does not necessarily contain any right or left neighbours of $e_s(k+1)$. Applying Lemma 4.1 as before, we find the two batches of $2s$ adjacent intersecting edges to the right and to the left of $e_s(k+1)$ on qv_s and we insert them into T. Then similarly for each $s < i \le 2s$ we find $e_i(k+1)$ and its corresponding window and we update T, W and E. Overall, finding a batch of s vertices, sorting and processing them, reporting the windows and the k-visible part of ∂P in the batch, take

$O(n + s \log s)$ time. Moreover, we run the selection subroutine in the first batch. We repeat this procedure for $O(n/s)$ iterations, until all the vertices are processed. Thus, the running time of the algorithm is $O(n/s(n + s \log s)) + T_{\text{selection}}$. Since $T_{\text{selection}}$ is dominated, Theorem 4.3 is follows:

Theorem 4.3. *Given a simple polygon P with n vertices in a read-only array, a point $q \in P$ and a parameter $k \in \mathbb{N}$, there is an algorithm which reports $\partial V_k(P, q)$ in $O(n^2/s + n \log s)$ time using $O(s)$ words of workspace.*

4.2   Algo 2: Processing only Critical Vertices

In this algorithm we process critical vertices in contiguous batches of size s in angular order. This algorithm works similarly as the algorithm in Sect. 4.1, but it differs in constructing and updating the data structure for finding the windows. In each iteration of this algorithm we find the next batch of s *critical* vertices with smallest angles and sort them in workspace memory in $O(s \log s)$ time. As in the previous algorithm, we construct a data structure T, which contains the possible candidates for the $(k + 1)^{th}$ intersecting edges of the rays to critical vertices of the batch. In each step, we process a critical vertex, and we use T to find its corresponding window and we update T. For updating T efficiently, we use another data structure, which is called T_θ; see below. After finding the windows of the batch, we report the k-visible part of ∂P in the interval of the batch. We repeat the same procedure for the next s critical vertices.

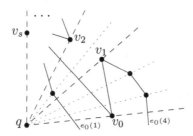

Fig. 5. The vertices $v_0, v_1, \ldots, v_s$ are the first batch of s critical vertices in angular order. The endpoint of the edge $e_0(1)$ is between qv_1 and qv_2, so $e_0(1)$ will be changed in T after processing v_1. The endpoint of $e_0(4)$ is between qv_0 and qv_1, so $e_0(4)$ will be changed in T after processing v_0.

In the first iteration, after computing $v_1, \ldots, v_s$, the angular sorted critical vertices after v_0, we find the two batches of $2s$ adjacent intersecting edges to the right and to the left of $e_0(k + 1)$ on qv_0. We sort them and insert them in a balanced search tree T, which takes $O(n + s \log s)$ time. Then for each edge in T we determine the larger angle of its endpoints. This angle shows the position of the endpoint between the rays from q to the critical vertices. Specifically, if the edge is incident to a non-critical vertex, this angle determines the step in which the edge in T should be updated to the other incident edge to the vertex; see Fig. 5. By traversing ∂P we determine these angles for the edges in T and we insert them in a balanced search tree T_θ, whose entries are connected through cross-pointers to their corresponding edges in T. We construct T_θ in $O(n+s \log s)$ time using $O(s)$ words of workspace.

Now for finding the $(k+1)^{th}$ intersecting edge of qv_1 we update T, so that it contains the edge list of qv_1: If there is any angle in T_θ which is smaller than the angle of v_1, we change the corresponding edge of the angle in T with its previous or next edge in P. In other words, we have found a non-critical vertex between qv_0 and qv_1, so we change its incident edge, which has been already in T, with its other incident edge. Then we find the larger angle of the endpoints of the new edge and insert it in T_θ. These two steps take $O(1)$ and $O(\log s)$ time for each angle that meets the condition. By doing these steps, changes in the edge list which are caused by non-critical vertices between qv_0 and qv_1 are handled. Then we update T and consequently T_θ according to the type of v_1, with the same procedure as in the previous algorithm: if v_1 is an end (start) critical vertex, we remove (insert) the two edges which are incident to v_1, this can be done in $O(\log s)$ time. Now T contains $2s$ intersecting edges of qv_1, and we can find $e_1(k+1)$ using the position of $e_0(k+1)$ and its neighbours in T in $O(1)$ time. We repeat this procedure for all critical vertices in this batch. In summary, updating T considering the changes that are caused by critical and non-critical vertices of the batch takes respectively $O(s \log s)$ and $O(n' \log s)$ time, where n' is the number of non-critical vertices that lie on the interval of the batch.

While processing the batch, we insert all $e_i(k+1)$, $1 \le i \le s$ into E. Also whenever we find a window we report it and we insert its endpoints, sorted according to the indices of the edges of P on which they lie, into a balanced search tree W in $O(\log s)$ time. After processing all the vertices of the batch, we use W (as a sorted array) and E to report the k-visible part of ∂P restricted to the interval of the batch. We know that, if there is no endpoint of a window on a chain, then the visibility of the chain is consistent on the entire chain. Using this, we walk on ∂P and for each chain C, we check if its endpoints, restricted to the interval of the batch, are k-visible or not (in $O(1)$ time using E). We also check whether there is any endpoint of windows on C (in $O(|w_e| + |C|)$ time using W, where $|w_e|$ is the number of windows' endpoints on the chain). Then we report the k-visible segments of C restricted to the interval of the batch. Since the endpoints of windows are sorted according to their positions on ∂P, we do not check any member of W more than one time. It follows that the procedure of reporting the k-visible part of ∂P takes $O(n)$ time in each batch.

In the next iteration, we repeat the same procedure for the next batch of critical vertices until all critical vertices are processed. Since the batches do not have any intersections, each non-critical vertex lies only on one batch. Thus, updating T in all batches takes $O(n \log s)$ time. All together, finding the batches of s sorted critical vertices, constructing and updating the data structures and reporting $\partial V_k(P, q)$ take $O(cn/s + n \log s)$ total time, in addition to $T_{\text{selection}}$ in the first batch. This leads to the following theorem:

Theorem 4.4. *Given a simple polygon P with n vertices in a read-only array, a point $q \in P$ and a parameter $k \in \mathbb{N}$, there is an algorithm which reports $\partial V_k(P, q)$ in $O(cn/s + n \log s + \min\{\lceil k/s \rceil n, n \log \log_s n\})$ expected time using $O(s)$ words of workspace, where c is the number of critical vertices of P.*

5   Variants and Extensions

Our results can be extended in several ways; for example, computing the k-visibility region of a point q inside a polygon P, when P may have some holes, or computing the k-visibility region of a point q in presence of a set of non-crossing segments inside a bounding box in the plane (the bounding box is only for bounding the k-visibility region). For the first problem, all the arguments in the algorithms for simple polygons hold for polygon with holes. The only remarkable issue is walking on ∂P to report the k-visible segments of ∂P. Here, after walking on the outer part of ∂P, we walk on the boundary of the holes one by one and we apply the same procedures for them. If there is no window on the boundary of a hole, then it is completely k-visible or completely non-k-visible. For such a hole, we check if it is k-visible and, if so, we report it completely. This leads to the following corollary:

Corollary 5.1. *Given a polygon P with $h \geq 0$ holes and n vertices in a read-only array, a point $q \in P$ and a parameter $k \in \mathbb{N}$, there is an algorithm which reports $\partial V_k(P, q)$ in $O(cn/s + n \log s + \min\{\lceil k/s \rceil n, n \log \log_s n\})$ expected time using $O(s)$ words of workspace. Here c is the number of critical vertices of P.*

In the second problem for a set of n non-crossing segments inside a bounding box in the plane, the output is the part of the segments which are k-visible. Here, the endpoints of all segments are critical vertices and should be processed. In the parts of the algorithm where a walk on the boundary is needed, reading the input sequentially leads to similar results. Similarly, there may be some segments with no windows' endpoints on. For these we only need to check visibility of an endpoint to decide whether they are completely k-visible or completely non-k-visible. This leads to the following corollary:

Corollary 5.2. *Given a set of n non-crossing segments S in a read-only array which lie in a bounding box B, a point $q \in B$ and a parameter $k \in \mathbb{N}$, there is an algorithm which reports $V_k(S, q)$ in $O(n^2/s + n \log s)$ time using $O(s)$ words of workspace, where $V_k(S, q)$ is the k-visible subsets of segments in S from q.*

6   Conclusion

We have proposed algorithms for a class of k-visibility problems in the constrained-memory model, which provide time-space trade-offs for these problems. We leave it as an open question whether the presented algorithms are optimal. Also, it would be interesting to see whether there exists an output sensitive algorithm whose running time depends on the number of windows in the k-visibility region, instead of the critical vertices in the input polygon.

References

1. Aichholzer, O., Fabila Monroy, R., Flores Peñaloza, D., Hackl, T., Huemer, C., Urrutia Galicia, J., Vogtenhuber, B.: Modem illumination of monotone polygons. In: Proceedings of 25th EWCG, pp. 167–170 (2009)
2. Asano, T., Buchin, K., Buchin, M., Korman, M., Mulzer, W., Rote, G., Schulz, A.: Memory-constrained algorithms for simple polygons. CGTA **46**(8), 959–969 (2013)
3. Bajuelos, A.L., Canales, S., Hernández-Peñalver, G., Martins, A.M.: A hybrid metaheuristic strategy for covering with wireless devices. J. UCS **18**(14), 1906–1932 (2012)
4. Ballinger, B., Benbernou, N., Bose, P., et al.: Coverage with k-transmitters in the presence of obstacles. In: Wu, W., Daescu, O. (eds.) COCOA 2010. LNCS, vol. 6509, pp. 1–15. Springer, Heidelberg (2010). doi:10.1007/978-3-642-17461-2_1
5. Barba, L., Korman, M., Langerman, S., Sadakane, K., Silveira, R.I.: Space-time trade-offs for stack-based algorithms. Algorithmica **72**(4), 1097–1129 (2015)
6. Barba, L., Korman, M., Langerman, S., Silveira, R.I.: Computing a visibility polygon using few variables. CGTA **47**(9), 918–926 (2014)
7. Chan, T.M.: Comparison-based time-space lower bounds for selection. TALG **6**(2), 26 (2010)
8. Chan, T.M., Chen, E.Y.: Multi-pass geometric algorithms. DCG **37**(1), 79–102 (2007)
9. Chan, T.M., Munro, J.I., Raman, V.: Selection and sorting in the restore model. In: Proceedings of 25th SODA, pp. 995–1004. SIAM (2014)
10. Dean, A.M., Evans, W., Gethner, E., Laison, J.D., Safari, M.A., Trotter, W.T.: Bar k-visibility graphs: bounds on the number of edges, chromatic number, and thickness. In: Healy, P., Nikolov, N.S. (eds.) GD 2005. LNCS, vol. 3843, pp. 73–82. Springer, Heidelberg (2006). doi:10.1007/11618058_7
11. Dean, J.A., Lingas, A., Sack, J.R.: Recognizing polygons, or how to spy. Vis. Comput. **3**(6), 344–355 (1988)
12. Eppstein, D., Goodrich, M.T., Sitchinava, N.: Guard placement for efficient point in-polygon proofs. In: Proceedings of 23rd SoCG, pp. 27–36. ACM (2007)
13. Fabila-Monroy, R., Vargas, A.R., Urrutia, J.: On modem illumination problems. In: Proceedings of 13th EGC (2009)
14. Felsner, S., Massow, M.: Parameters of bar k-visibility graphs. JGAA **12**(1), 5–27 (2008)
15. Fulek, R., Holmsen, A.F., Pach, J.: Intersecting convex sets by rays. DCG **42**(3), 343–358 (2009)
16. Ghosh, S.K.: Visibility Algorithms in the Plane. Cambridge University Press, New York (2007)
17. Hartke, S.G., Vandenbussche, J., Wenger, P.: Further results on bar k-visibility graphs. SIAM J. Discrete Math. **21**(2), 523–531 (2007)
18. Joe, B., Simpson, R.B.: Corrections to Lee's visibility polygon algorithm. BIT Numer. Math. **27**(4), 458–473 (1987)
19. Munro, J.I., Raman, V.: Selection from read-only memory and sorting with minimum data movement. TCS **165**(2), 311–323 (1996)
20. O'Rourke, J.: Computational geometry column 52. ACM SIGACT News **43**(1), 82–85 (2012)

Space-Efficient and Output-Sensitive Implementations of Greedy Algorithms on Intervals

Toshiki Saitoh[1(✉)] and David G. Kirkpatrick[2]

[1] Graduate School of Engineering, Kobe University, Kobe, Japan
saitoh@eedept.kobe-u.ac.jp
[2] Department of Computer Science, UBC, Vancouver, Canada
kirk@cs.ubc.ca

Abstract. We consider the space-efficient implementation of greedy algorithms for several fundamental problems on intervals. We assume a random access machine model with read-only access to input stored in $\Theta(n)$ words of memory, augmented with a random access memory (workspace) of size $\Theta(s)$ bits, where $\lg n \leq s \leq n$. Our implementations are based on the efficient realization of an abstract data structure that we call a *temporal priority queue* that supports *extract-min* and *advance-time* operations for a static collection of entities, each of which is active for some pre-specified interval of time. This realization is a generalization of the memory-adjustable navigation pile proposed by Asano et al. in studying time-space tradeoffs for sorting.

Using temporal priority queues we are able to implement familiar greedy algorithms for the maximum independent set problem and a variety of dominating set problems on intervals, using $O(m(\lg(sk/m) + n/s))$ time and $\Theta(s)$ bits of workspace, where k is the size of output and $m = \min(sk, n)$. Choosing $s = \Theta(n)$ this achieves $O(n \lg k)$ output-sensitive time complexity for the maximum independent set problem on intervals, previously realized using $\Omega(n)$ words of workspace.

1 Introduction

Motivated by the large and rapidly growing size of data sets, arising in applications involving biological data, social networks, etc., there has been a significant resurgence of interest in algorithms designed to run with a limited workspace. Various computational models that lend themselves to the analysis of time-space trade-offs have been the subject of study for several decades. Among the earliest of these was the *multi-pass streaming model* introduced by Munro and Paterson for selection and sorting problems [16]. In this model, the data is stored on a read-only sequential-access medium, which captures in a reasonable way the cost of accessing input data stored on a hard disk drive (HDD), since sequential access is fast compared with the random-access on a HDD. On the other hand, solid state drives (SSD) are widely used on computers or mobile devices. For SSDs, random-access time is almost same as the sequential-access, and the speed of the

© Springer International Publishing AG 2017
S.-H. Poon et al. (Eds.): WALCOM 2017, LNCS 10167, pp. 320–332, 2017.
DOI: 10.1007/978-3-319-53925-6_25

random-access is faster than that of sequential-access on HDD in general. For the reason, we consider a *read-only random-access machine model* in this paper [1,8]. Specifically, we assume a read-only random-access memory for input, a write-only memory for output, and a limited random-access memory for general computation (workspace). See [3,8] for a survey of read-only random-access machine models for sorting and selection.

This paper deals with several fundamental problems concerning sets of intervals on the real line, using this read-only random-access machine model. Sets of real-valued intervals are a fundamental geometric structure for which there are many important applications in scheduling, bioinfomatics, etc. In many instances optimization problems on intervals can be solved using simple greedy algorithms [15, Chap. 4]. The most direct implementations often start by sorting the input data, taking $\Theta(n \lg n)$ time[1] and $\Theta(n)$ words where n is the size of input. We are interested in understanding how these same greedy algorithms can be efficiently implemented with significantly less workspace.

Priority Queues. A general priority queue is an abstract data structure that stores a collection of elements and supports `find-min`, `insert`, and `extract` operations, efficiently. Building on earlier work of Pagter and Rauhe [17], Katajainen and Vitale [14] introduced a data structure called a *navigation pile* that provides, among other things, a space-efficient implementation of a general priority queue, with applications to sorting. Using the same structure, Darwish and Elmasry analyzed the time-space tradeoff for the 2D convex-hull problem [7].

Asano et al. [1] presented a memory-adjustable variant of this structure, called a *memory-adjustable navigation pile*, that serves to implement a priority queue restricted in a way that it produces only monotonically increasing extraction sequences. (This restriction, that we refer to as the *monotone extraction property* is consistent with important applications such as sorting).

A memory-adjustable navigation pile (or MANP) can be viewed as a complete binary tree, with $s \geq \lg n$ leaves, built on top of an equitable, but otherwise arbitrary, partition of the input into s buckets. Each node of a navigation pile provides efficient access to the smallest viable representative from among the buckets associated with the leaves of its subtree, where an element is *viable* as long as its value exceeds that of the most recently extracted element. To restrict the workspace to a total of $\Theta(s)$ auxiliary bits, this access is *indirect*: the specified access information suffices only to localize the element to an interval within one of its associated buckets. The monotone extraction property ensures that the selection of new bucket representatives can be done with the aid of a single filter value (the most recent output), thereby avoiding the need for $\Theta(n)$ bits to mark previously extracted elements.

Unfortunately the monotone extraction property alone does not adequately model situations, arising naturally in the implementation of greedy algorithms on intervals, where extraction of some elements makes others ineligible for extraction. In such situations the indirect association of representatives with nodes of the navigation pile creates complications in maintaining the invariant that

---

[1] The symbol lg denotes $\log_2$.

all representatives are viable. To address this problem we introduce another abstract data structure, which we call a *temporal priority queue*, that facilitates a direct implementation of several greedy algorithms on intervals, and can be implemented with a natural augmentation of the MANP of Asano et al.

Section 2 recalls some of the details of memory-adjustable navigation piles and explains their essential limitation for the implementation of greedy optimization algorithms on intervals. Temporal priority queues, and their realization as temporal memory-adjustable navigation piles, are described in Sect. 3. Applications to the implementation of a greedy algorithm for a maximum independent set in a collection of intervals, and several dominating set problems on intervals, are described in Sects. 4 and 5 respectively. The remainder of this section sets out some of the relevant background concerning optimizations on sets of intervals and then summarizes our results.

Optimization on Sets of Intervals. The MAXIMUM INDEPENDENT SET (MIS) problem on intervals, takes as input a collection of intervals and asks for a maximum size subset whose elements are pairwise disjoint. The MIS problem for intervals has a simple linear-time greedy solution assuming the input is sorted. Snoeyink [18] described an output-sensitive algorithm using divide-prune-and-conquer that runs in $O(n \lg k)$ time and $\Theta(n)$ words, where k is the output size of the solution. Recently, Bhattacharya et al. [2] studied this MIS problem with restrictions on workspace. They showed how to address the problem using a memory-adjustable heap implementation of a priority queue. This structure executes `find-min` operations in $O(1)$ time. However, when $O(s)$ words are available as a workspace, `extract` operations require updates to the structure whose total cost, for k extractions, is $O(m(\lg s + n/s))$, where $m = \min(sk, n)$.

A set of intervals on a straight line provides an implicit representation of the intersection graph of these intervals, called an *interval graph*. The MIS problem on interval graphs, and also on the more general graph classes, e.g., chordal graphs and circular-arc graphs, can be solved in linear time and space [9,12]. In a similar way, MIS algorithms for some of these more general graph classes have been formulated in terms of their implicit representations, e.g. a set of arcs on a circle [10].

A subset D of a set of intervals $\mathcal{I}$ is a *dominating set* if every interval in $\mathcal{I}$ intersects at least one interval in D. A minimum size dominating set can be constructed in linear time on interval and circular-arc graphs [13], and efficient algorithms have been constructed for a variety of other domination problems on these same graph classes [4]. Cheng et al. proposed efficient dominating set algorithms on a set of intervals [5]. These algorithms use dynamic programming taking linear space for a sorted intervals, and are not output sensitive. See [6,11] for a survey of results for a variety of dominating set problems on various graph classes.

Our Results. Our contributions in this paper are as follows:

- We propose an abstract data structure, called a temporal priority queue, that provides a natural generalization of priority queues with the monotone extraction property;
- We describe an efficient implementation of temporal priority queues, using temporal memory-adjustable navigation piles;
- We show how greedy algorithms for MAXIMUM INDEPENDENT SET (MIS), MINIMUM CONNECTED DOMINATING SET (MCDS), and MINIMUM DOMINATING SET (MDS) problems on intervals can all be implemented, using temporal memory-adjustable navigation piles, to run in $O(m(\lg(sk/m) + n/s))$ time, using $\Theta(s)$ bits for workspace, where $s \geq \lg n$. (Choosing $s = \Theta(n)$ this achieves the optimal $O(n \lg k)$ output-sensitive time complexity, previously realized [18] for the maximum independent set problem only with the aid of $\Omega(n)$ words of workspace.)

2   Navigation Piles

In this section, we first recall some of the essential details of the memory-adjustable navigation pile (MANP) structure, proposed by Asano et al. [1] as a space-efficient implementation of priority queues with the monotone extraction restriction. The original navigation pile structure, introduced by Katajainen and Fabio [14], stores $\Theta(n)$ bits where n is the size of an input. Asano et al. modified the navigation pile to a memory-adjustable structure using a workspace of $\Theta(s)$ bits for $\lg n \leq s$. The MANP structure supports the priority queue operations find-min and insert in $O(1)$ time, and extract in $O(n/s + \lg s)$ time. To avoid the complications associated with insert operations, that do not arise in our modifications of MANP structures, we describe a structure that has been initialized by batch insertion, and thereafter executes find-min and extract operations only.

An MANP is a complete binary tree T with s leaves built on top of an equitable partition of the input array $A = [e_1, \ldots, e_n]$ into s, $\lg n \leq s \leq n$ buckets $\{B_1, \ldots, B_s\}$ such that $B_i = [e_{(i-1)\lceil n/s \rceil+1}, \ldots, e_{i\lceil n/s \rceil}]$, for each i. The size of each bucket B_i is $\lceil n/s \rceil$ for $i \in \{1, \ldots, s-1\}$, and $|B_s| = n - (s-1)\lceil n/s \rceil$. (For ease of description, we assume hereafter that s is a power of 2.)

The leaves of T (nodes of height 0) are associated with individual buckets, and each internal node x with height h is associated with the 2^h consecutive buckets covered by the leaves in the subtree rooted at x. Each node x with height $h \leq \lceil \lg n \rceil/2$ stores $2h$ bits, called *navigation bits* that serve to describe implicitly the location of the minimum of the representatives from among the buckets covered by node x. Specifically, the interval of inputs contained in each of the 2^h buckets covered by a node x at height h is divided into $\lceil n/(s \cdot 2^h) \rceil$ blocks of contiguous elements, called *quantiles*, and the $2h$ bits associated with x specify the quantile that contains the minimum viable element in all of the buckets covered by x. (Recall that viable elements are those whose value exceed that of the most recently extracted element, initialized to $-\infty$). The first h bits represents the bucket that contains the specified quantile and the second h bits represents the index of the quantile within the bucket. If $h > \lceil \lg n \rceil / 2$, nodes with height h have $\lceil \lg n \rceil$ bits and point to the minimum element in the buckets, directly. Since the number of nodes with height h is $s/2^h$ and each node at height h stores $\min(2h, \lceil \lg n \rceil)$ bits, it follows that the total number of bits used in an MANP is $\sum_{h=0}^{\lg s} \left((s/2^h) \cdot \min(2h, \lceil \lg n \rceil) \right)$, which is $O(s)$.

An MANP supports `find-min` and `extract` operations. To support the `find-min` operation in $O(1)$ time, a pointer is kept to the minimum element in the navigation pile, by using $\lg n$ bits separately. This minimum pointer is updated following every `extract` operation.

To extract an element in $O(\lg s + n/s)$ time from a navigation pile, we start by locating the bucket containing the representative associated with the root of the pile. We then update the information of the $\lg s$ nodes on the path from the leaf associated with this bucket, to the root. To update each internal node x on the path, we access the representatives associated with the two children of x and compare their keys. We set the navigation information at x to point to the quantile containing the element with the smaller of these two keys. We can obtain the position of the quantile of an internal node in constant time because navigation bits are stored in a bit vector in breadth-first order. After getting to this position, we scan for the minimum viable element in the quantile. For a node of height h, the size of the quantile (and so the scan cost) is at most $\lceil n/(s \cdot 2^h) \rceil$. By summing on this cost over the updating path, the total update cost is $O(n/s + \lg s)$. See the analysis in [1] for additional details.

The monotone extraction property is exploited to avoid explicitly marking extracted elements. Instead the most recently extracted element is maintained as used as a filter in maintaining/certifying the viability of representatives. In our applications, unlike sorting which was the motivating application for the design of MANPs, it is expected that the extraction of some elements will *inactivate* others, i.e. render them ineligible for future extraction, independent of their viability. This leads to complications in updating an MANP following an extraction, specifically the reestablishment of the *viable-representatives invariant* which asserts that all representatives in the MANP are viable.

Directly coupling the inactivation of elements to their viability ensures that MANP updates accompanying an extraction can be confined to one leaf-to-root path in the MANP. A more relaxed extraction assumption, arising naturally in several priority queue applications, is that whenever an element becomes inactive, it must be the case that all elements with smaller keys are also inactive. What this assumption implies is that whenever the representative associated with a node x is inactive, so too is the representative associated with any ancestor of x in the MANP. This is easily implemented by maintaining a separate *inactivation threshold* that is increased with each extraction to some value at least as large as the most recently extracted key. With this in hand, we can modify the MANP by adding an operation `refresh` that efficiently reestablishes the viable-representatives invariant.

Although this relaxation makes it possible to model certain situations in which extraction of some elements will lead to the inactivation of others, there remain situations in which this hereditary property of inactivation does not hold. Thus we are led to the formulation of another even more relaxed priority queue model that we call a *temporal priority queue*. In this model, each element e has associated with them (i) a key $k(e)$, (ii) an *activation interval* $(l(e), r(e))$ and (iii) an *activation-time update value* $t(e)$. At any given *activation-time* t, an extraction operation (i) locates the viable element e, among those whose activation intervals intersect $t(l(e) < t < r(e))$, whose associated key $k(e)$ is smallest, (ii) returns the value $k(e)$ of this smallest key, and (iii) updates the activation-time to the activation-time update value $t(e)$ which is larger than t. Fixing the activation interval of elements with key $k(e)$ as $(-\infty, k(e))$ and choosing $k(e)$ as the associated activation-time update value, reduces the structure to a priority queue with the monotone extraction property. In general, however, temporal priority queues, while still producing monotonic extraction sequences, can exhibit inactivation behaviour fully decoupled from key values.

In the next section we first formulate and analyze the `refresh` operation for MANPs implementing priority queues that satisfy the following *hereditary inactivation assumption*:

Assumption 1. *If an element with key value x has become inactive then so too has every element with key value $y < x$.*

This restricted structure is an essential component of our temporal MANP implementation of more general temporal priority queues, the full description of which closes the section.

3   Temporal Memory-Adjustable Navigation Piles

Refreshing Memory-Adjustable Navigation Piles. We begin by assuming that we have a memory-adjustable navigation pile $\mathbb{P}$ whose elements satisfy the hereditary inactivation assumption. We propose a new operation `refresh` for such navigation piles to reestablish the viable representatives invariant. The standard priority-queue operation `extract` performs a bottom-up update, that is, `extract` updates elements from a leaf to the root. In contrast, `refresh` performs updates in a top-down recursive fashion. We describe the `refresh` procedure in Procedure 1 and analyze its correctness and efficiency in Lemmas 2 and 4 below.

The following lemma, which need not hold in general, follows directly from our assumption that the elements of $\mathbb{P}$ satisfy Assumption 1:

Lemma 1. *If the representative associated with a node x of $\mathbb{P}$ becomes inactive so too does the representative of all ancestors of x in $\mathbb{P}$.*

From the lemma above, if the representative of the root is inactive, we can follow the inactive nodes from the root to all inactive leaves.

Procedure 1. refresh(v: a node v in an MANP)

```
/* assume the hereditary inactivation assumption holds        */
```
1 **if** v *has height* $h > 0$ **then**
2   let v_l and v_r be left and right children of v, respectively;
3   **if** *the representative associated with* v_l *is inactive* **then** refresh (v_l);
4   **if** *the representative associated with* v_r *is inactive* **then** refresh (v_r);
5   Find the representatives e_l and e_r associated with v_l and v_r, respectively;
6   **if** *the key of* e_l *is smaller than that of* e_r **then**
7   set the navigation bits of v to point to the quantile associated with e_l
8   **else**
9   set the navigation bits of v to point to the quantile associated with e_r

Lemma 2. *Every initially inactive representative associated with some node in* $\mathbb{P}$ *is reset, by Procedure 1, to the smallest viable active element among the buckets covered by* v.

Proof. First note that, by the hierarchical inactivation assumption, if the representative associated with v is active to start this representative is unchanged by the procedure. Otherwise, we establish our claim by induction on h, the height of v in $\mathbb{P}$. If $h = 0$ the assertion holds trivially, since the quantile associated with v is the entire bucket associated with v.

For $h > 0$, we know, by the induction hypothesis, that following steps 4 and 5, the representatives associated with the children v_l and v_r of v are correctly set. Given this it is clear that step 6 sets the navigation bits associated with v correctly. $\qquad\square$

The time complexity of the refresh procedure obviously depends on the number of nodes (equivalently recursive calls) whose representatives are updated in the process. To estimate the number of such nodes, we take advantage of a straightforward counting result bounding the number of nodes in a binary tree with a specified height and number of leaves.

Lemma 3. *Let* T *be a binary tree with height* h *and* ℓ *leaves. The number of nodes in* T *is at most* $\ell \lg \left(2^h / \ell \right) + \ell - 1$.

We now turn to the analysis of the complexity of the refresh procedure on navigation piles.

Lemma 4. *For an MANP* $\mathbb{P}$ *using* $O(s)$ *bits, the* refresh *operation, when invoked at the root of* $\mathbb{P}$, *takes* $O\left(b\left(n/s + \lg\left(s/b\right)\right)\right)$ *time, where* b *is the number of buckets whose representatives are updated.*

Proof. In a given recursive call with a node at height $h > 0$ the refresh procedure accesses the representatives associated with both of its children, compares their keys, and reassigns the navigation bits associated with v. Since nodes do not have a direct pointer to their representatives, rather they point to a quantile containing the representative, it takes time proportional to the size of this quantile, specifically $O(\max\{1, \lceil n/(s \cdot 2^h) \rceil\})$ time for nodes at height h, to carry out these steps.

Now if b buckets have their representatives updated, recursive calls are made to at most b nodes at every height $h \leq \lg(n/s)$ and the quantiles are scanned at most four times for each node, for a total cost of $\sum_{h=0}^{\lg(n/s)} 4b\lceil n/(s \cdot 2^h) \rceil$, which is $O(bn/s)$. For nodes at height $h > \lg(n/s)$ the quantile size is $O(1)$ so the associated cost for all such nodes is bounded by the total number of nodes which, by Lemma 3 is at most $b\lg(s/b) + b - 1$. $\qquad\square$

Details of Temporal Memory-Adjustable Navigation Piles.

A temporal memory-adjustable navigation pile (T-MANP) can be viewed as a pair of MANPs built on top of the same partition of the input elements into s buckets. The first (*primary* MANP) is a standard MANP based on keys of representative bucket elements that are viable with respect to both the key value and activation interval (i.e. the key exceeds that of the most recently extracted element *and* the activation interval intersects the current activation time). The second *activation* MANP maintains *activation events* which are right endpoints $r(e)$ of activation intervals, and chooses as a representative for each bucket the smallest activation event, among its elements, that exceeds the current activation time. Namely, the elements in the activation MANP are most likely to be inactive in the buckets.

The role of the second MANP is to trigger a re-selection of the bucket representatives in the primary MANP, whenever any one of the elements in that bucket experiences an activation event (i.e. changes activation status). Of course, the bucket representative might not change with this re-selection, but this is a conservative strategy that ensures that once all of the representatives in the activation MANP are viable with respect to the current activation time, all of the representatives in the primary MANP are fully viable.

When an element e is extracted from the primary MANP, we (i) use its associated activation-time update value to update the current activation-time, (ii) refresh the representatives in the activation MANP, (iii) update the representatives in the primary MANP of all buckets whose activation MANP representatives have changed, and then, if there were no such changes, (iv) update the bucket representative in the primary MANP of the bucket containing the most recently extracted element. We note that, by design, extractions from the activation MANP satisfy the hereditary inactivation assumption and so its elements can be refreshed (step (ii)) in batch, as described above. We also note that the updates in step (iii) involve exactly the nodes in the primary MANP whose corresponding nodes in the activation MANP were undated in step (ii). Finally, the updates in step (iv) occur, as in the standard MANP, along a single leaf to root path.

Thus, the extraction cost for our T-MANP is dominated by the cost of refreshing its associated activation MANP, unless of course the latter experiences no updates, in which case the extract cost, like that of a standard MANP, is $O(n/s)$. We summarize this result in the following:

Theorem 1. *A T-MANP using $O(s)$ navigation bits, can be maintained at a cost of $O(b(n/s+\lg(s/b)))$ time per* extract-min *operation, where b denotes the number of buckets whose representatives are updated following the extraction.*

4   Algorithms for Maximum Independent Set on Intervals

Let $\mathcal{I}$ be an (unsorted) set of intervals on a straight line. A subset of $\mathcal{I}$ is *independent* if for any two distinct intervals I and I' in the subset, $I \cap I' = \emptyset$. The problem MAXIMUM INDEPENDENT SET (MIS) asks for an independent subset of $\mathcal{I}$ which has the maximum cardinality. We define $l(I)$ and $r(I)$ as the coordinate of the left and right endpoints of the interval $I \in \mathcal{I}$, respectively.

There is a well-known greedy algorithm for the MIS problem on intervals [15]. The algorithm selects intervals in an iterative fashion, where the next selection is the interval with the smallest right endpoint among those whose left endpoint is larger than the right endpoint of the most recently selected interval. Equivalently, if interval I is the most recently selected interval, we say that an interval I' is *active* if $l(I') > r(I)$, and make the next selection the active interval J that minimizes $r(J)$. In the most straightforward implementation, we first sort intervals with increasing order of the right endpoints. Then, we can find output intervals by sweeping the sorted array once. However, the algorithm of this implementation runs in $O(n \lg n)$ time and $\Theta(n)$ words for workspace. In this section, we propose a simple implementation of this MIS algorithm, using a T-MANP.

For each interval I, we choose $r(I)$ as its associated key, since our recurring goal is to find the interval with the minimum right endpoint among currently active intervals. The representatives of the primary MANP has the smallest right endpoint in active intervals. The activation interval associated with I is defined by $(-\infty, l(I))$, and associated the activation-time update value is $r(I)$. By this definition, an interval is active exactly when its activation interval intersects the current activation-time.

With these choices for key values and activation intervals, the T-MANP implements the greedy algorithm correctly. It follows that

Theorem 2. *Given a set of n intervals and $O(s)$ bits for workspace, the maximum independent set of the intervals can be constructed in $O\left(m\left(\lg\left(sk/m\right) + n/s\right)\right)$ time, where k is the size of the solution and $m = \min(n, sk)$.*

Proof. Finding an interval which has the minimum right endpoint is implemented by a single extract-min operation which, by Lemma 4, takes $O(b_i(n/s + \lg(s/b_i)))$ time where b_i is the number of buckets of intervals whose representatives on the activation MANP are refreshed in iteration i. Thus in total the algorithm takes $c\sum_{i=1}^{k} b_i n/s + c\sum_{i=1}^{k} b_i \lg(s/b_i)$ time, where c is some non-negative constant. We will show that (i) $\sum_{i=1}^{k} b_i n/s = O(mn/s)$ and (ii) $\sum_{i=1}^{k} b_i \lg(s/b_i) = O(m \lg(sk/m))$ where $m = \min(n, sk)$.

To show (i) above, we first observe that $\sum_{i=1}^{k} b_i = O(\min(n, sk))$. For each interval, the interval is chosen as a representative in the T-MANP a constant number of times, since the current activation-time increases monotonically. On the other hand, for each ith iteration, $b_i \leq s$ because the activation MANP maintains at most s intervals. Thus the total number of refreshed intervals is at most $\min(n, sk)$. Therefore, $\sum_{i=1}^{k} b_i n/s = O(mn/s)$.

To show (ii), we demonstrate that $\sum_{i=1}^{k} b_i \lg (s/b_i) \leq m \lg (sk/m)$. Let $n' = \sum_{i=1}^{k} b_i$ and $p_i = b_i/n'$, for each i. Note that, $n' \leq \min(n, sk)$ from previous paragraph and $\sum_{i=1}^{k} p_i = 1$. It follows that:

$$\sum_{i=1}^{k} b_i \lg \frac{s}{b_i} = n' \left(\sum_{i=1}^{k} p_i \lg \frac{s}{n'} + \sum_{i=1}^{k} p_i \lg \frac{1}{p_i} \right) \leq n' \left(\lg \frac{s}{n'} + \sum_{i=1}^{k} \frac{1}{k} \lg k \right) = n' \lg \frac{sk}{n'}$$

where the inequality follows from the fact that, $\sum_{i=1}^{k} p_i \lg (1/p_i)$ is maximized when $p_i = 1/k$, that is, $b_i = n'/k$ for each i. $\square$

5 Algorithms for Dominating Set Problems

Let I be an interval in a set of intervals $\mathcal{I}$. We define $N(I)$ to be a set of intervals that intersect I and $N[I] = N(I) \cup I$. For a set of intervals $\mathcal{I}'$, $N(\mathcal{I}') = \bigcup_{I \in \mathcal{I}'} N(I)$ and $N[\mathcal{I}'] = \bigcup_{I \in \mathcal{I}'} N[I]$. A set of intervals D is a *dominating set* if every interval not in D intersects to at least one interval in D, that is, $N[D] = \mathcal{I}$. We say that an interval I is *dominated* by $I' \in D$ if $I \in N[I']$, and is *undominated* if there is no interval $I' \in D$ such that $I \in N[I']$. The MINIMUM DOMINATING SET (MDS) problem is to find the minimum cardinality of a dominating set on a set of intervals. A set of intervals $\mathcal{I}'$ is *connected* if for any two intervals I and J in $\mathcal{I}'$, there is a sequence of intervals $I = I_1, I_2, \ldots, I_k = J$ such that $I_i \cap I_{i+1} \neq \emptyset$ for $1 \leq i \leq k - 1$. The MINIMUM CONNECTED DOMINATING SET PROBLEM (MCDS) is to find the minimum cardinality of a connected dominating set on intervals.

In this section, we first propose a simple implementation of an algorithm for MCDS. Following this we propose an algorithm for MDS that can be viewed as a combination of our algorithms for MIS and MCDS.

Minimum Connected Dominating Set. It is known that there is a greedy algorithm for MCDS [11]. Let J be a last outputted interval as a solution in the algorithm. We find an interval J' which has the maximum right endpoint in $N[J]$, and then, we output J' and set $J = J'$. We repeat the process until there is an undominated interval. In the initialization, we set an interval with the minimum right endpoint as J. In a simple implementation using $O(1)$ workspace of the algorithm, we scan the full input using $O(n)$ time for finding the interval J' and an undominated interval. The algorithm iterates k time in the while loop and it takes $O(n)$ time to find each J'. Thus, it takes $O(kn)$ time and $O(1)$ workspace. In this implementation, we scan all the blocks to find J'. We would like restrict our scan to blocks which include candidates of the next output interval. To decide whether the block may have a candidate, we maintain a T-MANP using $O(s)$ bits.

We define a T-MANP for our implementation as follows. We choose, for each interval I, a key-value matching its right endpoint $r(I)$, an activation interval $(l(I), r(I))$, and an activation-time update value $r(I)$. The next output interval J' has the maximum right endpoint such that $l(J') < r(J) < r(J')$, where J is the most recent interval selected by the algorithm. If the current activation time is set to $r(J)$, then the active element with the maximum key is precisely the next element to be selected. The time complexity of this implementation can be obtained by a direct modification of the proof of Theorem 2. Therefore, we have the following theorem.

Algorithm 2. Greedy Algorithm for MDS

---

1 Set $J = [-\infty, -\infty]$;
2 **while** *There is an undominated interval* **do**
3 $\quad$ Find an LUDI I which does not intersect J;
4 $\quad$ Find an interval J' with maximum right endpoint in intervals $N(I)$;
5 $\quad$ Output J' and set $J = J'$;

Theorem 3. *Our implementation of the greedy Minimum Connected Dominating Set algorithm using a T-MANP finds a minimum connected dominating set and runs in $O\left(m\left(\lg\left(sk/m\right) + n/s\right)\right)$ time using $O(s)$ bits, where $m = \min(n, sk)$ and k is the size of the optimal solution.*

Minimum Dominating Set. Let D be a subset of a set of intervals $\mathcal{I}$. We define $U(D)$ as the set of intervals in $\mathcal{I}$ that are undominated by intervals in D. An interval is called a *leftmost undominated interval*, *LUDI* for short, if the right endpoint of the interval is minimum among all intervals in $U(D)$. A greedy algorithm for minimum dominating set on intervals is presented in Algorithm 2. There is a straightforward implementation of this MDS algorithm using $O(n \lg n)$ time (and $\Theta(n)$ words) to pre-sort the intervals, followed by k iterations of time $O(n)$ each of which selects one more interval in the output.

We improve the implementation to output-sensitive time using a T-MANP with $O(s)$ bits of workspace. The idea is to use two cooperating T-MANPs, one to find the next LUDI and one to compute the next output interval. The first, like our MIS structure, associates with each interval I the key-value $r(I)$, the activation interval $(-\infty, l(I))$, and the activation-time update value $r(I)$. As before, the current activation time is maintained as $r(J)$, where J is the most recently output interval. The representative associated with its root is the desired LUDI interval.

The second structure uses the right endpoint of the current LUDI interval as its current activation time, and associates with each interval I the key-value $r(I)$, the activation interval $(l(I), r(I))$, and the activation-time update value $r(I)$. This ensures that active intervals are those that are intersected by $r(J)$, and the maximum one among these is the desired next output. Like our MCDS structure it extracts the active element with the maximum key.

In our implementation the outputs of the two T-MANP structures lead to updates on the other but, as before, intervals are bucket representatives only a constant number of times before they are inactive in all structures, and so our analysis from our MIS and MCDS implementations carry over directly. We summarize the properties of this implementation in the following:

Theorem 4. *Our implementation of the greedy Minimum Dominating Set algorithm using a T-MANP finds a minimum dominating set and runs in $O\left(m\left(\lg\left(sk/m\right)+n/s\right)\right)$ time, using $O(s)$ bits, where $m = \min(n, sk)$ and k is the size of the optimal solution.*

References

1. Asano, T., Elmasry, A., Katajainen, J.: Priority queues and sorting for read-only data. In: Chan, T.-H.H., Lau, L.C., Trevisan, L. (eds.) TAMC 2013. LNCS, vol. 7876, pp. 32–41. Springer, Heidelberg (2013). doi:10.1007/978-3-642-38236-9_4
2. Bhattacharya, B.K., De, M., Nandy, S.C., Roy, S.: Maximum independent set for interval graphs and trees in space efficient models. In: Canadian Conference on Computational Geometry (CCCG 2014) (2014)
3. Borodin, A.: Time space tradeos (getting closer to the barrier?). In: International Symposium on Algorithms and Computation (ISAAC 1993), pp. 209–220 (1993)
4. Chang, M.: Efficient algorithms for the domination problems on interval and circular-arc graphs. SIAM J. Comput. **27**(6), 1671–1694 (1998)
5. Cheng, S., Kaminski, M., Zaks, S.: Minimum dominating sets of intervals on lines. Algorithmica **20**(3), 294–308 (1998)
6. Corneil, D.G., Stewart, L.K.: Dominating sets in perfect graphs. Discrete Math. **86**(1–3), 145–164 (1990)
7. Darwish, O., Elmasry, A.: Optimal time-space tradeo for the 2D convex-hull problem. In: European Symposium on Algorithms (ESA 2014), pp. 284–295 (2014)
8. Frederickson, G.N.: Upper bounds for time-space trade-offs in sorting and selection. J. Comput. Syst. Sci. **34**(1), 19–26 (1987)
9. Golumbic, M.C.: Algorithmic Graph Theory and Perfect Graphs. Annals of Discrete Mathematics, vol. 57. Elsevier, Amsterdam (2004)
10. Golumbic, M.C., Hammer, P.L.: Stability in circular arc graphs. J. Algorithms **9**(3), 314–320 (1988)
11. Haynes, T., Hedetniemi, S., Slater, P.: Fundamentals of Domination in Graphs. Chapman & Hall/CRC Pure and Applied Mathematics. CRC Press, Boca Raton (1998)
12. Hsu, W., Spinrad, J.P.: Independent sets in circular-arc graphs. J. Algorithms **19**(2), 145–160 (1995)
13. Hsu, W., Tsai, K.: Linear time algorithms on circular-arc graphs. Inf. Process. Lett. **40**(3), 123–129 (1991)
14. Katajainen, J., Vitale, F.: Navigation piles with applications to sorting, priority queues, and priority deques. Nordic J. Comput. **10**(3), 238–262 (2003)
15. Kleinberg, J., Tardos, E.: Algorithm Design. Addison-Wesley Longman Publishing Co., Inc., Boston (2005)

16. Munro, J.I., Paterson, M.: Selection and sorting with limited storage. Theor. Comput. Sci. **12**, 315–323 (1980)
17. Pagter, J., Rauhe, T.: Optimal time-space trade-offs for sorting. In: Foundations of Computer Science (FOCS 1998), pp. 264–268 (1998)
18. Snoeyink, J.: Maximum independent set for intervals by divide and conquer with pruning. Networks **49**(2), 158–159 (2007)

Computational Complexity

Algorithms for Automatic Ranking of Participants and Tasks in an Anonymized Contest

Yang Jiao[✉], R. Ravi, and Wolfgang Gatterbauer

Tepper School of Business, Carnegie Mellon University,
5000 Forbes Avenue, Pittsburgh, PA 15213, USA
yangjiao@andrew.cmu.edu

Abstract. We introduce a new set of problems based on the *Chain Editing problem*. In our version of Chain Editing, we are given a set of anonymous participants and a set of undisclosed tasks that every participant attempts. For each participant-task pair, we know whether the participant has succeeded at the task or not. We assume that participants vary in their ability to solve tasks, and that tasks vary in their difficulty to be solved. In an ideal world, stronger participants should succeed at a superset of tasks that weaker participants succeed at. Similarly, easier tasks should be completed successfully by a superset of participants who succeed at harder tasks. In reality, it can happen that a stronger participant fails at a task that a weaker participants succeeds at. Our goal is to find a *perfect nesting of the participant-task relations* by flipping a minimum number of participant-task relations, implying such a "nearest perfect ordering" to be the one that is closest to the truth of participant strengths and task difficulties. Many variants of the problem are known to be NP-hard.

We propose six natural *k-near* versions of the Chain Editing problem and classify their complexity. The input to a *k-near* Chain Editing problem includes an initial ordering of the participants (or tasks) that we are required to respect by moving each participant (or task) at most *k* positions from the initial ordering. We obtain surprising results on the complexity of the six *k-near* problems: Five of the problems are polynomial-time solvable using dynamic programming, but one of them is NP-hard.

Keywords: Chain Editing · Chain Addition · Truth discovery · Massively open online classes · Student evaluation

1 Introduction

1.1 Motivation

Consider a contest with a set S of participants who are required to complete a set Q of tasks. Every participant either succeeds or fails at completing each

© Springer International Publishing AG 2017
S.-H. Poon et al. (Eds.): WALCOM 2017, LNCS 10167, pp. 335–346, 2017.
DOI: 10.1007/978-3-319-53925-6_26

task. The identities of the participants and the tasks are anonymous. We aim to obtain rankings of the participants' strengths and the tasks' difficulties. This situation can be modeled by an unlabeled bipartite graph with participants on one side, tasks on the other side, and edges defined by whether the participant succeeded at the task. From the edges of the bipartite graph, we can infer that a participant a_2 is stronger than a_1 if the neighborhood of a_1 is contained in (or is "nested in") that of a_2. Similarly, we can infer that a task is easier than another if its neighborhood contains that of the other. See Fig. 1 for a visualization of strengths of participants and difficulties of tasks. If all neighborhoods are nested, then this nesting immediately implies a ranking of the participants and tasks. However, participants and tasks are not perfect in reality, which may result in a bipartite graph with "non-nested" neighborhoods. In more realistic scenarios, we wish to determine a ranking of the participants and the tasks when the starting graph is not ideal, which we define formally in Sect. 1.2.

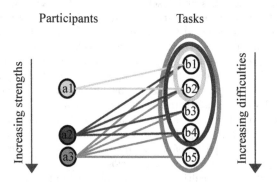

Fig. 1. An ideal graph is shown. Participants and tasks may be interpreted as students and questions, or actors and claims. Participant a_1 succeeds at b_1 to b_2; a_2 succeeds at b_1 to b_4; a_3 succeeds at b_1 to b_5. The nesting of neighborhoods here indicate that participant a_1 is weaker than a_2, who is weaker than a_3, and task b_1 and b_2 are easier than b_3 and b_4, which in turn are easier than b_5.

1.1.1   Relation to Truth Discovery

A popular application of unbiased rankings is computational "truth discovery." *Truth discovery* is the determination of trustworthiness of conflicting pieces of information that are observed often from a variety of sources [24] and is motivated by the problem of extracting information from networks where the trustworthiness of the actors are uncertain [15]. The most basic model of the problem is to consider a bipartite graph where one side is made up of actors, the other side is made up of their claims, and edges denote associations between actors and claims. Furthermore, claims and actors are assumed to have "trustworthiness" and "believability" scores, respectively, with known a priori values. According to a number of recent surveys [15,20,24], common approaches for truth discovery include iterative procedures, optimization methods, and probabilistic graphic models. Iterative methods [9, 13, 22, 27] update trust scores of actors to believability scores of claims, and vice versa, until convergence. Variants of these methods (such as Sums, Hubs and Authorities [18], AverageLog, TruthFinder, Investment, and PooledInvestment) have been extensively studied and proven in practice [2]. Optimization methods [3, 19] aim to find truths that minimize the total distance between the provided claims and the output truths for some specified continuous distance function; coordinate descent [5] is often used to obtain the solution. Probabilistic graphical models [23] of truth discovery are solved by expectation maximization. Other methods for truth discovery include those that leverage trust relationships between the sources [14]. Our study is conceptually closest to optimization approaches (we minimize the number of edge additions or edits), however we suggest a *discrete objective* for minimization, for which we need to develop new algorithms.

1.1.2   Our Context: Massively Open Online Courses

Our interest in the problem arises from trying to model the problem of automatic grading of large number of students in the context of MOOCs (massively open online courses). Our idea is to crowd-source the creation of automatically gradable questions (like multiple choice items) to students, and have all the students take all questions. From the performance of the students, we would like to quickly compute a roughly accurate ordering of the difficulty of the crowd-sourced questions. Additionally, we may also want to efficiently rank the strength of the students based on their performance. Henceforth, we refer to participants as students and tasks as questions in the rest of the paper.

1.1.3   Our Model

We cast the ranking problem as a discrete optimization problem of minimizing the number of changes to a given record of the students' performance to obtain nested neighborhoods. This is called the Chain Editing problem. It is often possible that some information regarding the best ranking is already known. For instance, if the observed rankings of students on several previous assignments are consistent, then it is likely that the ranking on the next assignment will be similar. We model known information by imposing an additional constraint that the changes made to correct the errors to an ideal ranking must result in a ranking that is near a given base ranking. By near, we mean that the output position of each student should be within at most k positions from the position in the base ranking, where k is a parameter. Given a nearby ranking for students, we consider all possible variants arising from how the question ranking is constrained. The question ranking may be constrained in one of the following three ways: the exact question ranking is specified (which we term the "constrained" case), it must be near a given question ranking (the "both near" case), or the question ranking is unconstrained (the "unconstrained" case). We provide the formal definitions of these problems next.

1.2   Problem Formulations

Here, we define all variants of the ranking problem. The basic variants of Chain Editing are defined first and the k-near variants are defined afterward.

1.2.1   Basic Variants of Chain Editing

First, we introduce the problem of recognizing an ideal input. Assume that we are given a set S of students, and a set Q of questions, and edges between S and Q that indicate which questions the students answered correctly - note that we assume that every student attempts every question. Denote the resulting bipartite graph by $G = (S \cup Q, E)$. For every pair $(s, q) \in S \times Q$, we are given an edge between s and q if and only if student s answered question q correctly. For a graph (V, E), denote the neighborhood of a vertex x by $N(x) := \{y \in V : xy \in E\}$.

Definition 1. We say that student s_1 is stronger than s_2 if $N(s_1) \supset N(s_2)$. We say that question q_1 is harder than q_2 if $N(q_1) \subset N(q_2)$. Given an ordering α on the students and β on the questions, $\alpha(s_1) \geq \alpha(s_2)$ shall indicate that s_1 is stronger than s_2, and $\beta(q_1) \geq \beta(q_2)$ shall indicate that q_1 is harder than q_2.

Definition 2. An ordering of the questions satisfies the interval property if for every s, its neighborhood $N(s)$ consists of a block of consecutive questions (starting with the easiest question) with respect to our ordering of the questions. An ordering α of the students is nested if $\alpha(s_1) \geq \alpha(s_2) \Rightarrow N(s_1) \supseteq N(s_2)$.

Definition 3. The objective of the Ideal Mutual Orderings (IMO) problem is to order the students and the questions so that they satisfy the nested and interval properties respectively, or output NO if no such orderings exist.

Observe that IMO can be solved efficiently by comparing containment relation among the neighborhoods of the students and ordering the questions and students according to the containment order.

Proposition 1. There is a polynomial time algorithm to solve IMO.

All missing proofs are in the full version of the paper [16]. Next, observe that the nested property on one side is satisfiable if and only if the interval property on the other side is satisfiable. Hence, we will require only the nested property in subsequent variants of the problem.

Proposition 2. A bipartite graph has an ordering of all vertices so that the questions satisfy the interval property if and only if it has an ordering with the students satisfying the nested property.

Next, we define several variants of IMO.

Definition 4. In the Chain Editing (CE) problem, we are given a bipartite graph representing student-question relations and asked to find a minimum set of edge edits that admits an ordering of the students satisfying the nested property.

A more restrictive problem than Chain Editing is Chain Addition. Chain Addition is variant of Chain Editing that allows only edge additions and no deletions. Chain Addition models situations where students sometimes accidentally give wrong answers on questions they know how to solve but never answer a hard problem correctly by luck, e.g. in numerical entry questions.

Definition 5. *In the* Chain Addition (CA) *problem, we are given a bipartite graph representing student-question relations and asked to find a minimum set of edge additions that admits an ordering of the students satisfying the nested property.*

Analogous to needing only to satisfy one of the two properties, it suffices to find an optimal ordering for only one side. Once one side is fixed, it is easy to find an optimal ordering of the other side respecting the fixed ordering.

Proposition 3. *In Chain Editing, if the best ordering (that minimizes the number of edge edits) for either students or questions is known, then the edge edits and ordering of the other side can be found in polynomial time.*

1.2.2 *k*-near Variants of Chain Editing

We introduce and study the nearby versions of Chain Editing or Chain Addition. Our problem formulations are inspired by Balas and Simonetti's [4] work on *k*-near versions of the TSP.

Definition 6. *In the k-near problem, we are given an initial ordering $\alpha : S \to [\|S\|]$ and a positive integer k. A feasible* solution *exhibits a set of edge edits (additions) attaining the nested property so that the associated ordering π, induced by the neighborhood nestings, of the students satisfies $\pi(s) \in [\alpha(s) - k, \alpha(s) + k]$.*

Next, we define three types of *k*-near problems. In the subsequent problem formulations, we bring back the interval property to our constraints since we will consider problems where the question side is not allowed to be arbitrarily ordered.

Definition 7. *In* Unconstrained *k*-near *Chain Editing (Addition), the student ordering must be k-near but the question side may be ordered any way. The objective is to minimize the number of edge edits (additions) so that there is a k-near ordering of the students that satisfies the nested property.*

Definition 8. *In* Constrained *k*-near *Chain Editing (Addition), the student ordering must be k-near while the questions have a fixed initial ordering that must be kept. The objective is to minimize the number of edge edits (additions) so that there is k-near ordering of the students that satisfies the nested property and respects the interval property according to the given question ordering.*

Definition 9. *In* Both *k*-near *Chain Editing (Addition), both sides must be k-near with respect to two given initial orderings on their respective sides. The objective is to minimize the number of edge edits (additions) so that there is a k-near ordering of the students that satisfies the nested property and a k-near ordering of the questions that satisfies the interval property.*

1.3 Main Results

In this paper, we introduce k-near models to the Chain Editing problem and present surprising complexity results. Our k-near model captures realistic scenarios of MOOCs, where information from past tests is usually known and can be used to arrive at a reliable initial nearby ordering.

We find that five of the k-near Editing and Addition problems have polynomial time algorithms while the Unconstrained k-near Editing problem is NP-hard. Our intuition is that the Constrained k-near and Both k-near problems are considerably restrictive on the ordering of the questions, which make it easy to derive the best k-near student ordering. The Unconstrained k-near Addition problem is easier than the corresponding Editing problem because the correct neighborhood of the students can be inferred from the neighborhoods of all weaker students in the Addition problem, but not for the Editing version.

Aside from restricting the students to be k-near, we may consider all possible combinations of whether the students and questions are each k-near, fixed, or unconstrained. The remaining (non-symmetric) combinations not covered by the above k-near problems are both fixed, one side fixed and the other side unconstrained, and both unconstrained. The both fixed problem is easy as both orderings are given in the input and one only needs to check whether the orderings are consistent with the nesting of the neighborhoods. When one side is fixed and the other is unconstrained, we have already shown that the ordering of the unconstrained side is easily derivable from the ordering of the fixed side via Proposition 3. If both sides are unconstrained, this is exactly the Chain Editing (or Addition) problem, which are both known to be NP-hard (see below). Figure 2 summarizes the complexity of each problem, including our results for the k-near variants, which are starred. Note that the role of the students and questions are symmetric up to flipping the orderings.

Questions \ Students	Unconstrained	k-near Editing	k-near Addition	Constrained
Unconstrained	NP-hard [26,10]	NP-hard	$O(n^3 k^{2k+2})$	$O(n^2)$
k-near Editing	NP-hard	$O(n^3 k^{4k+4})$		$O(n^3 k^{2k+2})$
k-near Addition	$O(n^3 k^{2k+2})$		$O(n^3 k^{4k+4})$	$O(n^3 k^{2k+2})$
Constrained	$O(n^2)$	$O(n^3 k^{2k+2})$	$O(n^3 k^{2k+2})$	$O(n^2)$

Fig. 2. All variants of the problems are shown with their respective complexities. The complexity of Unconstrained/Unconstrained Addition [26] and Editing [10] were derived before. All other results are given in this paper. Most of the problems have the same complexity for both Addition and Editing versions. The only exception is the Unconstrained k-near version where Editing is NP-hard while Addition has a polynomial time algorithm.

To avoid any potential confusion, we emphasize that our algorithms are not fixed-parameter tractable algorithms, as our parameter k is not a property of problem instances, but rather is part of the constraints that are specified for the outputs to satisfy.

The remaining sections are organized as follows. Section 2 discusses existing work on variants of Chain Editing that have been studied before. Section 3 shows the exact algorithms for five of the k-near problems and includes the NP-hardness proof for the last k-near problem. Section 4 summarizes our main contributions.

2   Related Work

The earliest known results on hardness and algorithms tackled Chain Addition. Before stating the results, we define a couple of problems closely related to Chain Addition. The *Minimum Linear Arrangement* problem considers as input a graph $G = (V, E)$ and asks for an ordering $\pi : V \to [|V|]$ minimizing $\sum_{vw \in E} |\pi(v) - \pi(w)|$. The *Chordal Completion* problem, also known as the *Minimum Fill-In* problem, considers as input a graph $G = (V, E)$ and asks for the minimum size set of edges F to add to G so that $(V, E \cup F)$ has no chordless cycles. A *chordless* cycle is a cycle $(v_1, \ldots, v_r, v_1)$ such that for every i, j with $|i - j| > 1$ and $\{i, j\} \neq \{1, r\}$, we have $v_i v_j \notin E$. Yannakakis [26] proved that Chain Addition is NP-hard by a reduction from Linear Arrangement. He also showed that Chain Addition is a special case of Chordal Completion on graphs of the form $(G = U \cup V, E)$ where U and V are cliques. Recently, Chain Editing was shown to be NP-hard by Drange et al. [10].

Another problem called *Total Chain Addition* is essentially identical to Chain Addition, except that the objective function counts the number of total edges in the output graph rather than the number of edges added. For Total Chain Addition, Feder et al. [11] give a 2-approximation. The total edge addition version of Chordal Completion has an $O(\sqrt{\Delta} \log^4(n))$-approximation algorithm [1] where Δ is the maximum degree of the input graph. For Chain Addition, Feder et al. [11] claim an $8d + 2$-approximation, where d is the smallest number such that every vertex-induced subgraph of the original graph has some vertex of degree at most d. Natanzon et al. [21] give an $8OPT$-approximation for Chain Addition by approximating Chordal Completion. However, no approximation algorithms are known for Chain Editing.

Modification to chordless graphs and to chain graphs have also been studied from a fixed-parameter point of view. A *fixed-parameter tractable (FPT)* algorithm for a problem of input size n and parameter p bounding the value of the optimal solution, is an algorithm that outputs an optimal solution in time $O(f(p)n^c)$ for some constant c and some function f dependent on p. For Chordal Completion, Kaplan et al. [17] give an FPT in time $O(2^{O(OPT)} + OPT^2 nm)$. Fomin and Villanger [12] show the first subexponential FPT for Chordal Completion, in time $O(2^{O(\sqrt{OPT} \log OPT)} + OPT^2 nm)$. Cao and Marx [7] study a generalization of Chordal Completion, where three operations are allowed: vertex deletion, edge addition, and edge deletion. There, they give an FPT in time $2^{O(OPT \log OPT)} n^{O(1)}$, where OPT is now the minimum total number of the three operations needed to obtain a chordless graph. For the special case of Chain Editing, Drange et al. [10] show an FPT in time $2^{O(\sqrt{OPT}\log OPT)} + \text{poly}(n)$. They also show the same result holds for a related problem called Threshold Editing.

On the other side, Drange et al. [10] show that Chain Editing and Threshold Editing do not admit $2^{o(\sqrt{OPT})}\text{poly}(n)$ time algorithms assuming the Exponential Time Hypothesis (ETH). For Chain Completion and Chordal Completion, Bliznets et al. [6] exclude the possibility of $2^{O(\sqrt{n}/\log n)}$ and $2^{O(OPT^{\frac{1}{4}}/\log^c k)}n^{O(1)}$ time algorithms assuming ETH, where c is a constant. For Chordal Completion, Cao and Sandeep [8] showed that no algorithms in time $2^{O(\sqrt{OPT}-\delta)}n^{O(1)}$ exist for any positive δ, assuming ETH. They also exclude the possibility of a PTAS for Chordal Completion assuming $P \neq NP$. Wu et al. [25] show that no constant approximation is possible for Chordal Completion assuming the Small Set Expansion Conjecture. Table 1 summarizes the known results for the aforementioned graph modification problems.

Table 1. Known results

	Chordal	Chain
Editing	Unknown approximation, FPT [9]	Unknown approximation, FPT [9]
Addition	$8OPT$-approx [21], FPT [9]	$8OPT$-approx [21], $8d + 2$-approx [11], FPT [9]
Total addition	$O(\sqrt{\Delta}\log^4(n))$-approx [1], FPT [9]	2-approx [11], FPT [9]

For the k-near problems, we show that the Unconstrained k-near Editing problem is NP-hard by adapting the NP-hardness proof for Threshold Editing from Drange et al. [9]. The remaining k-near problems have not been studied.

3   Polynomial Time Algorithms for k-near Orderings

We present our polynomial time algorithm for the Constrained k-near Addition problem and state similar results for the Constrained k-near Editing problem, the Both k-near Addition and Editing problems, and the Unconstrained k-near Addition problem. The algorithms and analyses for the other polynomial time results use similar ideas as the one for Constrained k-near Addition. They are provided in detail in the full paper [16]. We also state the NP-hardness of the Unconstrained k-near Editing problem and provide the proof in the full paper [16].

We assume correct orderings label the students from weakest (smallest label) to strongest (largest label) and label the questions from easiest (smallest label) to hardest (largest label). We associate each student with its initial label given by the k-near ordering. For ease of reading, we boldface the definitions essential to the analysis of our algorithm.

Theorem 1 (Constrained k-near Editing). *Constrained k-near Editing can be solved in time $O(n^3 k^{2k+2})$.*

Proof. Assume that the students are given in k-near order $1, \ldots, |S|$ and that the questions are given in exact order $1 \le \cdots \le |Q|$. We construct a dynamic program for Constrained k-near Editing. First, we introduce the subproblems that we will consider. Define $C(i, u_i, U_i, v_{j_i})$ to be the smallest number of edges incident to the weakest i positions that must be edited such that u_i is in position i, U_i is the set of students in the weakest $i - 1$ positions, and v_{j_i} is the hardest question correctly answered by the i weakest students. Before deriving the recurrence, we will define several sets that bound our search space within polynomial size of $n = |S| + |Q|$.

Search Space for U_i. Given position i and student u_i, define P_{i,u_i} to be the set of permutations on the elements in $\left[\max\{1, i - k\}, \min\{|S|, i + k - 1\}\right] \setminus \{u_i\}$. Let

$$F_{i,u_i} := \Big\{ \{\pi^{-1}(1), \ldots, \pi^{-1}(k)\} : \pi \in P_{i,u_i}, \pi(a) \in [a - k, a + k] \forall a \in \big[\max\{1, i - k\},$$

$$\min\{|S|, i + k - 1\}\big] \setminus \{u_i\} \Big\}.$$ The set P_{i,u_i} includes all possible permutations of the $2k$ students centered at position i, and the set F_{i,u_i} enforces that no student moves more than k positions from its label. We claim that every element of F_{i,u_i} is a candidate for $U_i \setminus \big[1, \max\{1, i - k - 1\}\big]$ given that u_i is assigned to position i. To understand the search space for U_i given i and u_i, observe that for all $i \ge 2$, U_i already must include all of $\big[1, \max\{1, i - k - 1\}\big]$ since any student initially at position $\le i - k - 1$ cannot move beyond position $i - 1$ in a feasible solution. If $i = 1$, we have $U_1 = \emptyset$. From now on, we assume $i \ge 2$ and treat the base case $i = 1$ at the end. So the set $U_i \setminus \big[1, \max\{1, i - k - 1\}\big]$ will uniquely determine U_i. We know that U_i cannot include any students with initial label $[k + i, |S|]$ since students of labels $\ge k + i$ must be assigned to positions i or later. So the only uncertainty remaining is which elements in $\big[\max\{1, i - k\}, \min\{|S|, i + k - 1\}\big] \setminus \{u_i\}$ make up the set $U_i \setminus \big[1, \max\{1, i - k - 1\}\big]$. We may determine all possible candidates for $U_i \setminus \big[1, \max\{1, i - k - 1\}\big]$ by trying all permutations of $\big[\max\{1, i - k\}, \min\{|S|, i + k - 1\}\big] \setminus \{u_i\}$ that move each student no more than k positions from its input label, which is exactly the set F_{i,u_i}.

Feasible and Compatible Subproblems. Next, we define $S_i = \Big\{ (u_i, U_i, v_{j_i}) :$ $u_i \in \big[\max\{1, i - k\}, \min\{|S|, i + k\}\big], U_i \setminus \big[1, \max\{1, i - k - 1\}\big] \in F_{i,u_i}, v_{j_i} \in Q \cup$ $\{0\} \Big\}$. The set S_i represents the search space for all possible vectors (u_i, U_i, v_{j_i}) given that u_i is assigned to position i. Note that u_i is required to be within k positions of i by the k-near constraint. So we encoded this constraint into S_i. To account for the possibility that the i weakest students answer no questions correctly, we allow v_{j_i} to be in position 0, which we take to mean that $U_i \cup \{u_i\}$ gave wrong answers to all questions.

Now, we define $R_{i-1, u_i, U_i, v_{j_i}} := \{(u_{i-1}, U_{i-1}, v_{j_{i-1}}) \in S_{i-1} : v_{j_{i-1}} \le v_{j_i}, U_i = \{u_{i-1}\} \cup U_{i-1}\}$. The set $R_{i-1, u_i, U_i, v_{j_i}}$ represents the search space for smaller subproblems that are compatible with the subproblem (i, u_i, U_i, v_{j_i}). More precisely, given that u_i is assigned to position i, U_i is the set of students assigned to the weakest $i - 1$ positions, and v_{j_i} is the hardest question correctly answered by $U_i \cup u_i$, the set of subproblems of the form $(i - 1, u_{i-1}, U_{i-1}, v_{j_{i-1}})$ which do not contradict the aforementioned assumptions encoded by (i, u_i, U_i, v_{j_i}) are exactly those whose $(u_{i-1}, U_{i-1}, v_{j_{i-1}})$ belongs to $R_{i-1, u_i, U_i, v_{j_i}}$. We illustrate compatibility in Fig. 3.

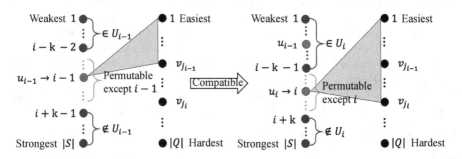

Fig. 3. Subproblem $(i - 1, u_{i-1}, U_{i-1}, v_{j_{i-1}})$ is compatible with subproblem (i, u_i, U_i, v_{j_i}) if and only if $v_{j_{i-1}}$ is no harder than v_{j_i} and $U_i = \{u_{i-1}\} \cup U_{i-1}$. The cost of (i, u_i, U_i, v_{j_i}) is the sum of the minimum cost among feasible compatible subproblems of the form $(i - 1, u_{i-1}, U_{i-1}, v_{j_{i-1}})$ and the minimum number of edits incident to u_i to make its neighborhood exactly $\{1, \ldots, v_{j_i}\}$.

The Dynamic Program. Finally, we define $c_{u_i, v_{j_i}}$ to be the smallest number of edge edits incident to u_i so that the neighborhood of u_i becomes exactly $\{1, \ldots, v_{j_i}\}$, i.e. $c_{u_i, v_{j_i}} := |N_G(u_i) \triangle \{1, \ldots, v_{j_i}\}|$. We know that $c_{u_i, v_{j_i}}$ is part of the cost within $C(i, u_i, U_i, v_{j_i})$ since v_{j_i} is the hardest question that $U_i \cup \{u_i\}$ is assumed to answer correctly and u_i is a stronger student than those in U_i who are in the positions before i. We obtain the following recurrence.

$$C(i, u_i, U_i, v_{j_i}) = \min_{(u_{i-1}, U_{i-1}, v_{j_{i-1}}) \in R_{i-1, u_i, U_i, v_{j_i}}} \{C(i-1, u_{i-1}, U_{i-1}, v_{j_{i-1}})\} + c_{u_i, v_{j_i}}$$

The base cases are $C(1, u_1, U_1, v_{j_1}) = |N_G(u_1) \triangle \{1, \ldots, v_{j_1}\}|$ if $v_{j_1} > 0$, and $C(1, u_1, U_1, v_{j_1}) = |N_G(u_1)|$ if $v_{j_1} = 0$ for all $u_1 \in [1, 1 + k], v_{j_1} \in Q \cup \{0\}$.

By definition of our subproblems, the final solution we seek is $\min_{(u_{|S|}, U_{|S|}, v_{j_{|S|}}) \in S_{|S|}} C(|S|, u_{|S|}, U_{|S|}, v_{j_{|S|}})$.

Running Time. Now, we bound the run time of the dynamic program. Note that before running the dynamic program, we build the sets P_{i, u_i}, F_{i, u_i}, S_i, $R_{i-1, u_i, U_i, v_{j_i}}$ to ensure that our solution obeys the k-near constraint and that the smaller subproblem per recurrence is compatible with the bigger subproblem it came from. Generating the set P_{i, u_i} takes $(2k)! = O(k^k)$ time per (i, u_i). Checking the k-near condition to obtain the set F_{i, u_i} while building P_{i, u_i} takes k^2 time per (i, u_i). So generating S_i takes $O(k \cdot k^k k^2 \cdot |Q|)$ time per i. Knowing S_{i-1}, generating $R_{i-1, u_i, U_i, v_{j_i}}$ takes $O(|S|)$ time. Hence, generating all of the sets is dominated by the time to build $\cup_{i \leq |S|} S_i$, which is $O(|S| k^3 k^k |Q|) = O(n^2 k^{k+3})$.

After generating the necessary sets, we solve the dynamic program. Each subproblem (i, u_i, U_i, v_{j_i}) takes $O(|R_{i-1,u_i,U_i,v_{j_i}}|)$ time. So the total time to solve the dynamic program is $O(\sum_{i \in S, (u_i, U_i, v_{j_i}) \in S_i} |R_{i-1,u_i,U_i,v_{j_i}}|) = O(|S||S_i||S_{i-1}|) = O(n(k \cdot k^k \cdot n)^2) = O(n^3 k^{2k+2})$. $\qquad\square$

Theorem 2 (Constrained k-near Addition). *Constrained k-near Addition can be solved in time $O(n^3 k^{2k+2})$.*

Theorem 3 (Unconstrained k-near Addition). *Unconstrained k-near Addition can be solved in time $O(n^3 k^{2k+2})$.*

Theorem 4 (Unconstrained k-near Editing). *Unconstrained k-near Editing is NP-hard.*

Theorem 5 (Both k-near Editing). *Both k-near Editing can be solved in time $O(n^3 k^{4k+4})$.*

Theorem 6 (Both k-near Addition). *Both k-near Addition can be solved in time $O(n^3 k^{4k+4})$.*

We present the proofs of the above theorems in the full paper [16].

4   Conclusion

We proposed a new set of problems that arise naturally from ranking participants and tasks in competitive settings and classified the complexity of each problem. First, we introduced six k-near variants of the Chain Editing problem, which capture the common scenario of having partial information about the final orderings from past rankings. Second, we provided polynomial time algorithms for five of the problems and showed NP-hardness for the remaining one.

Acknowledgments. This work was supported in part by the US National Science Foundation under award numbers CCF-1527032, CCF-1655442, and IIS-1553547.

References

1. Agrawal, A., Klein, P., Ravi, R.: Cutting down on fill using nested dissection: provably good elimination orderings. In: George, A., Gilbert, J.R., Liu, J.W.H. (eds.) Graph Theory and Sparse Matrix Computation, pp. 31–55. Springer, Heidelberg (1993)
2. Andersen, R., Borgs, C., Chayes, J., Feige, U., Flaxman, A., Kalai, A., Mirrokni, V., Tennenholtz, M.: Trust-based recommendation systems: an axiomatic approach. In: WWW, pp. 199–208. ACM (2008)
3. Aydin, B., Yilmaz, Y., Li, Y., Li, Q., Gao, J., Demirbas, M.: Crowdsourcing for multiple-choice question answering. In: IAAI, pp. 2946–2953 (2014)
4. Balas, E., Simonetti, N.: Linear time dynamic-programming algorithms for new classes of restricted TSPs: a computational study. INFORMS J. Comput. **13**(1), 56–75 (2000)

5. Bertsekas, D.P.: Non-linear Programming. Athena Scientific, Belmont (1999)
6. Bliznets, I., Cygan, M., Komosa, P., Mach, L., Pilipczuk, M.: Lower bounds for the parameterized complexity of minimum fill-in and other completion problems. In: SODA, pp. 1132–1151 (2016)
7. Cao, Y., Marx, D.: Chordal editing is fixed-parameter tractable. Algorithmica **75**(1), 118–137 (2016)
8. Cao, Y., Sandeep, R.B.: Minimum fill-in: inapproximability and almost tight lower bounds. CoRR abs/1606.08141 (2016). http://arxiv.org/abs/1606.08141
9. Dong, X.L., Berti-Equille, L., Srivastava, D.: Integrating conflicting data: the role of source dependence. PVLDB **2**(1), 550–561 (2009)
10. Drange, P.G., Dregi, M.S., Lokshtanov, D., Sullivan, B.D.: On the threshold of intractability. In: ESA, pp. 411–423 (2015)
11. Feder, T., Mannila, H., Terzi, E.: Approximating the minimum chain completion problem. Inf. Process. Lett. **109**(17), 980–985 (2009)
12. Fomin, F.V., Villanger, Y.: Subexponential parameterized algorithm for minimum fill-in. In: SODA, pp. 1737–1746 (2012)
13. Galland, A., Abiteboul, S., Marian, A., Senellart, P.: Corroborating information from disagreeing views. In: WSDM, pp. 131–140. ACM (2010)
14. Gatterbauer, W., Suciu, D.: Data conflict resolution using trust mappings. In: SIGMOD, pp. 219–230 (2010)
15. Gupta, M., Han, J.: Heterogeneous network-based trust analysis: a survey. ACM SIGKDD Explor. Newsl. **13**(1), 54–71 (2011)
16. Jiao, Y., Ravi, R., Gatterbauer, W.: Algorithms for automatic ranking of participants and tasks in an anonymized contest. CoRR abs/1612.04794 (2016). http://arxiv.org/abs/1612.04794
17. Kaplan, H., Shamir, R., Tarjan, R.E.: Tractability of parameterized completion problems on chordal, strongly chordal, and proper interval graphs. SIAM J. Comput. **28**(5), 1906–1922 (1999)
18. Kleinberg, J.M.: Authoritative sources in a hyperlinked environment. JACM **46**(5), 604–632 (1999)
19. Li, Q., Li, Y., Gao, J., Zhao, B., Fan, W., Han, J.: Resolving conflicts in heterogeneous data by truth discovery and source reliability estimation. In: SIGMOD, pp. 1187–1198 (2014)
20. Li, Y., Gao, J., Meng, C., Li, Q., Su, L., Zhao, B., Fan, W., Han, J.: A survey on truth discovery. ACM SIGKDD Explor. Newsl. **17**(2), 1–16 (2015)
21. Natanzon, A., Shamir, R., Sharan, R.: A polynomial approximation algorithm for the minimum fill-in problem. SIAM J. Comput. **30**(4), 1067–1079 (2000)
22. Pasternack, J., Roth, D.: Knowing what to believe (when you already know something). In: COLING, pp. 877–885 (2010)
23. Pasternack, J., Roth, D.: Latent credibility analysis. In: WWW, pp. 1009–1021 (2013)
24. Pasternack, J., Roth, D., Vydiswaran, V.V.: Information trustworthiness. AAAI Tutorial (2013)
25. Wu, Y.L., Austrin, P., Pitassi, T., Liu, D.: Inapproximability of treewidth, one-shot pebbling, and related layout problems. J. Artif. Int. Res. **49**(1), 569–600 (2014)
26. Yannakakis, M.: Computing the minimum fill-in is NP-complete. SIAM J. Algebr. Discret. Methods **2**(1), 77–79 (1981)
27. Yin, X., Han, J., Yu, P.S.: Truth discovery with multiple conflicting information providers on the web. TKDE **20**(6), 796–808 (2008)

The Complexity of (List) Edge-Coloring Reconfiguration Problem

Hiroki Osawa[1]([✉]), Akira Suzuki[1,2], Takehiro Ito[1,2], and Xiao Zhou[1]

[1] Graduate School of Information Sciences, Tohoku University, Aoba-yama 6-6-05, Aoba-ku, Sendai 980-8579, Japan
{osawa,a.suzuki,takehiro,zhou}@ecei.tohoku.ac.jp
[2] CREST, JST, 4-1-8 Honcho, Kawaguchi, Saitama 332-0012, Japan

Abstract. Let G be a graph such that each edge has its list of available colors, and assume that each list is a subset of the common set consisting of k colors. Suppose that we are given two list edge-colorings f_0 and f_r of G, and asked whether there exists a sequence of list edge-colorings of G between f_0 and f_r such that each list edge-coloring can be obtained from the previous one by changing a color assignment of exactly one edge. This problem is known to be PSPACE-complete for every integer $k \geq 6$ and planar graphs of maximum degree three, but any computational hardness was unknown for the non-list variant in which every edge has the same list of k colors. In this paper, we first improve the known result by proving that, for every integer $k \geq 4$, the problem remains PSPACE-complete even for planar graphs of maximum degree three and bounded bandwidth. Since the problem is known to be solvable in polynomial time if $k \leq 3$, our result gives a sharp analysis of the complexity status with respect to the number k of colors. We then give the first computational hardness result for the non-list variant: for every integer $k \geq 5$, the non-list variant is PSPACE-complete even for planar graphs of maximum degree k and bandwidth linear in k.

1 Introduction

Recently, reconfiguration problems [10] have been intensively studied in the field of theoretical computer science. The problem arises when we wish to find a step-by-step transformation between two feasible solutions of a combinatorial (search) problem such that all intermediate results are also feasible and each step conforms to a fixed reconfiguration rule, that is, an adjacency relation defined on feasible solutions of the original search problem. (See, e.g., the survey [9] and references in [6, 13].)

1.1 Our Problem

In this paper, we study the reconfiguration problem for (list) edge-colorings of a graph [11,12]. Let $C = \{1, 2, \ldots, k\}$ be the set of k colors. A (proper)

This work is partially supported by JSPS KAKENHI Grant Numbers JP26730001 (A. Suzuki), JP15H00849 and JP16K00004 (T. Ito), and JP16K00003 (X. Zhou).

S.-H. Poon et al. (Eds.): WALCOM 2017, LNCS 10167, pp. 347–358, 2017.
DOI: 10.1007/978-3-319-53925-6_27

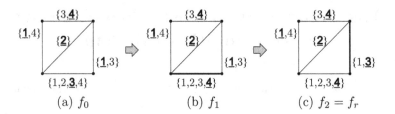

Fig. 1. A sequence of list edge-colorings of the same graph with the same list.

edge-coloring of a graph $G = (V, E)$ is a mapping $f : E \to C$ such that $f(e) \neq f(e')$ holds for every two adjacent edges $e, e' \in E$. In *list edge-coloring*, each edge $e \in E$ has a set $L(e) \subseteq C$ of colors, called the *list* of e. Then, an edge-coloring f of G is called a *list edge-coloring* of G if $f(e) \in L(e)$ holds for every edge $e \in E$. Figure 1 illustrates three list edge-colorings of the same graph G with the same list L; the list of each edge is attached to the edge, and the color assigned to each edge is written in bold with an underline. Clearly, an (ordinary) edge-coloring of G is a list edge-coloring of G for which $L(e) = C$ holds for every edge e of G, and hence list edge-coloring is a generalization of edge-coloring.

Ito et al. [12] introduced an adjacency relation defined on list edge-colorings of a graph, and define the LIST EDGE-COLORING RECONFIGURATION problem, as follows: Suppose that we are given two list edge-colorings of a graph G (e.g., the leftmost and rightmost ones in Fig. 1), and we are asked whether we can transform one into the other via list edge-colorings of G such that each differs from the previous one in only one edge color assignment; such a sequence of list edge-colorings is called a *reconfiguration sequence*. We call this problem the LIST EDGE-COLORING RECONFIGURATION problem. For the particular instance of Fig. 1, the answer is "yes" as illustrated in the figure, where the edge whose color assignment was changed from the previous one is depicted by a thick line.

For convenience, we call the problem simply the *non-list variant* (formally, EDGE-COLORING RECONFIGURATION) if $L(e) = C$ holds for every edge e of a given graph.

1.2 Known and Related Results

Despite recent intensive studies on reconfiguration problems (in particular, for graph colorings [1–5,7,11,12,14,16]), as far as we know, only one complexity result is known for LIST EDGE-COLORING RECONFIGURATION. Ito et al. [11] proved that LIST EDGE-COLORING RECONFIGURATION is PSPACE-complete even when restricted to $k = 6$ and planar graphs of maximum degree three. (Since the list of each edge is given as an input, this result implies that the problem is PSPACE-complete for every integer $k \geq 6$.) They also gave a sufficient condition for which any two list edge-colorings of a tree can be transformed into each other, which was improved by [12]; but these sufficient conditions do not clarify the complexity status for trees, and indeed it remains open.

As a related problem, the (LIST) VERTEX-COLORING RECONFIGURATION problem has been studied intensively. (LIST VERTEX-COLORING RECONFIGURATION and its non-list variant are defined analogously.) Bonsma and Cereceda [3] proved that VERTEX-COLORING RECONFIGURATION is PSPACE-complete for every integer $k \geq 4$. On the other hand, Cereceda et al. [5] proved that both LIST VERTEX-COLORING RECONFIGURATION and its non-list variant are solvable in polynomial time for any graph if $k \leq 3$. Thus, the complexity status of VERTEX-COLORING RECONFIGURATION is analyzed sharply with respect to the number k of colors.

Edge-coloring in a graph G can be reduced to vertex-coloring in the line graph of G. By this reduction, we can solve LIST EDGE-COLORING RECONFIGURATION for any graph if $k \leq 3$. However, the reduction does not work the other way, and hence this reduction does not yield any computational hardness result. Indeed, the complexity of EDGE-COLORING RECONFIGURATION was an open question proposed by [11].

1.3   Our Contribution

In this paper, we precisely analyze the complexity of (LIST) EDGE-COLORING RECONFIGURATION; in particular, we give the first complexity result for the non-list variant.

We first improve the known result for LIST EDGE-COLORING RECONFIGURATION by proving that, for every integer $k \geq 4$, the problem remains PSPACE-complete even for planar graphs of maximum degree three and bounded bandwidth. Recall that the problem is solvable in polynomial time if $k \leq 3$, and hence our result gives a sharp analysis of the complexity status with respect to the number k of colors. We then give the first complexity result for the non-list variant: for every integer $k \geq 5$, EDGE-COLORING RECONFIGURATION is PSPACE-complete even for planar graphs of maximum degree k and bandwidth linear in k.

We roughly explain our main technical contribution. Both our results can be obtained by constructing polynomial-time reductions from NONDETERMINISTIC CONSTRAINT LOGIC (NCL, for short), introduced by Hearn and Demaine [8]. This problem is often used to prove the computational hardness of puzzles and games, because a reduction from this problem requires to construct only two types of gadgets, called AND and OR gadgets. However, there is another difficulty for our problems: how the gadgets communicate with each other. We handle this difficulty by introducing the "neutral orientation" to NCL. In addition, our AND/OR gadgets are very complicated, and hence for showing the correctness of our reductions, we clarify the sufficient conditions (which we call "internally connected" and "external adjacency") so that the gadgets correctly work. We show that our gadgets indeed satisfy these conditions by a computer search.

2   Nondeterministic Constraint Logic

In this section, we define the NONDETERMINISTIC CONSTRAINT LOGIC problem [8].

An NCL "machine" is an undirected graph together with an assignment of weights from $\{1, 2\}$ to each edge of the graph. An (*NCL*) *configuration* of this machine is an orientation (direction) of the edges such that the sum of weights of in-coming arcs at each vertex is at least two. Figure 2(a) illustrates a configuration of an NCL machine, where each weight-2 edge is depicted by a thick (blue) line and each weight-1 edge by a thin (orange) line. Then, two NCL configurations are *adjacent* if they differ in a single edge direction. Given an NCL machine and its two configurations, it is known to be PSPACE-complete to determine whether there exists a sequence of adjacent NCL configurations which transforms one into the other [8].

An NCL machine is called an AND/OR *constraint graph* if it consists of only two types of vertices, called "NCL AND vertices" and "NCL OR vertices" defined as follows:

1. A vertex of degree three is called an *NCL* AND *vertex* if its three incident edges have weights 1, 1 and 2. (See Fig. 2(b).) An NCL AND vertex u behaves as a logical AND, in the following sense: the weight-2 edge can be directed outward for u if and only if both two weight-1 edges are directed inward for u. Note that, however, the weight-2 edge is not necessarily directed outward even when both weight-1 edges are directed inward.
2. A vertex of degree three is called an *NCL* OR *vertex* if its three incident edges have weights 2, 2 and 2. (See Fig. 2(c).) An NCL OR vertex v behaves as a logical OR: one of the three edges can be directed outward for v if and only if at least one of the other two edges is directed inward for v.

It should be noted that, although it is natural to think of NCL AND/OR vertices as having inputs and outputs, there is nothing enforcing this interpretation; especially for NCL OR vertices, the choice of input and output is entirely arbitrary because an NCL OR vertex is symmetric.

For example, the NCL machine in Fig. 2(a) is an AND/OR constraint graph. From now on, we call an AND/OR constraint graph simply an *NCL machine*, and call an edge in an NCL machine an *NCL edge*. NCL remains PSPACE-complete even if an input NCL machine is planar and bounded bandwidth [17].

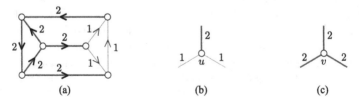

Fig. 2. (a) A configuration of an NCL machine, (b) NCL AND vertex u, and (c) NCL OR vertex v. (Color figure online)

3 Our Results

In this section, we give our results. Observe that LIST EDGE-COLORING RECONFIGURATION can be solved in (most conveniently, nondeterministic [15]) polynomial space, and hence it is in PSPACE. Therefore, we show the PSPACE-hardness. Our reductions from NCL take a similar construction, and hence we first give the common preparation in Sect. 3.1. Then, we prove the PSPACE-hardness of LIST EDGE-COLORING RECONFIGURATION in Sect. 3.2 and the non-list variant in Sect. 3.3.

3.1 Preparation for Reductions

Suppose that we are given an instance of NCL, that is, an NCL machine and two orientations of the machine.

We subdivide every NCL edge vw into a path $vv'w'w$ of length three by adding two new vertices v' and w'. (See Fig. 3(a) and (b).) We call the edge $v'w'$ a *link edge* between two NCL vertices v and w, and call the edges vv' and ww' *connector edges* for v and w, respectively. Notice that every vertex in the resulting graph belongs to exactly one of stars $K_{1,3}$ such that the center of each $K_{1,3}$ corresponds to an NCL AND/OR vertex. Furthermore, these stars are all mutually disjoint, and joined together by link edges. (See Fig. 3(c) as an example.)

Our reduction thus involves constructing three types of gadgets which correspond to link edges and stars of NCL AND/OR vertices; we will replace each of them with its corresponding gadget. In our reduction, assigning the color 1 to the connector edge vv' always corresponds to directing vv' from v' to v (i.e., the inward direction for v), while assigning the color 4 to vv' always corresponds to directing vv' from v to v' (i.e., the outward direction for v).

3.2 List Edge-Coloring Reconfiguration

In this subsection, we prove the following theorem.

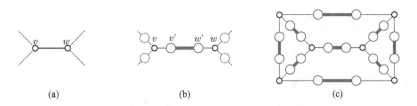

(a)   (b)   (c)

Fig. 3. (a) An NCL edge vw, (b) its subdivision into a path $vv'w'w$, and (c) the resulting graph which corresponds to the NCL machine in Fig. 2(a), where newly added vertices are depicted by (red) large circles, link edges by (green) thick lines and connector edges by (blue) thin lines. (Color figure online)

Theorem 1. *For every integer* $k \geq 4$, *the* LIST EDGE-COLORING RECONFIG-URATION *problem is PSPACE-complete for planar graphs of maximum degree three and bounded bandwidth.*

We prove the theorem in the remainder of this subsection. As we have mentioned in Sect. 3.1, it suffices to construct three types of gadgets which correspond to link edges and stars of NCL AND/OR vertices.

● **Link edge gadget**
Recall that, in a given NCL machine, two NCL vertices v and w are joined by a single NCL edge vw. Therefore, the link edge gadget between v and w should be consistent with the orientations of the NCL edge vw, as follows (see also Fig. 4): If we assign the color 1 to the connector edge vv' (i.e., the inward direction for v), then ww' must be colored with 4 (i.e., the outward direction for w); conversely, vv' must be colored with 4 if we assign 1 to ww'. In particular, the gadget must forbid a list edge-coloring which assigns 1 to both vv' and ww' (i.e., the inward directions for both v and w), because such a coloring corresponds to the direction which illegally contributes to both v and w at the same time. On the other hand, assigning 4 to both vv' and ww' (i.e., the outward directions for both v and w) corresponds to the *neutral* orientation of the NCL edge vw which contributes to neither v nor w, and hence we simply do not care such an orientation.

Figure 5 illustrates our link edge gadget between two NCL vertices v and w. Figure 6(b) illustrates the "reconfiguration graph" of this link edge gadget together with two connector edges vv' and ww': each rectangle represents a node of the reconfiguration graph, that is, a list edge-coloring of the gadget, where the underlined bold number represents the color assigned to the edge,

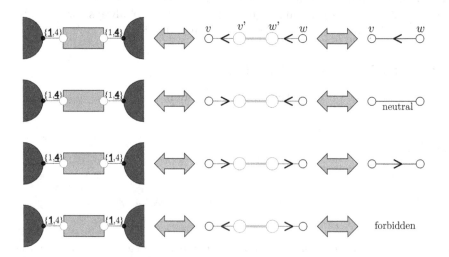

Fig. 4. (a) Color assignments to connector edges, (b) their corresponding orientations of the edges vv' and ww', and (c) the corresponding orientations of an NCL edge vw. (Color figure online)

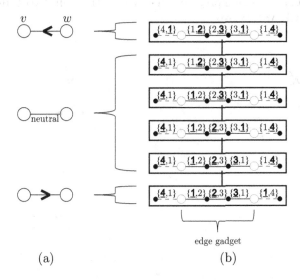

Fig. 5. Link edge gadget for LIST EDGE-COLORING RECONFIGURATION.

Fig. 6. (a) Three orientations of an NCL edge vw, and (b) all list edge-colorings of the link edge gadget with two connector edges.

and two rectangles are joined by an edge in the reconfiguration graph if their corresponding list edge-colorings are adjacent. Then, the reconfiguration graph is connected as illustrated in Fig. 6(b), and the link edge gadget has no list edge-coloring which assigns 4 to the two connector edges vv' and ww' at the same time, as required. Furthermore, the reversal of the NCL edge vw can be simulated by the path via the neutral orientation of vw, as illustrated in Fig. 6(a). Thus, this link edge gadget works correctly.

• AND **gadget**
Consider an NCL AND vertex v. Figure 7(a) illustrates all valid orientations of the three connector edges for v; each box represents a valid orientation of the three connector edges for v, and two boxes are joined by an edge if their orientations are adjacent. We construct our AND gadget so that it correctly simulates this reconfiguration graph in Fig. 7(a).

Figure 8 illustrates our AND gadget for each NCL AND vertex v, where e_1, e_2 and e_a correspond to the three connector edges for v such that e_1 and e_2 come from the two weight-1 NCL edges and e_a comes from the weight-2 NCL edge. Figure 7(b) illustrates the reconfiguration graph for all list edge-colorings of the AND gadget, where each large box surrounds all colorings having the same color assignments to the three connector edges for v. Then, we can see that these list edge-colorings are "internally connected," that is, any two list edge-colorings in the same box are reconfigurable with each other without recoloring any connector edge. Furthermore, this gadget preserves the "external adjacency" in the following sense: if we contract the list edge-colorings in the same box in Fig. 7(b) into a single vertex, then the resulting graph is exactly the graph depicted in Fig. 7(a). Therefore, we can conclude that our AND gadget correctly works as an NCL AND vertex.

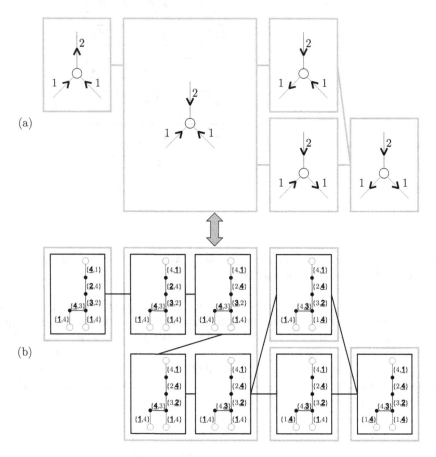

Fig. 7. (a) All valid orientations of the three connector edges for an NCL AND vertex v, and (b) all list edge-colorings of the AND gadget in Fig. 8.

• OR gadget

Figure 9 illustrates our OR gadget for each NCL OR vertex v, where e_1, e_2 and e_3 correspond to the three connector edges for v. To verify that this OR gadget correctly simulates an NCL OR vertex, it suffices to show that this gadget satisfies both the internal connectedness and the external adjacency. Since this gadget has 1575 list edge-colorings, we have checked these sufficient conditions by a computer search of all list edge-colorings of the gadget.

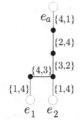

Fig. 8. AND gadget for LIST EDGE-COLORING RECONFIGURATION.

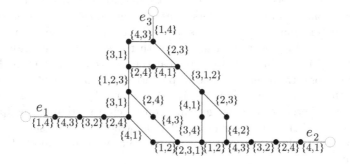

Fig. 9. OR gadget for LIST EDGE-COLORING RECONFIGURATION.

Reduction. As we have explained before, we replace each of link edges and stars of NCL AND/OR vertices with its corresponding gadget; let G be the resulting graph. Since NCL remains PSPACE-complete even if an input NCL machine is planar, bounded bandwidth and of maximum degree three, the resulting graph G is also planar, bounded bandwidth and of maximum degree three; notice that, since each gadget consists of only a constant number of edges, the bandwidth of G is also bounded.

In addition, we construct two list edge-colorings of G which correspond to two given NCL configurations C_0 and C_r of the NCL machine. Note that there are (in general, exponentially) many list edge-colorings which correspond to the same NCL configuration. However, by the construction of the three gadgets, no two distinct NCL configurations correspond to the same list edge-coloring of G. We thus arbitrarily choose two list edge-colorings f_0 and f_r of G which correspond to C_0 and C_r, respectively.

This completes the construction of our corresponding instance of LIST EDGE-COLORING RECONFIGURATION. Clearly, the construction can be done in polynomial time. Due to the page limitation, we omit the correctness proof of our reduction.

3.3   Edge-Coloring Reconfiguration

In this subsection, we prove the following theorem for the non-list variant.

Theorem 2. *For every integer* $k \geq 5$, *the* EDGE-COLORING RECONFIGURATION *problem is PSPACE-complete for planar graphs of maximum degree* k *and bandwidth linear in* k.

To prove the theorem, similarly as in the previous subsection, we will construct three types of gadgets corresponding to a link edge and stars of NCL AND/OR vertices. However, since we deal with the non-list variant, every edge has all k colors as its available colors. Thus, we construct one more gadget, called a *color gadget*, which restricts the colors available for the edge. The gadget is simply a star having k leaves, and we assign the k colors to the edges of the star in both f_0 and f_r. (See Fig. 10(a) as an example for $k = 5$.) Note that, since the color set consists of only k colors, these k edges must stay the same colors in any reconfiguration sequence. Thus, if we do not want to assign a color c to an edge e, then we connect the leaf edge with the color c to an endpoint v of e. (See Fig. 10(b).) In this way, we can treat the edge e as if it has the list $L(e)$ of available colors. However, we need to pay attention to the fact that all edges e' sharing the endpoint v cannot receive the color c by connecting such color gadgets to v. Therefore, our gadgets are constructed so that all endpoints of connector edges shared by other gadgets are attached with the same color gadgets which forbid colors 2 and $5, 6, \ldots, k$, and hence we can connect the gadgets consistently.

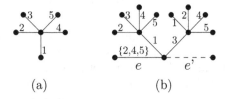

Fig. 10. (a) Gadget for restricting a color ($k = 5$), and (b) edge e whose available colors are restricted to $\{2, 4, 5\}$.

Figures 11 and 12 illustrate all gadgets for the non-list variant, where the available colors for each edge is attached as the list of the edge. Notice that the gadget in Fig. 11 forbids the colors i, j, and $6, 7, \ldots, k$. Then, the link edge gadget and AND/OR gadgets have 10, 40, 477192 edge-colorings, respectively. We have checked that all gadgets satisfy both the internal connectedness and the external adjacency by a computer search of all edge-colorings of the gadgets.

Recall that NCL remains PSPACE-complete even if an input NCL machine is planar, bounded bandwidth, and of maximum degree three. Thus, the resulting graph G is also planar. Notice that only the size of the color gadget depends on k, and the other (parts of) gadgets are of constant sizes. Since $k \geq 5$, the maximum degree of G is k, i.e., the degree of the center of each color gadget. In addition, since each of link edge and AND/OR gadgets contains only a constant number of color gadgets, the number of edges in each gadget can be bounded

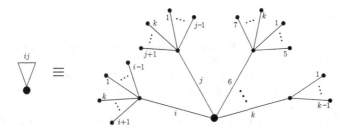

Fig. 11. Explanatory note for the gadgets in Fig. 12.

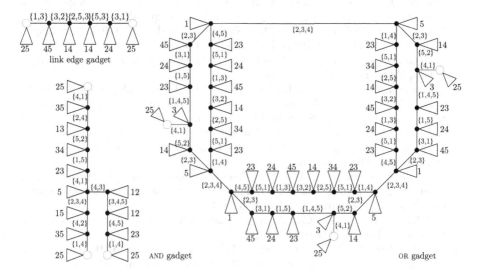

Fig. 12. Link edge, AND, and OR gadgets for the non-list variant.

by a linear in k. Since the bandwidth of the input NCL machine is a constant, that of G can be bounded by a linear in k.

4   Conclusion

In this paper, we have shown the PSPACE-completeness of LIST EDGE-COLORING RECONFIGURATION and its non-list variant. We emphasize again that our result for LIST EDGE-COLORING RECONFIGURATION gives a sharp analysis of the complexity status with respect to the number k of colors. In addition, our result is the first complexity hardness result for the non-list variant.

References

1. Bonamy, M., Bousquet, N.: Recoloring bounded treewidth graphs. Electron. Notes Discret. Math. **44**, 257–262 (2013)
2. Bonamy, M., Johnson, M., Lignos, I., Patel, V., Paulusma, D.: Reconfiguration graphs for vertex colourings of chordal and chordal bipartite graphs. J. Comb. Optim. **27**, 132–143 (2014)
3. Bonsma, P., Cereceda, L.: Finding paths between graph colourings: PSPACE-completeness and superpolynomial distances. Theoret. Comput. Sci. **410**, 5215–5226 (2009)
4. Bonsma, P., Mouawad, A.E., Nishimura, N., Raman, V.: The complexity of bounded length graph recoloring and CSP reconfiguration. In: Cygan, M., Heggernes, P. (eds.) IPEC 2014. LNCS, vol. 8894, pp. 110–121. Springer, Heidelberg (2014). doi:10.1007/978-3-319-13524-3_10
5. Cereceda, L., van den Heuvel, J., Johnson, M.: Finding paths between 3-colorings. J. Graph Theory **67**, 69–82 (2011)
6. Demaine, E.D., Demaine, M.L., Fox-Epstein, E., Hoang, D.A., Ito, T., Ono, H., Otachi, Y., Uehara, R., Yamada, T.: Linear-time algorithm for sliding tokens on trees. Theoret. Comput. Sci. **600**, 132–142 (2015)
7. Hatanaka, T., Ito, T., Zhou, X.: The list coloring reconfiguration problem for bounded pathwidth graphs. IEICE Trans. Fundam. Electron. Commun. Comput. Sci. **E98-A**, 1168–1178 (2015)
8. Hearn, R.A., Demaine, E.D.: PSPACE-completeness of sliding-block puzzles and other problems through the nondeterministic constraint logic model of computation. Theoret. Comput. Sci. **343**, 72–96 (2005)
9. van den Heuvel, J.: The Complexity of Change. Surveys in Combinatorics 2013. London Mathematical Society Lecture Notes Series 409 (2013)
10. Ito, T., Demaine, E.D., Harvey, N.J.A., Papadimitriou, C.H., Sideri, M., Uehara, R., Uno, Y.: On the complexity of reconfiguration problems. Theoret. Comput. Sci. **412**, 1054–1065 (2011)
11. Ito, T., Kamiński, M., Demaine, E.D.: Reconfiguration of list edge-colorings in a graph. Discret. Appl. Math. **160**, 2199–2207 (2012)
12. Ito, T., Kawamura, K., Zhou, X.: An improved sufficient condition for reconfiguration of list edge-colorings in a tree. IEICE Trans. Inf. Syst. **E95-D**, 737–745 (2012)
13. Ito, T., Ono, H., Otachi, Y.: Reconfiguration of cliques in a graph. In: Jain, R., Jain, S., Stephan, F. (eds.) TAMC 2015. LNCS, vol. 9076, pp. 212–223. Springer, Heidelberg (2015). doi:10.1007/978-3-319-17142-5_19
14. Johnson, M., Kratsch, D., Kratsch, S., Patel, V., Paulusma, D.: Finding shortest paths between graph colourings. Algorithmica **75**, 295–321 (2016)
15. Savitch, W.J.: Relationships between nondeterministic and deterministic tape complexities. J. Comput. Syst. Sci. **4**, 177–192 (1970)
16. Wrochna, M.: Reconfiguration in bounded bandwidth and treedepth (2014). arXiv:1405.0847
17. van der Zanden, T.C.: Parameterized complexity of graph constraint logic. In: Proceedings of IPEC 2015, LIPIcs, vol. 9076, pp. 282–293 (2015)

An Upper Bound for Resolution Size: Characterization of Tractable SAT Instances

Kensuke Imanishi[✉]

Department of Computer Science, The University of Tokyo,
7-3-1 Hongo, Bunkyo-ku, Tokyo 113-8656, Japan
japlj@is.s.u-tokyo.ac.jp

Abstract. We show the first upper bound for resolution size of a SAT instance by pathwidth of its incidence graph. Namely, we prove that if an incidence graph of an unsatisfiable CNF formula has pathwidth pw, the formula can be refuted by a resolution proof with at most $O^*(3^{pw})$ clauses. It is known that modern practical SAT-solvers run efficiently for instances which have small and narrow resolution refutations. Resolution size is one of the parameters which make SAT tractable, whereas it is shown that even linearly approximating the resolution size is NP-hard. In contrast, computing graph based parameters such as treewidth or pathwidth is fixed-parameter tractable, and also efficient FPT algorithms for SAT of bounded such parameters are widely researched. However, few explicit connection between these parameters and resolutions or SAT-solvers are known. In this paper, we provide an FPT algorithm for SAT on path decomposition of its incidence graph. The algorithm can construct resolution refutations for unsatisfiable formulas, and analyzing the size of constructed proof gives the new bound.

Keywords: Satisfiability · Parameterized complexity · Resolution size · Incidence graph · Pathwidth

1 Introduction

1.1 Background

SAT is an important problem both in theory and practice, since it is the most fundamental NP-complete problem and has many practical applications. Although performance of SAT-solvers has been improved and modern SAT-solvers can solve large instances of SAT with millions of variables, there is no theoretical explanation of their performance for general instances of SAT. Therefore studies about what kind of formulas can be solved efficiently and why modern SAT-solvers can solve such kind of formulas efficiently are needed.

Most of modern SAT-solvers are based on Davis-Putnam-Logemann-Loveland (DPLL) algorithm [10] which works by choosing a variable and assigning truth value to it. In actual implementations, many other heuristics and algorithms are used with DPLL. Conflict-Driven Clause Learning (CDCL) is one of

© Springer International Publishing AG 2017
S.-H. Poon et al. (Eds.): WALCOM 2017, LNCS 10167, pp. 359–369, 2017.
DOI: 10.1007/978-3-319-53925-6_28

such algorithms for solving SAT proposed by Marques-Silva and Sakallah in 1996 [16], which is a major breakthrough in SAT solving and widely used in modern SAT-solvers. Since Beame et al. [6] showed that the behavior of SAT-solvers with DPLL and CDCL can be regarded as searching resolution refutations for a given formula, it has been recognized that there is a strong connection between the performance of solvers and characteristics of formulas in resolution proof system (formal definitions of resolutions are given in Sect. 2). Indeed, the size of a resolution refutation (resolution size) is a lower bound for time complexity of solvers. In addition, it is shown by Atserias et al. [4] that, for an unsatisfiable formula with n variables and resolution width k, many of modern SAT-solvers can learn the empty clause with high probability in at most $\mathcal{O}(n^{2k+2})$ conflicts.

However, there are some difficulties to deal with resolution width and size directly. One of the difficulties is about general intractability of resolution; it is shown by Haken [12] that there are infinitely many CNF formulas ϕ such that the resolution size of ϕ cannot be bounded by a polynomial of the size of ϕ. For a relationship between SAT and resolution, Alekhnovich and Razborov [2] showed that SAT cannot be solved in polynomial time of the resolution size of an input formula unless $\mathbf{W[P]} \subseteq \mathbf{co\text{-}FPR}$, one of the hypotheses in computational complexity believed to be false, holds. Another difficulty is that to determine the value of resolution width or size of a formula ϕ is harder than to just solve SAT for ϕ, because, to be precise, resolution width and size are parameters defined by proofs of SAT instances, not by parameters of SAT instances (formulas itself). Actually, it is proven by Alekhnovich et al. [1] that even linearly approximating the minimum size of propositional proof is NP-hard for many proof systems, including resolution. Thus, although resolution width and size are key characteristics of SAT instances, it is nearly impossible to develop algorithms for SAT exploiting resolution width or size explicitly. On the other hand, by considering graph representation of formulas, we can develop algorithms using graph based parameters such as treewidth or branchwidth. This is because both treewidth and branchwidth can be computed in linear time for small width [8,17].

1.2   Graph Representations and Resolution

Several kinds of graph representations are considered for SAT. In terms of parameterized complexity, there are three kinds of graph representations mainly studied: primal, dual, and incidence graphs. There are FPT algorithms (dynamic programming on tree decompositions) for the treewidth of these three kinds of graphs known [15]. For the connection with resolution, Alekhnovich and Razborov [3] showed that the resolution width is bounded by the branchwidth of hypergraph representation which can be seen as more detailed primal graphs.

In this paper, we focus on incidence graphs. The reason is that incidence graphs have essentialy more information about structures of formulas than primal and dual graphs. Indeed, treewidth of primal and dual graphs can easily become large by only one variable or one clause. For example, primal graphs become complete graphs if there is a clause containing all the variables, even though such clauses are very weak since they forbid only one truth assignment out of 2^n possibilities. At the same time, branchwidth of hypergraphs becomes n, the maximum possible value, by a such large clause. In contrast, one clause with many variables generally does not immediately increase the treewidth of incidence graphs so much. Furthermore, the treewidth of incidence graphs tends to be smaller than that of primal and dual graphs. To be exact, if the treewidth of primal or dual graph is k, then the incidence graph has treewidth at most $k + 1$ [13].

We show a new connection between incidence pathwidth and resolution size by presenting an algorithm for SAT on path decompositions. Although pathwidth is not the same as treewidth, path decomposition can be regarded as a special case of tree decomposition, thus these parameters have deep connections. Pathwidth is lower bounded by treewidth, and also is upper bounded by treewidth times a logarithm of the number of vertices [14].

1.3 Our Results

The following theorem is our main result (formal definitions of resolution, pathwidth and treewidth are given in Sect. 2).

Theorem 1. *For an unsatisfiable formula ϕ with incidence pathwidth $\mathrm{pw}(\phi)$, there exists a resolution refutation of ϕ whose size is $\mathcal{O}^*(3^{\mathrm{pw}(\phi)})$.*

It is possible to construct resolution refutations of size $\mathcal{O}^*(3^{\mathrm{pw}(\phi)})$, since this result is obtained by developing a constructive algorithm. Descriptions of algorithms and proofs are in Sect. 3. We also can obtain a following connection between incidence treewidth and resolution size by using an upper bound of pathwidth [14].

Corollary 1. *For an unsatisfiable formula ϕ with incidence treewidth $\mathrm{tw}(\phi)$, there exists a resolution refutation of ϕ whose size is $|\phi|^{\mathcal{O}(\mathrm{tw}(\phi))}$.*

1.4 Resolution Size and Width

Although results of this paper focus on resolution size and do not mention about resolution width, there are some notable relationships between resolution width and size shown. In one direction, resolution width can be bounded by using the width of a formula and resolution size; $w(\phi \vdash 0) \leq w(\phi) + \mathcal{O}(\sqrt{n \ln S(\phi)})$ holds [7], where $w(\phi \vdash 0), S(\phi)$ are resolution width and size of a formula ϕ, $w(\phi)$ is the size of the largest clause in ϕ, and n is the number of variables. This result indicates that the resolution width becomes small when the resolution size is small and the formula has no large clauses. In another direction, for a bound of resolution size by resolution width, it is shown that an obvious upper bound is essentially optimal. The size of resolution proofs of resolution width w is obviously bounded by the number of distinct clauses $n^{\mathcal{O}(w)}$. Atserias et al. [5] proved that if $w = \mathcal{O}(n^c)$ for some constant $c < 1/2$, there exists 3-CNF formulas which have maximally long proofs, that is, formulas requiring $n^{\Omega(w)}$

clauses to refute by resolution. This result demonstrates that the bound of the number of rounds SAT-solvers take by resolution width and size [4] is optimal up to a constant in the exponent.

2   Preliminaries and Notations

2.1   Satisfiability and Resolution

A *literal* is a Boolean variable or its negation and a *clause* is a disjunction of literals. We say that a propositional formula ϕ is in *conjunctive normal form* (CNF) or ϕ is a CNF formula if ϕ is a conjunction of clauses. We regard a clause and a formula as a set of literals and clauses, respectively. For a clause C and a formula ϕ, we denote a set of variables occurring in it by $\text{Vars}(C)$ or $\text{Vars}(\phi)$. That is, $\text{Vars}(C) = \{x \mid x \in C\} \cup \{x \mid \overline{x} \in C\}$ and $\text{Vars}(\phi) = \bigcup_{C \in \phi} \text{Vars}(C)$.

A *truth assignment* for a formula ϕ is a function $\sigma : \text{Vars}(\phi) \to \{0, 1\}$. A truth assignment σ *satisfies* ϕ if a formula obtained by replacing occurrences of each variable x and its negation $\overline{x}$ by $\sigma(x)$ and $1 - \sigma(x)$, respectively, is evaluated to 1. We say that ϕ is *satisfiable* if there exists a truth assignment satisfying ϕ, otherwise ϕ is *unsatisfiable*. Satisfiability (SAT) is the problem to decide whether a given formula is satisfiable.

Resolution is one of the propositional proof systems and has only one inference rule (*resolution rule*):

$$\frac{C \vee x \qquad D \vee \overline{x}}{C \vee D}$$

Resolution rule takes two clauses and produces a new implied clause (*resolvent*). A *resolution refutation* of a formula ϕ is a sequence of clauses $C_1, C_2, \ldots, C_k$ such that

- for each $i \, (1 \leq i \leq k)$, C_i is a clause occurring in ϕ or a resolvent of two previous clauses, and
- the last clause C_k is an empty clause.

The *size* (or *length*) of the refutation is the number k of clauses. The *width* of the refutation is a maximum number of literals in clauses ($\max_{1 \leq i \leq k} |C_i|$).

Since an empty clause is inconsistent (always false), a resolution refutation of ϕ can be regarded as a proof of unsatisfiability of ϕ. Indeed, it is known that ϕ is unsatisfiable if and only if there exists a resolution refutation of ϕ. For an unsatisfiable formula ϕ, a resolution size and width of ϕ is the minimum size and width of resolution refutations of ϕ, respectively.

If a sequence of clauses (C_i) satisfies the first condition of a resolution refutation of a formula ϕ but does not satisfy the second condition (i.e., $C_k \neq \emptyset$), we can regard (C_i) as a proof of that C_k is a logical consequence of ϕ.

2.2 Treewidth and Pathwidth

Definition 1. *Let* $G = (V(G), E(G))$ *be an undirected graph. A tree decomposition of* G *is a pair of a tree* $T = (V(T), G(T))$ *and labels* $X : V(T) \rightarrow \mathcal{P}(V(G))$ *of tree vertices such that*

- $\bigcup_{v \in V(T)} X(v) = V(G)$,
- *for each vertex* $v \in V(G)$, *the subgraph of* T *induced by the vertices containing* v *in their label is connected, and*
- *for each edge* $(v_1, v_2) \in E(G)$, *there exists a vertex* $w \in V(T)$ *which has both* v_1 *and* v_2 *in its label.*

A *width* of a tree decomposition is defined as $\max_{v \in V(T)} |X(v)| - 1$. A *treewidth* of a graph is the minimum width of its tree decompositions.

Path decompositions and *pathwidth* are defined similarly. The difference of them is that in the definition of path decomposition, T is restricted to a simple path. For a graph G, we denote a treewidth and pathwidth of G by tw(G) and pw(G), respectively. Since a path decomposition is also a tree decomposition, pw(G) $\geq$ tw(G) holds. For upper bound of a pathwidth, it is shown that pw(G) = $\mathcal{O}(\log |V(G)| \cdot$ tw(G)) [14].

In this paper, the pathwidth of a SAT instance is defined as a pathwidth of its *incidence graph*.

Definition 2. *For a SAT instance* ϕ, *an* incidence graph *of* ϕ *is a bipartite graph* G_ϕ *whose vertices are variables and clauses in* ϕ. *There is an edge* (x, c) *if and only if a clause* $c \in \phi$ *contains* x *or* $\overline{x}$.

We denote a pathwidth of a SAT instance ϕ by pw(ϕ), which actually means pw(G_ϕ).

We assume that a path decomposition of ϕ with pathwidth pw(ϕ) is also given as input when we develop an algorithm later. For convenience, we also assume that given path decompositions are *nice*.

Definition 3. *Let* (T, X) *be a path decomposition of a graph* G *with vertices* $t_1, t_2, \ldots, t_{|V(T)|}$ *along the path. The path decomposition* (T, X) *is* nice *when each edge vertex* t_i $(2 \leq i \leq |V(T)|)$ *has one of following 2 characteristics:*

- *(introduce)* $\exists v \in V(G)$. $X(t_i) = X(t_{i-1}) \cup \{v\}$.
- *(forget)* $\exists v \in V(G)$. $X(t_i) = X(t_{i-1}) \setminus \{v\}$.

Nice decompositions are mainly considered for tree decompositions. The above definition of nice path decompositions are naturally obtained by restricting nice tree decompositions to path decompositions.

Remark 1. There is an fixed-parameter tractable algorithm for computing pathwidth and path decompositions [9]. Also we can transform a path decomposition into nice one with the same width in linear time. Thus the assumption can be made without loss of generality.

3   Results

3.1   Finding a Resolution Refutation on a Path Decomposition

This section is devoted to proof of the Theorem 1. Instead of repeating the theorem, we describe here the more detailed theorem which immediately implies Theorem 1.

Theorem 2. *There exists an algorithm which takes a CNF formula ϕ and the nice path decomposition (T, X) of its incidence graph with path-width $\mathrm{pw}(\phi)$ as inputs, runs in $2^{\mathcal{O}(\mathrm{pw}(\phi))}|\phi|^{\mathcal{O}(1)}$ time, and*

- *produces a satisfying assignment of ϕ if ϕ is satisfiable.*
- *constructs a resolution refutation of size $\mathcal{O}^*(3^{\mathrm{pw}(\phi)})$ if ϕ is unsatisfiable.*

Note that we assume that the nice path decomposition is also given for simplicity. There is a fixed-parameter tractable algorithm for computing path decompositions [9] and we can easily convert path decompositions into nice ones in polynomial time.

First we provide a detailed procedure of the algorithm and proof of its correctness to solve SAT on a path decomposition of a given formula, which can construct a resolution refutation for an unsatisfiable formula. Then we analyze the size of a resolution refutation constructed by the algorithm in the next subsection.

In the following description, we denote vertices of T as $t_1, t_2, ..., t_{|V(T)|}$ along the path, and each bag $X(t_i)$ as X_i for simplicity. We also assume following for given inputs and the algorithm:

- Both X_1 and $X_{|V(T)|}$ are empty. This means that along the path each variable in ϕ is introduced once and then finally forgotten. This assumption can be done by simple preprocessing to add some introduce nodes and forget nodes to a given path decomposition. The preprocessing increases the number of vertices in T by only $\mathcal{O}(\mathrm{pw}(\phi))$.
- Through the algorithm, clauses which have both of a variable and its negation are ignored. Such clauses are obviously always true, thus ignoring them does not affect the correctness of the algorithm. If an input ϕ contains such clauses, they also should be ignored.

Before presenting an algorithm, we introduce an operation to a set of clauses which will be used in the algorithm. This operation is the almost same as the one used in the algorithm for branch decompositions [3].

Definition 4. *For a set of clauses $\mathcal{C}$ and a variable x,*

$$\mathcal{C}^x = \{C \vee D \mid C \vee x, D \vee \overline{x} \in \mathcal{C}\} \cup \{C \mid C \in \mathcal{C}, x \notin \mathrm{Vars}(C)\}.$$

1: $\mathcal{C} \leftarrow \phi$ {note that ϕ is just a set of clause here}
2: $\mathcal{R} \leftarrow$ sequence of clauses of ϕ in arbitrary order
3: **for** $i = 1$ to $|V(T)|$ **do**
4:    **if** t_i is a forget node for a variable x **then**
5:       append clauses $\mathcal{C}^x \setminus \mathcal{C}$ to the end of $\mathcal{R}$
6:       $\mathcal{C} \leftarrow \mathcal{C}^x$
7:    **end if**
8: **end for**
9: **if** $\mathcal{C}$ contains the empty clause **then**
10:    ϕ is unsatisfiable, and $\mathcal{R}$ is its resolution refutation
11: **else**
12:    ϕ is satisfiable {a way to produce a satisfying assignment is described later}
13: **end if**

Fig. 1. Solving SAT on a path decomposition

This operation replaces clauses in $\mathcal{C}$ which has x or $\overline{x}$ with all possible resolvents for x. Clauses without x remain unchanged through the operation. One of the notable properties of this operation is that $\mathcal{C}^x$ no longer contains x (i.e., $x \notin \mathrm{Vars}(\mathcal{C}^x)$), since we assumed that no clauses have both x and $\overline{x}$. We can informally say that this operation removes a variable x, but keep all information about resolution for x.

Figure 1 shows the procedure to solve SAT on path decomposition and construct resolution refutations for unsatisfiable formulas. Although this algorithm is so simple that it just applies the operation introduced above to the input ϕ in the order in which variables are forgotten, this algorithm correctly decides whether the given formula is satisfiable or not and construct a resolution refutation.

The correctness of this algorithm can be confirmed by soundness and completeness of the Davis Putnam resolution [11]. We describe the proof of the correctness for completeness and to show how to produce a satisfying assignment for satisfiable formulas.

Lemma 1. ϕ *is unsatisfiable if and only if the algorithm ends with the empty clause.*

Proof. The "if" side of this lemma is rather easy to prove. Through the algorithm, every time $\mathcal{C}$ is updated, new clauses introduced to $\mathcal{C}$ are also introduced to $\mathcal{R}$. Thus all clauses in $\mathcal{C}$ are always contained by $\mathcal{R}$. Since the new clauses are all resolvents of the clauses in $\mathcal{C}$, $\mathcal{R}$ can be regarded as a resolution proof of a clause in $\mathcal{R}$. If $\mathcal{C}$ contains the empty clause at the end of the algorithm, $\mathcal{R}$ also contains the empty clause. Therefore, $\mathcal{R}$ is a resolution refutation of a formula ϕ.

To prove "only if" side, we use an induction on the for-loop in the algorithm. If ϕ is unsatisfiable, $\mathcal{C}$ is also unsatisfiable when the algorithm starts. In addition, if $\mathcal{C}$ is unsatisfiable when the algorithm ends, then $\mathcal{C}$ contains the empty clause, because applying the operation on variable x removes all occurrence of x in $\mathcal{C}$ and all variables were forgotten at the end. Thus it is sufficient to show that if

$\mathcal{C}$ is unsatisfiable, $\mathcal{C}^x$ is also unsatisfiable for any variable x. We can show this fact by proving its contraposition; if $\mathcal{C}^x$ is satisfiable, $\mathcal{C}$ is also satisfiable.

Suppose that there exists a truth assignment σ satisfying $\mathcal{C}^x$. Note that $\sigma(x)$ is undefined since $\text{Vars}(\mathcal{C}^x)$ does not contain x. Let us consider clauses in $\mathcal{C}$ under the assignment σ. We can split them into following three groups:

- Clauses without a variable x. These clauses remain unchanged in $\mathcal{C}^x$, thus the clauses are satisfied by σ.
- Clauses containing x. Let $A_i \vee x \in \mathcal{C}$ for $1 \leq i \leq a$ be such clauses.
- Clauses containing $\overline{x}$. Let $B_j \vee \overline{x} \in \mathcal{C}$ for $1 \leq j \leq b$ be such clauses.

By definition of the operation, $\mathcal{C}^x$ contains all clauses of the form $A_i \vee B_j$ for all i, j. If there exists a clause A_i that is not satisfied by σ, then clauses B_j for all j are satisfied by σ, since $A_i \vee B_j \in \mathcal{C}^x$ are all satisfied. Therefore, we can satisfy $\mathcal{C}$ by defining $\sigma(x)$ to be true if there exists a unsatisfied clause A_i, otherwise false.

We can use the argument in the above proof to construct a satisfying assignment for a satisfiable formula. If ϕ is satisfiable, then $\mathcal{C}^x$ is also satisfiable since clauses in $\mathcal{C}^x$ are all logical consequences of ϕ, that is there exists a truth assignment satisfying $\mathcal{C}^x$. From the argument in the above proof, we can construct a satisfying assignment of $\mathcal{C}$ by setting x be appropriate value. Thus we can construct a satisfying assignment for ϕ by looking operations performed backward and deciding the truth assignment for each variable.

Note that in the above proof of correctness we did not use the order of forgotten variables. This means that the algorithm correctly determines whether given formulas are satisfiable or not even if the order of operations performed on $\mathcal{C}$ changes arbitrarily. Performing operations in the order of forgotten variables is required and essential to bound the number of clauses in constructed proofs.

3.2   Bound of the Running Time and the Resolution Size

To conclude the proof for Theorem 2, we have to bound the size of resolution refutations constructed by the above algorithm; namely we show the following theorem.

Theorem 3. *For an unsatisfiable formula ϕ, at the end of the algorithm shown in Fig. 1, the size of $\mathcal{R}$ (the number of clauses in $\mathcal{R}$) is $\mathcal{O}^*(3^{\text{pw}(\phi)})$.*

Since $\mathcal{R}$ is a resolution refutation of ϕ, Theorem 1 follows from this theorem. To prove this theorem, it is sufficient to show nearly the same lemma for $\mathcal{C}$.

Lemma 2. *Through the algorithm shown in Fig. 1, the size of $\mathcal{C}$ is always less than or equal to $3^{\text{pw}(\phi)+1} + |\phi|$.*

Theorem 3 follows from Lemma 2 because the size of $\mathcal{R}$ is increased by at most $|\mathcal{C}^x|$ $(= \mathcal{O}^*(3^{\text{pw}(\phi)})$ if Lemma 2 holds) for at most $|V(T)|$ times through the algorithm. Then, let us go on to the proof of Lemma 2.

Proof. Clauses in $\mathcal{C}$ can be categorized into two types; a clause appears in ϕ or not. Obviously the number of clauses appearing in ϕ is bounded by $|\phi|$. We show that the number of clauses which do not appear in ϕ is bounded by $3^{\mathrm{pw}(\phi)+1}$.

Assume that we are just finished looking at a vertex v with variables $V(v)$ and clauses $C(v)$. Let F be a set of variables already forgotten at that moment, and D be a clause in $\mathcal{C} \setminus \phi$. It is easily confirmed that D may contain variables in $V(v)$ in arbitrary combinations (note that there are $3^{|V(v)|}$ possible combinations, since each variable can be taken or not, and taken variables can be negated or not), but does not contain variables in F. The problem is about the remaining part of D; what can we say about variables not in $V(v)$ and F? To answer this question we use the following fact: if D contains a literal l which is not a variable in $V(v)$ and F, then there exists a clause $L \in C(v)$ such that $l \in L$.

This fact is confirmed as follows. The first thing to be noted is that D can be derived with resolution by introduced clauses. Since all resolutions are performed on forgotten variables, clauses not introduced yet (in other words, clauses without forgotten variables) are not resolved at this moment. The next is that forgotten clauses, which are introduced and then already forgotten before the vertex v, do not have variables not in $V(v)$ and F. Thus variables not in $V(v)$ and F can only come from clauses in $C(v)$.

By this fact, the remaining part of D can be characterized by a subsequence of $\mathcal{R}$. Suppose $\mathcal{R} = (R_1, R_2, \ldots, R_{|\mathcal{R}|})$ and $D = R_k$. Then there exists a (not necessarily consecutive) subsequence of $\mathcal{R}$ such that it can be regarded as a proof of D under ϕ by resolution. The remaining part of D is determined by, for each clause C in $C(v)$, whether the proof of D under ϕ, a subsequence of $\mathcal{R}$, contains C or not.

For a clause C in $C(v)$, let C' be a clause obtained by removing variables in $V(v)$ and F from C. If the proof of D under ϕ contains C, then we can say that $C' \subseteq D$. A variable $x \notin V(v) \cup F$ is not in $\mathrm{Vars}(D)$ if we cannot say that D contains x or $\bar{x}$ from the above fact.

Finally we can bound the size of $\mathcal{C}$ by above arguments. If a combination of variables in $V(v)$ ($3^{|V(v)|}$ possibilities) and clauses in $C(v)$ contained by a proof under ϕ ($2^{|C(v)|}$ possibilities) are specified, we can uniquely determine a clause which may be in $\mathcal{C}$. That is, if a set of literals L_V over variables in $V(v)$ and a set of clauses $C_V \subseteq C(v)$ are given, we can construct a clause which possibly be in $\mathcal{C}$ as the union of L_V and clauses obtained by removing variables in $V(v)$ and F from clauses in C_V. In addition, all clauses in $\mathcal{C}$ (without ones appearing in ϕ) are characterized by above properties. Therefore $\mathcal{C}$ can contain at most $3^{|V(v)|} 2^{|C(v)|} + |\phi| \leq 3^{\mathrm{pw}(\phi)+1} + |\phi|$ clauses.

We proved Lemma 2 and thus Theorem 2 and Theorem 1 are also proved. By using the bound of path-width $\mathrm{pw}(\phi) = \mathcal{O}(\mathrm{tw}(\phi) \log |\phi|)$ [14], we immediately get the bound of resolution size by incidence treewidth, which we mentioned as Corollary 1 in the previous section.

Corollary 1. *For an unsatisfiable formula ϕ with incidence treewidth $\mathrm{tw}(\phi)$, there exists a resolution refutation of ϕ whose size is $|\phi|^{\mathcal{O}(\mathrm{tw}(\phi))}$.*

Since $3^{\mathrm{pw}(\phi)} = |\phi|^{\mathcal{O}(\mathrm{tw}(\phi))}$, the resolution size of an unsatisfiable formula ϕ is bounded by $|\phi|^{\mathcal{O}(\mathrm{tw}(\phi))}$.

Note that a computation of the operation $\mathcal{C}^x$ can be done in polynomial time of the size of $\mathcal{C}$, thus the algorithm runs in $2^{\mathcal{O}(\mathrm{pw}(\phi))}|\phi|^{\mathcal{O}(1)}$ time. That is, the algorithm is a fixed-parameter tractable algorithm for SAT on pathwidth of incidence graphs.

4   Conclusion

The result of this paper indicates that SAT instances of bounded pathwidth or treewidth can be rather easily solved by modern SAT-solvers. Although it is believed that we cannot solve SAT in polynomial time of resolution size [2], the paper [4] about connections between CDCL solvers and resolutions provides the more detailed bound $4\,km\ln(4knm)n^{k+1}$ of the number of restarts, where n is the number of variables, k and m are the resolution width and size, respectively. This means that shorter proofs are relatively easily found by modern solvers, and we showed that SAT instances with smaller pathwidth and treewidth have smaller resolution sizes. Thus, we may conclude that the pathwidth and treewidth of incidence graphs are one of the keys to grasp the structure of easily solved SAT for modern solvers.

There are several open questions about connections between resolution and graph based parameters. First, can we develop an algorithm on tree decompositions which can construct resolution refutations for unsatisfiable formulas, or we can bound the resolution size by the exponential factors of incidence treewidth with constant base? Although in this paper we showed an upper bound of the resolution size by incidence treewidth, that bound is exponential factor of incidence treewidth with non-constant base and is obtained indirectly from the algorithm on path decompositions.

Second, is there any lower bounds for resolution size by graph based parameters, especially pathwidth or treewidth? If we can bound resolution size from both upper and lower sides, then we can compare such bounds with other general (not necessary to involve resolution proofs) fixed-parameter tractable algorithms to study capabilities and limitations of resolution.

Acknowledgements. I would like to thank Hiroshi Imai for helpful advice for writing this paper, and Yoichi Iwata for motivating this work by providing a lot of information about SAT, CDCL, and other related researches.

References

1. Alekhnovich, M., Buss, S., Moran, S., Pitassi, T.: Minimum propositional proof length is NP-hard to linearly approximate. J. Symb. Log. **66**(1), 171–191 (2001)
2. Alekhnovich, M., Razborov, A.A.: Resolution is not automatizable unless W[P] is tractable. In: Proceedings of the 42nd Annual IEEE Symposium on Foundations of Computer Science (FOCS 2001), pp. 210–219. IEEE (2001)

3. Alekhnovich, M., Razborov, A.A.: Satisfiability, branch-width and Tseitin tautologies. In: Proceedings of the 43rd Annual IEEE Symposium on Foundations of Computer Science (FOCS 2002), pp. 593–603. IEEE (2002)
4. Atserias, A., Fichte, J.K., Thurley, M.: Clause-learning algorithms with many restarts and bounded-width resolution. J. Artif. Intell. Res. **40**, 353–373 (2011)
5. Atserias, A., Lauria, M., Nordström, J.: Narrow proofs may be maximally long. In: Proceedings of the 29th Annual IEEE Conference on Computational Complexity (CCC 2014), pp. 287–297. IEEE (2014)
6. Beame, P., Kautz, H., Sabharwal, A.: Understanding the power of clause learning. In: International Joint Conference on Artificial Intelligence (IJCAI), pp. 1194–1201 (2003)
7. Ben-Sasson, E., Wigderson, A.: Short proofs are narrow - resolution made simple. In: Proceedings of the 31st Annual ACM Symposium on Theory of Computing (STOC 1999), pp. 517–526 (1999)
8. Bodlaender, H.L.: A linear time algorithm for finding tree-decompositions of small treewidth. In: Proceedings of the 25th Annual ACM Symposium on Theory of Computing (STOC 1993), pp. 226–234. ACM (1993)
9. Bodlaender, H.L., Kloks, T.: Efficient and constructive algorithms for the pathwidth and treewidth of graphs. J. Algorithms **21**, 358–402 (1996)
10. Davis, M., Logemann, G., Loveland, D.: A machine program for theorem-proving. Commun. ACM **5**(7), 394–397 (1962)
11. Davis, M., Putnam, H.: A computing procedure for quantification theory. J. ACM **7**(3), 201–215 (1960)
12. Haken, A.: The intractability of resolution. Theoret. Comput. Sci. **39**, 297–308 (1985)
13. Kolaitis, P.G., Vardi, M.Y.: Conjunctive-query containment and constraint satisfaction. In: Proceedings of the 17th ACM SIGACT-SIGMOD-SIGART Symposium on Principles of Database Systems (PODS 1998), pp. 205–213 (1998)
14. Korach, E., Solel, N.: Tree-width, path-width, and cutwidth. Discret. Appl. Math. **43**, 97–101 (1993)
15. Samer, M., Szeider, S.: Algorithms for propositional model counting. J. Discret. Algorithms **8**(1), 50–64 (2010)
16. Silva, J.P.M., Sakallah, K.A.: GRASP - a new search algorithm for satisfiability. In: Proceedings of the 1996 IEEE/ACM International Conference on Computer-Aided Design, pp. 220–227 (1996)
17. Bodlaender, H.L., Thilikos, D.M.: Constructive linear time algorithms for branchwidth. In: Degano, P., Gorrieri, R., Marchetti-Spaccamela, A. (eds.) ICALP 1997. LNCS, vol. 1256, pp. 627–637. Springer, Heidelberg (1997). doi:10.1007/3-540-63165-8_217

Approximation Algorithms

Finding Triangles for Maximum Planar Subgraphs

Parinya Chalermsook[1]($\boxtimes$) and Andreas Schmid[2]($\boxtimes$)

[1] Aalto University, Espoo, Finland
parinya.chalermsook@aalto.fi
[2] Max Planck Institute for Informatics, Saarbrücken, Germany
ASchmid@mpi-inf.mpg.de

Abstract. In the MAXIMUM PLANAR SUBGRAPH (MPS) problem, we are given a graph G, and our goal is to find a planar subgraph H with maximum number of edges. Besides being a basic problem in graph theory, MPS has many applications including, for instance, circuit design, factory layout, and graph drawing, so it has received a lot of attention from both theoretical and empirical literature. Since the problem is NP-hard, past research has focused on approximation algorithms. The current best known approximation ratio is $\frac{4}{9}$ obtained two decades ago (Călinescu et al. SODA 1996) based on computing as many edge-disjoint triangles in an input graph as possible. The factor of $\frac{4}{9}$ is also the limit of this "disjoint triangles" approach.

This paper presents a new viewpoint that highlights the essences of known algorithmic results for MPS, as well as suggesting new directions for breaking the $\frac{4}{9}$ barrier. In particular, we introduce the MAXIMUM PLANAR TRIANGLES (MPT) problem: Given a graph G, compute a subgraph that admits a planar embedding with as many triangular faces as possible. Roughly speaking, any ρ-approximation algorithm for MPT can easily be turned into a $\frac{1}{3} + \frac{2\rho}{3}$ approximation for MPS. We illustrate the power of the MPT framework by "rephrasing" some known approximation algorithms for MPS as approximation algorithms for MPT (solving MPS as by-products). This motivates us to perform a further rigorous study on the approximability of MPT and show the following results:

- MPT is NP-hard, giving a simplified NP-hardness proof for MPS as a by-product.
- We propose a natural class of greedy algorithms that captures all known greedy algorithms that have appeared in the literature. We show that a very simple greedy rule gives better approximation ratio than all known greedy algorithms (but still worse than $\frac{4}{9}$).

Our greedy results, despite not improving the approximation factor, illustrate the advantage of overlapping triangles in the context of greedy algorithms. The MPT viewpoint offers various new angles that might be useful in designing a better approximation algorithm for MPS.

1 Introduction

In the maximum planar subgraph problem (MPS), we are given an input graph G, and our objective is to compute a planar subgraph H with the maximum number

© Springer International Publishing AG 2017
S.-H. Poon et al. (Eds.): WALCOM 2017, LNCS 10167, pp. 373–384, 2017.
DOI: 10.1007/978-3-319-53925-6_29

of edges. MPS has arisen in many distinct applications, including for instance, *architectural floor planning*, and *circuit design*. Besides the applications, the problem is fundamentally interesting and has often been used as a subroutine in solving other basic graph drawing problems: Graph drawing problems generally aim at computing an embedding of a graph G with respect to some optimization criterion. To draw G, one usually starts with a drawing of a planar subgraph H and then draws the remaining edges in $G - E(H)$, so it is natural to start with H that contains the maximum number of edges (for more detail, see [13] and references therein).

MPS is NP-hard, so past research has focused on approximation algorithms. Liu and Geldmacher showed that the problem is NP-hard [11], and later in SODA 1996, Călinescu et al. showed that it is in fact APX-hard [2]. By Euler's Formula, any planar graph has at most $3n - 6$ edges. This means that simply outputting a spanning tree gives immediately a $\frac{1}{3}$-approximation algorithm. Many heuristics were proposed [1,5,6], but they were shown to be unable to improve over the $\frac{1}{3}$-approximation guarantee.

To beat the "trivial" factor of $\frac{1}{3}$, Călinescu et al. (implicitly) proposed a framework of "augmenting" a spanning tree by *edge-disjoint* triangles. Indeed, adding one extra triangle to the spanning tree gives us one more edge, so it is natural to aim at adding as many triangles as possible. They showed that a simple algorithm based on greedily adding disjoint triangles gives a $\frac{7}{18}$-approximation and also devised a $\frac{4}{9}$-approximation by computing a maximum triangular structure[1] in G in polynomial time. The factor of $\frac{4}{9}$, however, is a limitation of this approach, as there is a graph G for which even a maximum triangular structure contains only a $\frac{4}{9}$-fraction of the number of edges in a maximum planar subgraph. Hence, the state of the art techniques on this problem have, more or less, reached their limitation.

1.1 Our Contributions

In this paper, we propose a new viewpoint that highlights the essences of known algorithmic results for MPS. We hope that, through this viewpoint, some promising directions towards improved approximation algorithms become visible. We not only give better explanations on previous results, but also suggest potential directions for breaking the $\frac{4}{9}$ barrier.

One message that was implicit in Călinescu et al.'s approach is that "the more triangles, the better the subgraph is as an approximate solution for MPS". They show nicely that if one only aims at maximizing disjoint triangles, then the problem can be reduced to the matroid parity problem, for which polynomial time algorithms exist. Some obvious questions arise:

> *Do we have any advantage in terms of approximating MPS if we want to maximize triangles that are allowed to overlap? Is the resulting optimization problem tractable?*

---

[1] A graph is a triangular structure if all cycles are of length three.

The goal of this paper is to address these two questions. We are able to provide partial answers in the restricted context of greedy algorithms.

*Our New Viewpoint: Maximum Planar Triangles (*Mpt*).* First, we quantify the connection between triangles and Mps by introducing a new optimization problem, that we call MAXIMUM PLANAR TRIANGLE (Mpt): Given a graph G, compute a subgraph H and its embedding with a maximum number of triangular faces.

Theorem 1. *If there is a β-approximation algorithm for* Mpt*, then there is* $\min(\frac{1}{2}, \frac{1}{3} + \frac{2\beta}{3} - O(\frac{1}{n}))$*-approximation algorithm for* Mps*.*

This shows, in particular, that a $\frac{1}{4}$-approximation for Mpt would immediately imply a $(\frac{1}{2} - O(1/n))$-approximation for Mps and that a $(\frac{1}{6} + \epsilon)$-approximation algorithm would suffice for improving the best known approximation factor.

We justify that Mpt is the right formulation to study by showing that many known algorithmic results can be obtained through approximating Mpt.

Theorem 2. *All greedy heuristics proposed in the literature are $\frac{1}{12}$ approximation for* Mpt*, thus yielding $\frac{7}{18} - O(1/n)$ approximation for* Mps*.*

The proof of this theorem is obtained via rather straightforward applications/modifications of the previous results. We only include them here for the sake of completeness and to provide exposition on the Mpt framework.

Unlike the question of finding disjoint triangles, overlapping triangles can be hard to compute, as shown in the following result:

Theorem 3. Mpt *is NP-hard.*

A New Greedy Framework: Match-and-Merge. The main goal of this paper is to study the advantages of overlapping triangles over the disjoint ones. To this end, we consider a class of greedy algorithms and analyze how greedy rules that allow overlapping triangles perform *strictly better* than the non-overlapping ones.

In particular, we introduce a systematic study of a greedy framework, that we call MATCH-AND-MERGE. Roughly speaking, the algorithms in this class iteratively find isomorphic copies of "pattern graphs" and merge connected components in the so far computed subgraph until no pattern can be applied. The algorithm in this class can be concisely described by a set of *merging rules* and the iterations to apply them. This class of algorithms is relatively rich: All known greedy algorithms can be cast concisely in this framework.

Theorem 4. *There is a greedy algorithm with approximation factor $\frac{1}{11}$ and $\frac{13}{33}$ for* Mpt *and* Mps *respectively.*

These ratios are better than the approximation ratios achieved by other algorithms of the same kind; all previous greedy algorithms did not perform better than $\frac{1}{12}$ for Mpt and $\frac{7}{18}$ for Mps respectively.

Related Results: More recently, [3] shows an approximation algorithm for weighted MPS in which we are given a weighted graph G, and the goal is to maximize the total weight of a planar subgraph of G. [4] considers maximum series-parallel subgraphs and gives a $\frac{7}{12}$-approximation.

Special cases of MPS also received attention, partly due to their connection to extremal graph theory. For instance, [7] shows that the problem is APX-hard even in cubic graphs. In [10], Kühn et al. showed a structural result that when the graph is dense enough (i.e. has large minimum degree), then there is a triangulated planar subgraph that can be computed in polynomial time. Therefore, MPS is polynomial-time solvable when the minimum degree is large. The proof of this result relies on Szemerédi's Regularity Lemma.

2   Preliminaries

Let $G = (V, E)$ be a graph. For any subset $S \subseteq V$, we use $G[S]$ to denote the induced subgraph of G on S. We denote by $V(G)$ and $E(G)$ the set of nodes and edges of G respectively. Moreover, if G is a plane graph we use $f(G)$ to denote the number of faces of G and by $f_j(G)$ the number of faces of G with a boundary containing exactly j edges. Let $t(G)$ denote the number of edges necessary to turn G into a maximal plane graph. By Euler's Formula it follows that $|E(G)| + t(G) = 3|V(G)| - 6$ and therefore $t(G)$ does not depend on the embedding of G. The following lemma was proven in [2].

Lemma 1 [2]. *For any plane graph G, $f_3(G) \geq 2|V(G)| - 4 - 2t(G)$.*

The following observation will be used in our proof of Theorem 1.

Lemma 2. *Let H be any connected subgraph of a connected plane graph. Then $|E(H)| \geq |V(H)| + f_3(H) - 2$.*

In the following sections, whenever we discuss MPS or MPT on a graph G, we will denote by OPT_{mps} the number of edges in a maximum planar subgraph H of G, and by OPT_{mpt} the maximum number of triangular faces in a plane subgraph H' of G.

3   Our New Concepts

In this section, we highlight our two new conceptual ideas: (i) the formulation of MPT and its connection to MPS and (ii) the MATCH-AND-MERGE framework. As discussed earlier, the MPT abstraction allows a cleaner analysis for algorithms in our framework, and therefore from this section on, we will focus on the case of MPT instead of MPS.

3.1 MPS ⇒ MPT

We show that an algorithm for MPT can be used for MPS. Let G be an input instance for MPS, and H be a planar subgraph of G that corresponds to the optimal solution. That is, $|E(H)| = \mathsf{OPT}_{mps}$. For simplicity, we abbreviate $|E(H)|$ and $|V(H)|$ by m and n respectively. We can write m in terms of $(1 + \gamma)n$ for some $\gamma \geq 0$.

By Euler's Formula $m = 3n - 6 - t(H)$, so $t(H) = (2 - \gamma)n - 6$. If we fix an embedding of H, then by Lemma 1, the number of triangular faces in H must be at least $2n - 4 - 2t(H) = 2n - 4 - 2(2 - \gamma)n + 12 > (2\gamma - 2)n$. This implies that $\mathsf{OPT}_{mpt} \geq (2\gamma - 2)n$. This term is only meaningful when $\gamma \geq 1$, so we distinguish between the following two cases that would imply Theorem 1.

- If $\mathsf{OPT}_{mps} < 2n$: This implies that any spanning tree is a $\frac{1}{2}$-approximation algorithm.
- Otherwise if $\mathsf{OPT}_{mps} \geq 2n$, then $\gamma \in [1, 2]$ (notice that γ can never be more than 2) and as argued above there are at least $(2\gamma - 2)n$ triangular faces in H. Then if we run a β-approximation algorithm for MPT, we will get a plane subgraph H' of G with $f_3(H') \geq \beta(2\gamma - 2)n$. We may assume that H' is connected: Otherwise, one can always add arbitrary edges to connect components without affecting planarity. By Lemma 2, $|E(H')| \geq \beta(2\gamma - 2)n + n - 2 = (1 + \beta(2\gamma - 2))n - 2$. The worst approximation factor is obtained by the infimum of the following term:

$$\inf_{\gamma \in [1,2]} \frac{1 + \beta(2\gamma - 2)}{1 + \gamma}.$$

To analyze this infimum, we first write a function $g(\gamma) = \frac{1 + \beta(2\gamma - 2)}{1 + \gamma}$. The derivative $\frac{dg}{d\gamma}$ can be written as $\frac{4\beta - 1}{(1 + \gamma)^2}$. As long as $\beta \in (0, 1/4]$, we have $\frac{dg}{d\gamma} < 0$, so this function is decreasing in γ. This means that the infimum is achieved at the maximum value of γ, i.e. at the boundary $\gamma = 2$. Plugging in $\gamma = 2$ gives the infimum as $\frac{1 + 2\beta}{3}$, leading to the approximation ratio of $\frac{1 + 2\beta}{3} - 2/n$, as desired.

3.2 MATCH-AND-MERGE

To achieve a $\frac{4}{9}$-approximation for MPS in [2] the authors reduce MPS to the linear matroid parity problem. The reduction is very simple but the process of picking the triangles is done by the black-box that solves the linear matroid parity problem. We introduce a class of simple greedy algorithms so that we can focus on studying the advantage of picking (potentially) overlapping triangles.

First, we formally define the term *merging rules*. Let G be an input graph. At any point of execution of the algorithm, let E' be a subset of edges in $E(G)$ that have been included so far and $\mathcal{C}$ be the connected components in $G' = (V(G), E')$.

Let H be a graph (that we refer to as pattern) and $\mathcal{P} = (V_1, V_2, \ldots, V_k)$ be a partition of $V(H)$. We say that an $(H, \mathcal{P})$-rule applies to G' if there is a subgraph H' in G that is isomorphic to H and such that, if we break H' into components based on $\mathcal{C}$ to obtain $U_1, \ldots, U_\ell$, then $\ell = k$ and $H'[U_i]$ is isomorphic to $H[V_i]$. When the rule is applied, all H-edges joining different components of $\mathcal{C}$ will be added. If $\mathcal{P}$ is a collection of singletons, we only use the abbreviation H-rule instead of $(H, \mathcal{P})$-rule: In this case, the rules would look for isomorphic copies of H where vertices come from different components in $\mathcal{C}$. Next we will show how previously proposed algorithms fit into this framework.

- **K_3-rule:** The K_3-rule, when applied to G', will merge three connected components $C_1, C_2, C_3 \in \mathcal{C}$ such that there are $v_1 \in C_1, v_2 \in C_2, v_3 \in C_3$ where $\{v_1, v_2, v_3\}$ induces K_3. This rule has been used in many algorithms. The CA_0 algorithm in [4] can be concisely described in our framework as follows: Iteratively apply K_3-rule until it cannot be applied any further.
- **Poranen's rule:** The $(K_3, \{\{1, 2\}, \{3\}\})$-rule would look for a triangle (v_1, v_2, v_3) such that an edge (v_1, v_2) belongs to one component $C_1 \in \mathcal{C}$ and vertex v_3 to another component $C_2 \in \mathcal{C}$. The purpose of this rule is obvious: It will create triangles that are not necessarily disjoint. This rule has been used in two algorithms, CA_1 and CA_2, suggested by Poranen [12]. Both CA_1 and CA_2 use the same set of rules, except that they differ in the conditions on which the rule is applied. Lemma 5 shows that having more rules does not necessarily improve the performance of a greedy algorithm.

Indeed, so far it was not clear whether there exists an $(H, \mathcal{P})$-rule that would improve over a $\frac{1}{12}$-approximation for MPT.

4   Analysis of Match-And-Merge Rules

We first analyze the approximation ratios of known algorithms in the context of MPT. We show that these algorithms give $\frac{1}{12}$-approximations, but not better. Next, we propose a new *diamond rule* that leads to an improvement.

4.1 K_3-rule Gives a $\frac{1}{12}$-approximation

The first algorithm we analyze for its performance in MPT was introduced in [2] as the first algorithm to exceed the trivial $\frac{1}{3}$-approximation ratio for MPS. Recall that the algorithm does the following: Repeatedly apply the K_3-rule until it cannot be applied anymore.

Lemma 3. *The approximation ratio of Algorithm CA_0 for MPT is $\frac{1}{12}$.*

Proof. Let $S_1 \subseteq E(G)$ denote the planar subgraph that CA_0 computed after the K_3-rule stops applying. Any component in S_1 is either a collection of triangular faces or just a single vertex. Let $\mathcal{C} = \{C_1, \ldots, C_r\}$ be a collection of all components in S_1 that contain at least one triangular face. Let p_i be the number of triangular faces found in component C_i, and p be the number of triangular faces found in S_1, so $p = \sum_{i=1}^{r} p_i$.

Let G^* be an optimal solution for MPT in G and G_i^* the plane subgraph of G^* induced on C_i. It is easy to make the following observation.

Proposition 1. *No triangle in G^* joins three different components of $\mathcal{C}$.*

Let $\Delta_{in}(C_i)$ denote the number of triangular faces in G^* that have all three vertices in $V(C_i)$ and let $\Delta_{out}(C_i)$ be the number of triangular faces in G^* with two vertices in $V(C_i)$ and a vertex not in $V(C_i)$. Then $\sum_{i=1}^{r}(\Delta_{in}(C_i) + \Delta_{out}(C_i)) = f_3(G^*)$, due to Proposition 1. Now notice that,

$$\frac{p}{f_3(G^*)} = \frac{\sum_{i=1}^{r} p_i}{\sum_{i=1}^{r}(\Delta_{in}(C_i) + \Delta_{out}(C_i))} \geq \min_i \frac{p_i}{\Delta_{in}(C_i) + \Delta_{out}(C_i)}.$$

Therefore, it suffices to show locally that $p_i/(\Delta_{in}(C_i) + \Delta_{out}(C_i)) \geq 1/12$.

Every edge in G_i^* can be incident to at most two triangular faces in G^*. By Euler's Formula there are at most $3|V(G_i^*)| - 6$ edges in G_i^*. Therefore $\Delta_{in}(C_i) + \Delta_{out}(C_i) \leq 6|V(G_i^*)| - 12$. Finally, the following simple lemma relates the number of triangles to the number of vertices in each component. The following lemma was proven in [2] as part of the analysis of CA_0.

Lemma 4 [2]. *Let X be a connected component constructed by iteratively merging three vertices in different connected components. Then we have $|V(X)| = 2p + 1$ where p is the number of triangles in X.*

This implies $|V(C_i)| = 2p_i + 1$ for all i. Therefore, $\Delta_{in}(C_i) + \Delta_{out}(C_i) \leq 6|V(C_i)| - 12 = 12p_i + 6 - 12 = 12p_i - 6$, and $\frac{p_i}{\Delta_{in}(C_i) + \Delta_{out}(C_i)} \geq \frac{1}{12}$ for every i.

Bad Example: We conclude this subsection by giving an example where CA_0 does not achieve a ratio better than $\frac{1}{12}$. To do this it suffices to bound $p_i/(\Delta_{in}(C_i) + \Delta_{out}(C_i))$ locally. The example is the same as the one used in [2] to show that CA_0 cannot exceed $\frac{7}{18}$ for MPS. Consider a sequence of k triangles, where the consecutive ones are joined by a vertex. Call this plane graph H. This is supposed to represent a component in the solution given by CA_0. Then the bad example H'' is obtained as follows: First triangulate H to get H'. Then for each face of H' add a vertex inside the face and connect it to the three bounding vertices of the face. The final graph is denoted by H''. Let $|V(H)| = n$. The number of triangles in H is exactly $k = (n-1)/2$. The number of triangular faces in H' is $2n - 4$, and each face of H' will give rise to three faces in H''. So $f_3(H'') = 3(2n - 4) = 6n - 12$. Clearly $\lim_{n \to \infty} \frac{(n-1)/2}{6n-12} = \frac{1}{12}$.

The second algorithm in [2], differs from CA_0 in the fact that it does not just find any triangular structure but a maximum triangular structure in G.

4.2   Other Proposed Rules Do Not Break $\frac{1}{12}$-Approximation

In this subsection we describe two algorithms CA_1 and CA_2 given in [12] by Poranen and analyze them for their performance in MPT as well as MPS. These algorithms are basically the following (in MATCH-AND-MERGE framework):

1. Check if $(K_3, \{\{1,2\}, \{3\}\})$-rule applies.
2. If not, check if K_3-rule applies.
3. If at least one of the rules applies, go back to (1).

It is easy to see that these algorithms perform at least as good as CA_0 for MPT (i.e. at least $\frac{1}{12}$-approximation for MPT). Based on their empirical successes, performed in [12], the author conjectured that they can even reach a $\frac{4}{9}$-approximation ratio in MPS matching the currently best known algorithm given in [2]; this would hint to a $\frac{1}{6}$-approximation for MPT.

We show a bad example where both algorithms can be as bad as a $\frac{1}{12}$-approximation for MPT and a $\frac{7}{18}$-approximation for MPS. This refutes the possibility of improvements over $\frac{7}{18}$ using CA_1 and CA_2 in MPS.

Lemma 5. *There is a graph G such that running CA_1 or CA_2 on G may yield at most $\frac{1}{12}OPT_{mpt}$ triangular faces, and $\frac{7}{18}OPT_{mps}$ edges.*

Due to space limitation, the bad example is deferred to the full version.

4.3 D_4 and K_3 Lead to a $\frac{1}{11}$-Approximation

We now propose a new rule that leads to a better approximation ratio. Let D_4 be the diamond graph (i.e. K_4 with one edge removed). This pattern graph intuitively captures the ideas of having two triangles sharing an edge. Our algorithm CA_3 proceeds in the following steps:

1. Keep applying D_4-rule until it cannot be applied any further.
2. Keep applying K_3-rule until it cannot be applied.

Now we analyze the performance of CA_3 in MPT. Let H be an optimal solution for MPT on a given graph G. Let $G' = (V, E')$ be the subgraph of G with E' as computed by CA_3 after leaving the second loop and $\mathcal{C} = \{C_1, \ldots, C_r\}$ be the collection of connected components in G'. Let $\mathcal{C}'$ be the connected components in G' formed after leaving the first loop; we call them *dense components*. (Notice that the components formed by diamonds are denser than those formed by adding triangles.) Notice that components in $\mathcal{C}$ are obtained by combining components in $\mathcal{C}'$. The following properties hold at the end of the algorithm.

Proposition 2.

- *For any four distinct dense components $X, Y, Z, W \in \mathcal{C}'$ and four vertices $x \in X, y \in Y, z \in Z, w \in W$, the induced subgraph $G[\{x,y,z,w\}]$ is not a diamond.*
- *For any three distinct components $X, Y, Z \in \mathcal{C}$ and three vertices $x \in X, y \in Y, z \in Z$, the induced subgraph $G[\{x,y,z\}]$ is not a triangle.*

For some connected component C in $\mathcal{C}$, we denote by $\Delta_{in}(C)$ the number of triangular faces in H whose three vertices belong to the induced subgraph $G[C]$. In addition we denote by $\Delta_{out}(C)$ the number of triangular faces in H that have an edge in $G[C]$ and one vertex in $V \setminus V(C)$. The following lemma follows easily.

Lemma 6. $f_3(H) = \sum_{C \in \mathcal{C}} (\Delta_{in}(C) + \Delta_{out}(C))$.

For a fixed component $C \in \mathcal{C}$, let $\Delta(C)$ denote the sum $\Delta_{in}(C) + \Delta_{out}(C)$.

$$\Delta(C) = \Delta_{in}(C) + \Delta_{out}(C) = (3\Delta_{in}(C) + \Delta_{out}(C)) - 2\Delta_{in}(C) \leq 2|E(H[C])| - 2\Delta_{in}(C).$$

The last inequality follows from the fact that each triangle contributing to $\Delta_{in}(C)$ uses three edges in C, while triangles in $\Delta_{out}(C)$ use only one edge.

Diamond Clusters and Triangular Cacti: Fix some component C of $\mathcal{C}$. We can break C into several parts based on the structure of $\mathcal{C}'$. Let $\mathcal{D}_C$ be the collection of non-singleton dense connected components in C, i.e. $\mathcal{D}_C = \{C' \in \mathcal{C}' : C' \subseteq C \text{ and } |V(C')| > 1\}$. Each non-singleton subcomponent $X \in \mathcal{D}_C$ is called a *diamond cluster* inside C; notice that $|V(X)| \geq 4$. Let $F = E(G'[C]) \setminus (\bigcup_{X \in \mathcal{D}_C} E(G'[X]))$ be the edges remaining after removing edges in induced subgraphs of components in $\mathcal{D}_C$. Observe that the graph (C, F) consists of connected components that are formed by applying the K_3-rule. Let $\mathcal{T}_C$ be such a collection of non-singleton connected components. Each $Y \in \mathcal{T}_C$ is a connected triangular cactus in the component C. Notice that the components in $\mathcal{D}_C$ are disjoint, and the same holds for $\mathcal{T}_C$. For each $X \in \mathcal{D}_C$ and $Y \in \mathcal{T}_C$, let $c(X)$ and $l(Y)$ be the number of triangles in $G'[X]$ and that in $G'[Y]$ respectively.

Proposition 3. $C \subseteq (\bigcup_{X \in \mathcal{D}_C} X) \cup (\bigcup_{Y \in \mathcal{T}_C} Y)$

Now we want to express the number of vertices $|V(C)|$ in terms of the sizes of the connected triangular cacti and diamond clusters in C. Let I be the set of vertices such that every vertex in I belongs to some diamond cluster X and to some connected triangular cactus Y. Then $|V(C)|$ is given by

$$|V(C)| = \sum_{X \in \mathcal{D}_C} |V(X)| + \sum_{Y \in \mathcal{T}_C} |V(Y)| - |I|$$

Lemma 7. For each diamond cluster $X \in \mathcal{D}_C$, we have $|V(X)| = 1 + \frac{3}{2}c(X)$.

Invoking Lemma 4, for each connected cactus $Y \in \mathcal{T}_C$, we have $|V(Y)| = 2l(Y) + 1$. Let $t = |\mathcal{T}_C|$ and $k = |\mathcal{D}_C|$. We now bound $|I|$ in terms of k and t.

Lemma 8. $|I| = k + t - 1$.

From these observations we derive an upper bound on the number of vertices in any component C, in terms of p (the number of triangles inside the diamond clusters) and l (the number of triangles inside the connected triangular cacti).

$$\begin{aligned}
|V(C)| &= \sum_{X \in \mathcal{D}_C} |V(X)| + \sum_{Y \in \mathcal{T}_C} |V(Y)| - |I| \\
&= \sum_{X \in \mathcal{D}_C} \left(1 + \frac{3}{2}c(X)\right) + \sum_{Y \in \mathcal{T}_C} (2l(Y) + 1) - k - t + 1 \\
&= \left(k + \frac{3}{2}p\right) + (t + 2l) - k - t + 1 = \frac{3}{2}p + 2l + 1.
\end{aligned}$$

The following is the main lemma that crucially exploits the new diamond rule.

Lemma 9. $\Delta_{out}(C) \leq 15p + 8l - 6k$.

Proof. Each triangle that contributes to $\Delta_{out}(C)$ must have an edge that appears in $H[C]$; we call them *supporting edges*. Let E^* be the set of such edges. Denote by E_1^* the set of supporting edges whose two endpoints belong to the same diamond cluster $X \in \mathcal{D}_C$. Let E_2^* denote $E^* \setminus E_1^*$.

Claim. The subgraph $(V(C), E_2^*)$ is triangle-free.

Now since $(V(C), E_2^*)$ is triangle-free, Euler's formula together with the upper bound on $|V(C)|$ imply that $|E_2^*| \leq 2|V(C)| - 4 \leq 3p + 4l - 2$. Moreover, we can bound the edges in E_1^* by applying Euler's formula to each diamond cluster $X \in \mathcal{D}_C$. That is, $|E_1^*| \leq \sum_{X \in \mathcal{D}_C}(3|V(X)| - 6) = \sum_{X \in \mathcal{D}_C} \left(\frac{9}{2}c(X) - 3\right) = \frac{9}{2}p - 3k$. Next, $\Delta_{out}(C) \leq 2|E^*|$ since each edge in E^* can only support at most two triangles. Plugging in the values of $|E_1^*|$ and $|E_2^*|$ gives

$$\Delta_{out}(C) \leq 2(|E_1^*| + |E_2^*|) \leq 2(\frac{9}{2}p - 3k + 3p + 4l - 2) \leq 15p + 8l - 6k.$$

Lemma 10. CA_3 *gives a* $\frac{1}{11}$-*approximation for* MPT.

Proof. We will bound the approximation ratio locally, i.e. for each connected component C, we argue that $p + l \geq \frac{1}{11}\Delta(C)$, which will imply that when summing over all components in $\mathcal{C}$ the number of triangles is at least $\frac{1}{11}f_3(H)$. Using Euler's formula, we get

$$\Delta \leq 2|E(H[C])| - 2\Delta_{in} \leq 6|V(H[C])| - 12 - 2\Delta_{in} \leq 9p + 12l - 6 - 2\Delta_{in} \quad (1)$$

The first inequality follows by a simple counting argument. Note that the last inequality follows from the fact that $|V(H[C])| \leq \frac{3}{2}p + 2l + 1$. From Lemma 9, we have that

$$\Delta = \Delta_{in} + \Delta_{out} \leq 15p + 8l - 6k + \Delta_{in}. \quad (2)$$

Adding (1) with twice of (2) gives us $3\Delta \leq 39p + 28l$, which implies that $\Delta \leq 13p + 10l$. Finally, we can combine this with (1) to get $\Delta \leq 11(p + l)$.

5   Hardness of Maximum Planar Triangles

In this section, we prove that MPT is NP-hard, as a by-product simplifying the MPS NP-hardness proof by Liu and Geldmacher [11]. Our reduction is from the Hamiltonian path problem in bipartite graphs. In [9], it is shown that the Hamiltonian cycle problem in bipartite graphs is NP-complete; it follows easily that the same holds for the Hamiltonian path problem.

Construction: Let G be an instance of the Hamiltonian path problem, i.e. G is a connected bipartite graph with n vertices. Note that G is triangle-free. Let G' be a copy of G, augmented with two vertices s and t, where s and t are both connected to every vertex in $V(G)$; we call the edges that connect vertices in G to $\{s, t\}$ *auxiliary edges*. More formally, $V(G') = V(G) \cup \{s, t\}$ and $E(G') = E(G) \cup \{(s, v) : v \in V(G)\} \cup \{(t, v), v \in V(G)\}$.

Analysis: We argue that there exists a spanning subgraph H of G' and an embedding ϕ_H of H with $2n - 2$ triangular faces, if and only if G has a Hamiltonian path. First assume that G has a Hamiltonian path P. We show how to construct a spanning subgraph H of G', that has an embedding ϕ_H with $2n - 2$ triangular faces. Let $V(H) = V(P) \cup \{s, t\}$ and $E(H) = E(P) \cup \{(s, v) : v \in V(P)\} \cup \{(t, v) : v \in V(P)\}$. For ϕ_H simply draw P on the plane on a vertical line, placing s and t on the left and right side of the line respectively.

To prove the converse, now assume that there exists a spanning subgraph H of G' and an embedding ϕ_H of H with at least $2n - 2$ triangular faces. Notice that each triangular face in H must be formed by an edge in $E(G)$ (called *supporting edge*) together with two auxiliary edges as G is triangle-free. Denote by $H' = H \setminus \{s, t\}$, which is a subgraph of G. We will show that there exists a Hamiltonian path in H' and therefore also in G.

Let E_s and E_t be the sets of edges in H' that support triangles formed with s and t in H respectively. Notice that the number of triangles in ϕ_H is $|E_s| + |E_t|$. We need the following structural lemma.

Lemma 11. *The subgraph $(V(G), E_s)$ (respectively $(V(G), E_t)$) of H' has the following properties:*

i. The maximum degree of a vertex in $(V(G), E_s)$ is at most two.
ii. If $(V(G), E_s)$ contains a cycle C, then $E_s \setminus E(C) = \emptyset$.

Proof. We first prove (i). Assume otherwise that some vertex v is adjacent to three supporting edges vv_1, vv_2, vv_3 for s. Suppose that the triangular faces (s, v, v_1) and (s, v, v_2) are adjacent in ϕ_H, sharing the edge sv. Then the triangle (s, v, v_3) cannot be a face, as it must contain one of the two faces in $\{(s, v, v_1), (s, v, v_2)\}$, a contradiction.

For (ii) note that every edge in E_s is incident to at least one triangular face in H. Assume now that E_s contains a cycle C and $E_s \setminus E(C) \neq \emptyset$. As $E_s \subseteq E(H') \subseteq E(G)$ and G is bipartite $|V(C)| \geq 4$. Note that by planarity s and the edges in $E_s \setminus E(C)$ must be embedded on the same side of C (inside or outside of C). Once we embed C, s and all auxiliary edges between C and s, every edge in $E(C)$ is incident to a triangular face (one of which is the outer face of the current graph) formed with the auxiliary edges and the face on the other side of C. Embedding any edge of $E_s \setminus E(C)$ on the same side as s and adding the auxiliary edges from its endpoints to s results in destroying one of these triangular faces. ∎

Lemma 11 implies that all subgraphs in H' induced by the endpoints of supporting edges for s (or t) must either be a disjoint union of paths or a cycle. Therefore E_s and E_t contribute at most n edges each to the triangular faces in H. At the same time we know that in order to form at least $2n - 2$ triangular faces in ϕ_H, one of them must have size at least $n - 1$. To complete the proof we consider the possible compositions of edges from E_s and E_t in H':

- If E_s or E_t induces a cycle C of length n, G contains a Hamiltonian path.
- If one of E_s and E_t has size at least $n - 1$ and at the same time induces a single path in H', then this path is also a Hamiltonian path in G.
- It remains to analyze the case where both E_s and E_t induce a cycle of length $n - 1$ in H'. Let C be the cycle induced by E_s in H' and u be the vertex in $V(G) \setminus V(C)$. As G is connected there is a vertex v in C that is a neighbor of u in G. Let P be a path starting in u and ending in one of the neighbors of v in C. Clearly, P is a Hamiltonian path in G.

Acknowledgement. We are grateful to an anonymous reviewer, whose detailed suggestions contributed to a clearer presentation of this work.

References

1. Cai, J., Han, X., Tarjan, R.E.: An $O(m \log n)$-time algorithm for the maximal planar subgraph problem. SIAM J. Comput. **22**(6), 1142–1162 (1993)
2. Călinescu, G., Fernandes, C.G., Finkler, U., Karloff, H.: A better approximation algorithm for finding planar subgraphs. J. Algorithms **27**(2), 269–302 (1998)
3. Călinescu, G., Fernandes, C.G., Karloff, H.J., Zelikovsky, A.: A new approximation algorithm for finding heavy planar subgraphs. Algorithmica **36**(2), 179–205 (2003)
4. Călinescu, G., Fernandes, C.G., Kaul, H., Zelikovsky, A.: Maximum series-parallel subgraph. Algorithmica **63**(1–2), 137–157 (2012)
5. Chiba, T., Nishioka, I., Shirakawa, I.: An algorithm of maximal planarization of graphs. In: Proceedings of IEEE Symposium on Circuits and Systems, pp. 649–652 (1979)
6. Cimikowski, R., Coppersmith, D.: The sizes of maximal planar, outerplanar, and bipartite planar subgraphs. Discret. Math. **149**(1–3), 303–309 (1996)
7. Faria, L., De Figueiredo, C.M.H., Mendonça, C.F.X.: On the complexity of the approximation of nonplanarity parameters for cubic graphs. Discret. Appl. Math. **141**(1), 119–134 (2004)
8. Gonzalez, T.F.: Handbook of Approximation Algorithms and Metaheuristics. CRC Press, Boca Raton (2007)
9. Krishnamoorthy, M.S.: An NP hard problem in bipartite graphs. ACM SIGACT News **7**(1), 26 (1975)
10. Kühn, D., Osthus, D., Taraz, A.: Large planar subgraphs in dense graphs. J. Comb. Theory, Ser. B **95**(2), 263–282 (2005)
11. Liu, P.C., Geldmacher, R.C.: On the deletion of nonplanar edges of a graph. In: Proceedings of 10th Southeastern Conference on Combinatorics, Graph Theory, and Computing, pp. 727–738 (1977)
12. Poranen, T.: Two new approximation algorithms for the maximum planar subgraph problem. Acta Cybern. **18**(3), 503–527 (2008)
13. Tamassia, R.: Handbook of Graph Drawing and Visualization. CRC Press, Boca Raton (2013)

An Approximation Algorithm for Maximum Internal Spanning Tree

Zhi-Zhong Chen[1]([✉]), Youta Harada[1], Fei Guo[2], and Lusheng Wang[3]

[1] Division of Information System Design, Tokyo Denki University,
Hatoyama, Saitama 350-0394, Japan
zzchen@mail.dendai.ac.jp
[2] School of Computer Science and Technology, Tianjin University,
No. 135, Yaguan Road, Tianjin, China
guofeieileen@163.com
[3] Department of Computer Science, City University of Hong Kong,
Tat Chee Avenue, Kowloon, Hong Kong SAR
cswangl@cityu.edu.hk

Abstract. Given a graph G, the *maximum internal spanning tree problem* (MIST for short) asks for computing a spanning tree T of G such that the number of internal vertices in T is maximized. MIST has possible applications in the design of cost-efficient communication networks and water supply networks and hence has been extensively studied in the literature. MIST is NP-hard and hence a number of polynomial-time approximation algorithms have been designed for MIST in the literature. The previously best polynomial-time approximation algorithm for MIST achieves a ratio of $\frac{3}{4}$. In this paper, we first design a simpler algorithm that achieves the same ratio and the same time complexity as the previous best. We then refine the algorithm into a new approximation algorithm that achieves a better ratio (namely, $\frac{13}{17}$) with the same time complexity. Our new algorithm explores much deeper structure of the problem than the previous best. As our recent $\frac{1}{2}$-approximation algorithm for the weighted version of the problem shows, the discovered structure may be used to design better algorithms for related problems.

1 Introduction

The *maximum internal spanning tree problem* (MIST for short) requires the computation of a spanning tree T in a given graph G such that the number of internal vertices in T is maximized. MIST has possible applications in the design of cost-efficient communication networks [7] and water supply networks [1]. Unfortunately, MIST is clearly NP-hard because the problem of finding a Hamiltonian path in a given graph is NP-hard and can be easily reduced to MIST. MIST is in fact APX-hard [5] and hence does not admit a polynomial-time approximation scheme.

L. Wang—Supported by NSFC (61373048) and GRF (CityU 123013) grants.

S.-H. Poon et al. (Eds.): WALCOM 2017, LNCS 10167, pp. 385–396, 2017.
DOI: 10.1007/978-3-319-53925-6_30

Since MIST is APX-hard, quite a number of polynomial-time approximation algorithms for MIST and its special cases have been designed (see [5] and a long list of references therein). The previously best polynomial-time approximation algorithm for MIST achieves a ratio of $\frac{3}{4}$ [5]. Unlike the other previously known approximation algorithms for MIST, the algorithm in [5] is based on a simple but crucial observation that the maximum number of internal vertices in a spanning tree of a graph G can be bounded from above by the maximum number of edges in a triangle-free path-cycle cover of G.

In the weighted version of MIST (WMIST for short), each vertex of the given graph G has a nonnegative weight and the objective is to find a spanning tree T of G such that the total weight of internal vertices in T is maximized. A number of polynomial-time approximation algorithms for WMIST and its special cases have been designed (see [2] and the references therein). The best known polynomial-time approximation algorithm for WMIST achieves a ratio of $\frac{1}{2}$ [2]. In the parameterized version of MIST (PMIST for short), we are asked to decide whether a given graph G has a spanning tree with at least a given number k of internal vertices. PMIST and its special cases and variants have also been extensively studied in the literature (see [6] and a long list of references therein). The best known kernel for PMIST is of size $2k$ and it leads to the fastest known algorithm for PMIST with running time $O(4^k n^{O(1)})$ [6].

In this paper, we first give a new approximation algorithm for MIST that is simpler than the one in [5] but achieves the same approximation ratio and time complexity. In more details, the time complexity is dominated by that of computing a maximum triangle-free path-cycle cover in a graph. We then show that the algorithm can be refined into a new approximation algorithm for MIST that has the same time complexity as the algorithm in [5] but achieves a better ratio (namely, $\frac{13}{17}$). To obtain our algorithm, we use three new main ideas. The first main idea is to bound the maximum number of internal vertices in a spanning tree of a graph G by the maximum number of edges in a *special* (rather than general) triangle-free path-cycle cover of G. Roughly speaking, we can figure out that certain vertices in G must be leaves in an optimal spanning tree of G, and hence we can require that the degrees of these vertices be at most 1 when computing a maximum triangle-free path-cycle cover $\mathcal{C}$ of G. In this sense, $\mathcal{C}$ is special and can have significantly fewer edges than a maximum (general) triangle-free path-cycle cover of G, and hence gives us a tighter upper bound. The second idea is to carefully modify $\mathcal{C}$ into a spanning tree T by local improvement. Unfortunately, we can not always guarantee that the number of internal vertices in T is at least $\frac{13}{17}$ times the number of edges in $\mathcal{C}$. Our third idea is to show that if this unfortunate case occurs, then an optimal spanning tree of G cannot have so many internal vertices. Although the improvement of approximation ratio may not look so significant, our ideas may be used to design even better approximation or parameterized algorithms for MIST and related problems in the future. Indeed, we have recently applied some of the ideas to obtain a polynomial-time approximation algorithm for WMIST that achieves a ratio of $\frac{1}{2}$ [2], significantly improving the previous best (namely, $\frac{1}{3} - \epsilon$ [4]).

2   Basic Definitions

Throughout this chapter, a graph means a simple undirected graph (i.e., it has neither parallel edges nor self-loops).

Let G be a graph. We denote the vertex set of G by $V(G)$, and denote the edge set of G by $E(G)$. For a subset U of $V(G)$, $G - U$ denotes the graph obtained from G by removing the vertices in U (together with the edges incident to them), while $G[U]$ denotes $G - (V(G) \setminus U)$. We call $G[U]$ the *subgraph of G induced by U*. For a subset F of $E(G)$, $G - F$ denotes the graph obtained from G by removing the edges in F. An edge e of G is a *bridge* of G if $G - \{e\}$ has more connected components than G, and is a *non-bridge* otherwise. A vertex v of G is a *cut-point* if $G - \{v\}$ has more connected components than G.

Let v be a vertex of G. The *neighborhood* of v in G, denoted by $N_G(v)$, is $\{u \mid \{v, u\} \in E(G)\}$. The *degree* of v in G, denoted by $d_G(v)$, is $|N_G(v)|$. If $d_G(v) = 0$, then v is an *isolated* vertex of G. If $d_G(v) \leq 1$, then v is a *leaf* of G; otherwise, v is a *non-leaf* of G. We use $L(G)$ to denote the set of leaves in G.

Let H be a subgraph of G. $N_G(H)$ denotes $\bigcup_{v \in V(H)} N_G(v) \setminus V(H)$. A *port* of H is a $u \in V(H)$ with $N_G(u) \setminus V(H) \neq \emptyset$. When H is a path, H is *dead* if neither endpoint of H is a port of H, while H is *alive* otherwise. H and another subgraph H' of G are *adjacent* in G if $V(H) \cap V(H') = \emptyset$ but $N_G(H) \cap V(H') \neq \emptyset$ (or equivalently, $N_G(H') \cap V(H) \neq \emptyset$).

A *cycle* in G is a connected subgraph of G in which each vertex is of degree 2. A *path* in G is either a single vertex of G or a connected subgraph of G in which exactly two vertices are of degree 1 and the others are of degree 2. A vertex v of a path P in G is an *endpoint* of P if $d_P(v) \leq 1$, and is an *internal vertex* of P if $d_P(v) = 2$. The *length* of a cycle or path C is the number of edges in C and is denoted by $|C|$. A *k-cycle* is a cycle of length k, while a *k-path* is a path of length k. A *tree* (respectively, *cycle*) *component* of G is a connected component of G that is a tree (respectively, cycle). In particular, if a tree component T of G is indeed a path (respectively, k-path), then we call T a *path* (respectively, *k-path*) *component* of G.

A *tree-cycle cover* (TCC for short) of G is a subgraph H of G such that $V(H) = V(G)$ and each connected component of H is a tree or cycle. Let H be a TCC of G. H is a *Hamiltonian path* (respectively, *cycle*) of G if H is a path (respectively, cycle), and is a *spanning tree* of G if H is a tree. H is a *path-cycle cover* (PCC for short) of G if each tree component of H is a path. H is a *path cover* of G if H has only path components. A *triangle-free* TCC (TFTCC for short) of G is a TCC without 3-cycles. Similarly, a *triangle-free* PCC (TFPCC for short) of G is a PCC without 3-cycles. A TFPCC of G is *maximum* if its number of edges is maximized over all TFPCCs of G. For convenience, let $t(n, m)$ denote the time complexity of computing a maximum TFPCC in a graph with n vertices and m edges. It is known that $t(n, m) = O(n^2 m^2)$ [3].

Suppose G is connected. The *weight* of a spanning tree T of G, denoted by $w(T)$, is the number of non-leaves in T. We use $opt(G)$ to denote the maximum weight of a spanning tree of G. An *optimal spanning tree* (OST for short) of G is a spanning tree T of G with $w(T) = opt(G)$.

3   A Simple $\frac{3}{4}$-Approximation Algorithm

Throughout the remainder of this paper, G means a connected graph for which we want to find an OST. Moreover, T denotes an OST of G. For convenience, let $n = |V(G)|$ and $m = |E(G)|$.

3.1   Reduction Rules

We want to make G smaller (say, by deleting one or more vertices or edges from G) without decreasing $opt(G)$. For this purpose, we define two *strongly safe* operations on G below. Here, an operation on G is *strongly safe* if performing it on G does not change $opt(G)$.

Operation 1. If $|V(G)| > 3$ and $E(G)$ contains two edges $\{u_1, v\}$ and $\{u_2, v\}$ such that both u_1 and u_2 are leaves of G, then delete u_2.

Operation 2. If for a non-bridge $e = \{u_1, u_2\}$ of G, $G - \{u_i\}$ has a connected component K_i with $u_{3-i} \notin V(K_i)$ for each $i \in \{1, 2\}$, then delete e. (*Comment:* When $|V(K_1)| = |V(K_2)| = 1$, Li and Zhu [5] showed that Operation 2 is strongly safe.)

Li and Zhu [5] showed that Operation 1 is strongly safe. We can show that Operation 2 is strongly safe.

An operation on G is *weakly safe* if performing it on G yields one or more graphs $G_1, \ldots, G_k$ such that (1) $|V(G)| \geq \sum_{i=1}^{k} |V(G_i)|$, $|E(G)| \geq \sum_{i=1}^{k} |E(G_i)|$, and $|V(G)| + |E(G)| > \sum_{i=1}^{k} |V(G_i)| + \sum_{i=1}^{k} |E(G_i)|$, (2) $opt(G) \leq \sum_{i=1}^{k} opt(G_i) + c$ for some nonnegative integer c, and (3) given a spanning tree T_i for each G_i, a spanning tree T of G with $w(T) \geq \sum_{i=1}^{k} w(T_i) + c$ can be computed in linear time. Note that the last two conditions in the definition imply that $opt(G) = \sum_{i=1}^{k} opt(G_i) + c$.

Operation 3. If G has a bridge $e = \{u_1, u_2\}$ such that for each $i \in \{1, 2\}$, u_i is a cut-point in the connected component G_i of $G - e$ with $u_i \in V(G_i)$, then obtain G_1 and G_2 as the connected components of $G - e$.

Operation 4. If G has a cut-point v such that one connected component K of $G - \{v\}$ has at least two but at most 8 vertices, then obtain G_1 from $G - V(K)$ by adding a new vertex u and a new edge $\{v, u\}$.

The number 8 in the definition of Operation 4 is not essential. It can be chosen at one's discretion as long as it is a constant. We here choose the number 8, because it will be the smallest number for the proofs of several lemmas in this paper to go through. We can show Operations 3 and 4 are weakly safe.

An operation on G is *safe* if it is strongly or weakly safe on G.

3.2   The Algorithm

As in [5], the algorithm is based on a lemma which says that G has a path cover $\mathcal{P}$ such that $opt(G)$ is bounded from above by the number of edges in $\mathcal{P}$. We next state the lemma in a stronger form and give an extremely simple proof.

Lemma 1. *Given a spanning tree $\tilde{T}$ of G, we can construct a path cover $\mathcal{P}$ of G such that $|E(\mathcal{P})| \geq w(\tilde{T})$ and $d_{\mathcal{P}}(v) \leq 1$ for each leaf v of $\tilde{T}$.*

Proof. We construct $\mathcal{P}$ from $\tilde{T}$ by rooting $\tilde{T}$ at an arbitrary non-leaf and then for each non-leaf u of $\tilde{T}$, deleting all but one edge between u and its children. ∎

Now, the outline of the algorithm is as follows.

1. Whenever there is an $i \in \{1,2\}$ such that Operation i can be performed on G, then perform Operation i on G.
2. Whenever there is an $i \in \{3,4\}$ such that Operation i can be performed on G, then perform the following steps:
   (a) Perform Operation i on G. Let $G_1, \ldots, G_k$ be the resulting graphs.
   (b) For each $j \in \{1, \ldots, k\}$, compute a spanning tree T_j of G_j recursively.
   (c) Combine $T_1, \ldots, T_k$ into a spanning tree $\tilde{T}$ of G such that $w(\tilde{T}) \geq \sum_{i=1}^{k} w(T_i) + c$.
   (d) Return $\tilde{T}$.
3. If $|V(G)| \leq 8$, then compute and return an OST of G in $O(1)$ time.
4. Compute a maximum TFPCC $\mathcal{C}$ of G. (*Comment:* By Lemma 1, $opt(G) \leq |E(\mathcal{C})|$).
5. Perform a preprocessing on $\mathcal{C}$ without decreasing $|E(\mathcal{C})|$.
6. Transform $\mathcal{C}$ into a spanning tree $\tilde{T}$ of G such that $w(\tilde{T}) \geq \frac{3}{4}|E(\mathcal{C})|$.
7. Return $\tilde{T}$.

Only Steps 5 and 6 are unclear. So, we detail them below. First, Step 5 is done by performing the next three operations until none of them is applicable.

Operation 5. If $\mathcal{C}$ has a dead path component P such that $2 \leq |P| \leq 4$ and $G[V(P)]$ has an alive Hamiltonian path Q, then replace P by Q.

Operation 6. If an endpoint u of a path component P of $\mathcal{C}$ is adjacent to a vertex v of a cycle C of $\mathcal{C}$ in G, then combine P and C into a single path by replacing one edge incident to v in C with the edge $\{u, v\}$.

Operation 7. If an endpoint u_1 of a path component P_1 of $\mathcal{C}$ is adjacent to an internal vertex u_2 of another path component P_2 in G such that one edge e' incident to u_2 in P_2 satisfies that combining P_1 and P_2 by replacing e' with the edge $\{u_1, u_2\}$ yields two paths Q_1 and Q_2 with $\max\{|Q_1|, |Q_2|\} > \max\{|P_1|, |P_2|\}$, then replace P_1 and P_2 by Q_1 and Q_2. (*Comment:* For each $i \in \{5,6,7\}$, Operation i does not change the maximality of $\mathcal{C}$. Due to the maximality, no endpoint of a path component P_1 of $\mathcal{C}$ is adjacent to an endpoint of another path component P_2 in G.)

Lemma 2. *Immediately after Step 5, the following hold:*

1. *C is a maximum TFPCC of G and hence has at least $opt(G)$ edges.*
2. *If a path component P of C is of length at most 3, then P is alive.*
3. *If an endpoint v of a path component P of C is a port of P, then each vertex in $N_G(v) \setminus V(P)$ is an internal vertex of a path component Q of C with $|Q| \geq 2|P| + 2$.*

We next detail Step 6. First, for each path component P of C with $1 \leq |P| \leq 3$, we select one edge $e_P \in E(G)$ connecting an endpoint of P to a vertex not in P, and add e_P to an initially empty set M. Such e_P exists by Statement 2 in Lemma 2. Moreover, by Statement 3 in Lemma 2, the endpoint of e_P not in P appears in a path component Q of C with $|Q| \geq 4$. So, for two path components P_1 and P_2 in C, $e_{P_1} \neq e_{P_2}$. Consider the graph H obtained from C by adding the edges in M. Each connected component of H is a cycle of length at least 4 or a tree. Suppose we modify H by performing the following three steps in turn:

– Whenever H has two cycles C_1 and C_2 such that some edge $e = \{u_1, u_2\} \in E(G)$ satisfies $u_1 \in V(C_1)$ and $u_2 \in V(C_2)$, delete one edge of C_1 incident to u_1 from H, delete one edge of C_2 incident to u_2 from H, and add e to H.
– Whenever H has a cycle C, choose an edge $e = \{u, v\} \in E(G)$ with $u \in V(C)$ and $v \notin V(C)$, delete one edge of C incident to u from H, and add e to H.
– Whenever H has two connected components C_1 and C_2 such that some edge $e = \{u_1, u_2\} \in E(G)$ satisfies $u_1 \in V(C_1)$ and $u_2 \in V(C_2)$, add e to H.

Step 6 is done by obtaining $\tilde{T}$ as the final modified H. Obviously, for each cycle C of C, at least $|C| - 1 \geq \frac{3}{4}|C|$ vertices of C are internal vertices of $\tilde{T}$. Moreover, for each path component P of C with $|P| \geq 4$, at least $|P| - 1 \geq \frac{3}{4}|P|$ vertices of P are internal vertices of $\tilde{T}$. Furthermore, for each path component P of C with $1 \leq |P| \leq 3$, at least $|P|$ vertices of P are internal vertices of $\tilde{T}$. So, $\tilde{T}$ has at least $\frac{3}{4}|E(C)|$ internal vertices. Obviously, all steps of the algorithm excluding Steps 2b and 4 can be done in $O(|E(G)|^2)$ time. Now, we have:

Theorem 1. *The algorithm achieves an approximation ratio of $\frac{3}{4}$ and runs in $O(m^2) + t(n, m)$ time.*

We improve the algorithm below. The first idea is to introduce more safe reduction rules (cf. Sect. 4). The second is to compute a better upper bound on $opt(G)$ than that given by a maximum TFPCC (cf. Sect. 5). The third is to perform a more sophisticated preprocessing on C (cf. Sect. 6). The last idea is to transform C into a spanning tree of G more carefully (cf. Sect. 7).

4   More Safe Reduction Rules

In addition to the four safe reduction rules in Sect. 3.1, we further introduce the following rules.

Operation 8. If for four vertices $u_1, \ldots, u_4$, $N_G(u_3) = N_G(u_4) = \{u_1, u_2\}$, $G - \{u_2\}$ has a connected component K with $u_1 \notin V(K)$, then delete the edge $e = \{u_2, u_3\}$.

Operation 9. If for five vertices $u_1, \ldots, u_5$, $N_G(u_3) = N_G(u_4) = N_G(u_5) = \{u_1, u_2\}$, then delete the edge $e = \{u_2, u_3\}$.

Operation 10. If for two vertices u and v of G, $G - \{u, v\}$ has a connected component K with $|V(K)| \leq 6$ such that $V(G) \neq V(K) \cup \{u, v\}$ and $G[V(K) \cup \{u, v\}]$ has a Hamiltonian path P from u to v, then delete all edges of $G[V(K) \cup \{u, v\}]$ that do not appear in P.

Operation 11. If G has an edge $e = \{u_1, u_2\}$ with $d_G(u_1) = d_G(u_2) = 2$, then obtain G_1 from G by merging u_1 and u_2 into a single vertex $u_1 u_2$.

We can show that Operations 8 through 10 are strongly safe and Operation 11 is weakly safe.

Accordingly, we need to modify Step 1 of the algorithm by choosing i from $\{1, 2, 8, 9, 10\}$ and also modify Step 2 by choosing i from $\{3, 4, 11\}$. Obviously, after the modification, Steps 1 and 2 can be done in $O(n^2 m)$ time.

5 Computing a Preferred TFPCC $\mathcal{C}$

In this section, we consider how to refine Step 4. Because of Steps 1 and 3, we hereafter assume that $|V(G)| \geq 9$ and there is no $i \in \{1, \ldots, 4, 8, \ldots, 11\}$ such that Operation i can be performed on G. Then, we can prove the next lemma:

Lemma 3. *Suppose C is a cycle of G with $|C| \leq 8$. Let A be the set of ports of C. Then, the following hold.*

1. *$|A| \geq 2$.*
2. *If $|A| = 2$, then the two vertices in A are not adjacent in C and $|C| \neq 5$.*
3. *If $|A| = 2$ and $|C| = 4$, then $G[V(C)]$ and C are the same graph.*

To refine Step 4, our idea is to compute $\mathcal{C}$ as a *preferred* TFPCC of G. Before defining what the word "preferred" means here, we need to prove a lemma. For ease of explanation, we assume, without loss of generality, that there is a linear order (denoted by $\prec$) on the vertices of G. Then, we can prove the next lemma:

Lemma 4. *Suppose u_1 and u_3 are two vertices of G such that $u_1 \prec u_3$ and Condition C1 below holds. Then, G has an OST in which u_1 or u_3 is a leaf. Consequently, G has an OST in which u_1 is a leaf.*

C1. *For two vertices u_2 and u_4 in $V(G) \backslash \{u_1, u_3\}$, $N_G(u_1) = N_G(u_3) = \{u_2, u_4\}$.*

If Condition C1 in Lemma 4 holds for u_1 and u_3, we refer to u_2 and u_4 as the *boundary points* of the pair $p = (u_1, u_3)$, and refer to the edges incident to u_1 or u_3 as the *supports* of p.

Let Π be the set of pairs (u_1, u_3) of vertices in G satisfying Condition C1. It is worth pointing out that for each $p \in \Pi$ and each boundary point u of p, $d_G(u) \geq 3$ because otherwise Operation 4 could be performed on G.

Lemma 5. *No two pairs in Π share a support.*

Lemma 6. *G has an OST in which u_1 is a leaf for each $(u_1, u_3) \in \Pi$.*

Now, we are ready to make two definitions. Let $\mathcal{C}$ be a TFPCC of G. $\mathcal{C}$ is *special* if for every pair $(u_1, u_3) \in \Pi$, $d_{\mathcal{C}}(u_1) \leq 1$. $\mathcal{C}$ is *preferred* if $\mathcal{C}$ is special and $|E(\mathcal{C})|$ is maximized over all special TFPCCs of G.

Lemma 7. *If $\mathcal{C}$ is a preferred TFPCC of G, then $opt(G) \leq |E(\mathcal{C})|$.*

Lemma 8. *We can compute a preferred TFPCC $\mathcal{C}$ of G in $t(2n, 2m)$ time.*

Recall $t(n, m) = O(n^2 m^2)$ [3]. By Lemma 8, after modifying Step 4 by computing $\mathcal{C}$ as a preferred TFPCC of G, Step 4 can still be done in $t(n, m)$ time.

6   Preprocessing $\mathcal{C}$

In this section, we consider how to refine Step 5. So, suppose that we have computed a preferred TFPCC $\mathcal{C}$ of G as in Lemma 8. To refine Step 5, we repeatedly perform not only Operations 5 through 7 but also the following three operations on $\mathcal{C}$ until none of the six is applicable.

Operation 12. If a cycle C_1 of $\mathcal{C}$ has an edge $e_1 = \{u_1, u_1'\}$ and another cycle or path component C_2 of $\mathcal{C}$ has an edge $e_2 = \{u_2, u_2'\}$ such that $e = \{u_1, u_2\} \in E(G)$ and $e' = \{u_1', u_2'\} \in E(G)$, then combine C_1 and C_2 into a single cycle or path by replacing e_1 and e_2 with e and e'.

Operation 13. If an endpoint u_1 of a path component P_1 of $\mathcal{C}$ is adjacent to an endpoint u_2 of another path component P_2 of $\mathcal{C}$ in G, then combine P_1 and P_2 into a single path by adding the edge $\{u_1, u_2\}$.

Operation 14. If $e = \{u, v\}$ is an edge of a path component of $\mathcal{C}$ such that for some isolated vertex x of $\mathcal{C}$, $\{u, x\} \in E(G)$ and $\{v, x\} \in E(G)$, then replace e by the edges $\{u, x\}$ and $\{v, x\}$.

Lemma 9. *Immediately after the refined preprocessing step, the following hold:*

1. *$\mathcal{C}$ is a TFPCC of G and has at least $opt(G)$ edges.*
2. *If a path component P of $\mathcal{C}$ is of length at most 3, then P is alive.*
3. *If an endpoint v of a path component P of $\mathcal{C}$ is a port of P, then each vertex in $N_G(v) \setminus V(P)$ is an internal vertex of a path component Q of $\mathcal{C}$ with $|Q| \geq 2|P| + 2$.*
4. *No pair $(u_1, u_3) \in \Pi$ satisfies that u_1 appears in a cycle of $\mathcal{C}$.*
5. *If a dead path component P of $\mathcal{C}$ is of length 4, then both endpoints of P are leaves in G.*
6. *Each 4-cycle C of $\mathcal{C}$ has at least three ports.*

Obviously, the refined preprocessing (i.e., Step 5) can be done in $O(nm)$ time.

7   Transforming $\mathcal{C}$ into a Spanning Tree

In this section, we consider how to refine Step 6. So, suppose that we have just performed the refined preprocessing on $\mathcal{C}$ as in Sect. 6. Let Γ be the set of (ordered) pairs (P, Q) of path components of $\mathcal{C}$ such that $|P| \geq 1$ and some endpoint v of P is adjacent to a vertex u of Q in G. Note that $d_{\mathcal{C}}(u) = 2$ and $2|P| + 2 \leq |Q|$ by Statement 3 in Lemma 9. Suppose we obtain a subset Γ' of Γ from Γ as follows.

– For each path component P of $\mathcal{C}$ such that there are two or more path components Q of $\mathcal{C}$ with $(P, Q) \in \Gamma$, delete all but one pair (P, Q) from Γ.

Now, consider an auxiliary digraph D such that the vertices of D one-to-one correspond to the path components P of $\mathcal{C}$ with $|P| \geq 1$ and the arcs of D one-to-one correspond to the pairs in Γ'. By Statement 3 in Lemma 9, D is a rooted forest (in which each leaf is of in-degree 0, each root is of out-degree 0, and each vertex is of out-degree at most 1).

To transform $\mathcal{C}$ into a spanning tree of G, the idea is to modify $\mathcal{C}$ in three stages. $\mathcal{C}$ is initially a TFPCC of G and we will always keep $\mathcal{C}$ being a TFTCC of G. For each $i \in \{1, 2, 3\}$, we use $\mathcal{C}_i$ to denote the $\mathcal{C}$ immediately after the i-th stage. For convenience, we use $\mathcal{C}_0$ to denote the $\mathcal{C}$ immediately before the first stage. Moreover, for each $i \in \{1, 2, 3\}$ and each connected component C of $\mathcal{C}_i$, we use $b(C)$ to denote the number of edges $\{u, v\} \in E(\mathcal{C}_0)$ such that $\{u, v\} \subseteq V(C)$.

In the first stage, we modify $\mathcal{C}$ by performing the following step:

1. For each pair $(P, Q) \in \Gamma'$, add an arbitrary $\{u, v\} \in E(G)$ to $\mathcal{C}$ such that u is an endpoint of P and v appears in Q.

Lemma 10. *Each connected component of $\mathcal{C}_1$ that is not a path or cycle is a tree $\hat{T}$ satisfying Condition C2 below:*

C2. $b(\hat{T}) \geq 5$, $|L(\hat{T})| \leq b(\hat{T}) - 2$, and $w(\hat{T}) \geq \frac{4}{5} b(\hat{T})$.

Hereafter, a connected component of $\mathcal{C}$ is *good* if it is a tree $\hat{T}$ satisfying Condition C2 in Lemma 10 or Condition C3 below, while it is *bad* otherwise.

C3. $w(\hat{T}) \geq b(\hat{T}) = 4$ and $|L(\hat{T})| = 3$.

Lemma 11. *Suppose C is a bad connected component of $\mathcal{C}_1$. Then, C is a cycle of length at least 4, a 0-path, or a 4-path whose endpoints are leaves of G. Moreover, if C is a 0-path, then the unique vertex $u \in V(C)$ satisfies that each $v \in N_G(u)$ is an internal vertex of a tree component of $\mathcal{C}_1$ and no two vertices in $N_G(u)$ are adjacent in $\mathcal{C}_1$.*

We next want to define several operations on $\mathcal{C}$ none of which will produce a new cycle or a new bad connected component in $\mathcal{C}$. An operation on $\mathcal{C}$ is *good* if it either just connects two or more connected components of $\mathcal{C}$ into a single good connected component, or modifies a good connected component of $\mathcal{C}$ so that it has more internal vertices (and hence remains good).

In the second stage, we modify $\mathcal{C}$ by repeatedly performing the following operations on $\mathcal{C}$ until none of them is applicable.

Operation 15. If $\mathcal{C}$ has two cycles C_1 and C_2 such that $|C_1| + |C_2| \geq 10$ and some edge $e = \{v_1, v_2\}$ of G satisfies $v_1 \in V(C_1)$ and $v_2 \in V(C_2)$, then connect C_1 and C_2 into a single path T by deleting one edge incident to v_1 in C_1, deleting one edge incident to v_2 in C_2, and adding the edge e.

Operation 16. If $\mathcal{C}$ has a cycle C_1 of length at least 5 and a good connected component C_2 such that some edge $e = \{v, u\}$ of G satisfies $v \in V(C_1)$ and $u \in V(C_2)$, then connect C_1 and C_2 into a single tree T by deleting one edge incident to v in C_1 and adding the edge e.

Operation 17. If $\mathcal{C}$ has a cycle C of length at least 6 and a 4-path component P such that some edge $e = \{v, u\}$ of G satisfies $v \in V(C)$ and $u \in V(P)$, then connect C and P into a single tree T by deleting one edge incident to v in C and adding the edge e.

Operation 18. If $\mathcal{C}$ has a 0-path component P whose unique vertex u has two neighbors v_1 and v_2 in G such that v_1 and v_2 fall into different connected components C_1 and C_2 of $\mathcal{C}$, then connect P, C_1, and C_2 into a single connected component T by adding the edges $\{u, v_1\}$ and $\{u, v_2\}$.

Operation 19. If $\mathcal{C}$ has a good connected component C_1 and another connected component C_2 such that some leaf u of C_1 is adjacent to a vertex v of C_2 in G, then connect C_1 and C_2 into a single tree component T by deleting one edge incident to v in C_2 if C_2 is a cycle, and further adding the edge $\{u, v\}$.

Operation 20. If a cycle C of $\mathcal{C}$ has an edge $e = \{v_1, v_2\}$ such that some $u_1 \in N_G(v_1) \backslash V(C)$ and some $u_2 \in N_G(v_2) \backslash V(C)$ fall into different connected components C_1 and C_2 of $\mathcal{C}$ other than C, then connect C, C_1, and C_2 into a single tree component T by deleting e, deleting one edge incident to u_1 if C_1 is a cycle, deleting one edge incident to u_2 if C_2 is a cycle, and adding the edges $\{v_1, u_1\}$ and $\{v_2, u_2\}$.

Operation 21. If a good connected components C of $\mathcal{C}$ is not a Hamiltonian path of G but is a dead path whose endpoints are adjacent in G, then choose an arbitrary port u of C, modify C by adding the edge of G between the endpoints of C and deleting one edge incident to u in C, and further perform Operation 19.

Operation 22. If a good connected component C of $\mathcal{C}$ is not a path but has two leaves u and v with $\{u, v\} \in E(G)$, then modify C by first finding an arbitrary vertex x on the path P between u and v in C with $d_C(x) \geq 3$, then deleting one edge incident to x in P, and further adding the edge $\{u, v\}$.

Operation 23. If $\mathcal{C}$ has a 0-path component C_1, a 4-path component P, and a connected component C_2 other than C_1 and P such that the center vertex u_3 of P is adjacent to a vertex x of C_2 in G and the unique vertex v of C_1 is adjacent to the other two internal vertices u_2 and u_4 of P (than u_3) in G, then connect C_1, P, and C_2 into a single connected component T by deleting the edge $\{u_2, u_3\}$, deleting one edge incident to x if C_2 is a cycle, and adding the edges $\{v, u_2\}$, $\{v, u_4\}$, $\{u_3, x\}$.

We can show that Operations 15 through 23 are good.

We next show that the above operations lead to a number of useful properties of $\mathcal{C}_2$. The properties are stated in the next five lemmas:

Lemma 12. *Each 4-cycle of C_2 is adjacent to at most one other connected component of C_2 in G.*

Lemma 13. *No two 4-cycles of C_2 are adjacent in G.*

Lemma 14. *No 4-cycle C of C_2 is adjacent to a 4-path component of C_2 in G.*

Lemma 15. *No 4-cycle of C_2 is adjacent to a 5-cycle of C_2 in G.*

Lemma 16. *Let C be a connected component of C_2. Then, C is a 4-cycle, 5-cycle, 0-path, 4-path, or good connected component. Moreover, the following hold:*

1. *If C is a 0-path, then its unique vertex u satisfies that for a single tree component C' of C_2, each $v \in N_G(u)$ is an internal vertex of C', and u is a leaf of G if C' is bad.*
2. *If C is a 4-path component of C_2, then its endpoints are leaves of G and each internal vertex u of C satisfies that each neighbor of u in G is a leaf of G, a vertex of a 5-cycle of C_2, or an internal vertex of a 4-path component or a good connected component of C.*
3. *If C is a 4-cycle of C_2, then each vertex u of C satisfies that each neighbor of u in G is an internal vertex of a good connected component of C_2.*
4. *If C is a 5-cycle of C_2, then each vertex u of C satisfies that each neighbor of u in G is an internal vertex of a 4-path component of C_2.*
5. *If C is a good connected component but not a Hamiltonian path of G, then for each leaf u of C, each neighbor of u in G is an internal vertex of C.*

Finally, in the third stage, we complete the transformation of C into a spanning tree of G by further modifying C by performing the following steps:

1. For each cycle C of C, first select an arbitrary edge $e = \{u, v\} \in E(G)$ such that $u \in V(C)$ and $v \in V(G) \setminus V(C)$, then delete one edge incident to u in C, and further add e. (*Comment:* Since no two cycles in C_2 are adjacent in G, v appears in a tree component of C. Moreover, after this step, C has only tree components.)
2. Arbitrarily connect the connected components of C into a tree by adding some edges of G.

It is easy to see that for each $i \in \{15, \ldots, 23\}$, Operation i can be done in $O(m)$ time. So, the second stage takes $O(nm)$ time. Since the other two stages can be easily done in $O(m)$ time, the refined Step 6 can be done $O(nm)$ time.

8 Performance Analysis

Let g_2 (respectively, g_3) be the number of internal vertices in connected components of C_2 satisfying Condition C2 (respectively, C3), b_2 (respectively, b_3) be the total number of edges in C_0 whose endpoints appear in the same connected components of C_2 satisfying Condition C2 (respectively, C3), c_4 (respectively, c_5) be the number of 4-cycles (respectively, 5-cycles) in C_2, and p_4 be the number of 4-path components in C_2. We can show the next lemma and theorem:

Lemma 17. *Let T_{apx} be the spanning tree of G outputted by the refined algorithm. Then, the following hold:*

1. $w(T_{apx}) \geq 3c_4 + 4c_5 + 3p_4 + g_2 + g_3 \geq 3c_4 + 4c_5 + 3p_4 + \frac{4}{5}b_2 + b_3$.
2. $opt(G) \leq 4c_4 + 5c_5 + 4p_4 + b_2 + b_3$.
3. $opt(G) \leq 3c_4 + 5c_5 + 3p_4 + 2g_2 + 2g_3$.

Theorem 2. *The algorithm achieves an approximation ratio of $\frac{13}{17}$ and runs in $O(n^2 m) + t(2n, 2m)$ time.*

Proof. Let T_{apx} be as in Lemma 17, and $r = w(T_{apx})/opt(G)$. By Lemma 17, $r \geq \max\{r_1, r_2\}$, where $r_1 = \frac{3c_4 + 4c_5 + 3p_4 + g_2 + g_3}{4c_4 + 5c_5 + 4p_4 + b_2 + b_3}$ and $r_2 = \frac{3c_4 + 4c_5 + 3p_4 + g_2 + g_3}{3c_4 + 5c_5 + 3p_4 + 2g_2 + 2g_3}$. Note that $r_1 \geq \min\left\{\frac{4}{5}, r_1'\right\}$ and $r_2 \geq \min\left\{\frac{4}{5}, r_2'\right\}$, where $r_1' = \frac{3c_4 + 3p_4 + g_2 + g_3}{4c_4 + 4p_4 + b_2 + b_3}$ and $r_2' = \frac{3c_4 + 3p_4 + g_2 + g_3}{3c_4 + 3p_4 + 2g_2 + 2g_3}$. So, it suffices to show that $\max\{r_1', r_2'\} \geq \frac{13}{17}$. This is done if $r_1' \geq \frac{13}{17}$. Thus, we assume that $r_1' < \frac{13}{17}$. Then, $c_4 + p_4 > 17g_2 + 17g_3 - 13b_2 - 13b_3$. Hence, $r_2' > \frac{52g_2 + 52g_3 - 39b_2 - 39b_3}{53g_2 + 53g_3 - 39b_2 - 39b_3} \geq \min\left\{\frac{52g_2 - 39b_2}{53g_2 - 39b_2}, \frac{52g_3 - 39b_3}{53g_3 - 39b_3}\right\}$. Now, since $g_2 \geq \frac{4}{5}b_2$, $\frac{52g_2 - 39b_2}{53g_2 - 39b_2} \geq \frac{13}{17}$. Moreover, since $g_3 \geq b_3$, $\frac{52g_3 - 39b_3}{53g_3 - 39b_3} \geq \frac{13}{14}$. Therefore, $r_2' > \frac{13}{17}$. The running is clearly as claimed. ∎

Recall that $t(n, m) = O(n^2 m^2)$ [3]. So, the algorithm takes $O(n^2 m^2)$ time.

References

1. Binkele-Raible, D., Fernau, H., Gaspers, S., Liedloff, M.: Exact and parameterized algorithms for max internal spanning tree. Algorithmica **65**(1), 95–128 (2013)
2. Chen, Z.-Z., Lin, G., Wang, L., Chen, Y.: Approximation Algorithms for the Maximum Weight Internal Spanning Tree Problem. CoRR, abs/1608.03299 (2016)
3. Hartvigsen, D.: Extensions of matching theory. Ph.D. thesis, Carnegie-Mellon University (1984)
4. Knauer, M., Spoerhase, J.: Better approximation algorithms for the maximum internal spanning tree problem. In: Dehne, F., Gavrilova, M., Sack, J.-R., Tóth, C.D. (eds.) WADS 2009. LNCS, vol. 5664, pp. 459–470. Springer, Heidelberg (2009). doi:10.1007/978-3-642-03367-4_40
5. Li, X., Zhu, D.: Approximating the maximum internal spanning tree problem via a maximum path-cycle cover. In: Ahn, H.-K., Shin, C.-S. (eds.) ISAAC 2014. LNCS, vol. 8889, pp. 467–478. Springer, Heidelberg (2014). doi:10.1007/978-3-319-13075-0_37
6. Li, W., Wang, J., Chen, J., Cao, Y.: A $2k$-vertex kernel for maximum internal spanning tree. In: Dehne, F., Sack, J.-R., Stege, U. (eds.) WADS 2015. LNCS, vol. 9214, pp. 495–505. Springer, Heidelberg (2015). doi:10.1007/978-3-319-21840-3_41
7. Salamon, G., Wiener, G.: On finding spanning trees with few leaves. Inf. Process. Lett. **105**(5), 164–169 (2008)

Approximation Algorithm for Cycle-Star Hub Network Design Problems and Cycle-Metric Labeling Problems

Yuko Kuroki$^{(\boxtimes)}$ and Tomomi Matsui$^{(\boxtimes)}$

Department of Industrial Engineering and Economics, Graduate School
of Engineering, Tokyo Institute of Technology, Tokyo, Japan
{kuroki.y.aa,matsui.t.af}@m.titech.ac.jp

Abstract. We consider a single allocation hub-and-spoke network design problem which allocates each non-hub node to exactly one of given hub nodes so as to minimize the total transportation cost. This paper deals with a case in which the hubs are located in a cycle, which is called a cycle-star hub network design problem. The problem is essentially equivalent to a cycle-metric labeling problem. The problem is useful in the design of networks in telecommunications and airline transportation systems. We propose a $2(1 - 1/h)$-approximation algorithm where h denotes the number of hub nodes. Our algorithm solves a linear relaxation problem and employs a dependent rounding procedure. We analyze our algorithm by approximating a given cycle-metric matrix by a convex combination of Monge matrices.

1 Introduction

In this paper, we propose a $2(1 - 1/h)$-approximation algorithm for cycle-star hub network design problems with h hubs and/or a cycle-metric labeling problem with h labels.

Hub-and-spoke networks arise in airline transportation systems, delivery systems and telecommunication systems. Hub networks have an important role when there are many origins and destinations. Hub facilities work as switching points for flows. In order to reduce transportation costs and set-up costs in a large network, each non-hub node is allocated to exactly one of the hubs instead of assigning every origin-destination pair directly.

Hub location problems (HLPs) consist of locating hubs and of designing hub networks so as to minimize the total transportation cost. HLPs are formulated as quadratic integer programming problem by O'Kelly [25], first. Since O'Kelly proposed HLPs, many researches on HLPs have been done in various applications (see [3, 9, 12, 16, 20, 26] for example).

In this paper, we discuss the situation in which the locations of the hubs are given, and deal with a problem, called a *single allocation hub-and-spoke network design problem*, which finds a connection of the non-hubs to the given hubs minimizing total transportation cost. Sohn and Park [27, 28] proposed a

© Springer International Publishing AG 2017
S.-H. Poon et al. (Eds.): WALCOM 2017, LNCS 10167, pp. 397–408, 2017.
DOI: 10.1007/978-3-319-53925-6_31

polynomial time exact algorithm for a problem with 2 hubs and proved NP-completness of the problem even if the number of hubs is equal to 3. Iwasa et al. [18] proposed a simple 3-approximation algorithm and a randomized 2-approximation algorithm under the assumptions of triangle inequality. They also proposed a $(5/4)$-approximation algorithm for the special case where the number of hubs is 3. Ando and Matsui [5] deal with the case in which all the nodes are embedded in a 2-dimensional plane and the transportation cost of an edge per unit flow is proportional to the Euclidean distance between the end nodes of the edge. They proposed a randomized $(1 + 2/\pi)$-approximation algorithm. Saito et al. [29] discussed some facets of polytopes corresponding to the convex hull of feasible solutions of the problem.

Fundamental HLPs assume a full interconnection between hubs. Recently, several researches consider incomplete hub networks which arise especially in telecommunication systems (see [4,8,10,11] for example). These models are useful when set-up costs of hub links are considerably large or full interconnection is not required. There are researches which assume that hub networks constitute a particular structure such as a tree [14,15,19,23], a star [22,30,31], a path [24] or a cycle [17].

In this paper, we consider a single allocation hub-and-spoke network design problem where the given hubs are located in a cycle. We call this problem the *cycle-star hub network design problem*. When the number of hubs is 3, the hub network becomes a 3-cycle and constitutes a complete graph. Thus, the 4/3-approximation algorithm for a complete 3-hub network proposed in [18] is valid for this special case. In this paper we propose a $2(1 - 1/h)$-approximation algorithm when a set of h hubs forms an h-cycle. To the best of our knowledge, our algorithm is the first approximation algorithm for this problem.

A single allocation hub-and-spoke network design problem is essentially equivalent to a metric labeling problem introduced by Kleinberg and Tardos in [21], which has connections to Markov random field and classification problems that arise in computer vision and related areas. They proposed an $O(\log h \log \log h)$-approximation algorithm where h is the number of labels (hubs). Chuzhoy and Naor [13] showed that there is no polynomial time approximation algorithm with a constant ratio for the problem unless P = NP. We deal with a cycle-metric labeling problem where a given metric matrix is defined by an undirected cycle and non-negative edge length. Thus, our results give an important class of the metric labeling problem, which has a polynomial time approximation algorithm with a constant approximation ratio.

2   Problem Formulation

Let $H = \{1, 2, \ldots, h\}$ be a set of hub nodes and N be a set of non-hub nodes where $|H| \geq 3$ and $|N| = n$. This paper deals with a single assignment hub network design problem which assigns each non-hub node to exactly one hub node. We discuss the case in which the set of hubs forms an undirected cycle, and the corresponding problem is called the *cycle-star hub network design problem*.

More precisely, we are given an undirected cycle $\Gamma = (H, T)$ defined by a vertex-set H and an edge-set $T = \{\{1,2\}, \{2,3\}, \ldots, \{h-1,h\}, \{h,1\}\}$. In the rest of this paper, we identify hub i and hub $h + i$ when there is no ambiguity. For each edge $e = \{i, i+1\} \in T$, the corresponding length, denoted by c_e or $c_{i\,i+1}$ represents a non-negative cost per unit of flow on the edge. For each ordered pair $(p, i) \in N \times H$, c_{pi} also denotes a non-negative cost per unit flow on an undirected edge $\{p, i\}$. We denote a given non-negative amount of flow from a non-hub p to another non-hub q by $w_{pq} (\geq 0)$. Throughout this paper, we assume that $w_{pp} = 0$ ($\forall p \in N$). We discuss the problem for finding an assignment of non-hubs to hubs which minimizes the total transportation cost defined below.

When non-hub nodes p and q ($p \neq q$) are assigned hubs i and j, respectively, an amount of flow w_{pq} is sent along a path $((p, i), \Omega_{ij}, (j, q))$ where Ω_{ij} denotes a shortest path in $\Gamma = (H, T)$ between i and j. For each pair of hub nodes $(i, j) \in H^2$, c_{ij} denotes the length of a shortest path Ω_{ij}. More precisely, cycle Γ contains exactly two paths between i and j and c_{ij} denotes the minimum of the lengths of these two paths. It is easy to see that $c_{ij} = c_{ji}$. In the rest of this paper, a matrix $C = (c_{ij})$ defined above is called a *cost matrix* and/or a *cycle-metric matrix*. The transportation cost corresponding to a flow from p to q is defined by $w_{pq}(c_{pi} + c_{ij} + c_{qj})$.

Now we describe our problem formally. First, we introduce a 0–1 variable x_{pi} for each pair $\{p, i\} \in N \times H$ as follows:

$$x_{pi} = \begin{cases} 1 & (p \in N \text{ is assigned to } i \in H), \\ 0 & (\text{otherwise}). \end{cases}$$

We have a constraint $\sum_{i \in H} x_{pi} = 1$ for each $p \in N$, since each non-hub is connected to exactly one hub. Then, the cycle-star hub network design problem can be formulated as follows:

$$\text{SAP:} \quad \min. \quad \sum_{(p,q) \in N^2} w_{pq} \left(\sum_{i \in H} c_{pi} x_{pi} + \sum_{j \in H} c_{jq} x_{qj} + \sum_{(i,j) \in H^2} c_{ij} x_{pi} x_{qj} \right)$$

$$\text{s. t.} \quad \sum_{i \in H} x_{pi} = 1 \qquad (\forall p \in N),$$

$$x_{pi} \in \{0, 1\} \qquad (\forall (p, i) \in N \times H).$$

The above formulation also appears in [18,28]. In case $h = 3$, 3-cycle Γ is a complete graph, and thus the corresponding problem is NP-complete [28].

Next we describe an integer linear programming problem proposed in [18], which is derived from SAP by employing the linearization technique introduced by Adams and Sherali [1]. We replace $x_{pi} x_{qj}$ with y_{piqj}. We have a new constraint $\sum_{i \in H} y_{piqj} = x_{qj}$ from the equation $\sum_{i \in H} x_{pi} = 1$ by multiplying both sides by x_{qj}. We also obtain a constraint $\sum_{j \in H} y_{piqj} = x_{pi}$ in a similar way. Then we obtain the following 0–1 integer linear programming problem:

$$\text{SAPL:} \quad \text{min.} \quad \sum_{(p,q)\in N^2} w_{pq} \left(\sum_{i\in H} c_{pi}x_{pi} + \sum_{j\in H} c_{jq}x_{qj} + \sum_{(i,j)\in H^2} c_{ij}y_{piqj} \right)$$

$$\text{s. t.} \quad \sum_{i\in H} x_{pi} = 1 \qquad (\forall p \in N),$$

$$\sum_{j\in H} y_{piqj} = x_{pi} \qquad (\forall(p,q)\in N^2, \forall i \in H, p < q),$$

$$\sum_{i\in H} y_{piqj} = x_{qj} \qquad (\forall(p,q)\in N^2, \forall j \in H, p < q),$$

$$x_{pi} \in \{0,1\} \qquad (\forall(p,i)\in N \times H),$$

$$y_{piqj} \in \{0,1\} \qquad (\forall(p,q)\in N^2, \forall(i,j)\in H^2).$$

By substituting non-negativity constraints of all the variables for 0–1 constraints in SAPL, we obtain a linear relaxation problem, denoted by LRP. We can solve LRP in polynomial time by employing an interior point algorithm.

3   Monge Property and Dependent Rounding Procedure

First, we give the definition of a Monge matrix. A comprehensive research on the Monge property appears in a recent survey [7].

Definition 1. *An $m \times n$ matrix C' is a Monge matrix if and only if C' satisfies the so-called Monge property*

$$c'_{ij} + c'_{i'j'} \le c'_{ij'} + c'_{i'j} \quad \text{for all} \quad 1 \le i < i' \le m, 1 \le j < j' \le n.$$

Although the Monge property depends on the orders of the rows and columns, in this paper, we say that a matrix is a Monge matrix when there exist permutations of rows and columns which yield the Monge property.

For each edge $e \in T$, we define a path $\Gamma^e = (H, T \setminus \{e\})$ obtained from cycle Γ by deleting the edge e. Let $C^e = (c^e_{ij})$ be a cost matrix where c^e_{ij} denotes the length of the unique subpath of Γ^e connecting i and j.

Lemma 1. *For any edge $e = \{\ell, \ell + 1\} \in T$, a Monge matrix is obtained from C^e by permuting rows and columns simultaneously in the ordering $(\ell + 1, \ell + 2, \ldots, h, 1, 2, \ldots, \ell)$.*

Proof is omitted (see [7] for example).

Next, we approximate a given cost matrix (cycle-metric matrix) C by a convex combination of h Monge matrices $\{C^e \mid e \in T\}$. Alon et al. [2] considered approximating a cycle-metric matrix by a probability distribution over path-metric matrices, and showed a simple distribution such that the expected length of each edge is no more than twice its original length. The following theorem improves their result especially when the size of the cycle (number of hubs) is small.

Theorem 1. *Let C be a cost matrix obtained from a cycle $\Gamma = (H, T)$ and non-negative edge lengths $(c_e \mid e \in T)$. Then, there exists a vector of coefficients $(\theta_e \mid e \in T)$ satisfying*

$$\theta_e \geq 0 \ (\forall e \in T), \quad \sum_{e \in T} \theta_e = 1, \quad and \quad C \leq \sum_{e \in T} \theta_e C^e \leq 2 \left(1 - \frac{1}{h}\right) C.$$

Proof. When there exists an edge $e° \in T$ satisfying $c_{e°} \geq (1/2) \sum_{f \in T} c_f$, it is easy to see that for every pair $(i, j) \in H^2$, there exists a shortest path Ω_{ij} on cycle $\Gamma = (H, T)$ between i and j excluding edge $e°$. Thus, a given cost matrix C is equivalent to the Monge matrix $C^{e°}$. In this case, the desired result is trivial.

We assume that $2c_e < L = \sum_{f \in T} c_f$ $(\forall e \in T)$ and introduce a positive coefficient θ_e for each $e \in T$ defined by

$$\theta_e = \frac{c_e}{K} \prod_{f \in T \setminus \{e\}} (L - 2c_f)$$

where K is a normalizing constant which yields the equality $\sum_{e \in T} \theta_e = 1$. Let $\Omega_{ij} \subseteq T$ be a set of edges in a shortest path in Γ between i and j. The definition of the coefficients $(\theta_e \mid e \in T)$ directly implies that for each pair $(i, j) \in H^2$,

$$\sum_{e \in T} \theta_e c_{ij}^e = \sum_{e \notin \Omega_{ij}} \theta_e c_{ij} + \sum_{e \in \Omega_{ij}} \theta_e (L - c_{ij}) = \sum_{e \in T} \theta_e c_{ij} + \sum_{e \in \Omega_{ij}} \theta_e (L - 2c_{ij})$$

$$\leq c_{ij} \sum_{e \in T} \theta_e + \sum_{e \in \Omega_{ij}} \theta_e (L - 2c_e)$$

$$= c_{ij} + \sum_{e \in \Omega_{ij}} \left((L - 2c_e) \frac{c_e}{K} \prod_{f \in T \setminus \{e\}} (L - 2c_f) \right)$$

$$= c_{ij} + \frac{\prod_{f \in T} (L - 2c_f)}{K} \sum_{e \in \Omega_{ij}} c_e = c_{ij} + \frac{\prod_{f \in T} (L - 2c_f)}{K} c_{ij}.$$

From the assumption, the last term appearing above is positive. Then, we have

$$\frac{K}{\prod_{f \in T} (L - 2c_f)} = \frac{\sum_{e \in T} \left(c_e \prod_{f \in T \setminus \{e\}} (L - 2c_f) \right)}{\prod_{f \in T} (L - 2c_f)} = \sum_{e \in T} \frac{c_e}{L - 2c_e}.$$

Now we introduce a function $f(z_1, \ldots, z_h) = \sum_{\ell=1}^{h} z_\ell / (L - 2z_\ell)$ defined on a domain $\{z \in [0, L/2)^h \mid z_1 + \cdots + z_h = L\}$. From the convexity and symmetry of variables of f, the minimum of f is attained at $z_1 = z_2 = \cdots = z_h = L/h$, and $f(L/h, \ldots, L/h) = 1/(1 - 2/h)$, which gives the following inequality

$$\sum_{e \in T} \theta_e c_{ij}^e \leq c_{ij} + \frac{\prod_{f \in T} (L - 2c_f)}{K} c_{ij} \leq c_{ij} + \left(1 - \frac{2}{h}\right) c_{ij} = 2 \left(1 - \frac{1}{h}\right) c_{ij}.$$

Since $C \leq C^e$ $(\forall e \in T)$, it is obvious that $C \leq \sum_{e \in T} \theta_e C^e$.

Dependent Rounding $(\boldsymbol{x}, \boldsymbol{y}; \pi)$

Input: A feasible solution $(\boldsymbol{x}, \boldsymbol{y})$ of LRP and a total order π of the hubs.
Step 1: Generate a random variable U which follows a uniform distribution defined on $[0, 1)$.
Step 2: Assign each non-hub node $p \in N$ to a hub $\pi(i)$, where $i \in \{1, 2, \ldots h\}$ is the minimum number that satisfies $U < x_{p\pi(1)} + \cdots + x_{p\pi(i)}$.

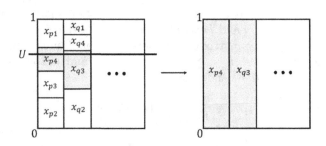

Fig. 1. Dependent rounding π_ℓ where $(\pi(1), \pi(2), \pi(3), \pi(4)) = (2, 3, 4, 1)$.

Next, we describe a rounding technique proposed in [18]. We will describe a connection between the Monge matrix and the rounding technique, later.

The above procedure can be explained roughly as follows (see Fig. 1). For each non-hub $p \in N$, we subdivide a rectangle of height 1 with horizontal segments into smaller rectangles whose heights are equal to the values of the given feasible solution $x_{p\pi(1)}, x_{p\pi(2)}, \ldots, x_{p\pi(h)}$. Here we note that $\sum_{i \in H} x_{p\pi(i)} = 1 \ (\forall p \in N)$. We assume that the smaller rectangles are heaped in the order π. We generate a horizontal line whose height is equal to the random variable U and round a variable x_{pi} to 1 if and only if the corresponding rectangle intersects the horizontal line.

Given a feasible solution $(\boldsymbol{x}, \boldsymbol{y})$ of LRP and a total order π of H, a vector of random variables $\boldsymbol{X}^\pi$, indexed by $N \times H$, denotes a solution obtained by **Dependent Rounding** $(\boldsymbol{x}, \boldsymbol{y}; \pi)$. In the following, we discuss the probability $\Pr[X_{pi}^\pi X_{qj}^\pi = 1]$.

Lemma 2 [18]. *Let $(\boldsymbol{x}, \boldsymbol{y})$ be a feasible solution of LRP and π a total order of H. A vector of random variables $\boldsymbol{X}^\pi$ obtained by* **Dependent Rounding** $(\boldsymbol{x}, \boldsymbol{y}; \pi)$ *satisfies that*

(1) $\mathrm{E}[X_{pi}^\pi] = x_{pi} \qquad (\forall (p, i) \in N \times H)$,

(2) $\mathrm{E}[X_{pi}^\pi X_{qj}^\pi] = y_{piqj}^\pi (\forall (p, q) \in N^2, \forall (i, j) \in H^2)$,

where $\boldsymbol{y}^{\pi}$ is a unique solution of the following system of equalities

$$\sum_{i=1}^{i'}\sum_{j=1}^{j'} y_{p\pi(i)q\pi(j)}^{\pi} = \min\left\{\sum_{i=1}^{i'} x_{p\pi(i)}, \sum_{i=1}^{j'} x_{q\pi(j)}\right\} \left(\forall (p,q) \in N^2, \forall (i',j') \in H^2\right).$$

(1)

Proof is omitted (see [18]).

In the rest of this paper, a pair of vectors $(\boldsymbol{x}, \boldsymbol{y}^{\pi})$ defined by (1) is called a *north-west corner rule solution* with respect to $(\boldsymbol{x}, \boldsymbol{y}; \pi)$. When $\boldsymbol{x}$ is non-negative and $\sum_{i\in H} x_{pi} = \sum_{i\in H} x_{qi}$ holds, the unique solution of (1) gives the so-called *north-west corner rule solution* for a Hitchcock transportation problem (Fig. 2 shows an example of a north-west corner rule solution where details are given in [18]). Here we note that the above definition is different from the ordinary definition of the north-west corner rule solution, which is known as a result of a procedure for finding a feasible solution of a Hitchcock transportation problem. In the rest of this section, we describe Hitchcock transportation (sub)problems contained in LRP.

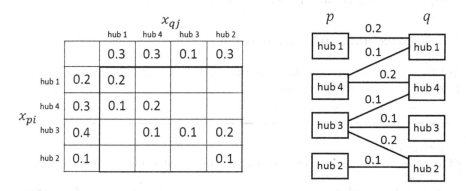

Fig. 2. North-west corner rule solution where $(\pi(1), \pi(2), \pi(3), \pi(4)) = (2, 3, 4, 1)$. In this case, $\mathrm{E}[X_{p1}^{\pi}X_{q1}^{\pi}] = 0.2$, $\mathrm{E}[X_{p4}^{\pi}X_{q1}^{\pi}] = 0.1$, $\mathrm{E}[X_{p4}^{\pi}X_{q4}^{\pi}] = 0.2$, $\mathrm{E}[X_{p3}^{\pi}X_{q4}^{\pi}] = 0.1$, $\mathrm{E}[X_{p3}^{\pi}X_{q3}^{\pi}] = 0.1$, $\mathrm{E}[X_{p3}^{\pi}X_{q2}^{\pi}] = 0.2$, and $\mathrm{E}[X_{p2}^{\pi}X_{q2}^{\pi}] = 0.1$.

Let $(\boldsymbol{x}^{\circ}, \boldsymbol{y}^{\circ})$ be a feasible solution of LRP. For any $p \in N$, $\boldsymbol{x}_p^{\circ}$ denotes a subvector of $\boldsymbol{x}^{\circ}$ defined by $(x_{p1}^{\circ}, x_{p2}^{\circ}, \ldots, x_{ph}^{\circ})$. When we fix variables $\boldsymbol{x}$ in LRP to $\boldsymbol{x}^{\circ}$, we can decompose the obtained problem into n^2 Hitchcock transportation problems $\{HTP(\boldsymbol{x}_p^\circ, \boldsymbol{x}_q^\circ, C) \mid (p,q) \in N^2\}$ where

$$HTP\,(\boldsymbol{x}_p^\circ, \boldsymbol{x}_q^\circ, C) : \text{min.} \quad \sum_{i \in H} \sum_{j \in H} c_{ij} y_{piqj}$$

$$\text{s. t.} \quad \sum_{j \in H} y_{piqj} = x_{pi}^\circ \qquad (\forall i \in H),$$

$$\sum_{i \in H} y_{piqj} = x_{qj}^\circ \qquad (\forall j \in H),$$

$$y_{piqj} \geq 0 \qquad (\forall (i,j) \in H^2).$$

Next, we describe a well-known relation between a north-west corner rule solution of a Hitchcock transportation problem and the Monge property.

Theorem 2 *If a given cost matrix $C = (c_{ij})$ is a Monge matrix with respect to a total order π of hubs, then the north-west corner rule solution $\boldsymbol{y}^\pi$ defined by (1) gives optimal solutions of all the Hitchcock transportation problems $\{HTP(\boldsymbol{x}_p^\circ, \boldsymbol{x}_q^\circ, C) \mid (p,q) \in N^2\}$.*

Proof is omitted here (see [6,7] for example).

4   Approximation Algorithm

In this section, we propose an algorithm and discuss its approximation ratio. First, we describe our algorithm.

Algorithm 4
Step 1: Solve the linear relaxation problem LRP and obtain an optimal solution $(\boldsymbol{x}^*, \boldsymbol{y}^*)$.
Step 2: For each edge $e \in T$, execute **Dependent Rounding** $(\boldsymbol{x}^*, \boldsymbol{y}^*; \pi^e)$ where π^e denotes a total order $(\pi^e(1), \pi^e(2), \ldots, \pi^e(h)) = (\ell+1, \ell+2, \ldots, h, 1, 2, \ldots, \ell-1, \ell)$.
Step 3: Output a best solution obtained in Step 2.

In the rest of this section, we discuss the approximation ratio. We define

$$W_1^* = \sum_{(p,q) \in N^2} w_{pq} \left(\sum_{i \in H} c_{pi} x_{pi}^* + \sum_{j \in H} c_{jq} x_{qj}^* \right)$$

and

$$W_2^* = \sum_{(p,q) \in N^2} w_{pq} \left(\sum_{(i,j) \in H^2} c_{ij} y_{piqj}^* \right)$$

where $(\boldsymbol{x}^*, \boldsymbol{y}^*)$ is an optimal solution of LRP.

Theorem 3 *Algorithm 4 is a $2(1 - 1/h)$-approximation algorithm.*

Proof Let z^{**} be the optimal value of the original problem SAP and $(\theta_e \mid e \in T)$ be a vector of coefficients defined in Theorem 1. For each $e \in T$, $(X^{\pi(e)})$ denotes a solution obtained by `Dependent Rounding`$(x^*, y^*; \pi^e)$ and $y^{\pi(e)}$ be the northwest corner rule solution defined by (1) (where π is set to π^e). Then we have that

$$2\left(1 - \frac{1}{h}\right)z^{**} \geq 2\left(1 - \frac{1}{h}\right)(\text{optimal value of LRP}) = 2\left(1 - \frac{1}{h}\right)(W_1^* + W_2^*)$$

$$\geq W_1^* + 2\left(1 - \frac{1}{h}\right)W_2^* = W_1^* + \sum_{(p,q)\in N^2} w_{pq}\left(\sum_{(i,j)\in H^2} 2\left(1 - \frac{1}{h}\right)c_{ij}y_{piqj}^*\right)$$

$$\geq \sum_{e\in T}\theta_e W_1^* + \sum_{(p,q)\in N^2} w_{pq}\left(\sum_{(i,j)\in H^2}\sum_{e\in T}\theta_e c_{ij}^e y_{piqj}^*\right) \qquad \text{(Theorem 1)}$$

$$= \sum_{e\in T}\theta_e W_1^* + \sum_{e\in T}\theta_e\left(\sum_{(p,q)\in N^2} w_{pq}\left(\sum_{(i,j)\in H^2} c_{ij}^e y_{piqj}^*\right)\right)$$

$$\geq \sum_{e\in T}\theta_e W_1^* + \sum_{e\in T}\theta_e\left(\sum_{(p,q)\in N^2} w_{pq}\,(\text{optimal value of HTP}(x_p^*, x_q^*, C^e))\right)$$

$$= \sum_{e\in T}\theta_e W_1^* + \sum_{e\in T}\theta_e\left(\sum_{(p,q)\in N^2} w_{pq}\left(\sum_{(i,j)\in H^2} c_{ij}^e y_{piqj}^{\pi(e)}\right)\right) \qquad \text{(Theorem 2)}$$

$$= \sum_{e\in T}\theta_e\left(\sum_{(p,q)\in N^2} w_{pq}\left(\sum_{i\in H} c_{pi}x_{pi}^* + \sum_{j\in H} c_{jq}x_{qj}^* + \sum_{(i,j)\in H^2} c_{ij}y_{piqj}^{\pi(e)}\right)\right)$$

$$= \sum_{e\in T}\theta_e\left(\sum_{(p,q)\in N^2} w_{pq}\left(\begin{matrix}\sum_{i\in H} c_{pi}\mathrm{E}[X_{pi}^{\pi(e)}] + \sum_{j\in H} c_{jq}\mathrm{E}[X_{qj}^{\pi(e)}] \\ + \sum_{(i,j)\in H^2} c_{ij}\mathrm{E}[X_{pi}^{\pi(e)}X_{qj}^{\pi(e)}]\end{matrix}\right)\right)$$

$$= \mathrm{E}\left[\sum_{e\in T}\theta_e\left(\sum_{(p,q)\in N^2} w_{pq}\left(\begin{matrix}\sum_{i\in H} c_{pi}X_{pi}^{\pi(e)} + \sum_{j\in H} c_{jq}X_{qj}^{\pi(e)} \\ + \sum_{(i,j)\in H^2} c_{ij}X_{pi}^{\pi(e)}X_{qj}^{\pi(e)}\end{matrix}\right)\right)\right]$$

$$\geq \mathrm{E}\left[\min_{e\in T}\left\{\sum_{(p,q)\in N^2} w_{pq}\left(\begin{matrix}\sum_{i\in H} c_{pi}X_{pi}^{\pi(e)} + \sum_{j\in H} c_{jq}X_{qj}^{\pi(e)} \\ + \sum_{(i,j)\in H^2} c_{ij}X_{pi}^{\pi(e)}X_{qj}^{\pi(e)}\end{matrix}\right)\right\}\right]$$

$$= \mathrm{E}[Z]$$

where Z denotes the objective value of a solution obtained by Algorithm 4. The last inequality in the above transformation is obtained from the equality $\sum_{e\in T}\theta_e = 1$ and the non-negativity of coefficients $(\theta_e \mid e \in T)$.

5 Discussions

In this paper, we proposed a polynomial time $2(1-1/h)$-approximation algorithm for a cycle-star hub network design problem with h hubs. Our algorithm solves a linear relaxation problem and employs a dependent rounding procedure. The attained approximation ratio is based on an approximation of a cycle-metric matrix by a convex combination of Monge matrices.

Lastly we discuss a simple independent rounding technique which independently connects each non-hub node $p \in N$ to a hub node $i \in H$ with probability x^*_{pi}, where $(\boldsymbol{x}^*, \boldsymbol{y}^*)$ is an optimal solution of LRP. Iwasa et al. [18] showed that the independent rounding technique gives a 2-approximation algorithm for a single allocation hub-and-spoke network design problem under the following assumption.

Assumption 1 A given symmetric non-negative cost matrix C satisfies $c_{ij} \leq c_{ik} + c_{kj}$ $(\forall(i,j,k) \in H^3)$ and $c_{ij} \leq c_{pi} + c_{pj}$ $(\forall(i,j,p) \in H^2 \times N)$.

We also have the following result.

Lemma 3 *Under an assumption that $c_{ij} \leq c_{pi} + c_{pj}$ $(\forall(i,j,p) \in H^2 \times N)$, a $\left(\frac{3}{2} - \frac{1}{2(h-1)}\right)$-approximation algorithm (for a cycle-star hub network design problem) is obtained by choosing the better of the two solutions given by Algorithm 4 and the independent rounding technique.*

Proof Let a random variable Z_2 be an objective function value with respect to a solution obtained by the independent rounding technique. Iwasa et al. [18] showed that $E[Z_2] \leq 2W_1^* + W_2^*$. In the proof of Theorem 3, we have shown that $E[Z] \leq W_1^* + 2(1 - 1/h)W_2^*$. By combining these results, we obtain that

$$
\begin{aligned}
E[\min\{Z_2, Z\}] &\leq E\left[\frac{h-2}{2(h-1)}Z_2 + \frac{h}{2(h-1)}E[Z]\right] \\
&= \frac{h-2}{2(h-1)}E[Z_2] + \frac{h}{2(h-1)}E[Z] \\
&\leq \frac{h-2}{2(h-1)}(2W_1^* + W_2^*) + \frac{h}{2(h-1)}\left(W_1^* + 2\left(1 - \frac{1}{h}\right)W_2^*\right) \\
&= \left(\frac{3}{2} - \frac{1}{2(h-1)}\right)(W_1^* + W_2^*) \leq \left(\frac{3}{2} - \frac{1}{2(h-1)}\right)(\text{opt. val. of SAP}).
\end{aligned}
$$

References

1. Adams, M.P., Sherali, H.D.: A tight linearization and an algorithm for zero-one quadratic programming problems. Manage. Sci. **32**, 1274–1290 (1986)
2. Alon, N., Karp, R.M., Peleg, D., West, D.: A graph-theoretic game and its application to the k-server problem. SIAM J. Comput. **24**, 78–100 (1995)
3. Alumur, S.A., Kara, B.Y.: Network hub location problems: the state of the art. Eur. J. Oper. Res. **190**, 1–21 (2008)

4. Alumur, S.A., Kara, B.Y., Karasan, O.E.: The design of single allocation incomplete hub networks. Transport. Res. B-Methodol. **43**, 951–956 (2009)
5. Ando, R., Matsui, T.: Algorithm for single allocation problem on hub-and-spoke networks in 2-dimensional plane. In: Asano, T., Nakano, S., Okamoto, Y., Watanabe, O. (eds.) ISAAC 2011. LNCS, vol. 7074, pp. 474–483. Springer, Heidelberg (2011). doi:10.1007/978-3-642-25591-5_49
6. Bein, W.W., Brucker, P., Park, J.K., Pathak, P.K.: A Monge property for the d-dimensional transportation problem. Discret. Appl. Math. **58**, 97–109 (1995)
7. Burkard, R.E., Klinz, B., Rudolf, R.: Perspectives of Monge properties in optimization. Discret. Appl. Math. **70**, 95–161 (1996)
8. Calik, H., Alumur, S.A., Kara, B.Y., Karasan, O.E.: A tabu-search based heuristic for the hub covering problem over incomplete hub networks. Comput. Oper. Res. **36**, 3088–3096 (2009)
9. Campbell, J.F.: A survey of network hub location. Stud. Locat. Anal. **6**, 31–47 (1994)
10. Campbell, J.F., Ernst, A.T., Krishnamoorthy, M.: Hub arc location problems: part I: introduction and results. Manage. Sci. **51**, 1540–1555 (2005)
11. Campbell, J.F., Ernst, A.T., Krishnamoorthy, M.: Hub arc location problems: part II: formulations and optimal Algorithms. Manage. Sci. **51**, 1556–1571 (2005)
12. Campbell, J.F., O'Kelly, M.E.: Twenty-five years of hub location research. Transport. Sci. **46**, 153–169 (2012)
13. Chuzhoy, J., Naor, J.: The hardness of metric labeling. SIAM J. Comput. **36**, 1376–1386 (2007)
14. Contreras, I., Fernández, E., Marín, A.: Tight bounds from a path based formulation for the tree of hub location problem. Comput. Oper. Res. **36**, 3117–3127 (2009)
15. Contreras, I., Fernández, E., Marín, A.: The tree of hubs location problem. Eur. J. Oper. Res. **202**, 390–400 (2010)
16. Contreras, I.: Hub location problems. In: Laporte, G., Nickel, S., Da Gama, F.S. (eds.) Location Science, pp. 311–344. Springer, New York (2015). (Chap. 12)
17. Contreras, I., Tanash, M., Vidyarthi, N.: Exact and heuristic approaches for the cycle hub location problem. Ann. Oper. Res. 1–23 (2016). doi:10.1007/s10479-015-2091-2
18. Iwasa, M., Saito, H., Matsui, T.: Approximation algorithms for the single allocation problem in hub-and-spoke networks and related metric labeling problems. Discret. Appl. Math. **157**, 2078–2088 (2009)
19. Kim, J.G., Tcha, D.W.: Optimal design of a two-level hierarchical network with tree-star configuration. Comput. Ind. Eng. **22**, 273–281 (1992)
20. Klincewicz, J.G.: Hub location in backbone/tributary network design: a review. Locat. Sci. **6**, 307–335 (1998)
21. Kleinberg, J., Tardos, E.: Approximation algorithms for classification problems with pairwise relationships. J. ACM **49**, 616–630 (2002)
22. Labbé, M., Yaman, H.: Solving the hub location problem in a star-star network. Networks **51**, 19–33 (2008)
23. de Sá, E.M., de Camargo, R.S., de Miranda, G.: An improved Benders decomposition algorithm for the tree of hubs location problem. Eur. J. Oper. Res. **226**, 185–202 (2013)
24. Martins de Sá, E., Contreras, I., Cordeau, J.F., de Camargo, R.S., de Miranda, G.: The hub line location problem. Transp. Sci. **49**, 500–518 (2015)
25. O'Kelly, M.E.: A quadratic integer program for the location of interacting hub facilities. Eur. J. Oper. Res. **32**, 393–404 (1987)

26. O'Kelly, M.E., Miller, H.J.: The hub network design problem: a review and synthesis. J. Transp. Geogr. **2**, 31–40 (1994)
27. Sohn, J., Park, S.: A linear program for the two-hub location problem. Eur. J. Oper. Res. **100**, 617–622 (1997)
28. Sohn, J., Park, S.: The single allocation problem in the interacting three-hub network. Networks **35**, 17–25 (2000)
29. Saito, H., Fujie, T., Matsui, T., Matuura, S.: A study of the quadratic semi-assignment polytope. Discret. Optim. **6**, 37–50 (2009)
30. Yaman, H.: Star p-hub median problem with modular arc capacities. Comput. Oper. Res. **35**, 3009–3019 (2008)
31. Yaman, H., Elloumi, S.: Star p-hub center problem and star p-hub median problem with bounded path lengths. Comput. Oper. Res. **39**, 2725–2732 (2012)

Improved Approximation for Two Dimensional Strip Packing with Polynomial Bounded Width

Klaus Jansen and Malin Rau[✉]

Institute of Computer Science, University of Kiel, 24118 Kiel, Germany
{kj,mra}@informatik.uni-kiel.de

Abstract. We study the well-known two-dimensional strip packing problem. Given is a set of rectangular axis-parallel items and a strip of width W with infinite height. The objective is to find a packing of these items into the strip, which minimizes the packing height. Lately, it has been shown that the lower bound of $3/2$ of the absolute approximation ratio can be beaten when we allow a pseudo-polynomial running-time of type $(nW)^{f(1/\varepsilon)}$, for any function f. If W is polynomially bounded by the number of items, this is a polynomial running-time. We present a pseudo-polynomial algorithm with approximation ratio $4/3 + \varepsilon$ and running time $(nW)^{1/\varepsilon^{\mathcal{O}(2^{1/\varepsilon})}}$.

Keywords: Strip packing · Pseudo polynomial running time · Structure analysis

1 Introduction

An instance of the strip packing problem consists of a strip of width $W \in \mathbb{N}$ and infinite height, and a set of items I, where each item $i \in I$ has width $w_i \in \mathbb{N}$ and height $h_i \in \mathbb{N}$, such that all items fit into the strip (i.e. $w_i \leq W$ f.a. $i \in I$).

A packing of the items is a mapping $\rho : I \to \mathbb{N} \times \mathbb{N}, i \mapsto (x_i, y_i)$, where $x_i \leq W - w_i$. We say an *inner point* of a placed item i is a point $(x, y) \in \mathbb{N} \times \mathbb{N}$, with $y_i \leq y < y_i + h_i$ and $x_i \leq x < x_i + w_i$. We say two items i and j *overlap* if there exists a point $(x, y) \in \mathbb{N} \times \mathbb{N}$, such that (x, y) is an inner point of i and an inner point of j. A packing is *feasible* if no two items overlap. The objective is to find a feasible packing, which minimizes its height $\max_{i \in I} y_i + h_i$. For a set of items S we denote its area by $A(S) := \sum_{i \in S} h_i w_i$. We denote the packing area by $W \times \max_{i \in I} y_i + h_i$.

Strip packing is one of the classical two-dimensional packing problems, which received a high research interest [2–4,6,7,10–16]. It arises naturally in many practical applications as manufacturing and logistics as well as in computer science. In computer science strip packing can be used to model scheduling parallel jobs on consecutive addresses. Here the width W of the strip equals the number of given processors.

Research was supported by German Research Foundation (DFG) project JA 612/14-2.

© Springer International Publishing AG 2017
S.-H. Poon et al. (Eds.): WALCOM 2017, LNCS 10167, pp. 409–420, 2017.
DOI: 10.1007/978-3-319-53925-6_32

If W occurs polynomially in the running time, it is called pseudo-polynomial. If $W \leq poly(n)$ the running time can be considered polynomial. The algorithm with the so far best absolute approximation ratio using pseudo-polynomial running time is the algorithm by Nadiradze and Wiese [13]. Their algorithm has an absolute approximation ratio of $1.4 + \varepsilon$.

Results and Methodology. We present an algorithm with absolute approximation ratio $4/3 + \varepsilon$, which has a pseudo-polynomial running time. The main difficulty arises when placing items which have a small width and a large height. If the considered algorithm can not place all these items into the area $W \times$ OPT, it would have to place it above this area, adding its height to the height of the packing. Since these items can have a height up to OPT this can double the height of the packing.

In [13] Nadiradze and Wiese presented a new technique to handle tall items, which have small width and height larger than 0.4OPT. They managed to place all these items into the area $W \times$ OPT. In this packing some of the items with height up to 0.4OPT are shifted upwards and are placed above OPT. These shifted items are responsible for adding 0.4OPT to the absolute approximation ratio.

We present a new structural result, leading to an algorithm, that can place all items with height at least $\frac{1}{3}$OPT in $W \times$ OPT. By this optimization just items with height up to $\frac{1}{3}$OPT have to be placed above this area, which results in an approximation algorithm with absolute approximation ratio $4/3 + \varepsilon$. This is possible since we could reduce the area of the items, that have to be shifted on top of the optimal packing area. The key to this better approximation lies in Lemma 5.

The second improvement to the algorithm in [13] lies in the running time. The main idea in [13] is to divide the packing area into a constant number of rectangular areas. The number of these areas depends on ε and can be quite large (i.e. $\Omega(6^{1/\delta})$). Since the number of these boxes influences the choice of δ this induces a very large running time, i.e. $\mathcal{O}(W^{1/\delta})$, where in the worst case $\delta \in \Omega(1/\exp_6^{1/\varepsilon}(1/\varepsilon))$, where $\exp_6^{1/\varepsilon}(1/\varepsilon) := 6^{\cdots^{6^{1/\varepsilon}}}$ and the 6 occurs $1/\varepsilon$ times (tower of exponents). We manage to reduce the number of these areas dramatically (i.e. $\mathcal{O}(1/\varepsilon^3\delta^2)$ which implies $\delta \in \Omega(\varepsilon^{\mathcal{O}(2^{1/\varepsilon})})$). How we find this better partition is described in the proof of Lemma 7. The result of our research is summarized in the following theorem:

Theorem 1. *For each $\varepsilon > 0$ there is an algorithm that finds a solution for each instance of the strip packing problem with height at most $(4/3 + \varepsilon)$OPT. The algorithm needs at most $(nW)^{1/\varepsilon^{\mathcal{O}(2^{1/\varepsilon})}}$ operations.*

An algorithm with the same approximation ratio was developed independently and at the same time by Gálvez et al. [5]. They extended their approach to strip packing with rotations, but did not improve the running time.

Related Work. The first algorithm for the strip packing problem was described by Baker and Coffman [3] in 1980. If the rectangles are ordered by descending width, this algorithm has an asymptotic approximation ratio of 3. The first algorithms with proven absolute approximation ratios of 3 and 2.7 were given by Coffman et al. [4]. After that Sleator [15] presented an algorithm which generates a schedule of height $2OPT(I) + h_{\max}(I)/2$, where $h_{\max}$ is the largest height of the items. So this algorithm has an asymptotic approximation ratio 2. Schiermeyer [14] and Steinberg [16] improved this algorithm independently to an algorithm with absolute approximation ratio 2. Harren and van Stee were the first to beat the barrier of 2. They presented an algorithm with an absolute approximation ratio of 1.9396 [8]. The so far best absolute approximation is given by the algorithm by Harren et al. [7], which has an absolute approximation ratio of $(5/3 + \epsilon)OPT(I)$. A reduction from the partition problem gives a lower bound on the absolute approximation ratio of $3/2 \cdot \text{OPT}$ for any polynomial approximation algorithm.

In the asymptotic case, the barrier of $3/2$ can be beaten. Golan [6] presented the first algorithm with asymptotic approximation ratio smaller than $3/2$. It has an asymptotic approximation ratio of $4/3$. Next Baker et al. [2] gave an algorithm with asymptotic ratio $5/4$. After that Kenyon and Rémila [12] presented an AFPTAS which has an approximation ratio of $(1 + \epsilon)\text{OPT}$ and an additive constant $\mathcal{O}(h_{\max}/\epsilon^2)$. Later the additive constant was improved by Jansen and Solis-Oba [10] at the expense of the processing time of the algorithm. They presented an APTAS, which generates a schedule of height $(1 + \epsilon)\text{OPT} + h_{\max}$.

If we allow pseudo-polynomial processing time, there are better approximations possible. This is thanks to the fact that the underlying partition problem is solvable in pseudo-polynomial time. Jansen and Thöle [11] presented an algorithm with approximation ratio $3/2 + \epsilon$. Recently Nadiradze and Wiese [13] have presented an algorithm which beats the bound $3/2$. It has an approximation ratio of $1.4 + \epsilon$. On the negative side Adamaszek et al. [1] have shown that there is no pseudo-polynomial algorithm with approximation ratio smaller than $\frac{12}{11}\text{OPT}$.

2 Simplifying the Input Instance

Let $\varepsilon > 0$, such that $1/\varepsilon \in \mathbb{N}$. Further, let an instance of the strip packing problem be given and consider an optimal solution to it, which has a packing height of OPT. We will show in the following that it is possible to transform this optimal solution into a solution which has a particular structure. By this transformation we add at most $(1/3 + \varepsilon)\text{OPT}$ to the height of the packing, resulting in a packing with height $(4/3 + \varepsilon)\text{OPT}$. When we know the structure of the optimal packing, we can use the same techniques as in [13] to describe an algorithm. Notice that we can find the height of the optimal packing by a binary search framework in $\mathcal{O}(\log(\text{OPT}))$ steps, which is polynomial in the input size. The described algorithm would also work if we would approximate the optimal packing height within the range of $(1 + \mathcal{O}(\varepsilon))$, which would result in $\mathcal{O}(\log(1/\varepsilon))$ steps of the framework, but for the simplification of the notation we use the exact height.

The first step in the transformation as well as in the algorithm is to partition the set of items I. Let $\delta = \delta(\varepsilon) > \mu = \mu(\varepsilon)$ be two suitable constants depending on ε. We define the set of large items $L := \{i \in I | h_i \geq \delta\mathrm{OPT}, w_i \geq \delta W\}$, tall items $T := \{i \in I \setminus L | h_i \geq (1/3 + \varepsilon)\mathrm{OPT}\}$, vertical items $V := \{i \in I \setminus T | h_i \geq \delta\mathrm{OPT}, w_i \leq \mu W\}$, medium sized vertical items $M_V := \{i \in I \setminus T | h_i \geq \delta\mathrm{OPT}, \mu W < w_i < \delta W\}$, horizontal items $H := \{i \in I | h_i \leq \mu\mathrm{OPT}, w_i \geq \delta W\}$, small items $S := \{i \in I | h_i \leq \mu\mathrm{OPT}, w_i \leq \mu W\}$ and medium sized horizontal items $M_H := I \setminus (L \cup T \cup V \cup M_V \cup H \cup S)$.

The medium sized items will be placed outside the optimal packing area. To guarantee that these items do not occupy to much space there, their total area has to be small. We achieve this by finding appropriate values for δ and μ, whose existence is stated in Lemma 1. It is a standard argument which follows by the pigeon-hole principle and is often used in packing algorithms, e.g. in [10]. The proofs not stated in this short version can be found in the full version [9].

Lemma 1. *Consider the sequence $\sigma_0 = \varepsilon^6$, $\sigma_{i+1} = \sigma_i^2 \varepsilon^6$. There is a value $j \in \{0, \ldots, 6/\varepsilon - 1\}$ such that when defining $\delta = \sigma_j$ and $\mu = \sigma_{j+1}$ the total area of the items in $M_V \cup M_H$ is at most $\varepsilon/6 \cdot \mathrm{OPT} \cdot W$.*

Note that $\sigma_i = \varepsilon^{6(2^{i+1}-1)}$. Since σ is strictly monotonic decreasing we have $\delta \geq \sigma_{6/\varepsilon-1} = \varepsilon^k$, for $k = 6 \cdot 2^{6/\varepsilon}$. The next step in our transformation is to round the heights of the items in $L \cup T \cup V$ and shift them such that they start and end at certain heights. Our rounding strategy is similar to the strategy in [13] but we manage to reduce the number of different heights.

Lemma 2. *Let $\delta = \varepsilon^k$ for some value $k \in \mathbb{N}$. At a loss of at most a factor $1 + 2\varepsilon$ in the approximation ratio we can ensure that each item $i \in L \cup T \cup V$ with $\varepsilon^{l-1}\mathrm{OPT} \geq h_i \geq \varepsilon^l\mathrm{OPT}$ for some $l \in \mathbb{N}_{\leq k}$ has height $h_i = k_i\varepsilon^{l+1}\mathrm{OPT} = k_i/\varepsilon^{k-l} \cdot \varepsilon^{k+1}\mathrm{OPT}$ for some $k_i \in \{0, \ldots, 1/\varepsilon - 1\}$. Furthermore the items' y-coordinates can be placed at multiples of $\varepsilon^{l+1}\mathrm{OPT}$.*

3   Partitioning the Packing Area

We apply the rounding according to Lemma 2, obtaining a packing where each item in $L \cup T \cup V$ starts and ends at multiples of $\delta\varepsilon$. The packing has now a height of $(1 + 2\varepsilon)\mathrm{OPT}$. We will show that we can partition its packing area into a constant number of rectangular areas, such that each of these areas contains items just from one of the following sets: L, $H \cup S$, or $T \cup V \cup S$. We allow items from $H \cup S$ or $T \cup V \cup S$ to be positioned into more than one area. While the placement of the items L or H is simple, it is difficult to place the items from $T \cup V$, without increasing the height of the packing too much. Note that there are at most $1/\delta^2$ large items since each of them covers an area of at least $\delta^2 W\mathrm{OPT}$.

Lemma 3. *We can partition the area $W \times (1 + 2\varepsilon)\mathrm{OPT}$ into at most $4(1 + 2\varepsilon)/(\varepsilon\delta^2)$ rectangular boxes. These boxes can be partitioned into sets of boxes $\mathcal{B}_L, \mathcal{B}_H$ and $\mathcal{B}_{T \cup V}$ such that*

- boxes in $\mathcal{B}_L$ are identified by items $i \in L$, i.e. they have box height h_i and box width w_i,
- $\mathcal{B}_H$ consists of at most $(1 + 2\varepsilon)/(\varepsilon\delta^2) - |L|/\delta$ many boxes of height $\varepsilon\delta$OPT, each of them containing at least some item in H but only items in $H \cup S$,
- $\mathcal{B}_{T \cup V}$ consists of at most $3(1+2\varepsilon)/(\varepsilon\delta^2)$ many boxes, each of them containing only items in $T \cup V \cup S$,
- no item in H is intersected vertically by any box border,
- no item in $T \cup V$ is intersected horizontally by any box border,
- no item in L is intersected by any box border.

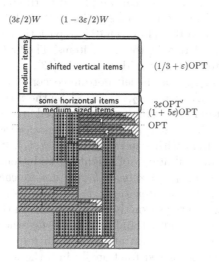

Fig. 1. Structure of the simplified packing. Boxes in $\mathcal{B}_H$ are hatched and boxes in $\mathcal{B}_{T \cup V}$ are dotted. Above the packing area we need an area for medium sized items, some horizontal and some vertical items, that have to be shifted up.

An overview over the simplified partition can be found in Fig. 1. To simplify the argumentation in the following, we forget about the items in S. We are going to show that it is possible to partition the area in each box in $\mathcal{B}_{T \cup V}$ into a constant number of subboxes. By this partition, we will get subboxes for vertical and small items $\mathcal{B}_V$ and other subboxes $\mathcal{B}_T$, which will contain only tall items, which have the height of the subbox. To achieve this property, we have to rearrange the items in the boxes $B \in \mathcal{B}_{T \cup V}$. In the following, we will focus on the more interesting case, where the height of the box $h(B)$ is at least $(2/3 + 2\varepsilon)$OPT. Let $\hat{\mathcal{B}}_{T \cup V} \subseteq \mathcal{B}_{T \cup V}$ be the set of boxes with height at least $(2/3 + 2\varepsilon)$OPT and $\check{\mathcal{B}}_{T \cup V} := \mathcal{B}_{T \cup V} \setminus \hat{\mathcal{B}}_{T \cup V}$.

Let us, for now, assume that we are allowed to slice all vertical items vertically as often as we desire. If we consider a packing of items, where some of the vertical items are sliced, we call it a fractional packing. We call all tall items, which are crossed (vertically) by a box border unmovable and the other tall items movable.

The first step is to shift tall items up or down respectively such that all movable tall items either touch the top or the bottom of the box. That items can be rearranged this way was shown in Lemma 1.4 in [13].

Lemma 4 [13]. *If we are allowed to slice the items in V vertically, we can ensure the following: In each box $B \in \mathcal{B}_{T \cup V}$ there is a packing where all movable tall items are either touching the top or the bottom of the box.*

Let us from now on assume that all movable tall items are touching the top or the bottom of the boxes in $\mathcal{B}_{T \cup V}$. We now want to reorder the tall items to find a partition into subboxes $\mathcal{B}_V$ and $\mathcal{B}_T$. If we reorder the tall items, it can happen that not all vertical items can be placed in the box. All vertical items that can not be placed in the box have to be shifted above the packing area. Since we have just the area $W \times (1/3 + \varepsilon)\mathrm{OPT}$ to pack the shifted items and M_V, we have to be careful, not to shift too many items. The first step is to introduce pseudo and dummy items similar as described in [13].

For $B \in \hat{\mathcal{B}}_{T \cup V}$ let (x_l, y_b) be the left bottom corner and (x_r, y_t) the top right corner respectively. Let $X = \{x_1, x_2, \ldots, x_{k-1}\}$ be the x-coordinates of the left bottom corner of the tall items in the packing, ordered in increasing order. Define $x_0 := x_l$ and $x_k := x_r$. If $[x_{j-1}, x_j) \times [y_b, y_t)$ does not overlap any tall item, we introduce one pseudo item with size $[x_{j-1}, x_j) \times [y_b, y_t)$. If $[x_{j-1}, x_j) \times [y_b, y_t)$ overlaps with exactly one tall item i of height h_i, we introduce one pseudo item which covers the area $[x_{j-1}, x_j) \times [y_b + h_i, y_t)$ if i touches the bottom, or a pseudo item which covers the area $[x_{j-1}, x_j) \times [y_b, y_t - h_i)$ otherwise. In the case $[x_{j-1}, x_j) \times [y_b, y_t)$ overlaps exactly two tall items. In this case we introduce $x_j - x_{j-1}$ dummy items of width 1 and height $h_t - h_b$, where h_t is the height of the item at the top and h_b the item at the bottom respectively. Obviously, all vertical items in B can be placed fractionally into the area of the pseudo and the dummy items.

Let P be the set of introduced pseudo items for the box B. A reordering of the tall and pseudo items in B is a rearrangement of the items, such that we just change the x-coordinate of the bottom-left corner, but not the y-coordinate. It is feasible if there are no two items in $T \cup P$ that overlap. For each reordering we have an individual set of dummy items D. If we reorder the tall and pseudo items, it can happen, that it is not possible to place all the items which are contained fractionally in the original set of dummy items into the new set of dummy items. Unlike in [13] we have to ensure that at least a constant amount of these items can be placed in B. We are now going to analyze which amount of these items can be placed in any reordering.

Lemma 5. *Let $B \in \hat{\mathcal{B}}_{T \cup V}$ with width w, D be its set of dummy items, and P its set of pseudo items. Let $\beta := \min\{|h_i - h_j| : i, j \in T \cup P, h_i \neq h_j\}$ be the minimal difference between the heights of two items in $T \cup P$. Let $t \in T \cup P$ be the shortest item touching the top and $b \in T \cup P$ be the shortest item touching the bottom, with $h_t > 0$ and $h_b > 0$. Define $h := (h(B) - h_t - h_b)$. For each feasible reordering of the tall and pseudo items and each $\alpha \leq \beta/(\beta + h)$, we can find a subset $S \subseteq D$, with $|S| \geq \alpha w$ that can be placed in the reordered packing.*

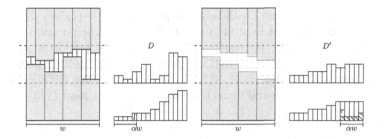

Fig. 2. Two orderings of the items in $T \cup P$.

Proof. Let D' be the set of dummy items in a given feasible reordering. We sort both sets D and D' in ascending order and index them from 1 to w. We will show that the $\lceil \alpha w \rceil$ smallest items in D fit into the $\lceil \alpha w \rceil$ largest items in D'. Let l be the item with index $\lceil \alpha w \rceil$ in the set D and let r be the item with index $w - \lceil \alpha w \rceil + 1$ in the set D'. If $h_l \leq h_r$ the $\lceil \alpha w \rceil$ shortest items in D can be placed into the $\lceil \alpha w \rceil$ longest items in D, see Fig. 2.

Assume for contradiction that $h_l > h_r$. We know that $A(D) = A(D')$, since the set of tall and pseudo items is unchanged. Since each dummy item with index $\geq l$ has height at least h_l, we know that $A(D) \geq h_l(w - \lceil \alpha w \rceil + 1)$. Furthermore we know that $A(D') \leq h_r(w - \lceil \alpha w \rceil + 1) + h(\lceil \alpha w \rceil - 1)$, since each dummy item i with $i \leq r$ has height at most h_r and each dummy item i with $i > r$ has height at most h. So in total we have

$$h_l(w - \lceil \alpha w \rceil + 1) \leq A(D) = A(D') \leq h_r(w - \lceil \alpha w \rceil + 1) + h(\lceil \alpha w \rceil - 1).$$

Since $h_l > h_r$ and the difference between two items out of $T \cup P$ is at least β we have $h_l \geq h_r + \beta$. This leads to

$$(h_r + \beta)(w - \lceil \alpha w \rceil + 1) \leq h_l(w - \lceil \alpha w \rceil + 1) \leq h_r(w - \lceil \alpha w \rceil + 1) + h(\lceil \alpha w \rceil - 1).$$

It follows that $w\beta \leq (\beta + h)(\lceil \alpha w \rceil - 1)$. Since $\lceil \alpha w \rceil - 1 < \alpha w$ this leads to $\beta < (\beta + h)\alpha$, which is a contradiction for each $\alpha \leq \beta/(h + \beta)$.

All the dummy items that can not be placed into the rearranged packings will be placed in an extra box V_0 of height $(1/3 + \varepsilon)$OPT and width $(1 - \alpha)W$. Since next to this box we have to place a box for medium sized items we want α to be as large as possible. The following Lemma states some useful properties, we can assume when reordering the items in $T \cup P$.

Lemma 6. *If we increase the height of the packing area $W \times (1 + 2\varepsilon)$OPT by 3εOPT, we can assume that each item in $T \cup P$ has a height, which is a multiple of εOPT. Furthermore, at each side of a box, there is at most one tall item overlapping its border, that touches either its bottom or top and has a height, that is a multiple of εOPT.*

Since now each item height in $T \cup P$ is a multiple of εOPT we have $\beta \geq \varepsilon$OPT. Let us take a look at items that are very tall with respect to the size of a box

$B \in \hat{\mathcal{B}}_{T \cup V}$. Consider an item i with height larger than $h(B) - (1/3 + \varepsilon)$OPT. Since each tall item has height at least $(1/3 + \varepsilon)$OPT, there can be no tall item placed above or below this item. By construction, there is one pseudo item directly above or below i. We combine i and the pseudo item to one new pseudo item which has height $h(B)$ and width w_i. Now it holds that the distance between items touching the bottom and items touching the top is at most $(1/3 + 3\varepsilon)$OPT. So we can choose $\alpha = 2\varepsilon < \varepsilon/(1/3 + 4\varepsilon)$ for $\varepsilon < 1/24$.

Note that there are at most $R := (1/3 + 3\varepsilon)/\varepsilon + 2 = (1/3 + 5\varepsilon)/\varepsilon$ possible item heights in $P \cup T$ with respect to a box $B \in \mathcal{B}_{T \cup V}$. In the next step, we show how it is possible to rearrange the tall and pseudo items touching the top and the bottom of a box, such that we need few boxes for tall and vertical items.

Lemma 7. *Let $B \in \hat{\mathcal{B}}_{T \cup V}$. We can find a rearrangement of tall and pseudo items in B, such that we can partition B into at most $\mathcal{O}(1/\varepsilon^2)$ subboxes $\mathcal{B}_V$ and $\mathcal{B}_T$, such that each subbox B_T contains just tall items with height $h(B_T)$, and all vertical items in B can be packed fractionally into the subboxes $\mathcal{B}_V$ and one extra box V_B of size $(1/3 + \varepsilon)$OPT $\times (1 - 2\varepsilon)w(B)$.*

Proof. We consider two cases. In the first no tall item overlaps the left or right border of B. For this case it is shown in [13] that we can simply sort the items from $T \cup P$ touching the top of B in descending order of heights and the items touching the bottom in ascending order. We sort tall and pseudo items of the same height such that pseudo items are positioned left to the tall items. By this reordering no two items overlap. And we have at most one group of items of the same size on each side of the box. For each group of tall items of the same size we introduce one subbox for $\mathcal{B}_T$, which contains exactly these group of tall items and for each group of pseudo item we introduce one subbox in $\mathcal{B}_V$. Furthermore for each group of dummy items of the same size, we introduce one subbox in $\mathcal{B}_V$ analogously. Since we have at most R different item sizes, we introduce at most $2R$ boxes for tall items $\mathcal{B}_T$. Since the height of the items touching the bottom and touching the top are changing at most $2R$ times totally, we introduce at most $4R$ subboxes for vertical items $\mathcal{B}_V$.

In the second case, it is possible that tall items overlap the box B on the left or right side. By Lemma 6 we can assume that there is at most one tall item per box side. Now we reorder the items differently from [13]. We reduce the number of boxes from an exponential to a quadratic function in the number of different heights in $T \cup P$.

Let h_b be the height of a tallest item touching the bottom of the box and b_l be the leftmost and b_r the rightmost item of height h_b. Similarly choose h_t, t_l, t_r with respect to the top. Further, let i_l be the item in $\{t_l, b_l\}$ which is further left and i_r the item which in $\{t_r, b_r\}$, which touches the other border. W.l.o.g let $i_l = b_l$. We draw a vertical line at the left border of b_l. The item cut by this line defines a new unmovable item b'. We do the same on the right side of t_r and name the cut item t' (see Fig. 3). Now the movable items touching the top are sorted in ascending order with respect to their height, while the movable items touching the bottom are sorted in descending order.

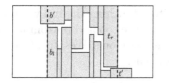

Fig. 3. A packing before and after the reordering of the items.

We show, that in this reordering no two items overlap. There is no tall item touching the bottom that overlaps b' since each item touching the bottom has height at most h_b. Since b' was placed above b_l this means b' fits above each item in the box B. Similarly one can see that no item overlaps t'. Assume there is an item i_b touching the bottom that overlaps an item i_t touching the top. Let $p = (x_p, y_p)$ be a point, which is overlapped by the item i_b and i_t. Let (x_l, y_b) denote the left bottom corner of b_l and (x_r, y_t) the right top corner of t_l. The reordering guarantees the existents of a set of items I_b touching the bottom with total width greater than $x_p - x_l$, which is placed between x_l and x_r and has height at least $y_p - y_b$. Furthermore there must be a set of items I_t with total width greater than $x_r - x_p$ touching the bottom and having height at least $y_t - y_p$. Since the area the items can be placed in has a width of $x_r - x_l$ and the sets I_b and I_t have a total width of $w(I_t \cup I_b) > x_p - x_l + x_r - x_p = x_r - x_l$ by the pidgin hole principle there must be an item in I_b that overlaps an item in I_t in the original packing, which is a contradiction.

Now for each group of tall, pseudo and dummy items between i_l and i_r having the same hight, we introduce subboxes as described above. We have introduced at most $R + 1$ boxes to $\mathcal{B}_T$ and at most $4R + 2$ boxes to $\mathcal{B}_T$. The total number of different heights touching the bottom and touching the top, on the left of i_l is at least one smaller than in the whole box. Same holds for the number on the right side of i_r.

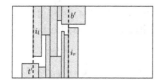

Fig. 4. A recursive rearrangement of the tall and pseudo items with updated labels.

We repeat the following step until a break condition occur. In each step, we will reduce the total number of different heights of the items touching the top and bottom by at least one. We look on the left side of i_l. W.l.o.g. let i_l touching the bottom of the box. Let b' be the item, which was intersected by the vertical line at the left border of i_l. Let h_t be the height of the largest item touching the top left of i_l. We rename the item i_l as i_r and redefine i_l as the left most item touching the top, which has height h_t. Between the items i_l and i_r we proceed as above (Fig. 4).

By choosing i_l as the leftmost tallest item touching the top we have reduced the total number of different heights touching the top and bottom in the remaining area, which has to be reordered, by at least one. We repeat the described step until one of the following conditions occur:

1. The tallest item touching the top and the tallest item touching the bottom have a summed height of at most h.
2. the item i_r is the unmovable item, which overlaps the left border.

If condition 1 occur in any reordering of the items it can not happen that a tall or pseudo item touching the bottom overlaps any tall or pseudo item touching the top, since their height is not large enough. So at this point we simply sort the items touching the top in ascending order and the items touching the bottom just as well.

If condition 2 occur we repeat the normal reordering step once again. When we draw the vertical line, it will be placed exactly on the box border, and we are finished.

We repeat this steps analog on the right side of the initial i_r. As seen before in each of the reordering steps we introduce at most $2R + 1$ subboxes to $\mathcal{B}_T$ and at most $4R + 2$ subboxes to $\mathcal{B}_V$. In each of the partition steps, we reduce the total number of different heights touching the bottom and the top by one. If the tallest item at the top and the tallest item at the bottom are both smaller than $h(B)/2$ condition 1 is fulfilled. Since in each partitioning step the number of tall item sizes is reduced by one, we need at most R steps until the tallest item on the bottom and the tallest item on the top both have a height of at most $h(B)/2$. By all these steps we have created at most one box with height $h(B)$, since this item is the largest item in the first step and it touches the bottom as well as the top. So if we sort the tall items in this box, we generate at most R further boxes. So we partition B into at most $2R + 2$ parts, each containing at most $2R$ subboxes for tall items and at most $4R$ subboxes for vertical items. So in totat we have $|\mathcal{B}_T| \leq (2R+2)2R = (2(1/3+5\varepsilon)/\varepsilon + 2)2(1/3+5\varepsilon)/\varepsilon \leq 1.3/\varepsilon^2$ and $|\mathcal{B}_V| \leq (2R+2)4R = (2(1/3+5\varepsilon)/\varepsilon + 2)4(1/3+5\varepsilon)/\varepsilon \leq 2.2/\varepsilon^2$, if $\varepsilon \leq 24$.

Considering Lemma 5 clearly all vertical items fit into the boxes $\mathcal{B}_V$ and the additional box V_B.

Lemma 8. *We can find a rearrangement of the items in each box $B \in \check{\mathcal{B}}_{T \cup V}$ such that we can partition the area in B into at most $\mathcal{O}(1/\varepsilon)$ subboxes $\mathcal{B}_V$ and $\mathcal{B}_T$, such that all vertical items can be packed fractionally in the subboxes $\mathcal{B}_V$, and each subbox in $B' \in \mathcal{B}_T$ contains just tall items with height $h(B')$.*

Since the boxes in $\check{\mathcal{B}}_{T \cup V}$ can be partitioned into less boxes than the boxes $\hat{\mathcal{B}}_{T \cup V}$, the following Lemma follows from Lemmas 5, 6 and 7.

Lemma 9. *We can partition the packing area $W \times (1 + 5\varepsilon)\mathrm{OPT}$ such that we introduce at most $\mathcal{O}(1/\varepsilon^3\delta^2)$ boxes for tall items $\mathcal{B}_T$, each containing just items with the same height, and at most $\mathcal{O}(1/\varepsilon^3\delta)$ boxes $\mathcal{B}_V$ for vertical items, such that all vertical items can be packed fractionally into the boxes $\mathcal{B}_V$ and an additional* box V_0 with height $(1/3 + \varepsilon)$OPT and width $(1 - 2\varepsilon)W$. Furthermore at most $|L|$ boxes $\mathcal{B}_L$ containing large items, and at most $\mathcal{O}(1/\varepsilon\delta^2)$ boxes $\mathcal{B}_H$, such that all horizontal items fit fractionally into $\mathcal{B}_H$. Furthermore the area of the small items fit into the free spaces in $\mathcal{B}_H \cup \mathcal{B}_V$.

4 Algorithm

We define boxes H_0 with size $W \times 2\varepsilon$OPT, B_{MH} with size $W \times \varepsilon$OPT, B_{MV} with size $3\varepsilon W/2 \times (1/3 + \varepsilon)$OPT and B_V with size $(1 - 3\varepsilon/2)W \times (1/3 + \varepsilon)$OPT.

Lemma 10. *Let a partition into boxes $\mathcal{B}_L$, $\mathcal{B}_H$, $\mathcal{B}_T$, and $\mathcal{B}_V$ be given, such that $|\mathcal{B}_L \cup \mathcal{B}_H \cup \mathcal{B}_T \cup \mathcal{B}_V| \in \mathcal{O}(1/\varepsilon^3\delta^2)$. There is an algorithm with running time $\mathcal{O}(n \log n + W^{(1/\varepsilon\delta)^{\mathcal{O}(1)}})$ that places all the items in I into the boxes $\mathcal{B}_L \cup \mathcal{B}_H \cup \mathcal{B}_T \cup \mathcal{B}_V \cup \{H_0, B_{MH}, B_{MV}, B_V\}$ or decides that such packing does not exist.*

Let us summarize what the structure of the adjusted optimal packing looks like (see Fig. 1): We have stretched the optimal packing area, such that it has a height of $(1 + 5\varepsilon)$OPT. We have an extra box H_0 for horizontal items, which has height 2εOPT and width W. We place this box exactly above the packing area of height $(1 + 5\varepsilon)$OPT. For the medium sized items, we have introduced two boxes. One has height εOPT and can be placed above the box for horizontal items. The other has height $(1/3 + \varepsilon)$OPT and width $(3\varepsilon/2)W$. We will place this box next to the extra box for vertical items, which has height $(1/3 + \varepsilon)$OPT and width $(1 - 3\varepsilon/2)W$. So the total height of the current packing is $(4/3 + 9\varepsilon)$OPT. So if we substitute ε with $\varepsilon' := \varepsilon/9$ the simplified packing has a height of at most $(4/3 + \varepsilon)$OPT.

The algorithm works as follows: First we set $\varepsilon' := \min\{\varepsilon/9, 1/24\}$. After that we have to find the height of the optimal packing OPT with a binary search framework, which takes $\mathcal{O}(\log(\text{OPT}))$ steps. For each guessed packing height we find the correct values for δ and μ and round the items in $T \cup V \cup L$. This can be done in $\mathcal{O}(n)$. Now we guess the structure of the packing. There are at most $W^{1/\varepsilon^{\mathcal{O}(2^{1/\varepsilon})}}$ possibilities. For each of the guessed partitions, we check with the algorithm from Lemma 10 if we can place the items in I into that partition. The total running time is bounded by $\log(\text{OPT})(nW)^{1/\varepsilon^{\mathcal{O}(2^{1/\varepsilon})}}$. If we use just an approximation of the value OPT the binary search needs just $\log(1/\varepsilon)$ steps, which results in a running time of $(nW)^{1/\varepsilon^{\mathcal{O}(2^{1/\varepsilon})}}$.

References

1. Adamaszek, A., Kociumaka, T., Pilipczuk, M., Pilipczuk, M.: Hardness of approximation for strip packing. CoRR abs/1610.07766 (2016)
2. Baker, B., Brown, D., Katseff, H.: A 5/4 algorithm for two-dimensional packing. J. Algorithms **2**(4), 348–368 (1981)
3. Baker, B., Coffman Jr., E.G., Rivest, R.: Orthogonal packings in two dimensions. SIAM J. Comput. **9**(4), 846–855 (1980)

4. Coffman Jr., E., Garey, M., Johnson, D., Tarjan, R.: Performance bounds for level-oriented two-dimensional packing algorithms. SIAM J. Comput. **9**(4), 808–826 (1980)

5. Gálvez, W., Grandoni, F., Ingala, S., Khan, A.: Improved pseudo-polynomial-time approximation for strip packing. In: 36th IARCS Annual Conference on Foundations of Software Technology and Theoretical Computer Science, FSTTCS, 13–15 December 2016, Chennai, India, pp. 9:1–9:14 (2016)

6. Golan, I.: Performance bounds for orthogonal oriented two-dimensional packing algorithms. SIAM J. Comput. **10**(3), 571–582 (1981)

7. Harren, R., Jansen, K., Prädel, L., Van Stee, R.: A $(5/3 + \varepsilon)$-approximation for strip packing. Comput. Geom. **47**(2), 248–267 (2014)

8. Harren, R., van Stee, R.: Improved absolute approximation ratios for two-dimensional packing problems. In: Dinur, I., Jansen, K., Naor, J., Rolim, J. (eds.) APPROX/RANDOM 2009. LNCS, vol. 5687, pp. 177–189. Springer, Heidelberg (2009). doi:10.1007/978-3-642-03685-9_14

9. Jansen, K., Rau, M.: Improved approximation for two dimensional strip packing with polynomial bounded width. CoRR abs/1610.04430 (2016)

10. Jansen, K., Solis-Oba, R.: Rectangle packing with one-dimensional resource augmentation. Discrete Optim. **6**(3), 310–323 (2009)

11. Jansen, K., Thöle, R.: Approximation algorithms for scheduling parallel jobs. SIAM J. Comput. **39**(8), 3571–3615 (2010)

12. Kenyon, C., Rémila, E.: A near-optimal solution to a two-dimensional cutting stock problem. Math. Oper. Res. **25**(4), 645–656 (2000)

13. Nadiradze, G., Wiese, A.: On approximating strip packing with a better ratio than 3/2. In: 27th Annual ACM-SIAM Symposium on Discrete Algorithms (SODA), pp. 1491–1510. SIAM (2016)

14. Schiermeyer, I.: Reverse-fit: a 2-optimal algorithm for packing rectangles. In: Leeuwen, J. (ed.) ESA 1994. LNCS, vol. 855, pp. 290–299. Springer, Heidelberg (1994). doi:10.1007/BFb0049416

15. Sleator, D.: A 2.5 times optimal algorithm for packing in two dimensions. Inf. Process. Lett. **10**(1), 37–40 (1980)

16. Steinberg, A.: A strip-packing algorithm with absolute performance bound 2. SIAM J. Comput. **26**(2), 401–409 (1997)

An FPTAS for Computing the Distribution Function of the Longest Path Length in DAGs with Uniformly Distributed Edge Lengths

Ei Ando[(✉)]

Sojo University, 4-22-1, Ikeda, Nishi-ku, Kumamoto 860-0082, Japan
ando-ei@cis.sojo-u.ac.jp

Abstract. Given a directed acyclic graph (DAG) $G = (V, E)$ with n vertices and m edges, we consider random edge lengths. That is, as the input, we have $a \in \mathbb{Z}_{>0}^m$, whose components are given for each edges $e \in E$. Then, the random length Y_e of edge e is a mutually independent random variable that obeys a uniform distribution on $[0, a_e]$. In this paper, we consider the probability that the longest path length is at most a certain value $x \in \mathbb{R}_{\geq 0}$, which is equal to the probability that all paths in G have length at most x. The problem can be considered as the computation of an m-dimensional polytope $K_G(a, x)$ that is a hypercube truncated by exponentially many hyperplanes that are as many as the number of paths in G. This problem is $\#P$-hard even if G is a directed path. In this paper, motivated by the recent technique of deterministic approximation of $\#P$-hard problems, we show that there is a *deterministic* FPTAS for the problem of computing $\text{Vol}(K_G(a, x))$ if the pathwidth of G is bounded by a constant p. Our algorithm outputs a value V' satisfying that $1 \leq V'/\text{Vol}(K_G(a, x)) \leq 1 + \epsilon$ and finishes in $O(p^4 2^{1.5p} n (\frac{2mnp}{\epsilon})^{3p} L)$ time, where L is the number of bits in the input. If the pathwidth p is a constant, the running time is $O(n(\frac{mn}{\epsilon})^{3p} L)$.

1 Introduction

In this paper, we consider the longest path problem in directed acyclic graphs (DAGs) with random edge lengths. We consider a DAG $G = (V, E)$ with vertex set $V = \{v_1, \ldots, v_n\}$ and edge set $E \subseteq V \times V$ where $|E| = m$. We assume that the vertex set $V = \{v_1, \ldots, v_n\}$ are topologically ordered. Then, we consider m mutually independent random variables. The distribution of the random edge lengths are uniform distribution given by a vector $a \in \mathbb{Z}_{>0}^m$, where each component a_{ij} of a is given for each edge $(v_i, v_j) \in E$. Let X_{ij} for $e = (v_i, v_j) \in E$ be mutually independent random variables uniformly distributed over $[0, 1]$. For each edge $e = (v_i, v_j) \in E$, $Y_e = a_{ij} X_{ij}$ is the random edge length with its distribution function $F_{ij}(x) = \Pr[a_{ij} X_{ij} \leq x]$. Let Π be the set of all paths from the sources to the terminals in G, where a source (resp. a terminal) is a vertex with indegree (resp. outdegree) 0. We are to compute the probability that the longest path length $X_{\text{MAX}} = \max_{\pi \in \Pi} \{\sum_{e \in \pi} a_e X_e\}$ is at most a certain value $x \in \mathbb{R}_{\geq 0}$.

© Springer International Publishing AG 2017
S.-H. Poon et al. (Eds.): WALCOM 2017, LNCS 10167, pp. 421–432, 2017.
DOI: 10.1007/978-3-319-53925-6_33

This problem is well studied in the field of VLSI design (see e.g., [4]). The time difference (signal delay) between the input and the output of each logical circuit product may be different even though they are produced in the same line of the same design. The signal delay fluctuates because the signal delay of each logical gate fluctuates. Therefore, the VLSI makers would like to know if sufficient number of their new chips are going to perform as expected before they start costly mass-production. To estimate the signal delay of a logical circuit, we consider the longest path length in a DAG by considering each of gates and lines as an edge and each fluctuating signal delay as a random edge length.

When the edge lengths are mutually independent and uniformly distributed, the distribution function $\Pr[X_{\mathrm{MAX}} \leq x]$ of the longest path length is equal to the volume of a polytope

$$K_G(\boldsymbol{a}, x) \stackrel{\text{def}}{=} \left\{ \boldsymbol{x} \in [0,1]^m \ \middle| \ \bigwedge_{\pi \in \Pi} \sum_{e \in \pi} a_e x_e \leq x \right\}.$$

If G is a directed path, $K_G(\boldsymbol{a}, x)$ is called a $0-1$ knapsack polytope. Computing the volume of $K_G(\boldsymbol{a}, x)$ is #P-hard even if G is a directed path (see [7])[1]. However, we show that there is a deterministic FPTAS (fully polynomial time approximation scheme) if the pathwidth of G is bounded by a constant p. In this paper, we show the following theorem.

Theorem 1.1. *Suppose that the pathwidth of G is bounded by a constant p. There is an algorithm that approximates* $\mathrm{Vol}(K_G(\boldsymbol{a}, x))$ *in* $O(p^4 2^{1.5p} n(\frac{2mnp}{\epsilon})^{3p} L)$ *time satisfying* $1 \leq V'/\mathrm{Vol}(K_G(\boldsymbol{a}, x)) \leq 1 + \epsilon$, *where V' is the output of the algorithm, and L is the number of bits in the input. If p is bounded by a constant, the running time is* $O(n(\frac{mn}{\epsilon})^{3p} L)$.

The running time of our algorithm depends on the number L of input bits because we use the linear programming [17] as the subroutine.

In n-dimensional space, computing the volume of a polytope is hard if the randomness is not available. In 1986, Lovász [14] considered an n-dimensional convex body that is accessible by membership oracle, and showed that no deterministic polynomial time algorithm can achieve the approximation ratio of 1.999^n (See also Elekes [9]). The bound is updated by Bárány and Füredi [3] up to $(cn/\log n)^{n/2}$, where c does not depend on n. Dyer and Frieze [7] showed that computing the volume of the $0-1$ knapsack polytope K is #P-hard.

Then, the randomized approximation has been studied. Dyer et al. [8] showed the first FPRAS (fully polynomial time randomized approximation scheme) that finishes in $O^*(n^{23})$ time for volume of the general n-dimensional convex body. Here O^* ignores the factor of poly($\log n$) and $1/\epsilon$ factor. There are faster FPRASes [6,15]. The current fastest FPRAS [6] runs in $O^*(n^3)$ time for well-rounded convex bodies.

---

[1] Intuitively, the breakpoints of the function $F(x) = \mathrm{Vol}(K_G(\boldsymbol{a}, x))$ increases exponentially with respect to n. For example, consider the case where each component a_i of $\boldsymbol{a}$ is $a_i = 2^i$ for $i = 1, \ldots, n$.

The above results lead us to a challenge in algorithm design: Is it possible to approximate the volume of the convex body K if we can access K not only by the membership oracle?

Recently, there are some deterministic approximation algorithms for the volume of the knapsack polytope. Li and Shi [13] showed an FPTAS for distribution function of the sum of the discrete random variables. Their algorithm can be used to approximate $\mathrm{Vol}(K_G(a, x))$ if G is a directed path. Their algorithm is based on the dynamic programming due to Štefankovič et al. [16] (See also [10,11]).

Ando and Kijima [1], motivated by the deterministic approximation technique of the above results, showed another FPTAS that is based on the approximate convolution integral. Their algorithm runs in $O(n^3/\epsilon)$ time. Especially, the FPTAS in [1] is extensible to the volume of the multiple constraint knapsack polytope. Given $m \times n$ matrix $A \in Z_{\geq 0}^{mn}$ and a vector $b \in Z_{\geq 0}^m$, the multiple constraint knapsack polytope $K_m(A, b)$ is

$$K_m(A, b) \stackrel{\text{def}}{=} \{x \in [0,1]^n | Ax \leq b\}.$$

Their algorithm finishes in $O((n^2\epsilon^{-1})^{m+1}nm \log m)$ time. Thus, there is an FPTAS for $\mathrm{Vol}(K_m(A, b))$ if the number of constraints m is bounded by a constant. We show, in this paper, that the result in [1] can be extended to the volume of the knapsack polytopes with $\Omega(2^n)$ constraints. That is, the volume of $K_G(a, x)$ in case the pathwidth of G is at most p.

We here note some results about the pathwidth for undirected graphs and directed graphs. As for the undirected pathwidth, Bodlaender [5] found an algorithm that finds the path decomposition of an undirected graph G with its width at most p in linear time with respect to the graph size if the pathwidth of G is bounded by a constant p. Johnson et al. [12] defined the directed treewidth. They proved that many NP-hard problems, including computing the directed treewidth, can be solved in polynomial time on the directed graphs with at most constant directed treewidth. To make the argument easier, we use the pathwidth of the underlying undirected graph of G.

Definition 1.2. A path decomposition of G is a sequence $\mathcal{B} = \{B_1, \ldots, B_b\}$ of subsets of V satisfying the following three conditions. We call $B \in \mathcal{B}$ a bag.

1. $\bigcup_{B \in \mathcal{B}} B = V$;
2. $(u, v) \in E \Rightarrow \exists B \in \mathcal{B}$ s.t. $\{u, v\} \subseteq B$;
3. if $1 \leq i \leq j \leq k \leq b$, then $B_i \cap B_k \subseteq B_j$.

The width of path decomposition $\mathcal{B}$ is $\max_{B \in \mathcal{B}} |B| - 1$. The pathwidth of G is the minimum of the width of all possible path decomposition.

This paper is organized as follows. In Sect. 2, we explain our notations for multiple integrals. In Sect. 3, we show how we can compute the volume of $K_G(a, x)$ by a repetition of definite integrals. In Sect. 4, we show our approximation algorithm for the volume of $K_G(a, x)$. In Sect. 5, we prove the approximation ratio and the running time so that our algorithm is an FPTAS for the volume of $K_G(a, x)$. We finish this paper with the conclusion and the future work in Sect. 6.

2   Preliminaries About Notations

We would like to introduce some notations about vector components. Let C be a set and consider a vector $\boldsymbol{x} \in \mathbb{R}^{|C|}$. Each component of $\boldsymbol{x}$ is specified by an element of C. That is, $\boldsymbol{x}(c)$ is a component of $\boldsymbol{x}$ for $c \in C$. Let C_1 and C_2 be two sets. Let $\boldsymbol{x} \in \mathbb{R}^{|C_1|}$ and $\boldsymbol{y} \in \mathbb{R}^{|C_2|}$ be two vectors. Then we write $\boldsymbol{w} = (\boldsymbol{x}, \boldsymbol{y})$ meaning that $\boldsymbol{w}$ is a concatenation of $\boldsymbol{x}$ and $\boldsymbol{y}$. That is, $\boldsymbol{w} \in \mathbb{R}^{|C_1 \cup C_2|}$ so that $\boldsymbol{w}(c) = \boldsymbol{x}(c)$ for $c \in C_1$ and $\boldsymbol{w}(c') = \boldsymbol{y}(c')$ for $c' \in C_2$. We consider this concatenation of vectors only in cases where $\boldsymbol{x}$ and $\boldsymbol{y}$ has the same component for the same element of $C_1 \cup C_2$ (i.e., $\boldsymbol{x}(c) = \boldsymbol{y}(c)$ holds for any $c \in C_1 \cup C_2$).

Throughout this paper, we are concerned with the integrals with respect to dummy variables $z_1, \ldots, z_n$. We define $\boldsymbol{v}$ as an n-dimensional vector of variables $\boldsymbol{v} = (z_1, \ldots, z_n)$, where each component z_i is associated to $v_i \in V$ (i.e., $\boldsymbol{v}(v_i) = z_i$). We write $\boldsymbol{u} = \boldsymbol{v}(C)$ for some $C \subseteq V$, $\boldsymbol{u}$ is a vector with $|C|$ components where $\boldsymbol{u}(v_i) = z_i$ for $v_i \in C$. We consider a multiple integral with $|C|$ dummy variables of some $|C|$ variables function $F(\boldsymbol{z})$, where $\boldsymbol{z} = \boldsymbol{v}(C)$. We write

$$\int_{\boldsymbol{z} \in \mathbb{R}^{|C|}} F(\boldsymbol{z}) \mathrm{d}\boldsymbol{z} = \int_{\boldsymbol{z}(v_1) \in \mathbb{R}} \cdots \int_{\boldsymbol{z}(v_{|C|}) \in \mathbb{R}} F(\boldsymbol{z}) \, \mathrm{d}\boldsymbol{z}(v_1) \cdots \mathrm{d}\boldsymbol{z}(v_{|C|}),$$

for $C = \{v_1, \ldots, v_{|C|}\}$.

3   Exact Computation Using Path Decomposition

We show a variant of Theorem 1 in [2] so that the running time for computing the volume of $K_G(\boldsymbol{a}, x)$ is bounded by using the width of the path decomposition of G. We first give the definition of terms and symbols. We compute a partial computation result for each bag and put them together so that we obtain $\mathrm{Vol}(K_G(\boldsymbol{a}, x))$.

Here, we define some terms and symbols. Let $\mathcal{B} = \{B_1, \ldots, B_b\}$ be the path decomposition of G. Let $G_{i,j}$ be the subgraph induced in G by $B_i \cup \cdots \cup B_j$. We define that a *source* of $G_{i,j}$ is a vertex that has no incoming edge from the vertices in $G_{i,j}$ or, a vertex that has one or more incoming edge from outside of $G_{i,j}$. Also, a *terminal* of $G_{i,j}$ is a vertex that has no outgoing edge to the vertices in $G_{i,j}$ or, a vertex that has one or more outgoing edge to outside of $G_{i,j}$. Let $S_{i,j}$ and $T_{i,j}$ be the sets of the sources and the terminals of $G_{i,j}$. We call $S_{i,i}$ (resp. $T_{i,i}$) as the set of the sources (resp. the terminals) of bag B_i. Let $\Pi_{i,j}(v_s, v_t)$ be the set of paths from a source v_s to a terminal v_t of $G_{i,j}$. Let $\boldsymbol{s}_{ij} = \boldsymbol{v}(S_{i,j})$ and $\boldsymbol{t}_{ij} = \boldsymbol{v}(T_{i,j})$, where $\boldsymbol{v}$ is a $|V|$-dimensional vector with each of its component given as a variable z_i for each $v_i \in V$. Then, $\Phi_{i,j}(\boldsymbol{s}_{ij}, \boldsymbol{t}_{ij})$ is an $|S_{i,j} \cup V_{i,j}|$ variables function

$$\Phi_{i,j}(\boldsymbol{s}_{ij}, \boldsymbol{t}_{ij}) \stackrel{\text{def}}{=} \Pr\left[\bigwedge_{\substack{v_s \in S_{i,j} \\ v_t \in T_{i,j}}} \bigwedge_{\pi \in \Pi_{i,j}(v_s, v_t)} \sum_{e \in \pi} a_e X_e \leq z_s - z_t \right].$$

Roughly speaking, z_i means the longest path length from v_i to the terminal (e.g., $z_1 = x$, $z_n = 0$) in this formula. The following is the Theorem 1 in [2].

Theorem 3.1. [2] *Let $H(x)$ be the step function satisfying $H(x) = 1$ for $x > 0$ and $H(x) = 0$ for $x \leq 0$. Then we have*

$$\Pr[X_{\text{MAX}} \leq x] = \int_{\mathbb{R}^{n-1}} H(x - z_1) \prod_{1 \leq i \leq n-1} \left(\frac{\partial}{\partial z_i} \prod_{i+1 \leq j \leq n} F_{ij}(z_i - z_j) \right) dz_i. \quad (1)$$

Here, $F_{ij}(x) = \Pr[a_{ij}X_{ij} \leq x]$ if $(v_i, v_j) \in E$; otherwise $F_{ij}(x) = 1$ for $x \in \mathbb{R}$.

Corollary 3.2. *Let $I_{i,j} = (B_i \cup \cdots \cup B_j) \setminus (S_{i,j} \cup T_{i,j})$. For $s_{ij} = v(S_{i,j})$ and $t_{ij} = v(T_{i,j})$, we have*

$$\Phi_{i,j}(s_{ij}, t_{ij}) = \int_{\substack{z \in \mathbb{R}^{|I_{i,j}|} \\ \substack{v_s \in S_{i,j} \\ (v_s, v_k) \in E_{i,j}}}} \prod H(z_s - z_k) \frac{\partial}{\partial z_k} \prod_{(v_k, v_\ell) \in E_{i,j}} F_{k\ell}(z_k - z_\ell) dz, \quad (2)$$

where $F_{k\ell}(x) = \Pr[a_{k\ell}X_{k\ell} \leq x]$ and $z_i = v(I_{i,j})$.

Proof. We consider a DAG $G'_{i,j}$ with vertex set $B_i \cup \cdots \cup B_j \cup \{v_+, v_-\}$, where v_+ and v_- are the source and the terminal, respectively. $G'_{i,j}$ is obtained by connecting v_+ to each source $v_s \in S_{i,j}$ of $G_{i,j}$ by an edge with constant length $x - z_s$. Also we connect each terminal $v_t \in T_{i,j}$ of $G_{i,j}$ to v_- by an edge with constant length z_t. Then by Theorem 3.1 with sufficiently large x, we have the claim. □

We further transform (2) into the following lemma. We omit the proof due to the space limit.

Lemma 3.3. *Let $J_i = (T_{1,i-1} \cap S_{i,i}) \cup (T_{i,i} \cap S_{1,i-1})$. Let $z_i = v(J_i)$, $s_i = v(S_{i,i} \setminus J_i)$, and $t_i = v(T_{i,i} \setminus J_i)$. For $i = 2, \ldots, b$, let $\phi_i(s_i, z_i, t_i)$ be the derivative of $\Phi_{i,i}(s_i, z_i, t_i)$ with respect to all z_j's for $v_j \in J_i$, that is*

$$\phi_i(s_i, z_i, t_i) \stackrel{\text{def}}{=} \left(\prod_{v_j \in J_i} \frac{\partial}{\partial z_j} \right) \Phi_{i,i}(s_i, z_i, t_i).$$

Let $\tilde{s}_i = v(S_{1,i})$, $\tilde{t}_i = v(T_{1,i})$, $s'_{i-1} = v(S_{1,i-1} \setminus J_i)$ and $t'_{i-1} = v(T_{1,i-1} \setminus J_i)$. We have

$$\Phi_{1,i}(\tilde{s}_i, \tilde{t}_i) = \int_{z_i \in [0,x]^{|J_i|}} \Phi_{1,i-1}(s'_{i-1}, z_i, t'_{i-1}) \phi_i(s_i, z_i, t_i) dz_i. \quad (3)$$

Here J_i is the set of vertices by which the bags B_i and B_{i-1} are joined. We consider the paths that goes backward the bags of the path decomposition by considering $T_{1,i} \cap S_{1,i-1}$ as a part of J_i.

We can prove the following.

Lemma 3.4. *Let* $\tilde{s}_i = v(S_{1,i})$, $\tilde{t}_i = v(T_{1,i})$, $s'_{i-1} = v(S_{1,i-1} \setminus J_i)$, $t'_{i-1} = v(T_{1,i-1} \setminus J_i)$, $s_i = v(S_{i,i} \setminus J_i)$, $t_i = v(T_{i,i} \setminus J_i)$, *and* $z_i = v(J_i)$. *We have*

$$\Phi_{1,i}(\tilde{s}_i + (c_1 + c_2)\mathbf{1}, \tilde{t}_i) \geq \int\limits_{z_i \in [0,x]^{|J_i|}} \Phi_{1,i-1}(s'_{i-1} + c_1\mathbf{1}, z_i, t'_{i-1})\phi_i(s_i, z_i + c_2\mathbf{1}, t_i)\mathrm{d}z_i,$$

where $\mathbf{1}$ *is the vector whose components are all* 1.

The volume of $K_G(\boldsymbol{a}, x)$ is $\Phi_{1,b}(x\mathbf{1}, \mathbf{0})$. The computation of $\Phi_{1,b}(x\mathbf{1}, \mathbf{0})$ is hard because there may be exponentially many breakpoints in the derivative of some order of $\Phi_{1,b}(x\mathbf{1}, \mathbf{0})$ with respect to x.

4   Approximation Algorithm

The idea is an extension of the algorithm in [1] for the volume of multiple constraint knapsack polytope. For $\overline{s_i} = v(S_{i,i})$ and $\overline{t_i} = v(T_{i,i})$, we first approximate $\Phi_{i,i}(\overline{s_i}, \overline{t_i})$ by a staircase function $A_i(M, \overline{s_i}, \overline{t_i})$ using parameter M. Then, as an approximation of $\phi_i(\overline{s_i}, \overline{t_i})$, we compute the discrete difference of $A(M, \overline{s_i}, \overline{t_i})$. We compute an extended form of the convolution of the differences for $i = 1, \ldots, b$ so that we have the approximation of $\Phi_{1,b}(\tilde{s}_b, \tilde{t}_b)$ for $\tilde{s}_b = v(S_{1,b})$ and $\tilde{t}_b = v(T_{1,b})$.

We approximate $\Phi_{i,i}(\overline{s_i}, \overline{t_i})$ by counting. Consider a rectangular parallelepiped P_i given by

$$P_i \stackrel{\text{def}}{=} \{\boldsymbol{x} \in \mathbb{R}^{|E_{i,i}|} \mid 0 \leq x_e \leq \min\{1, x/a_e\} \text{ for } e \in E_{i,i}\},$$

where $E_{i,i}$ is the edge set of $G_{i,i}$. Let $M = \lceil 2mnp/\epsilon \rceil$ be an integer parameter of our algorithm. We divide P_i into $M^{|E_{i,i}|}$ *cells* whose size in x_e axis direction is $\min\{1, x/a_e\}/M$ for $e \in E_{i,i}$. Then, we count the number $N_i(M, \overline{s_i}, \overline{t_i})$ of cells that intersects with $K'_{G_{i,}}(\overline{s_i}, \overline{t_i})$, where

$$K'_{G_{i,i}}(\overline{s_i}, \overline{t_i}) \stackrel{\text{def}}{=} \left\{ \boldsymbol{x} \in P_i \; \middle| \; \bigwedge_{\substack{v_s \in S_{i,i} \\ v_t \in T_{i,i}}} \bigwedge_{\pi \in \Pi_{i,i}(v_s, v_t)} \sum_{e \in \pi} a_e x_e \leq z_s - z_t \right\}.$$

We can decide if the intersection is empty by using the linear programming. That is, the decision problem is the linear programming with $|E_{i,i}|$ $(\leq p^2)$ variables and $|\Pi(i, i)|$ $(\leq 2^p)$ inequalities, where $p = \max_{1 \leq i \leq b} |B_i| - 1$ is the width of the path decomposition. We have an approximation of $\Phi_{i,i}(\overline{s_i}, \overline{t_i})$ as

$$A_i(M, \overline{s_i}, \overline{t_i}) = N_i\left(M, \left\lceil \frac{\overline{s_i}M}{x} \right\rceil \frac{x}{M}, \left\lceil \frac{\overline{t_i}M}{x} \right\rceil \frac{x}{M}\right) \frac{\prod_{e \in E_i} \min\left\{1, \frac{x}{a_e}\right\}}{M^{|E_{i,i}|}}, \quad (4)$$

where $\left\lceil \frac{\overline{s_i}M}{x} \right\rceil$ (resp. $\left\lceil \frac{\overline{t_i}M}{x} \right\rceil$) is a vector with each element $\left\lceil \frac{z_jM}{x} \right\rceil$ corresponding to $v_j \in S_{i,i}$ (resp. $v_j \in T_{i,i}$).

Let $\mathrm{LP}(n', m')$ be the time to solve the linear programming with n' variables and m' constraints. Vaidya's algorithm [17] can solve the linear programming in $O(((m' + n')n'^2 + (m' + n')^{1.5}n')L')$ time, where L' is bounded by the number of bits in the input of the linear programming. Since the number of constraints is the number of paths in a bag, we have $m' = O(2^p)$ and $L' \leq mpL$, where L is the number of the bits of our problem (i.e., L is the number of bits necessary for $\boldsymbol{a}, x$ and G). We have the following observation.

Observation 4.1. Let $\overline{s_i} = v(S_{i,i})$ and $\overline{t_i} = v(T_{i,i})$. we can compute $A_i(M, \overline{s_i}, \overline{t_i})$ in $O(\mathrm{LP}(p^2, 2^p)M^p) = O(p^2 2^{1.5p}LM^p)$ time, where p is the width of the path decomposition of G, and L is the number of bits necessary for $\boldsymbol{a}, x$ and G. If p is bounded by a constant, the running time is $O(LM^p)$.

Here, we have $L = O(\sum_{e \in E} \log a_e + \log x + \log n + 2m \log n)$.

It remains to, for $\boldsymbol{s}_i = v(S_{i,i} \setminus J_i)$, $\boldsymbol{z}_i = v(J_i)$ and $\boldsymbol{t}_i = t(T_{i,i} \setminus J_i)$, show how we put $A_i(M, \overline{s_i}, \overline{t_i})$'s together for $i = 1, \ldots, b$ into the approximation of $\Phi_{1,b}(\tilde{\boldsymbol{s}}_b, \tilde{\boldsymbol{t}}_b)$.

We approximate (3) using $A_i(M, \overline{s_i}, \overline{t_i})(= A_i(M, \boldsymbol{s}_i, \boldsymbol{z}_i, \boldsymbol{t}_i))$. Since $A_i(M, \boldsymbol{s}_i, \boldsymbol{z}_i, \boldsymbol{t}_i)$ is a staircase function, we consider discrete difference instead of the derivative of $A_i(M, \boldsymbol{s}_i, \boldsymbol{z}_i, \boldsymbol{t}_i)$. We define the difference operator $\Delta(z_j)$, i.e.,

$$\Delta(z_j)A_i(M, \boldsymbol{s}_i, \boldsymbol{z}_i, \boldsymbol{t}_i) \stackrel{\text{def}}{=} A_i\left(M, \boldsymbol{s}_i, \boldsymbol{z}_i + \frac{x}{M}\boldsymbol{e}_j, \boldsymbol{t}_i\right) - A_i(M, \boldsymbol{s}_i, \boldsymbol{z}_i, \boldsymbol{t}_i).$$

Then, by assuming an order in the vertices in J_i, we repeatedly take the difference for z_j's corresponding to $v_j \in J_i$. For the simplicity, we write

$$\Delta(\boldsymbol{z}_i)A_i(M, \boldsymbol{s}_i, \boldsymbol{z}_i, \boldsymbol{t}_i) \stackrel{\text{def}}{=} \Delta(z_{j_1}) \cdots \Delta(z_{j_{|J_i|}})A_i(M, \boldsymbol{s}_i, \boldsymbol{z}_i, \boldsymbol{t}_i),$$

where $J_i = \{v_{j_1}, \ldots, v_{j_{|J_i|}}\}$. To obtain the values of $\Delta(\boldsymbol{z}_i)A(M, \boldsymbol{s}_i, \boldsymbol{z}_i), \boldsymbol{t}_i)$, we first compute the value of $A_i(M, \boldsymbol{s}_i, \boldsymbol{z}_i, \boldsymbol{t}_i)$ for the gridpoints of $(\boldsymbol{s}_i, \boldsymbol{z}_i, \boldsymbol{t}_i)$. That is, the values of $A_i(M, \boldsymbol{s}_i, \boldsymbol{z}_i, \boldsymbol{t}_i)$'s of points $(\boldsymbol{s}_i, \boldsymbol{z}_i, \boldsymbol{t}_i) = v(S_{i,i} \cup T_{i,i}) = (x/M)\boldsymbol{i}$ for $\boldsymbol{i} \in \{0, \ldots, M\}^{|S_{i,i} \cup T_{i,i}|}$. Then, we compute $\Delta(z_j)A_i(M, (x/M)\boldsymbol{i})$ for all $\boldsymbol{i} \in \{0, \ldots, M\}^{|S_{i,i} \cup T_{i,i}|}$. We store the values in an array with $(M + 1)^{|S_{i,i} \cup T_{i,i}|}$ elements and then compute $\Delta(z_{j'})\Delta(z_j)A_i(M, (x/M)\boldsymbol{i})$ similarly for $v_{j'} \in J_i$. This way we repeat computing the differences. When we obtain the values of $\Delta(\boldsymbol{z}_i)A_i(M, (x/M)\boldsymbol{i})$ for $\boldsymbol{i} \in \{0, \ldots, M\}^{|S_{i,i} \cup T_{i,i}|}$, we will have $|J_i|$ arrays with $(M + 1)^{|S_{i,i} \cup T_{i,i}|}$ elements each. We have the following observation.

Observation 4.2. Given $\boldsymbol{z}_i = v(J_i), \boldsymbol{s}_i = v(S_{i,i} \setminus J_i)$, $\boldsymbol{t}_i = v(T_{i,i} \setminus J_i)$, and $A_i(M, \lceil x/M \rceil \boldsymbol{i})$ for all $\boldsymbol{i} \in \{0, \ldots, M\}^{|S_{i,i} \cup T_{i,i}|}$, we can compute $\Delta(\boldsymbol{z}_i)A_i(M, \boldsymbol{s}_i, \boldsymbol{z}_i, \boldsymbol{t}_i)$ in $O(pM^p)$ time.

We set $\Psi_1(M, \boldsymbol{s}_1, \boldsymbol{t}_1) \stackrel{\text{def}}{=} A_1(M, \boldsymbol{s}_1, \boldsymbol{t}_1)$. Let $\boldsymbol{i}' = \{i'_1, i'_2, \ldots, i'_{|J_i|}\}$ be an integer vector with $|J_i|$ components, where $0 \leq i'_j \leq M$ for each $v_j \in J_i$. We define the approximation $\Psi_i(M, \boldsymbol{s}_i, \boldsymbol{z}_i, \boldsymbol{t}_i)$ of $\Phi_{i,i}(\boldsymbol{s}_i, \boldsymbol{z}_i, \boldsymbol{t}_i)$ by

$$\Psi_i(M, \boldsymbol{s}_i, \boldsymbol{z}_i, \boldsymbol{t}_i) \stackrel{\text{def}}{=} \sum_{\boldsymbol{i}' \in \{0, \ldots, M\}^{|J_i|}} \Psi_{i-1}\left(M, \boldsymbol{s}_{i-1}, \frac{x}{M}\boldsymbol{i}', \boldsymbol{t}_{i-1}\right)\Delta(\boldsymbol{z}_i)A_i\left(M, \boldsymbol{s}_i, \frac{x}{M}\boldsymbol{i}', \boldsymbol{t}_i\right), \quad (5)$$

where $s_{i-1} = v(S_{1,i-1} \setminus J_i)$ and $t_{i-1} = v(T_{1,i-1} \setminus J_i)$.

The following pseudocode shows our algorithm A.

Algorithm $A(G, a, x, \mathcal{B})$

Input: DAG G, edge lengths parameter $a \in \mathbb{Z}_{>0}^m$, $x \in \mathbb{R}$,

   and bags of a path decomposition $\mathcal{B} = \{B_1, \ldots, B_b\}$ of G;

1. For $i = 1, \ldots, b$ do
2.   Compute $A_i(M, (x/M)i)$ for all $i \in \{0, \ldots, M\}^{|S_{i,i} \cup T_{i,i}|}$ by (4);
3.   Compute $\Delta(z_i)A_i(M, (x/M)i)$ for all $i \in \{0, \ldots, M\}^{|S_{i,i} \cup T_{i,i}|}$;
4. done;
5. Set $\Psi_1(M, (x/M)i) := A_1(M, (x/M)i$ for $i \in \{0, \ldots, M\}^{|S_{1,1} \cup T_{1,1}|}$;
6. For $i = 1, \ldots b$ do
7.   Compute $\Psi_i(M, (x/M)i')$ for $i' \in \{0, \ldots, M\}^{|S_{1,i} \cup T_{1,i}|}$ by (5);
8. done;
9. Output $\Psi_b(M, x\mathbf{1}, \mathbf{0})$.

As for the running time, we have the following observation. We actually set $M = \lceil 2mnp/\epsilon \rceil$ to bound the approximation ratio at most $1 + \epsilon$.

Observation 4.3. *If a path decomposition of G with width p is given, our algorithm outputs a value $V' = \Psi_b(x\mathbf{1}, \mathbf{0})$ satisfying $V'/\Phi_{1,b}(x\mathbf{1}, \mathbf{0}) \leq 1 + \epsilon$ in $O(\text{LP}(p^2, 2^p)bp^2(M+1)^{3p}) = O(p^4 2^{1.5p} n(\frac{2mnp}{\epsilon})^{3p}L)$ time, where L is the number of bits in the input. If p is bounded by a constant, the running time is $O(n(\frac{mn}{\epsilon})^{3p}L)$.*

5   Analysis

In this section, we bound the approximation ratio. We first bound the approximation error "horizontally". We show that $\Psi_b(M, x\mathbf{1}, \mathbf{0})$ is bounded by $\Phi_{1,b}(x\mathbf{1}, \mathbf{0})$ and its translation from below and from above. Then, we prove the "vertical" approximation ratio by showing that the upper bound and the lower bound are not too far away from each other.

To "horizontally" bound the approximation error, we prove the following lemmas. After that, we obtain the "vertical" approximation ratio.

Lemma 5.1. *Let $s_i = v(S_{i,i} \setminus J_i)$, $t_i = v(T_{i,i} \setminus J_i)$, and $z_i = v(J_i)$. Let $I_i = B_i \setminus (S_{i,i} \cup T_{i,i})$. Then,*

$$\Phi_{i,i}(s_i, z_i, t_i) \leq A_i(M, s_i, z_i, t_i) \leq \Phi_{i,i}\left(s_i, z_i + \frac{(|I_i| - 1)x}{M}\mathbf{1}, t_i\right).$$

Proof. Since we count all the cells that intersects with $K'_{G_{i,i}}(\overline{s_i}, \overline{t_i})$, $(= K'_{G_{i,i}}(s_i, z_i, t_i), \overline{s_i} = v(S_{i,i}), \overline{t_i} = v(T_{i,i}))$, the earlier inequality is obvious.

The latter inequality is shown as follows. Let $C \subseteq [0,1]^{|E_{i,i}|}$ be a cell on the border (i.e., $C \neq C \setminus K'_{G_{i,i}}(\overline{s_i}, \overline{t_i}) \neq \emptyset$). Then, we claim $C \subseteq K'_{G_{i,i}}\left(\overline{s_i} + \frac{(|I_i|-1)x}{M}\mathbf{1}, \overline{t_i}\right)$. Let $c_1, c_2 \in C$ be two vertices of C such that $c_1 \in K'_{G_{i,i}}(s, t)$ and $c_2 \notin K'_{G_{i,i}}(s, t)$. Let $c_1(e) = x_e$ and $c_2(e) = x_e + 1/M$ for

$e \in E_{i,i}$. Since $|\pi| \le |I_i| - 1$ for $v_s \in S_{i,i}, v_t \in T_{i,i}$ and $\pi \in \Pi_{i,i}(v_s, v_t)$, we have that

$$z_s - z_t \le \sum_{e \in \pi} \min\{a_e, x\} x_e + \sum_{e \in \pi} \frac{\min\{a_e, x\}}{M} \le z_s - z_t + \frac{(|I_i| - 1)x}{M},$$

by definition of $K'_{G_{i,i}}(\overline{s_i}, \overline{t_i})$. This shows the lemma. $\square$

By Lemma 5.1, we have the following lemma.

Lemma 5.2. Let $\tilde{s}_{i-1} = v(S_{1,i-1} \setminus J_i)$, $\tilde{t}_{i-1} = v(T_{1,i-1} \setminus J_i)$ and $z_i = v(J_i)$. Let $F(s_{i-1}, z_i, t_{i-1})$ be an $|S_{1,i} \cup T_{1,i}|$ variables staircase function such that for any $z_i \in \mathbb{R}^{|J_i|}$,

$$F(s_{i-1}, z_i, t_{i-1}) = F\left(s_{i-1}, \left\lceil \frac{z_i M}{x} \right\rceil \frac{x}{M}, t_{i-1}\right),$$

where $\lceil z_i M/x \rceil$ is a vector such that its element for $v_j \in V$ is $\lceil z_j M/x \rceil$. For $s_i = v(S_{i,i} \setminus J_i)$ and $t_i = v(T_{i,i} \setminus J_i)$, we have

$$\int_{z_i \in [0,x]^{|J_i|}} F(s_{i-1}, z_i, t_{i-1}) \phi_i(s_i, z_i, t_i) dz$$

$$\le \sum_{i=\{0,\ldots,M\}^{|J_i|}} F\left(s_{i-1}, \frac{x}{M} i, t_{i-1}\right) \Delta(z_i) A_i\left(M, s_i, \frac{x}{M} i, t_i\right)$$

$$\le \int_{z_i \in [0,x]^{|J_i|}} F(s_{i-1}, z_i, t_{i-1}) \phi_i\left(s_i, z_i + \frac{(|I_i| - 1)px}{M} 1, t_i\right) dz.$$

Proof. Since $F(s_{i-1}, z_i, t_{i-1})$ is a staircase function that is a constant in each cell, the lemma is clear from Lemma 5.1. $\square$

The following lemma shows the "horizontal" error bound.

Lemma 5.3. Let $\tilde{s}_i = v(S_{1,i})$, $\tilde{t}_i = v(T_{1,i})$. For $i = 1, \ldots, b$,

$$\Phi_{1,i}(\tilde{s}_i, \tilde{t}_i) \le \Psi_i(M, \tilde{s}_i, \tilde{t}_i) \le \Phi_{1,i}\left(\tilde{s}_i + \frac{|\bigcup_{1 \le j \le i} B_j| px}{M} 1, \tilde{t}_i\right).$$

Proof. Since the inequality on the left is obvious by the definition, we prove the inequality on the right in the following. The proof is induction on i. As for the base case, we have that, for $\tilde{s}_1 = v(S_{1,1})$ and $\tilde{t}_1 = v(T_{1,1})$.

$$\Psi_1(M, \tilde{s}_1, \tilde{t}_1) = A_1(M, \tilde{s}_1, \tilde{t}_1) \le \Phi_{1,1}\left(\tilde{s}_1 + \frac{|I_1|x}{M} 1, \tilde{t}_1\right),$$

by Lemma 5.1. Let $s'_{i-1} = v(S_{1,i-1} \setminus J_i)$, $t'_{i-1} = v(T_{1,i-1} \setminus J_i)$, $s_i = v(S_{i,i} \setminus J_i)$, $t_i = v(T_{i,i} \setminus J_i)$, and $z_i = v(J_i)$. As the induction hypothesis, we assume that

$$\Psi_{i-1}(M, s'_{i-1}, z_i, t'_{i-1}) \le \Phi_{1,i-1}\left(s'_{i-1} + \frac{|\bigcup_{1 \le j \le i-1} B_j| px}{M} 1, z_i, t'_{i-1}\right).$$

Then, we have

$$\Psi_i(M, \tilde{s}_i, \tilde{t}_i)$$

$$= \sum_{i \in \{0,\dots,M\}^{|J_i|}} \Psi_{i-1}\left(M, s'_{i-1}, \frac{x}{M}i, t'_{i-1}\right) \Delta(z_i) A_i\left(M, s_i, \frac{x}{M}i, t_i\right) \quad (\because \text{Definition})$$

$$\leq \int_{z_i \in [0,x]^{|J_i|}} \Psi_{i-1}\left(M, s'_{i-1}, z_i, t'_{i-1}\right) \phi_i\left(s_i, z_i + \frac{|I_i|px}{M}1, t_i\right) dz_i \quad (\because \text{Lemma 5.2})$$

$$\leq \int_{z_i \in [0,x]^{|J_i|}} \Phi_{1,i-1}\left(s'_{i-1} + \frac{\left|\bigcup_{1\leq j\leq i-1} B_j\right| px}{M}1, z_i, t'_{i-1}\right) \phi_i\left(s_i, z_i + \frac{|I_i|px}{M}1, t_i\right) dz_i$$

$$(\because \text{Induction hypo.})$$

$$\leq \Phi_{1,i}\left(\tilde{s}_i + \left(|I_i|p + \left|\bigcup_{1\leq j\leq i-1} B_j\right| p\right)\frac{x}{M}1, \tilde{t}_i\right) \quad (\because \text{Lemma 3.4})$$

$$\leq \Phi_{1,i}\left(\tilde{s}_i + \frac{|\bigcup_{1\leq j\leq i} B_j|px}{M}1, \tilde{t}_i\right) \quad (\because \Phi_{1,i}(s,t) \leq \Phi_{1,i}(s',t) \text{ if } s \leq s').$$

$$\square$$

The following shows the "vertical" approximation ratio of our algorithm.

Lemma 5.4. *Let* $M = \lceil 2mnp/\epsilon \rceil$ *where* $n = |V|, m = |E|$. *Then*

$$1 \leq \Psi_b(M, x1, 0)/\Phi_{1,b}(x1, 0) \leq 1 + \epsilon.$$

Proof. Since the earlier inequality is clear by the definition. We prove the latter inequality. Since we have that $\Psi_b(M, x1, 0)$ is at most

$$\Phi_{1,b}\left(x1 + \frac{|\bigcup_{1\leq j\leq b} B_j|px}{M}1, 0\right) \leq \Phi_{1,b}\left(x1 + \frac{npx}{M}1, 0\right)$$

by Lemma 5.3, we bound, from below,

$$\frac{\Phi_{1,b}(x1, 0)}{\Psi_b(M, x1, 0)} \geq \frac{\Phi_{1,b}(x1, 0)}{\Phi_{1,b}\left(x1 + \frac{npx}{M}1, 0\right)}.$$

Then, we claim that

$$\frac{\Phi_{1,b}(x1, 0)}{\Phi_{1,b}\left(x1 + \frac{npx}{M}1, 0\right)} \geq \left(\frac{1}{1 + np/M}\right)^m = \left(\frac{1}{1 + \frac{1}{2m\epsilon}}\right)^m.$$

This claim is verified as follows. By definition of $\Phi_{1,b}(x1, 0)$, is the volume of

$$K_G(a, x) = \left\{ x \in [0,1]^m \;\middle|\; \bigwedge_{\pi \in \Pi} \sum_{e \in \pi} a_e x_e \leq x \right\}.$$

Consider another polytope $\overline{K_G}(\boldsymbol{a}, x)$ that is obtained by scaling $K_G(\boldsymbol{a}, x)$ by $1 + np/M$, that is,

$$\overline{K_G}(\boldsymbol{a}, x) \overset{\text{def}}{=} \{\boldsymbol{x} \in \mathbb{R}^m | \exists \boldsymbol{y} \in K_G(\boldsymbol{a}, x) \text{ s.t. } \boldsymbol{x} = (1 + np/M)\boldsymbol{y}\}$$

$$= \left\{ \boldsymbol{x} \in [0, 1 + np/M]^m \,\middle|\, \bigwedge_{\pi \in \Pi_{1,i}} \sum_{e \in \pi} a_e x_e \leq x(1 + np/M) \right\}.$$

Then, it is clear that $K_G(\boldsymbol{a}, x + npx/M) \subseteq \overline{K_G}(\boldsymbol{a}, x)$. This implies that

$$\frac{\Phi_{1,b}(x\mathbf{1}, 0)}{\Phi_{1,b}\left(x\mathbf{1} + \frac{npx}{M}\mathbf{1}, 0\right)} = \frac{\text{Vol}(K_G(\boldsymbol{a}, x))}{\text{Vol}(K_G(\boldsymbol{a}, x + npx/M))} \geq \frac{\text{Vol}(K_G(\boldsymbol{a}, x))}{\text{Vol}(\overline{K_G}(\boldsymbol{a}, x))}$$

$$= \left(\frac{1}{1 + np/M}\right)^m = \left(\frac{1}{1 + \frac{1}{2m\epsilon}}\right)^m,$$

which shows the claim. Now we have

$$\left(\frac{1}{1 + \frac{1}{2m\epsilon}}\right)^m \geq \left(1 - \frac{1}{2m\epsilon}\right)^m \geq 1 - \frac{m\epsilon}{2m} = 1 - \frac{\epsilon}{2}.$$

The first inequality is because $((1 + \frac{1}{2m\epsilon})(1 - \frac{1}{2m\epsilon}))^m \leq 1$. Then,

$$\frac{\Phi_{1,b}(x\mathbf{1} + \frac{npx}{M}, 0)}{\Phi_{1,b}(x\mathbf{1}, 0)} \leq \frac{1}{1 - \epsilon/2} \leq 1 + \epsilon,$$

for $0 \leq \epsilon \leq 1$, we have the lemma. The restriction $\epsilon \leq 1$ is not essential. We set $\epsilon = 1$ instead of larger approximation ratio.   □

Now Theorem 1 is clear from Lemma 5.4 and Observation 4.3.

6   Conclusion and Discussion

In this paper, we showed that there is an FPTAS for the distribution function of the longest path length of a DAG G if the pathwidth of G is bounded by a constant p. Though the problem is #P-hard for a directed path (see [7]), there are cases where there are exponentially many paths in G. The idea of our algorithm may be extended to the DAGs with constant treewidth by making the description slightly more complex.

For the future work, the cases in which the pathwidth of G is large would be interesting. Also, it would be interesting if we can find some #P-hard volumes such that there is an FPTAS that is based on the staircase approximation.

Acknowledgment. This work was supported by JSPS KAKENHI Grant Number 15K15945. The author thanks the anonymous referees for the helpful suggestions.

References

1. Ando, E., Kijima, S.: An FPTAS for the volume computation of $0 - 1$ knapsack polytopes based on approximate convolution. Algorithmica **76**, 1245–1263 (2016)
2. Ando, E., Ono, H., Sadakane, K., Yamashita, M.: Computing the exact distribution function of the stochastic longest path length in a DAG. In: Chen, J., Cooper, S.B. (eds.) TAMC 2009. LNCS, vol. 5532, pp. 98–107. Springer, Heidelberg (2009). doi:10.1007/978-3-642-02017-9_13
3. Bárány, I., Füredi, Z.: Computing the volume is difficult. Discrete Comput. Geom. **2**, 319–326 (1987)
4. Blaauw, D., Chopra, K., Srivastava, A., Scheffer, L.: Statistical timing analysis: from basic principles to state of the art. IEEE Trans. Comput.-Aided Des. Integr. Circ. Syst. **27**, 589–606 (2008)
5. Bodlaender, H.: A linear-time algorithm for finding tree-decompositions of small treewidth. SIAM J. Comput. **25**, 1305–1317 (1996)
6. Cousins, B., Vempala, S.: Bypassing KLS: Gaussian cooling and an $O^*(n^3)$ volume algorithm. In: 47th Annual ACM Symposium on the Theory of Computing, pp. 539–548. ACM, New York (2015)
7. Dyer, M., Frieze, A.: On the complexity of computing the volume of a polyhedron. SIAM J. Comput. **17**, 967–974 (1988)
8. Dyer, M., Frieze, A., Kannan, R.: A random polynomial-time algorithm for approximating the volume of convex bodies. J. ACM **38**, 1–17 (1991)
9. Elekes, G.: A geometric inequality and the complexity of computing volume. Discrete Comput. Geom. **1**, 289–292 (1986)
10. Gopalan, P., Klivans, A., Meka, R.: Polynomial-time approximation schemes for knapsack and related counting problems using branching programs. arXiv:1008.3187v1 (2010)
11. Gopalan, P., Klivans, A., Meka, R., Štefankovič, D., Vempala, S., Vigoda, E.: An FPTAS for #knapsack and related counting problems. In: 52nd Annual IEEE Symposium on Foundations of Computer Science, pp. 817–826. IEEE Publications, New York (2011)
12. Johnson, T., Robertson, N., Seymour, P.D., Thomas, R.: Directed tree-width. J. Combin. Theory Ser. B **82**, 138–155 (2001)
13. Li, J., Shi, T.: A fully polynomial-time approximation scheme for approximating a sum of random variables. Oper. Res. Lett. **42**, 197–202 (2014)
14. Lovász, L.: An Algorithmic Theory of Numbers, Graphs and Convexity. SIAM Society for Industrial and Applied Mathematics, Philadelphia (1986)
15. Lovász, L., Vempala, S.: Simulated annealing in convex bodies and an $O^*(n^4)$ volume algorithm. J. Comput. Syst. Sci. **72**, 392–417 (2006)
16. Štefankovič, D., Vempala, S., Vigoda, E.: A deterministic polynomial-time approximation scheme for counting knapsack solutions. SIAM J. Comput. **41**, 356–366 (2012)
17. Vaidya, P.: An algorithm for linear programming which requires $O(((m + n)n^2 + (m + n)^{1.5}n)L)$ arithmetic operations. Math. Program. **47**, 175–201 (1990)

Graph Algorithms II

Sequentially Swapping Colored Tokens on Graphs

Katsuhisa Yamanaka[1]([✉]), Erik D. Demaine[2], Takashi Horiyama[3],
Akitoshi Kawamura[4], Shin-ichi Nakano[5], Yoshio Okamoto[6], Toshiki Saitoh[7],
Akira Suzuki[8], Ryuhei Uehara[9], and Takeaki Uno[10]

[1] Iwate University, Morioka, Japan
yamanaka@cis.iwate-u.ac.jp
[2] Massachusetts Institute of Technology, Cambridge, USA
edemaine@mit.edu
[3] Saitama University, Saitama, Japan
horiyama@al.ics.saitama-u.ac.jp
[4] The University of Tokyo, Meguro, Japan
kawamura@graco.c.u-tokyo.ac.jp
[5] Gunma University, Kiryu, Japan
nakano@cs.gunma-u.ac.jp
[6] The University of Electro-Communications, Chofu, Japan
okamotoy@uec.ac.jp
[7] Kobe University, Kobe, Japan
saitoh@eedept.kobe-u.ac.jp
[8] Tohoku University, Sendai, Japan
a.suzuki@ecei.tohoku.ac.jp
[9] Japan Advanced Institute of Science and Technology, Nomi, Japan
uehara@jaist.ac.jp
[10] National Institute of Informatics, Chiyoda, Japan
uno@nii.jp

Abstract. We consider a puzzle consisting of colored tokens on an n-vertex graph, where each token has a distinct starting vertex and a set of allowable target vertices for it to reach, and the only allowed transformation is to "sequentially" move the chosen token along a path of the graph by swapping it with other tokens on the path. This puzzle is a variation of the Fifteen Puzzle and is solvable in $O(n^3)$ token-swappings. We thus focus on the problem of minimizing the number of token-swappings to reach the target token-placement. We first give an inapproximability result of this problem, and then show polynomial-time algorithms on trees, complete graphs, and cycles.

1 Introduction

In this paper, we consider the following puzzle on graphs. Let $G = (V, E)$ be an undirected unweighted graph with vertex set V and edge set E. Suppose that each vertex in G has a color in $C = \{1, 2, \ldots, |C|\}$, $|C| \leq |V|$, and has a token of a color in C. Then, we wish to transform the current token-placement into the

© Springer International Publishing AG 2017
S.-H. Poon et al. (Eds.): WALCOM 2017, LNCS 10167, pp. 435–447, 2017.
DOI: 10.1007/978-3-319-53925-6_34

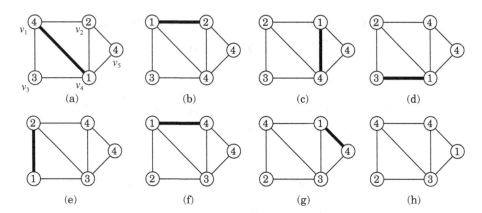

Fig. 1. An example of SEQUENTIAL TOKEN SWAPPING. (a) An input graph and its initial token-placement. (b)–(g) Intermediate token-placements. (h) The target token-placement. Token colors on vertices are written inside circles. We sequentially swap the token on v_4 along the walk $\langle v_4, v_1, v_2, v_4, v_3, v_1, v_2, v_5 \rangle$. For each solid line, the two tokens on its endpoints are swapped.

one such that a token of color i is placed on a vertex of color i for all vertices by a "sequential" token swapping. For a walk $W = \langle w_1, w_2, \ldots, w_k \rangle$[1] of G, a *sequential swapping* is to swap the two tokens on w_i and w_{i+1} in the order of $i = 1, 2, \ldots, k - 1$. Intuitively, the token on w_1 is moved to w_k along W and for each $i = 2, 3, \ldots, k$, the token on w_i is shifted to w_{i-1}. Figure 1 shows an example of a sequential swapping. If there exists a color i such that the number of vertices of color i is not equal to the number of tokens of color i in a current token-placement, then we cannot transform the token-placement into the target one. Thus, without loss of generality, we assume that the number of vertices of color i for each $i = 1, 2, \ldots, |C|$ is equal to the number of tokens of the same color.

Our problem is regarded as a variation of the Fifteen Puzzle or 15 Puzzle [1] if we assume that vertices and tokens are labeled. There are some generalizations for the Fifteen Puzzle. Ratner and Warmuth [5] considered the Fifteen Puzzle on a $N \times N$ board. They demonstrated that the problem of finding the shortest solution of the Fifteen Puzzle on $N \times N$ board is NP-complete. Goldreich [3] generalized the problem to a game on graphs. He demonstrated that the problem of finding the shortest solution of the Fifteen Puzzle on graphs is NP-complete. Kornhauser et al. [4] and Wilson [6] also considered the problem of the Fifteen Puzzle on graphs. See Demaine and Hearn's survey [2] on the Fifteen Puzzle and its related puzzles for further details.

Recently, Yamanaka et al. [7,8] considered the same problem with the different swapping rule which is to swap any two tokens on adjacent vertices. Yamanaka et al. [7] dealt with the case where the number of colors is equal

---

[1] In this paper, we denote a walk of a graph by a sequence of vertices.

to the number of vertices, and showed a polynomial-time 2-approximation algorithm for trees and a polynomial-time exact algorithm for complete bipartite graphs. However, the complexity of the problem was not proved in the paper. On the other hands, Yamanaka et al. [8] considered the more general case in which the number of colors is equal to or smaller than the number of vertices. They demonstrated that the problem is NP-complete when the number of colors is 3 or more, and otherwise the problem is polynomially solvable.

In this paper, we consider the sequential token swapping problem which asks to find the shortest walk W such that the sequential swapping along W gives the target token-placement. We first demonstrate an inapproximability of our problem even if the number of colors is 2. This result shows a difference on complexity between the problem in [8] and our problem in the sense that, when the number of colors is 2, the former is polynomially solvable, however the latter is computationally hard. Then, we present some positive results for restricted graph classes: trees, complete graphs, and cycles.

Due to space limitations, several proofs are omitted.

2   Preliminaries

2.1   Notations

In this paper, we assume without loss of generality that graphs are simple and connected. Let $G = (V, E)$ be an undirected unweighted graph with vertex set V and edge set E. We sometimes denote by $V(G)$ and $E(G)$ the vertex set and the edge set of G, respectively. We always denote $|V|$ by n. For a vertex v in G, let $N_G(v)$ be the set of all neighbors of v. that is, $N_G(v) = \{w \in V(G) \mid (v, w) \in E(G)\}$. Let $N_G[v] = N_G(v) \cup \{v\}$. Each vertex of a graph G has a color in $C = \{1, 2, \ldots, |C|\}$. We denote by $c(v) \in C$ the color of a vertex $v \in V$. A token is placed on each vertex in G. Each token also has a color in C. For a vertex v, we denote by $f(v)$ the color of the token placed on v. Then, we call the surjective function $f : V \to C$ a *token-placement* of G. Note that, since c is also a function from V to C, it can be regarded as a token-placement of G. Let f and f' be two token-placements of G. For a walk $W = \langle w_1, w_2, \ldots, w_h \rangle$ of G, a sequence $\mathcal{S} = \langle f_1, f_2, \ldots, f_h \rangle$ of token-placements is a *swapping sequence* of W between f and f' if the following three conditions (1)–(3) hold:

(1) $f_1 = f$ and $f_h = f'$;
(2) f_k is a token-placement of G for each $k = 1, 2, \ldots, h$; and
(3) f_k is obtained from f_{k-1} by swapping the two tokens on w_{k-1} and w_k for each $k = 2, 3, \ldots, h$.

Intuitively, a swapping sequence of W represents to move the token on w_1 to w_h along W.

Let S be a sequence. Then, the length of S, denoted by $\mathrm{len}(S)$, is defined to be the number of elements in S minus one. The length of a swapping sequence $\mathcal{S}$, $\mathrm{len}(\mathcal{S})$, indicates the number of token-swappings in $\mathcal{S}$. For two token-placements

f and c of G, we denote by $\mathrm{OPT}_{\mathrm{STS}}(G, f, c)$ the minimum length of a swapping sequence between f and c. Given two token-placements f and c of a graph G and a nonnegative integer ℓ, the SEQUENTIAL TOKEN SWAPPING problem is to determine whether or not $\mathrm{OPT}_{\mathrm{STS}}(G, f, c) \leq \ell$ holds. We call f and c the *initial* and *target* token-placements of G, respectively.

2.2   Polynomial-Length Upper Bound

We prove that, if there exists a swapping sequence between two token-placements, then the length of the sequence is polynomial.

Theorem 1. *For any graph G and an initial token-placement f and a target token-placement c, if there exists a swapping sequence between the two token-placements, $OPT_{STS}(G, f, c)$ is at most n^3.*

Proof. We show that, if the instance (G, f, c) has a swapping sequence, one can construct a swapping sequence of length at most n^3 using an existing result [4]. Let P_i be a permutation from the set of tokens of color i in f to the set of tokens of color i in c, for each $i = 1, 2, \ldots, |C|$. We denote by $\mathcal{P} = \{P_1, P_2, \ldots, P_{|C|}\}$ the set of the permutations. Then, by using the permutations in $\mathcal{P}$, we can create an initial token-placement f' and a target token-placement c' such that, using a color set $C' = \{1, 2, \ldots, n\}$, the token colors of any two vertices in f' are distinct and the token colors of any two vertices in c' are also distinct. Note that, for f' and c', the target place (vertex) of each token is unique and a swapping sequence between f' and c' is also a swapping sequence between f and c. Then, we can observe that, among all the pairs of token-placements which are created from all possible permutation sets, there exists at least one swapping sequence of length at most n^3 if and only if, for f and c, there exists a swapping sequence of length at most n^3.

Kornhauser *et al.* [4] showed that, for two token-placement of C' and a token t, one can decide whether there exists a swapping sequence between the two token-placements in which t is moved, and if a swapping sequence exists one can find a swapping sequence of length at most n^3. By applying their method to all the pairs of the token-placements which are created from all possible permutation sets and all tokens, one can obtain a swapping sequence of length at most n^3 if a swapping sequence exists. $\square$

3   Inapproximability

In this section, we demonstrate an inapproximability of SEQUENTIAL TOKEN SWAPPING problem. To show the hardness result, we give a gap-preserving reduction from the following problem:

Problem: MAXIMUM VERTEX-DISJOINT PATH COVER ON UNDIRECTED BIPARTITE GRAPHS

Instance: An undirected bipartite graph $G = (V, E)$ with partite sets X and Y such that $|X| = |Y|$.

Question: Find a set of vertex-disjoint paths that cover all the vertices in G such that the paths contain the maximum number of edges.

If an input graph G is hamiltonian, then a hamiltonian path in the graph is an optimal solution and the number of edges in the path is $n - 1$, where n is the number of vertices in G. We can show that $(1 - \varepsilon, 1)$-gap MAXIMUM VERTEX-DISJOINT PATH COVER ON UNDIRECTED BIPARTITE GRAPHS problem is NP-hard (We omit the proof of this claim). Thus, we give a gap-preserving reduction from the problem to SEQUENTIAL TOKEN SWAPPING problem with only 2 colors. Let $\mathrm{OPT_{U\text{-}MVDPC}}(G)$ denote the optimal value, which is the number of edges in paths in an optimal path cover, of MAXIMUM VERTEX-DISJOINT PATH COVER ON UNDIRECTED BIPARTITE GRAPHS problem for an input graph G.

Theorem 2. *There is a gap preserving reduction from* MAXIMUM VERTEX-DISJOINT PATH COVER ON UNDIRECTED BIPARTITE GRAPHS *problem to* SEQUENTIAL TOKEN SWAPPING *problem that transforms a bipartite graph* $G = (V, E)$ *with partite sets* X *and* Y *such that* $|X| = |Y|$ *to a graph* $H = (V_H, E_H)$ *and its two token-placements* f, c *with 2 colors, where* $n = |V|$ *such that*

(1) if $\mathrm{OPT_{U\text{-}MVDPC}}(G) = n - 1$, *then* $\mathrm{OPT_{STS}}(H, f, c) = n - 1$ *and*
(2) if $\mathrm{OPT_{U\text{-}MVDPC}}(G) < (1 - \varepsilon)(n - 1)$, *then* $\mathrm{OPT_{STS}}(H, f, c) > (1 + \varepsilon)(n - 1)$.

Proof. Let G be an instance of MAXIMUM VERTEX-DISJOINT PATH COVER ON UNDIRECTED BIPARTITE GRAPHS problem. Now we construct an instance of SEQUENTIAL TOKEN SWAPPING problem, that is a graph, an initial token-placement, and a target one. We first construct a copy $G' = (V', E')$ of G, and denote its two partite sets by X' and Y'. We set $f(u) = 1$ and $c(u) = 2$ for every vertex $u \in X'$, and set $f(v) = 2$ and $c(v) = 1$ for every vertex $v \in Y'$. We then insert "connection gadgets" for non-adjacent vertex pairs between X' and Y', as follows. Let $A = \{(u, v) \mid u \in X', v \in Y', \text{and } (u, v) \notin E'\}$. The *connection gadget* for $(u, v) \in A$ consists of two paths of length 2 connecting $u \in X'$ and $v \in Y'$ (see Fig. 2). For the two intermediate vertices w and z in the paths, we set $f(w) = 1$ and $c(w) = 1$, and $f(z) = 2$ and $c(z) = 2$. We denote the obtained graph and its initial and target token-placements by H, f, and c, respectively. Figure 3 shows an example of the reduction graph. In the figure, connection gadgets are represented as dotted lines for convenience. The reduction graph, its initial token-placement, and its target token-placement can be constructed in polynomial time.

Now we show that, an optimal solution of the reduced instance of SEQUENTIAL TOKEN SWAPPING can be obtained from an optimal solution of an instance of MAXIMUM VERTEX-DISJOINT PATH COVER ON UNDIRECTED BIPARTITE GRAPHS, as follows. Let $\mathcal{P}$ be a maximum vertex-disjoint cover of G and let $\mathrm{cost}(\mathcal{P}) = \sum_{P \in \mathcal{P}} \mathrm{len}(P)$, where $\mathrm{len}(P)$ is the number of edges in the path $P \in \mathcal{P}$. We construct a swapping sequence between f and c from $\mathcal{P}$. Let P_1 and P_2 be two paths in $\mathcal{P}$ such that P_1 contains an endpoint v_1 in X' and P_2 contains

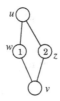

Fig. 2. A connection gadget.

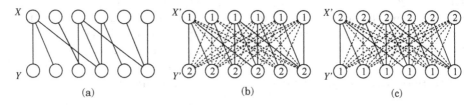

Fig. 3. An example of reduction. (a) A bipartite graph $G = (V, E)$ with partite sets X and Y such that $|X| = |Y|$. (b) The reduction graph H and its initial token-placement f. Connection gadgets are represented as dotted lines. (c) The target token-placement.

an endpoint v_2 in Y'. Then, we connect $v_1 \in X'$ and $v_2 \in Y'$ with a path of the connection gadget between v_1 and v_2. Note that v_1 and v_2 are not adjacent from optimality of $\mathcal{P}$. When a token is sequentially swapped from $v_1 \in X'$ to $v_2 \in Y'$, we choose the path with the intermediate vertex of color 2. Otherwise, we choose the path with the intermediate vertex of color 1. Since $|X| = |Y|$, by repeating the above process, we can find a path spanning all vertices in $X' \cup Y'$. Note that, since $|X| = |Y|$ holds again, the number of paths whose endpoints are both in X' are equal to the number of paths whose endpoints are both in Y'. Then, by swapping the token on an endpoint to the other endpoint along the obtained path, we have the target token-placement.

Let $\mathcal{S}$ be the swapping sequence obtained from $\mathcal{P}$ by the above process. Now, we show that $\mathcal{S}$ is optimal. We assume for a contradiction that $\mathcal{S}'$ is a better solution than $\mathcal{S}$. Let W be the walk corresponding to $\mathcal{S}$, and let m be the the number of edges in W in connection gadgets. Similarly, let W' be the walk corresponding to $\mathcal{S}'$, and let m' be the number of edges in W' in connection gadgets. Then, W' is a path spanning vertices in $X' \cup Y'$. More precisely, a vertex in $X' \cup Y'$ appears once in W' and an intermediate vertex in a connection gadget appears at most once in W'. Note that every vertex v in H does not appear twice or more, since to visit v twice or more produces redundant token-swappings. Hence, we can construct a path cover from W', as follows. First, we split W' into subsequences by regarding intermediate vertices as boundaries and removing the intermediate vertices. Then we obtain the set of subsequences. Since W' spans the vertices in $X' \cup Y'$ and visits each vertex at most once, the set is a path cover of G. Let $\mathcal{P}'$ denote the path cover obtained from W'.

Now, to derive a contradiction, we first focus on the two path covers $\mathcal{P}$ and $\mathcal{P}'$, and then we derive an inequality between $|\mathcal{P}|$ and $|\mathcal{P}'|$, more specifically the number of paths in $\mathcal{P}$ is smaller than or equal to the number of paths in $\mathcal{P}'$. Since each of $\mathcal{P}$ and $\mathcal{P}'$ visits every vertex in $X' \cup Y'$ exactly once, we have $cost(\mathcal{P}) = (n-1) - |\mathcal{P}| + 1$ and $cost(\mathcal{P}') = (n-1) - |\mathcal{P}'| + 1$. Then, we also have $cost(\mathcal{P}) \geq cost(\mathcal{P}')$, since $\mathcal{P}'$ is an optimal path cover. Thus, we obtain $|\mathcal{P}| \leq |\mathcal{P}'|$.

Next we focus on the two walks W and W' and we also derive an inequality between $|\mathcal{P}|$ and $|\mathcal{P}'|$. Since $\mathcal{S}'$ is a shorter swapping sequence than $\mathcal{S}$, $len(W) > len(W')$ holds. Each of W and W' visits every vertex in $X' \cup Y'$ exactly once. Therefore, the number of edges in connection gadgets in W is greater than the number of edges in connection gadgets in W', that is $m > m'$. Since W and W' include exactly two edges in each connection gadget in W and W', we have $|\mathcal{P}| = \frac{m}{2} + 1$ and $|\mathcal{P}'| = \frac{m'}{2} + 1$, respectively. Therefore, $|\mathcal{P}| > |\mathcal{P}'|$ holds, which contradicts to $|\mathcal{P}| \leq |\mathcal{P}'|$. Therefore, $\mathcal{S}$ is an optimal swapping sequence of H, f, and c.

Now, we demonstrate the claims (1) and (2).

If $\text{OPT}_{\text{U-MVDPC}}(G) = n - 1$, then G has a hamiltonian path P. Note that an endpoint of P is in X and the other is in Y since $|X| = |Y|$. By sequentially swapping the token on an endpoint of P in H to the other endpoint, we obtain the target token-placement. Hence, $\text{OPT}_{\text{STS}}(H, f, c) \leq n - 1$. Since we must visit every vertex v in H such that $f(v) \neq c(v)$, the number of such vertices minus one is a lower bound for $\text{OPT}_{\text{STS}}(H, f, c)$. Therefore $\text{OPT}_{\text{STS}}(H, f, c) = n - 1$.

Let $\mathcal{S}$ be an optimal swapping sequence obtained from an optimal path cover $\mathcal{P}$ of G by the transformation above. The length of $\mathcal{S}$ is equal to the sum of the number of edges in $\mathcal{P}$ and the number of edges in the connection gadgets which used in $\mathcal{S}$. (Recall that at most two edges in each connection gadget are used in $\mathcal{S}$.) Therefore we have the following equation:

$$len(\mathcal{S}) = cost(\mathcal{P}) + 2(|\mathcal{P}| - 1)$$
$$= (n-1) + (|\mathcal{P}| - 1) \tag{1}$$

If $\text{OPT}_{\text{U-MVDPC}}(G) < (1 - \varepsilon)(n-1)$, then $|\mathcal{P}|$ is bounded from below:

$$|\mathcal{P}| = (n-1) - cost(\mathcal{P}) + 1$$
$$> (n-1) - (1-\varepsilon)(n-1) + 1$$
$$= \varepsilon(n-1) + 1. \tag{2}$$

From Equality (1) and Inequality (2), we have $\text{OPT}_{\text{STS}}(H, f, c) > (1 + \varepsilon)(n-1)$. □

From Theorem 2, even if the number of colors is 2, there is no polynomial-time $(1 + \varepsilon)$-approximation algorithm, unless $P = NP$.

4 Polynomial-Time Algorithms

In the previous section, we showed the inapproximability of SEQUENTIAL TOKEN SWAPPING problem even if the number of colors is 2. On the other hand, if graph classes are restricted, the problem can be solved in polynomial time for any number of colors. In this section, we consider the problem of computing the minimum numbers of token-swappings for trees, complete graphs, and cycles.

4.1 Trees

Let $G = (V, E)$ be a tree, and let f and c be an initial and a target token-placements of G, respectively. We show below that SEQUENTIAL TOKEN SWAPPING problem on trees can be solved in linear time.

First we repeat to remove a leaf v with $f(v) = c(v)$ until the tree has no such leaf. If all the vertices are removed, then we return zero as a solution. We assume otherwise. Let G' be the obtained tree. Note that G' is a tree with two or more vertices, and hence G' has two or more leaves. If G' has three or more leaves, the input tree and its token-placements f and c is a no-instance. Recall that only one token is sequentially moved. Hence we assume that G' have exactly two leaves. For the token on each leaf v, which satisfies $f(v) \neq c(v)$, we sequentially move the token on v to its target vertex. We finally check whether or not the obtained token-placement is the target one. If the answer is yes for at least one of them, then the sequential token-swapping above is optimal. Otherwise, the input tree and its token-placements f and c is a no-instance. Hence we have the following theorem.

Theorem 3. *For a tree G, an initial token-placement f, and a target token-placement c one can compute $OPT_{STS}(G, f, c)$ in $O(n)$ time.*

4.2 Complete Graphs

Let $G = (V, E)$, f, and c be a complete graph, an initial token-placement and a target token-placement, respectively. Let $C = \{1, 2, \ldots, |C|\}$, $|C| \leq n$, be the set of colors. For a token-placement f, let $V'(f) \subseteq V$ be the set of vertices v such that $f(v) \neq c(v)$.

We first introduce a multiple digraph $D(f) = (V_D(f), E_D(f))$ called the *destination graph* as follows:

- $V_D(f) = \{i \mid i \in C \text{ and } i = f(v) \text{ for some } v \in V'(f)\}$.
- $E_D(f) = \{(f(v), c(v)) \mid v \in V'(f)\}$

Note that $V_D(f)$ corresponds to the set of the colors each of which is a color of at least one token placed in a vertex in $V'(f)$ and each arc $(f(v), c(v)) \in E_D(f)$ corresponds to a vertex $v \in V'(f)$. Since, for each color in C, the number of vertices in G of the color is equal to the number of tokens of the same color, each node in destination graph $D(f)$ has the same numbers of incoming edges and outgoing edges. Thus, each connected component of $D(f)$ has a directed Euler cycle. Therefore, $D(f)$ consists of only strongly connected components. Let $s(f)$ be the number of strongly connected components in $D(f)$. Then we claim the following:

Claim. For a complete graph, an initial token-placement f, and a target token-placement c, $OPT_{STS}(G, f, c) = |V'(f)| + s(f) - 2$.

The claim above immediately implies the following theorem, since we can calculate the values of $|V'(f)|$ and $s(f)$ in $O(n)$ time.

Theorem 4. *For a complete graph G, an initial token-placement f, and a target token-placement c one can compute $OPT_{STS}(G, f, c)$ in $O(n)$ time.*

In the rest of this section, we prove the above claim. First we show that $OPT_{STS}(G, f, c) \leq |V'(f)| + s(f) - 2$ by constructing a swapping sequence of length $|V'(f)| + s(f) - 2$, then we show that $OPT_{STS}(G, f, c) \geq |V'(f)| + s(f) - 2$ by using a potential function.

Upper Bound. We present an algorithm that finds a swapping sequence of length $|V'(f)| + s(f) - 2$. Let C_i, $i = 1, 2, \ldots, s(f)$, be a strongly connected component of $D(f)$. Recall that C_i has a directed Euler cycle. We here denote a directed Euler cycle as a sequence of directed edges: For each C_i, $i = 1, 2, \ldots, s(f)$, let $\langle e_{i,1}, e_{i,2}, \ldots, e_{i,t_i} \rangle$ denote a directed Euler cycle of C_i, where t_i is the number of edges in C_i. For each $e_{i,j}$, let $v_{i,j}$ be the corresponding vertex in $V'(f)$. Let

$$
\begin{aligned}
W = \langle\ & v_{1,1}, v_{1,2}, \ldots, v_{1,t_1}, \\
& v_{2,1}, v_{2,2}, \ldots, v_{2,t_2}, v_{1,t_1}, \\
& v_{3,1}, v_{3,2}, \ldots, v_{3,t_3}, v_{1,t_1}, \\
& \ldots, \\
& v_{s(f),1}, v_{s(f),2}, \ldots, v_{s(f),t_{s(f)}}, v_{1,t_1} \rangle.
\end{aligned}
$$

be a walk in $V'(f)$. Now we show that the length of W is $|V'(f)| + s(f) - 2$ and the target token-placement c is obtained by swapping along W. This immediately implies that we have a swapping sequence between f and c of length $|V'(f)| + s(f) - 2$.

Since each vertex in $V'(f) \setminus \{v_{1,t_1}\}$ appears exactly once and v_{1,t_1} appears $s(f)$ times, $\text{len}(W) = |V'(f)| + s(f) - 2$. Let f' be the token-placement obtained by sequentially swapping the token on $v_{1,1}$ along W, and let $v_{i,j}$, $1 \leq j \leq t_i - 1$, denote a vertex in $V'(f) \setminus \{v_{1,t_1}\}$. Since $v_{i,j}$ appears exactly once, $f'(v_{i,j}) = f(v_{i,j+1})$ holds. Recall that $v_{i,j}$ and $v_{i,j+1}$ correspond to $e_{i,j}$ and $e_{i,j+1}$ in the directed Euler cycle of C_i, respectively. Thus, $f(v_{i,j+1}) = c(v_{i,j})$ holds. Next, let us consider the vertex v_{i,t_i} in $V'(f) \setminus \{v_{1,t_1}\}$. It can be observed that, while we traverse from $v_{i,1}$ to v_{i,t_i}, v_{1,t_1} has $f(v_{i,1})$, since the sequence $\langle e_{i,1}, e_{i,2}, \ldots, e_{i,t_i} \rangle$ is an Euler cycle, $f(v_{i,1}) = c(v_{i,t_i})$ holds. Thus, v_{i,t_i} has its expected token after the token-swappings on vertices of C_i. Finally, we have $f'(v_{1,t_1}) = c(v_{1,t_1})$, since $f(v_{1,1}) = c(v_{1,t_1})$ holds.

Lower Bound. Now we show that the length of any swapping sequence is at least $|V'(f)| + s(f) - 2$. Let $\mathcal{S} = \langle f_1, f_2, \ldots, f_h \rangle$ be an arbitrary swapping sequence between f and c for a walk $W = \langle w_1, w_2, \ldots, w_h \rangle$. Note that $f_1 = f$ and $f_h = c$. We call the token x on w_1 in f the *moving token* of $\mathcal{S}$. Let $D(f_i)$ be the destination graph for each token-placement.

First, we define a potential function $p(f_i)$. Let $p_1(f_i)$ be the number of vertices in $V'(f_i)$ except the vertex with the moving token x, and let $p_2(f_i)$ be the number of strongly connected components in $D(f_i)$ that do not include x. We define the potential function as $p(f_i) = p_1(f_i) + p_2(f_i)$. Then,

$$p(f_i) \geq (|V'(f_i)| - 1) + (s(f_i) - 1)$$

and

$$p(c) = 0.$$

Note that for any token-placement $f_i \neq c$,

$$p(f_i) \geq 1.$$

Now we can show the following lemma.

Lemma 1. $p(f_{i+1}) \geq p(f_i) - 1$ *holds.*

Hence, the length of any swapping sequence f is at least $p(f)$. This completes the proof of Theorem 4.

4.3 Cycles

In this section, we present two algorithms for cycles. The first algorithm runs in $O(n^4)$ time, while the second one is faster and runs in $O(n^2)$ time. Let $G = (V, E)$ be a cycle with n vertices, and let f and c be an initial and a target token-placements of G. For cycles, the moving token goes clockwise or counterclockwise. In the shortest sequential swapping, the moving token does not turn back, since changing the direction produces redundant token-swappings. Hence, the moving token always goes either clockwise or counterclockwise. The optimal solution is the shortest sequential swapping among both directions. Thus, in this section, we suppose that the moving token always goes clockwise, since the same discussion can be applied to the other direction.

Naïve Algorithm

We denote vertices in clockwise order on the cycle by $\langle v_1, v_2, \ldots, v_n \rangle$. We first define the following $n \times n$ table $T[x][k]$:

$$T[x][k] = \begin{cases} 1 & f(v_x) = c(v_{(x-k) \bmod n}) \\ 0 & \text{otherwise.} \end{cases}$$

The value of $T[x][k]$ represents whether or not the token on v_x is placed on its expected vertex after the token goes counterclockwise by k token-swappings. Using this table, we make sure the token-placement after a sequential swapping is identical to the target one.

If we move the token on a vertex v_x clockwise with a sequential swapping of length $n-1$, then all other tokens are shifted once counterclockwise. Similarly, if we move the token on v_x clockwise with a sequential swapping of length $i(n-1)$, for $i = 1, 2, \ldots, n-1$, then all other tokens are shifted i times counterclockwise. Thus, a sequential token-swapping of length $i(n-1) + j$ moves each token on v_w, $w = x+1, x+2, \ldots, x+j \pmod{n}$, $i+1$ times counterclockwise and each token on v_w, $w = x-1, x-2, \ldots, x+j+1 \pmod{n}$, i times counterclockwise. Therefore, we have the following observation.

Observation 1. *The token-placement obtained by a sequential swapping of length $i(n-1) + j$ with the moving token on v_x is identical to the target one if and only if*

(1) $T[w][i+1] = 1$ for each $w = x+1, x+2, \ldots, x+j \pmod{n}$
(2) $T[w][i] = 1$ for each $w = x-1, x-2, \ldots, x+j+1 \pmod{n}$
(3) $T[x][i-j] = 1$.

From this observation, we have a naïve algorithm. We denote a candidate of a solution by a triple (x, i, j) for $1 \le x \le n$, $1 \le i \le n-1$, and $1 \le j \le n-1$. A triple (x, i, j) is *feasible* if it satisfies the above three conditions. The naïve algorithm simply investigates whether or not every triple (x, i, j) is feasible, then returns the triple that minimizes the value of $(n-1)i + j$ among all the feasible triples. This algorithm runs in $O(n^4)$ time.

Theorem 5. *For a cycle G, an initial token-placement f, and a target token-placement c one can compute $OPT_{STS}(G, f, c)$ in $O(n^4)$ time.*

Improvement

In this subsection, we improve the running time of the naïve algorithm. We construct three other tables that store auxiliary information to efficiently check the conditions in Observation 1.

First we define the table T'. For a vertex v_x, $1 \le x \le n$, and an integer k, $1 \le k \le n-1$, $T'[x][k]$ stores the maximum index s such that $T[w][k] = 1$ for all $w = x+1, x+2, \ldots, x+s \pmod{n}$. Intuitively, the entry $s = T'[x][k]$ means that, after k token-swappings, all consecutive tokens $f(w)$, $w = x+1, x+2, \ldots, x+s \pmod{n}$, are placed on their expected vertices, $f(x+s+1 \bmod n)$ is placed on an unexpected vertex. Similarly, we define the table T'', as follows. For a vertex v_x, $1 \le x \le n$, and an integer k, $1 \le k \le n-1$, $T''[x][k]$ stores the maximum index s such that $T[w][k] = 1$ for any $w = x-1, x-2, \ldots, x-s \pmod{n}$. The table T'' focuses on the consecutive tokens from v_x in the opposite direction of T'. We also define the table T'''. The table $T'''[x][k]$ stores the maximum index s such that $T[x][\ell] = 0$ for any $\ell = k, k-1, \ldots, k-s+1 \pmod{n}$. Intuitively, this entry means how many token-swappings we need to place the moving token, which is placed on v_x in f, on its expected vertex, after the token is swapped k times counterclockwise.

Our goal is to find the feasible triple (x, i, j) that minimizes the value of $(n-1)i + j$. To find such a triple, for every pair of (x, i), we find the smallest j such that the triple (x, i, j) is feasible. Among them, the triple that minimizes the value of $(n-1)i + j$ is a desired solution.

Now we describe the algorithm. Suppose we are given a pair of (x, i), $1 \leq x \leq n$ and $1 \leq i \leq n-1$. First, we investigate a range of j using the tables. Since a feasible triple needs to satisfy the first and second conditions in Observation 1, we have two ranges $j \leq T'[x][i+1]$ and $j \geq n - T''[x][i] - 1$. Let $j_{min} \leq j \leq j_{max}$ be the range of j which satisfies the above two inequalities. Note that, if the range is empty, it implies that there is no feasible triple for the given pair (x, i) and thus the algorithm returns false. Then, we investigate whether there is j, $j_{min} \leq j \leq j_{max}$, with the third condition in Observation 1. This can be checked by $j_{min} + T'''[x][i - j_{min}] \leq j_{max}$. If the inequality is true, then the algorithm returns $j = j_{min} + T'''[x][i - j_{min}]$. Note that this value is the minimum j for the given pair (x, i) from the definition of T'''. Otherwise, the algorithm returns false. Therefore, we have the following theorem.

Theorem 6. *For a cycle G, an initial token-placement f, and a target token-placement c, one can compute $OPT_{STS}(G, f, c)$ in $O(n^2)$ time.*

Acknowledgment. This work is partially supported by MEXT/JSPS KAKENHI Grant Numbers JP24106002, JP24106004, JP24106005, JP24106007, JP24220003, JP24700008, JP26330004, JP26330009, JP26730001, JP15K00008, JP15K00009, JP16K00002, and JP16K16006, the Asahi Glass Foundation, JST, CREST, Foundations of Innovative Algorithms for Big Data, and JST, CREST, Foundations of Data Particlization for Next Generation Data Mining.

References

1. Berlekamp, E.R., Conway, J.H., Guy, R.K.: Winning Ways for Your Mathematical Plays, 2nd edn. AK Peters, Natick (2004)
2. Demaine, E.D., Hearn, R.A.: Playing games with algorithms: algorithmic combinatorial game theory. In: Games of No Chance 3, vol. 56, pp. 3–56. Cambridge University Press, Cambridge (2009)
3. Goldreich, O.: Finding the shortest move-sequence in the graph-generalized 15-puzzle is NP-hard. In: Goldreich, O. (ed.) Studies in Complexity and Cryptography. Miscellanea on the Interplay between Randomness and Computation. LNCS, vol. 6650, pp. 1–5. Springer, Heidelberg (2011). doi:10.1007/978-3-642-22670-0_1
4. Kornhauser, D.M., Miller, G., Spirakis, P.: Coordinating pebble motion on graphs, the diameter of permutation groups, and applications. In: Proceedings of the 25th Annual Symposium on Foundations of Computer Science (FOCS), pp. 241–250 (1984)
5. Ratner, D., Warmuth, M.: The $(n^2 - 1)$-puzzle and related relocation problems. J. Symb. Comput. **10**, 111–137 (1990)

6. Wilson, R.M.: Graph puzzles, homotopy, and the alternating group. J. Comb. Theory (B) **16**, 86–96 (1974)
7. Yamanaka, K., Demaine, E.D., Ito, T., Kawahara, J., Kiyomi, M., Okamoto, Y., Saitoh, T., Suzuki, A., Uchizawa, K., Uno, T.: Swapping labeled tokens on graphs. Theoret. Comput. Sci. **586**, 81–94 (2015)
8. Yamanaka, K., Horiyama, T., Kirkpatrick, D., Otachi, Y., Saitoh, T., Uehara, R., Uno, Y.: Swapping colored tokens on graphs. In: Dehne, F., Sack, J.-R., Stege, U. (eds.) WADS 2015. LNCS, vol. 9214, pp. 619–628. Springer, Heidelberg (2015). doi:10.1007/978-3-319-21840-3_51

The Time Complexity of the Token Swapping Problem and Its Parallel Variants

Jun Kawahara[1], Toshiki Saitoh[2], and Ryo Yoshinaka[3(✉)]

[1] Graduate School of Information Science, NAIST, Ikoma, Japan
[2] Graduate School of Engineering, Kobe University, Kobe, Japan
[3] Graduate School of Information Sciences, Tohoku University, Sendai, Japan
ry@ecei.tohoku.ac.jp

Abstract. The token swapping problem (TSP) and its colored version are reconfiguration problems on graphs. This paper is concerned with the complexity of the TSP and two new variants; namely parallel TSP and parallel colored TSP. For a given graph where each vertex has a unique token on it, the TSP requires to find a shortest way to modify a token placement into another by swapping tokens on adjacent vertices. In the colored version, vertices and tokens are colored and the goal is to relocate tokens so that each vertex has a token of the same color. Their parallel versions allow simultaneous swaps on non-incident edges in one step. We investigate the time complexity of several restricted cases of those problems and show when those problems become tractable and remain intractable.

1 Introduction

Yamanaka et al. [14] have introduced a kind of reconfiguration problem on graphs, called the *token swapping problem (TSP)*[1]. Suppose that we have a simple graph where each vertex is assigned a token. Each token is labeled with its unique goal vertex, which may be different from where the token is currently placed. We want to relocate every misplaced token to its goal vertex. What we can do is to swap the two tokens on the ends of an arbitrary edge. The problem is to decide how many swaps are needed to realize the goal token placement. The upper half of Fig. 1 illustrates a problem instance and a solution. The graph has 4 vertices $1, 2, 3, 4$ and 4 edges $\{1, 2\}, \{1, 3\}, \{2, 4\}, \{3, 4\}$. Each token i is initially put on the vertex $5 - i$. By swapping the tokens on the edges $\{3, 4\}, \{1, 3\}, \{2, 4\}, \{3, 4\}$ in this order, we can match the indices of the tokens and vertices.

Yamanaka et al. have presented several positive results on the TSP in addition to classical results which can be seen as special cases of the TSP [7]. Namely, graph classes for which the TSP can be solved in polynomial-time are paths, cycles, complete graphs and complete bipartite graphs. They showed that the TSP for general graphs belongs to NP. The NP-hardness is recently shown

---

[1] No salesman is traveling in this paper.

© Springer International Publishing AG 2017
S.-H. Poon et al. (Eds.): WALCOM 2017, LNCS 10167, pp. 448–459, 2017.
DOI: 10.1007/978-3-319-53925-6_35

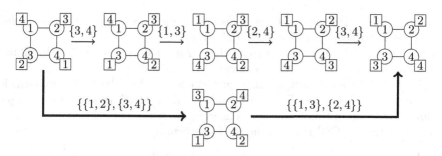

Fig. 1. Vertices and tokens are shown by circles and squares, respectively. Optimal solutions for the TSP and the PTSP are shown by small and big arrows, respectively.

in the preliminary version [9] of this paper and by Miltzow et al. [11] and Bonnet et al. [2] independently. On the other hand, some polynomial-time approximation algorithms are known for different classes of graphs including the general case [6,11,14]. For more backgrounds of the problem, the reader is referred to [14,15].

A variant of the TSP is the *c-colored token swapping problem (c-CTSP)*. Tokens and vertices in the c-CTSP are colored by one of the c admissible colors. The c-CTSP is to decide how many swaps are required to relocate the tokens so that each vertex has a token of the same color. Yamanaka et al. [15] have investigated the c-CTSP and shown that the 3-CTSP is NP-complete while the 2-CTSP is solvable in polynomial time. This problem and a further generalization are also studied in [2].

This paper is concerned with the TSP and variants of it. First, we give a proof of the NP-hardness of the TSP.

– The TSP is NP-complete even when graphs are restricted to bipartite graphs where every vertex has degree at most 3 (Theorem 1).

The result is tight with respect to the maximum vertex degree as the problem is in P if an input graph is a path or a cycle. In addition, we present two polynomial-time solvable subcases of the TSP. One is of lollipop graphs, which are combinations of a complete graph and a path. The other is the class of graphs which are combinations of a star and a path.

Variants of the TSP we will consider in this paper are the parallel versions of the TSP and c-CTSP. While in the TSP just one pair of tokens is swapped at once, the *parallel token swapping problem (PTSP)* allows us to swap token pairs on unadjacent edges simultaneously. We call a set of compatible swaps a *parallel swap*. The PTSP is to estimate how many parallel swaps are needed to achieve a goal token configuration. Figure 1 compares optimal solutions for the same instance of the TSP and the PTSP, where two parallel swaps are enough to relocate all the tokens to the goal vertices. Our main results concerning those problems include the following.

- The PTSP is NP-complete even to decide whether an instance admits a solution consisting of 3 parallel swaps (Theorem 4).
- One can decide in polynomial time whether an instance of the PTSP admits a solution consisting of 2 parallel swaps (Theorem 6).
- A polynomial-time algorithm that approximately solves the PTSP on paths is presented. It gives a parallel swap sequence whose length is at most one larger than that of an optimal solution (Theorem 7).
- The parallel 2-CTSP is NP-complete (Theorem 9).

The last result contrasts the fact that the 2-CTSP is solvable in polynomial-time [15].

One may consider the TSP and PTSP as special cases of the *minimum generator sequence problem (MGSP)* [4]. The MGSP is to determine whether one can obtain a permutation f on a finite set X by multiplying at most k permutations from a finite permutation set Π, where all of X, f, k and Π are input. The problem is known to be PSPACE-complete if k is specified in binary notation [7], while it becomes NP-complete if k is given in unary representation [4]. In the TSP and PTSP, permutation sets Π are restricted to the ones that have a graph representation. However, this does not necessarily mean that the NP-hardness of the PTSP implies the hardness of the MGSP, since the description size of all the admissible parallel swaps on a graph is exponential in the graph size.

2   Time Complexity of the Token Swapping Problem

We denote by $G = (V, E)$ an undirected graph whose vertex set is V and edge set is E. More precisely, elements of E are subsets of V consisting of exactly two distinct elements. A *configuration* f (on G) is a permutation on V, i.e., bijection from V to V. By $[u]_f$ we denote the orbit $\{f^i(u) \mid i \in \mathbb{N}\}$ of $u \in V$ under f. We call each element of V a *token* when we emphasize the fact that it is in the range of f. We say that a token v *is on a vertex* u *in* f if $v = f(u)$. A *swap* on G is a synonym for an edge of G, which behaves as a transposition. For a configuration f and a swap $e \in E$, the configuration obtained by applying e to f, which we denote by fe, is defined by

$$fe(u) = \begin{cases} f(v) & \text{if } e = \{u, v\}, \\ f(u) & \text{otherwise.} \end{cases}$$

For a sequence $\vec{e} = \langle e_1, \ldots, e_m \rangle$ of swaps, the length m is denoted by $|\vec{e}|$. For $i \leq m$, by $\vec{e}_{|\leq i}$ we denote the prefix $\langle e_1, \ldots, e_i \rangle$. The configuration $f\vec{e}$ obtained by applying $\vec{e}$ to f is $(\ldots((fe_1)e_2)\ldots)e_m$. We say that the token $f(u)$ on u *is moved to* v by $\vec{e}$ if $f\vec{e}(v) = f(u)$. We count the total moves of each token $u \in V$ in the application as

$$\mathsf{move}(f, \vec{e}, u) = |\{\, i \in \{1, \ldots, m\} \mid (f\vec{e}_{|\leq i-1})^{-1}(u) \neq (f\vec{e}_{|\leq i})^{-1}(u) \,\}|.$$

Clearly $\mathsf{move}(f, \vec{e}, u) \geq \mathsf{dist}(f^{-1}(u), (f\vec{e})^{-1}(u))$, where $\mathsf{dist}(u_1, u_2)$ denotes the length of a shortest path between u_1 and u_2, and $\sum_{u \in V} \mathsf{move}(f, \vec{e}, u) = 2|\vec{e}|$.

We denote the set of *solutions* for a configuration f by

$$\mathsf{SOL}(G, f) = \{\vec{e} \mid \vec{e} \text{ is a swap sequence on } G \text{ such that } f\vec{e} \text{ is the identity}\}.$$

A solution $\vec{e}_0 \in \mathsf{SOL}(G, f)$ is said to be *optimal* if $|\vec{e}_0| = \min\{|\vec{e}| \mid \vec{e} \in \mathsf{SOL}(G, f)\}$. The length of an optimal solution is denoted by $\mathsf{OPT}(G, f)$.

Problem 1 (Token Swapping Problem, TSP).

Instance: A graph G, a configuration f on G and a natural number k.
Question: $\mathsf{OPT}(G, f) \leq k$?

2.1   TSP Is NP-complete

This subsection proves the NP-hardness of the TSP by a reduction from the 3DM, which is known to be NP-complete [8].

Problem 2 (Three dimensional matching problem, 3DM).

Instance: Three disjoint sets A_1, A_2, A_3 such that $|A_1| = |A_2| = |A_3|$ and a set $T \subseteq A_1 \times A_2 \times A_3$.
Question: Is there $M \subseteq T$ such that $|M| = |A_1|$ and every element of $A_1 \cup A_2 \cup A_3$ occurs just once in M?

An instance of the 3DM is denoted by (A, T) where $A = A_1 \cup A_2 \cup A_3$ assuming that the partition is understood. Let $A_k = \{a_{k,1}, \ldots, a_{k,n}\}$ for $k \in \{1, 2, 3\}$ and $T = \{t_1, \ldots, t_m\}$. For notational convenience we write $a \in t$ if $a \in A$ occurs in $t \in T$ by identifying t with the set of the elements of t. We construct an instance (G_T, f) of the TSP as follows. The vertex set of G_T is $V_A \cup V_T$ with

$$V_A = \{u_{k,i}, u'_{k,i} \mid k \in \{1, 2, 3\} \text{ and } i \in \{1, \ldots, n\}\},$$
$$V_T = \{v_{j,k}, v'_{j,k} \mid j \in \{1, \ldots, m\} \text{ and } k \in \{1, 2, 3\}\}.$$

The edge set E_T is given by

$$E_T = \{\{u_{k,i}, v'_{j,k}\}, \{u'_{k,i}, v_{j,k}\} \mid a_{k,i} \in A_k \text{ occurs in } t_j \in T\}$$
$$\cup \{\{v_{j,k}, v'_{j,l}\} \subseteq V_T \mid j \in \{1, \ldots, n\} \text{ and } k \neq l\}.$$

We call the subgraph induced by $\{v_{j,1}, v'_{j,2}, v_{j,3}, v'_{j,1}, v_{j,2}, v'_{j,3}\}$ the t_j-cycle. The initial configuration f is defined by

$$f(u_{k,i}) = u'_{k,i} \text{ and } f(u'_{k,i}) = u_{k,i} \text{ for all } a_{k,i} \in A_k \text{ and } k \in \{1, 2, 3\},$$
$$f(v_{j,k}) = v_{j,k} \text{ and } f(v'_{j,k}) = v'_{j,k} \text{ for all } t_j \in T \text{ and } k \in \{1, 2, 3\}.$$

In the initial configuration f, all and only the tokens in V_A are misplaced. Each token $u_{k,i} \in V_A$ on the vertex $u'_{k,i}$ must be moved to $u_{k,i}$ via (a part of) t_j-cycle for some $t_j \in T$ in which $a_{k,i}$ occurs. To design a short solution for (G_T, f), it is desirable to have swaps at which both of the swapped tokens get closer to the destination. If (A, T) admits a solution, then one can find an optimal solution for (G_T, f) of length $21n$, where $9n$ of the swaps satisfy this property as we will see in Lemma 1. On the other hand, such an "efficient" solution is possible only when (A, T) admits a solution as shown in Lemma 2.

Lemma 1. *If (A,T) has a solution then $\mathsf{OPT}(G_T, f) \leq 21n$ with $n = |A_1|$.*

Proof. We show in the next paragraph that for each $t_j \in T$, there is a sequence σ_j of 21 swaps such that $g\sigma_j$ is identical to g except $(g\sigma_j)(u_{k,i}) = g(u'_{k,i})$ and $(g\sigma_j)(u'_{k,i}) = g(u_{k,i})$ if $a_{k,i}$ occurs in t_j for any configuration g. If $M \subseteq T$ is a solution, by collecting σ_j for all $t_j \in M$, we obtain a swap sequence σ_M of length $21n$ such that $f\sigma_M$ is the identity.

Let $t_j = (a_{1,i_1}, a_{2,i_2}, a_{3,i_3})$. We first move each of the tokens u_{k,i_k} on the vertex u'_{k,i_k} to the vertex $v_{j,k}$ and the tokens u'_{k,i_k} on u_{k,i_k} to $v'_{j,k}$. We then move the tokens u_{k,i_k} on $v_{j,k}$ to the opposite vertex $v'_{j,k}$ of the t_j-cycle for each $k \in \{1, 2, 3\}$ while moving u'_{k,i_k} on $v'_{j,k}$ to $v_{j,k}$ in the opposite direction simultaneously. At last we make swaps on the same 6 edges we used in the first phase. The above procedure consists of 21 swaps and gives the desired configuration. □

Lemma 2. *If $\mathsf{OPT}(G_T, f) \leq 21n$ with $n = |A_1|$ then (A,T) has a solution.*

Proof. We first show that $21n$ is a lower bound on $\mathsf{OPT}(G_T, f)$. Suppose that $f\sigma$ is the identity. For each token $u_{k,i} \in V_A$, we have

$$\mathsf{move}(f, \sigma, u_{k,i}) \geq \mathsf{dist}(u_{k,i}, f^{-1}(u_{k,i})) = \mathsf{dist}(u_{k,i}, u'_{k,i}) = 5\,.$$

The adjacent vertices of the vertex $u'_{k,i}$ are $v_{j,k}$ such that $a_{k,i} \in t_j$. Among those, let $\tau(u_{k,i}) \in V_T$ be the vertex to which $u_{k,i}$ goes for its first step, i.e., the first occurrence of $u'_{k,i}$ in σ is as $\{u'_{k,i}, \tau(u_{k,i})\}$. This means that $\mathsf{move}(f, \sigma, \tau(u_{k,i})) \geq 2$, since the token $\tau(u_{k,i})$ must once leave from and later come back to the vertex $\tau(u_{k,i})$. The symmetric discussion holds for all tokens $u'_{k,i}$. Therefore, noting that τ is an injection, we obtain

$$|\sigma| = \frac{1}{2} \sum_{x \in V_A \cup V_T} \mathsf{move}(f, \sigma, x) \geq \frac{1}{2} \sum_{x \in V_A} \left(\mathsf{move}(f, \sigma, x) + \mathsf{move}(f, \sigma, \tau(x))\right) \geq 21n\,.$$

This has shown that if $f\sigma$ is the identity and $|\sigma| \leq 21n$, then

(1) $\mathsf{move}(f, \sigma, x) = 5$ for all $x \in V_A$,
(2) $\mathsf{move}(f, \sigma, y) \neq 0$ for $y \in V_T$ if and only if $y = \tau(x)$ for some $x \in V_A$.

Let $M_\sigma = \{y \in V_T \mid \mathsf{move}(f, \sigma, y) \neq 0\} = \{\tau(x) \in V_T \mid x \in V_A\}$. We are now going to prove that if $v_{j,1} \in M_\sigma$ then $\{v_{j,2}, v_{j,3}, v'_{j,1}, v'_{j,2}, v'_{j,3}\} \subseteq M_\sigma$, which implies that $\widetilde{M_\sigma} = \{t_j \in T \mid v_{j,1} \in M_\sigma\}$ is a solution for (A,T).

Suppose $v_{j,1} \in M_\sigma$ and let $t_j \cap A_1 = \{a_{1,i}\}$. This means that $\tau(u_{1,i}) = v_{j,1}$ and $u_{1,i}$ goes from $u'_{1,i}$ to $u_{1,i}$ through $(u'_{1,i}, v_{j,1}, v'_{j,2}, v_{j,3}, v'_{j,1}, u_{1,i})$ or $(u'_{1,i}, v_{j,1}, v'_{j,3}, v_{j,2}, v'_{j,1}, u_{1,i})$ by (2) and (1). In either case, $v'_{j,1} \in M_\sigma$. Suppose that $u_{1,i}$ takes the former $(u'_{1,i}, v_{j,1}, v'_{j,2}, v_{j,3}, v'_{j,1}, u_{1,i})$. Then $v'_{j,2}, v_{j,3} \in M_\sigma$. Just like $v_{j,1} \in M_\sigma$ implies $v'_{j,1} \in M_\sigma$, we now see $v_{j,2}, v'_{j,3} \in M_\sigma$. □

It is known that the 3DM is still NP-complete if each $a \in A$ occurs at most three times in T [5]. Assuming that T satisfies this constraint, it is easy to see that G_T is a bipartite graph with maximum vertex degree 3.

Theorem 1. *The TSP is NP-complete even on bipartite graphs with maximum vertex degree 3.*

2.2 PTIME Subcases of TSP

In this subsection, we present two graph classes on which the TSP can be solved in polynomial time. One is that of *lollipop graphs*, which are obtained by connecting a path and a complete graph with a bridge. That is, a lollipop graph is $L_{m,n} = (V, E)$ where $V = \{-m, \ldots, -1, 0, 1, \ldots, n\}$ and

$$E = \{\{i, j\} \subseteq V \mid i < j \leq 0 \text{ or } j = i + 1 > 0\}.$$

The other class consists of graphs obtained by connecting a path and a star. A *star-path graph* is $Q_{m,n} = (V, E)$ such that $V = \{-m, \ldots, -1, 0, 1, \ldots, n\}$ and

$$E = \{\{i, 0\} \subseteq V \mid i < 0\} \cup \{\{i, i + 1\} \subseteq V \mid i \geq 0\}.$$

Algorithms 1 and 2 give optimal solutions for the TSP on lollipop and star-path graphs in polynomial time, respectively. Proofs are found in [10].

Algorithm 1. TSP Algorithm for Lollipop Graphs

Input: A lollipop graph $L_{m,n}$ and a configuration f on $L_{m,n}$
for $k = n, \ldots, 1, 0, -1, \ldots, -m$ **do**
    Move the token k to the vertex k directly;
end for

Algorithm 2. TSP Algorithm for Star-Path Graphs

Input: A star-path graph $Q_{m,n}$ and a configuration f on $Q_{m,n}$
for $k = n, \ldots, 1, 0, -1, \ldots, -m$ **do**
    while the token on the vertex 0 has an index less than 0 **do**
        Move the token on the vertex 0 to its goal vertex;
    end while
    Move the token k to the vertex k;
end for

---

3 Parallel Token Swapping Problem

The *parallel token swapping problem (PTSP)* is the parallel version of the TSP. Definitions and notation for the TSP are straightfordwardly generalized for the PTSP. A *parallel swap S on G* is a synonym for an involution which is a subset of E, or for a matching of G, i.e., $S \subseteq E$ such that $\{u, v_1\}, \{u, v_2\} \in S$ implies $v_1 = v_2$. For a configuration f and a parallel swap $S \subseteq E$, the configuration obtained by applying S to f is defined by $fS(u) = f(v)$ if $\{u, v\} \in S$ and $fS(u) = f(u)$ if $u \notin \bigcup S$. Let

$$\text{P-SOL}(G, f) = \{\vec{S} \mid \vec{S} \text{ is a parallel swap sequence s.t. } f\vec{S} \text{ is the identity}\}$$
$$\text{P-OPT}(G, f) = \min\{|\vec{S}| \mid \vec{S} \in \text{P-SOL}(G, f)\}.$$

Problem 3 (Parallel Token Swapping Problem, PTSP).

Instance: A graph G, a configuration f on G and a natural number k.
Question: P-OPT$(G, f) \leq k$?

It is trivial that P-OPT$(G, f) \leq$ OPT$(G, f) \leq$ P-OPT$(G, f)|V|/2$, since any parallel swap S consists of at most $|V|/2$ (single) swaps. Since OPT$(G, f) \leq |V|(|V| - 1)/2$ holds [14], the PTSP belongs to NP.

Yamanaka et al. [14] discussed the relation between the TSP and parallel sorting on an SIMD machine consisting of several processors with local memory which are connected by a network [1]. The relation to the PTSP is more direct.

Theorem 2. *If there is a parallel sorting algorithm with r rounds for an interconnection network G, then* P-OPT$(G, f) \leq r$ *for any configuration f on G.*

3.1   PTSP Is NP-complete

We show the NP-hardness of the PTSP by a reduction from a restricted kind of the satisfiability problem, which we call *PPN-Separable 3SAT* (*Sep-SAT* for short). For a set X of *(Boolean) variables*, $\neg X$ denotes the set of their negative literals. A *3-clause* is a subset of $X \cup \neg X$ whose cardinality is at most 3. An instance of the Sep-SAT consists of three finite collections F_1, F_2, F_3 of 3-clauses such that for each variable $x \in X$, the positive literal x occurs just once in each of F_1, F_2 and the negative literal $\neg x$ occurs just once in F_3. We will simply denote a Sep-SAT instance as $F = F_1 \cup F_2 \cup F_3$, from which one can find the right partition in polynomial time.

Theorem 3 [10]. *The Sep-SAT is NP-complete.*

We give a reduction from the Sep-SAT to the PTSP. For a given instance $F = \{C_1, \ldots, C_n\}$ over a variable set $X = \{x_1, \ldots, x_m\}$ of the Sep-SAT, we define a graph $G_F = (V_F, E_F)$ in the following manner. Let F be partitioned into F_1, F_2, F_3 where each of F_1 and F_2 has just one occurrence of each variable as a positive literal and F_3 has just one occurrence of each negative literal. Define

$$V_F = \{u_i, u_i', u_{i,1}, u_{i,2}, u_{i,3}, u_{i,4} \mid 1 \leq i \leq m\}$$
$$\cup \{v_j, v_j' \mid 1 \leq j \leq n\} \cup \{v_{j,i} \mid x_i \in C_j \text{ or } \neg x_i \in C_j\}.$$

The edge set E_F is the least set that makes G_F contain the following paths of length 3:

$$(u_i, u_{i,1}, u_{i,2}, u_i') \text{ and } (u_i, u_{i,3}, u_{i,4}, u_i') \text{ for each } i \in \{1, \ldots, m\},$$
$$(v_j, v_{j,i}, u_{i,k}, v_j') \text{ if } x_i \in C_j \in F_k \text{ or } \neg x_i \in C_j \in F_k .$$

It is not hard to see that G_F is a bipartite graph. Vertices v_j and v_j' have degree at most 3 for $j \in \{1, \ldots, n\}$, while $u_{i,k}$ has degree 4 for $i \in \{1, \ldots, m\}$ and $k \in \{1, 2, 3\}$. The initial configuration f is defined to be the identity except

$$f(u_i) = u_i', \ f(u_i') = u_i, \ f(v_j) = v_j', \ f(v_j') = v_j,$$

for each $i \in \{1, \ldots, m\}$ and $j \in \{1, \ldots, n\}$. Since $\mathsf{dist}(w, f(w)) = 3$ if $w \neq f(w)$, obviously $\mathsf{P\text{-}OPT}(G_F, f) \geq 3$.

Here we describe an intuition behind the reduction by giving the following observation between a 3-step solution for (G_F, f) and a solution for F:

- tokens u_i and u_i' pass vertices $u_{i,1}$ and $u_{i,2}$ iff x_i should be assigned 0, while they pass over $u_{i,3}$ and $u_{i,4}$ iff x_i should be assigned 1,
- if tokens v_j and v_j' pass a vertex $u_{i,k}$ for some $k \in \{1, 2\}$ then $C_j \in F_k$ is satisfied thanks to x_i, while if they pass over $u_{i,3}$ then $C_j \in F_3$ is satisfied thanks to $\neg x_i$.

Of course it is contradictory that a clause $C_j \in F_1$ is satisfied by $x_i \in C_j$ which is assigned 0. This impossibility corresponds to the fact that there are no i, j such that both u_i and v_j with $C_j \in F_1$ go to their respective goals via $u_{i,1}$ in a 3-step solution.

Theorem 4. *To decide whether* $\mathsf{P\text{-}OPT}(G, f) \leq 3$ *is NP-complete even when* G *is restricted to be a bipartite graph with maximum vertex degree 4.*

One can modify the above reduction so that every vertex has degree at most 3 by dividing vertices $u_{i,k}$ into two vertices of degree at most 3. Let

$$V_F = \{u_i, u_i', u_{i,1}, u_{i,1}', u_{i,2}, u_{i,2}', u_{i,3}, u_{i,3}', u_{i,4}, u_{i,4}' \mid 1 \leq i \leq m\}$$
$$\cup \{v_j, v_j' \mid 1 \leq j \leq n\} \cup \{v_{j,i}, v_{j,i}' \mid x_i \in C_j \text{ or } \neg x_i \in C_j\}.$$

Our graph G_F contains the following paths of length 5:

$(u_i, u_{i,1}, u_{i,1}', u_{i,2}, u_{i,2}', u_i')$ and $(u_i, u_{i,3}, u_{i,3}', u_{i,4}, u_{i,4}', u_i')$ for each $i \in \{1, \ldots, m\}$,
$(v_j, v_{j,i}, u_{i,k}, u_{i,k}', v_{j,i}', v_j')$ if $x_i \in C_j \in F_k$ or $\neg x_i \in C_j \in F_k$.

The initial configuration f is defined similarly to the previous construction.

Theorem 5. *To decide whether* $\mathsf{P\text{-}OPT}(G, f) \leq 5$ *is NP-complete even when* G *is restricted to be a bipartite graph with maximum vertex degree 3.*

3.2 PTIME Subcases of PTSP

In this subsection we discuss tractable subcases of the PTSP. In contrast to Theorem 4, the 2-step PTSP is decidable in polynomial time. In addition, we present an approximation algorithm for finding a solution for the PTSP on paths whose length can be at most one larger than that of an optimal solution.

2-Step PTSP. It is well-known that any permutation can be expressed as a product of 2 involutions, which means that any problem instance of the PTSP on a complete graph has a 2-step solution. Graphs we treat are not necessarily complete but the arguments by Petersen and Tenner [12] on involution factorization lead to the following observation, which is useful to decide whether $\mathsf{P\text{-}OPT}(G, f) \leq 2$ for general graphs G.

Corollary 1. $\langle S, T \rangle \in$ P-SOL(G, f) *if and only if the set of orbits under f is partitioned as* $\{\{[u_1]_f, [v_1]_f\}, \ldots, \{[u_k]_f, [v_k]_f\}\}$ *(possibly $[u_j]_f = [v_j]_f$ for some $j \in \{1, \ldots, k\}$) so that for every $j \in \{1, \ldots, k\}$,*

$$\{f^i(u_j), f^{-i}(v_j)\} \in \check{S} \quad and \quad \{f^{i+1}(u_j), f^{-i}(v_j)\} \in \check{T} \quad for\ all\ i \in \mathbb{Z},$$

where $\check{S} = S \cup \{\{v\} \mid v \in V - \bigcup S\}$ for a parallel swap S.

Theorem 6. *It is decidable in polynomial time if* P-OPT$(G, f) \leq 2$ *for any G and f.*

Proof. Suppose G and f are given. One can compute in polynomial time all the orbits $[\cdot]_f$. Let us denote the subgraph of G induced by a vertex set $U \subseteq V$ by G_U and the sub-configuration of f restricted to $[u]_f \cup [v]_f$ by $f_{u,v}$. The set

$$\Gamma_f = \{\{[u]_f, [v]_f\} \mid \text{P-OPT}(G_{[u]_f \cup [v]_f}, f_{u,v}) \leq 2\}$$

can be computed in polynomial time by Corollary 1. It is clear that P-OPT$(G, f) \leq 2$ if and only if there is a subset $\Gamma \subseteq \Gamma_f$ in which every orbit occurs exactly once. This problem is a very minor variant of the problem of finding a perfect matching on a graph, which can be solved in polynomial time [3]. □

One can calculate the number of 2-step solutions in P-SOL(K_n, f) for any configuration on the complete graph K_n using Petersen and Tenner's formula [12]. On the other hand, it is a #P-complete problem to calculate $|$P-SOL$(G, f)|$ for general graphs G. This can be shown by a reduction from the problem of calculating the number of perfect matchings in a bipartite graph, which is known to be #P-complete [13]. For $H = (V, E)$, let the vertex set of G be $V' = \{u_i \mid u \in V \text{ and } i \in \{1, 2\}\}$ and the edge set $E' = \{(u_i, v_j) \mid (u, v) \in E \text{ and } i, j \in \{1, 2\}\}$. The initial configuration is defined by $f(u_1) = u_2$ and $f(u_2) = u_1$ for all $u \in V$. Then it is easy to see that $|$P-SOL$(G, f)| = 2^m$ for the number m of perfect matchings in H. Note that if H is bipartite, then so is G.

Approximation Algorithm for the PTSP on Paths. We present an approximation algorithm for the PTSP on paths which outputs a parallel swap sequence whose length is no more than P-OPT$(P_n, f) + 1$, where $P_n = (\{1, \ldots, n\}, \{\{i, i+1\} \mid 1 \leq i < n\})$ and f is a configuration on P_n. We say that a swap $\{i, i+1\} \in E$ is *reasonable w.r.t. f* if $f(i) > f(i+1)$, and moreover, a parallel swap sequence $\vec{S} = \langle S_1, \ldots, S_m \rangle$ is *reasonable w.r.t. f* if every $e \in S_j$ is reasonable w.r.t. $f\langle S_1, \ldots, S_{j-1} \rangle$ for all $j \in \{1, \ldots, m\}$. The parallel swap sequence $\langle S_1, \ldots, S_m \rangle$ output by Algorithm 3 is reasonable and satisfies the condition which we call the *odd-even condition*: for each odd number j, all swaps in S_j are of the form $\{2i-1, 2i\}$ for some $i \geq 1$, and for each even number j, all swaps in S_j are of the form $\{2i, 2i+1\}$ for some $i \geq 1$.

Lemma 3. *Suppose that $g = fS$ for a reasonable parallel swap S w.r.t. f. For any $\langle S_1, \ldots, S_m \rangle \in$ P-SOL(P_n, f), there is $\langle S'_1, \ldots, S'_m \rangle \in$ P-SOL(P_n, g) such that $S'_j \subseteq S_j$ for all $j \in \{1, \ldots, m\}$.*

The lemma implies that we may assume without loss of generality that an optimal solution $\langle S_1, \ldots, S_m \rangle$ is reasonable and moreover if $f\langle S_1, \ldots, S_j \rangle(i) > f\langle S_1, \ldots, S_j \rangle(i+1)$ then $\{i, i+1\} \cap \bigcup S_{j+1} \neq \emptyset$ for $j < m$.

Algorithm 3. Approximation algorithm for PTSP on paths

Input: A configuration f_0 on P_n
Output: A solution $\vec{S} \in$ P-SOL(P_n, f_0)
Let $j = 0$;
while f_j is not identity **do**
  Let $j = j + 1$, $S_j = \{\{i, i+1\} \mid f_{j-1}(i) > f_{j-1}(i+1)$ and $i+j$ is even$\}$ and $f_j = f_{j-1}S_j$;
end while
return $\langle S_1, \ldots, S_j \rangle$;

Let us denote the output of Algorithm 3 by AP(P_n, f_0).

Theorem 7. AP$(P_n, f_0) \in$ P-SOL(P_n, f_0) $and |$AP$(P_n, f_0)| \leq$ P-OPT$(P_n, f_0)+1$.

Proof. Let $\vec{T} = $ AP(P_n, f_0). It is obvious that $\vec{T} \in$ P-SOL(P_n, f_0) and it is odd-even. It is easy to see by Lemma 3 that $|\vec{T}| \leq |\vec{S}|$ for any odd-even solution $\vec{S} \in$ P-SOL(P_n, f_0).

We next show that every swap sequence $\vec{S} = \langle S_1, \ldots, S_m \rangle$ admits an equivalent odd-even sequence that is not much longer than the original. Without loss of generality we assume that $S_j \cap S_{j+1} = \emptyset$ for any j (in fact, any reasonable parallel swap sequence meets this condition). For a parallel swap sequence $\vec{S} = \langle S_1, \ldots, S_m \rangle$, define $Œ(\vec{S}) = \langle S'_1, \ldots, S'_{m+1} \rangle$ by delaying swaps which do not meet the odd-even condition, that is,

$$S'_j = \{\{i, i+1\} \in S_j \cup S_{j-1} \mid i+j \text{ is even}\}$$

for $j = 1, \ldots, m + 1$ assuming that $S_0 = S_{m+1} = \emptyset$. By the parity restriction, each S'_j is a parallel swap. It is easy to show by induction on j that

$$f\langle S'_1, \ldots, S'_j \rangle(i) = \begin{cases} f\langle S_1, \ldots, S_{j-1} \rangle(i) & \text{if } \{i, i+1\} \in S_j \text{ and } i+j \text{ is odd,} \\ f\langle S_1, \ldots, S_j \rangle(i) & \text{otherwise,} \end{cases}$$

for each $j \in \{1, \ldots, m+1\}$, which implies that $f\vec{S} = f Œ(\vec{S})$. Therefore, for an optimal reasonable solution $\vec{S}_0$, we have $|\vec{S}_0| + 1 = |Œ(\vec{S}_0)| \geq |\vec{T}|$. $\square$

4 Parallel Colored Token Swapping Problem

The *colored token swapping problem* (CTSP) is a generalization of the TSP, where each token is colored and different tokens may have the same color. By swapping tokens on adjacent vertices, the goal coloring configuration should be realized. More formally, a *coloring* is a map f from V to $\mathbb{N}$. The definition of a swap application to a configuration can be applied to colorings with no change. We say that two colorings f and g are *consistent* if $|f^{-1}(i)| = |g^{-1}(i)|$ for all $i \in \mathbb{N}$. Since the problem is a generalization of the TSP, obviously it is NP-hard. Yamanaka et al. [15] have investigated subcases of the CTSP called the c-CTSP where the codomain of colorings is restricted to $\{1, \ldots, c\}$. We discuss the parallel version of the c-CTSP in this section.

Problem 4 (Parallel c-Colored Token Swapping Problem, c-PCTSP).

Instance: A graph G, two consistent c-colorings f and g, and a number $k \in \mathbb{N}$.
Question: Is there $\vec{S}$ with $|\vec{S}| \leq k$ such that $f\vec{S} = g$?

Define $\mathsf{P\text{-}OPT}(G, f, g) = \min\{|\vec{S}| \mid f\vec{S} = g\}$ for two consistent colorings f and g. Since $\mathsf{P\text{-}OPT}(G, f, g)$ can be bounded by $\mathsf{P\text{-}OPT}(G, h)$ for some configuration h, the c-PTSP belongs to NP.

Yamanaka et al. have shown that the 3-CTSP is NP-hard by a reduction from the 3DM. It is not hard to see that their reduction works to prove the NP-hardness of the 3-PCTSP. We then obtain the following theorem as a corollary to their discussion.

Theorem 8. *To decide whether* $\mathsf{P\text{-}OPT}(G, f, g) \leq 3$ *is NP-hard even if G is restricted to be a planar bipartite graph with maximum vertex degree 3 and f and g are 3-colorings.*

Yamanaka et al. have shown that the 2-CTSP is solvable in polynomial time on the other hand. In contrast, we prove that the 2-PCTSP is still NP-hard.

Theorem 9. *To decide whether* $\mathsf{P\text{-}OPT}(G, f, g) \leq 3$ *is NP-hard even if G is restricted to be a bipartite graph with maximum vertex degree 4 and 2-colorings f and g.*

Proof. We prove the theorem by a reduction from the Sep-SAT. We use the same graph used in the proof of Theorem 4. The initial and goal colorings f and g are defined to be $f(w) = 1$ and $g(w) = 1$ for all w but $f(u_i) = g(u_i') = 2$ for each $x_i \in X$, $f(v_j) = g(v_j') = 2$ for each $C_j \in F_1 \cup F_3$ and $f(v_j') = g(v_j) = 2$ for each $C_j \in F_2$. The claim that F is satisfiable if and only if $\mathsf{P\text{-}OPT}(G_F, f, g) = 3$ can be established by the same manner as the proof of Theorem 4. $\square$

We can also show the following using the ideas for proving Theorems 5 and 8.

Theorem 10. *To decide whether* $\mathsf{P\text{-}OPT}(G, f, g) \leq 5$ *is NP-hard even if G is restricted to be a bipartite graph with maximum vertex degree 3 and f and g are 2-colorings.*

References

1. Bitton, D., DeWitt, D.J., Hsiao, D.K., Menon, J.: A taxonomy of parallel sorting. ACM Comput. Surv. **16**(3), 287–318 (1984)
2. Bonnet, É., Miltzow, T., Rzążewski, P.: Complexity of Token Swapping and Its Variants. CoRR, abs/1607.07676 (2016)
3. Edmonds, J.: Paths, trees, and flowers. Canad. J. Math. **17**, 449–467 (1965)
4. Even, S., Goldreich, O.: The minimum-length generator sequence problem is NP-hard. J. Algorithms **2**(3), 311–313 (1981)
5. Garey, M.R., Johnson, D.S.: Computers and Intractability: A Guide to the Theory of NP-Completeness. W.H. Freeman, San Francisco (1979)
6. Heath, L.S., Vergara, J.P.C.: Sorting by short swaps. J. Comput. Biol. **10**(5), 775–789 (2003)
7. Jerrum, M.: The complexity of finding minimum-length generator sequences. Theor. Comput. Sci. **36**, 265–289 (1985)
8. Karp, R.M.: Reducibility among combinatorial problems. In: Miller, R.E., Thatcher, J.W., Bohlinger, J.D. (eds.) The Complexity of Computer Computations. The IBM Research Symposia Series, pp. 85–103. Springer, Heidelberg (1972)
9. Kawahara, J., Saitoh, T., Yoshinaka, R.: A note on the complexity of the token swapping problem. IPSJ SIG Technical report, 2015-AL-156(3), 1–7 January 2016 (in Japanese)
10. Kawahara, J., Saitoh, T., Yoshinaka, R.: The time complexity of the token swapping problem and its parallel variants. CoRR, abs/1612.02948 (2016)
11. Miltzow, T., Narrins, L., Okamoto, Y., Rote, G., Thomas, A., Uno, T.: Approximation, hardness of token swapping. In: ESA, pp. 66:1–66:15 (2016)
12. Petersen, T.K., Tenner, B.E.; How to write a permutation as a product of involutions (and why you might care). Integers **13** (2013). Paper No. A63
13. Valiant, L.G.: The complexity of computing the permanent. Theoret. Comput. Sci. **8**, 189–201 (1979)
14. Yamanaka, K., Demaine, E.D., Ito, T., Kawahara, J., Kiyomi, M., Okamoto, Y., Saitoh, T., Suzuki, A., Uchizawa, K., Uno, T.: Swapping labeled tokens on graphs. In: Ferro, A., Luccio, F., Widmayer, P. (eds.) Fun with Algorithms. LNCS, vol. 8496, pp. 364–375. Springer, Heidelberg (2014). doi:10.1007/978-3-319-07890-8_31
15. Yamanaka, K., Horiyama, T., Kirkpatrick, D., Otachi, Y., Saitoh, T., Uehara, R., Uno, Y.: Swapping colored tokens on graphs. In: Dehne, F., Sack, J.-R., Stege, U. (eds.) WADS 2015. LNCS, vol. 9214, pp. 619–628. Springer, Heidelberg (2015). doi:10.1007/978-3-319-21840-3_51

Sliding Tokens on Block Graphs

Duc A. Hoang[1(✉)], Eli Fox-Epstein[2], and Ryuhei Uehara[1]

[1] JAIST, Nomi, Japan
{hoanganhduc,uehara}@jaist.ac.jp
[2] Brown University, Providence, USA
ef@cs.brown.edu

Abstract. Let I, J be two given independent sets of a graph G. Imagine that the vertices of an independent set are viewed as tokens (coins). A token is allowed to move (or slide) from one vertex to one of its neighbors. The SLIDING TOKEN problem asks whether there exists a sequence of independent sets of G starting from I and ending with J such that each intermediate member of the sequence is obtained from the previous one by moving a token according to the allowed rule. In this paper, we claim that this problem is solvable in polynomial time when the input graph is a block graph—a graph whose blocks are cliques. Our algorithm is developed based on the characterization of a non-trivial structure that, in certain conditions, can be used to indicate a NO-instance of the problem. Without such a structure, a sequence of token slidings between any two independent sets of the same cardinality exists.

1 Introduction

Recently, motivated by the purpose of understanding the solution space of a problem, many theoretical computer scientists have focused on the study of *reconfiguration problems*. *Reconfiguration problems* are the set of problems in which we are given a collection of *feasible solutions*, together with some *reconfiguration rule(s)* that defines an *adjacency relation* on the set of feasible solutions of the original problem. The question is, using a reconfiguration rule, whether there is a step-by-step transformation which transforms one feasible solution to another, such that each intermediate result is also feasible. A simple example is the famous Rubik's cube puzzle. The reconfigurability of several well-known problems, including SATISFIABILITY, INDEPENDENT SET, VERTEX-COLOURING, MATCHING, CLIQUE, etc. have been studied extensively. For more information about this research area, see the survey [10].

As the INDEPENDENT SET problem is one of the most important problems in the computational complexity theory, its reconfiguration variants have been well-studied [5,7,8]. Recall that an *independent set* of a graph is a set of pairwise non-adjacent vertices. Among these variants, the SLIDING TOKEN problem (first introduced by Hearn and Demaine [5]) is of particular interest (see [8] for the other variants). Given two independent sets I and J of a graph G, and imagine that a token is placed on each vertex in I. Then, the SLIDING TOKEN problem

© Springer International Publishing AG 2017
S.-H. Poon et al. (Eds.): WALCOM 2017, LNCS 10167, pp. 460–471, 2017.
DOI: 10.1007/978-3-319-53925-6_36

asks whether there exists a sequence (called a TS-*sequence*) $\mathcal{S} = \langle I_1, I_2, \ldots, I_\ell \rangle$ of independent sets of G such that

(a) $I_1 = I$, $I_\ell = J$, and $|I_i| = |I| = |J|$ for all i, $1 \leq i \leq \ell$; and
(b) for each i, $1 \leq i \leq \ell - 1$, there is an edge uv in G such that $I_i \setminus I_{i+1} = \{u\}$ and $I_{i+1} \setminus I_i = \{v\}$.

If such a sequence $\mathcal{S}$ exists, we say that $\mathcal{S}$ *reconfigures* I to J in G and write $I \overset{G}{\rightsquigarrow} J$. An example of a TS-sequence is given in Fig. 1. Observe that "$\overset{G}{\rightsquigarrow}$" is indeed an equivalence relation. SLIDING TOKEN is PSPACE-complete even for planar graphs [5] and bounded-treewidth graphs [9]. On the positive side, polynomial-time algorithms have been designed recently for claw-free graphs [1], cographs [8], trees [2], bipartite permutation graphs [4], and cactus graphs [6].

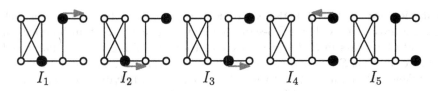

Fig. 1. Example of a TS-sequence $\langle I_1, I_2, \ldots, I_5 \rangle$ in a given graph that reconfigures I_1 to I_5. The vertices in independent sets are depicted by black circles (tokens).

A *block* of a graph G is a maximal connected subgraph with no cut vertex. A *block graph* is a graph whose blocks are cliques (for example, see the graph in Fig. 1). Note that, in order to preserve the independence property of the set of tokens, a token sometimes needs to make "detours". This restriction indeed makes SLIDING TOKEN more complicated (recall that the problem is PSPACE-complete even for bounded-treewidth graphs), even when the input graph is a tree (see [2]). As there might be exponential number of paths between any two vertices of a block graph (while in a tree, there is a unique path), for each token, we may have exponentially many choices of "routes" to slide and possibly super polynomial detours in general. Thus, in this case, the problem becomes more difficult. In this paper, we design a polynomial-time algorithm for solving the SLIDING TOKEN problem for block graphs.

Our algorithm is designed based on the following observations. Given a block graph G and an independent set I of G, one can characterize the properties of a non-trivial structure, called (G, I)-*confined clique* (Sect. 4). More precisely, we claim that one can find all (G, I)-confined cliques in polynomial time (Lemma 3), and, in certain conditions, we can easily derive if an instance of SLIDING TOKEN is a NO-instance (Lemma 5). Without such a structure, we claim that for any pair of independent sets I, J, I is reconfigurable to J (and vice versa) if and only if they are of the same cardinality (Lemma 9).

Due to the limitation of space, some proofs are omitted.

2   Preliminaries

Graph Notation. We define some notation that is commonly used in graph theory. For the notation that is not mentioned here, see [3]. Let G be a given graph, with edge set $E(G)$ and vertex set $V(G)$.

We sometimes denote by $|G|$ the size of $V(G)$. For a vertex v, we define $N_G(v) = \{w \in V(G) : vw \in E(G)\}$, $N_G[v] = N_G(v) \cup \{v\}$ and $\deg_G(v) = |N_G(v)|$. For two vertices u, v, we denote by $\mathrm{dist}_G(u, v)$ the *distance* between u and v in G. For a graph G, sometimes we write $I \cap G$ and $I - G$ to indicate the sets $I \cap V(G)$ and $I \setminus V(G)$, respectively.

For $X \subseteq V(G)$, we denote by $G[X]$ the subgraph of G *induced* by vertices of X. We write $G - X$ to indicate the graph $G[V(G) \setminus X]$. Similarly, for an induced subgraph H of G, $G - H$ indicates the graph $G[V(G) \setminus V(H)]$, and we say that the graph $G - H$ is obtained by *removing H from G*.

Notation for SLIDING TOKEN. We now define some useful notation for tackling SLIDING TOKEN. For a TS-sequence $\mathcal{S}$, we write $I \in \mathcal{S}$ if an independent set I of G appears in $\mathcal{S}$. For a vertex v, if there exists $I \in \mathcal{S}$ such that $v \in I$, then we say that $\mathcal{S}$ *involves* v. We say that $\mathcal{S} = \langle I_1, I_2, \ldots, I_\ell \rangle$ slides (or *moves*) the token t placed at $u \in I_1$ to $v \notin I_1$ in G if after applying the sliding steps described in $\mathcal{S}$, the token t is placed at $v \in I_\ell$. For convenience, we sometimes identify the token placed at a vertex with the vertex itself, and simply say "a token in an independent set I."

Let $W \subseteq V(G)$ and assume that $I \cap W \neq \emptyset$. We say that a token t placed at some vertex $u \in I \cap W$ is (G, I, W)-*confined* if for every J such that $I \overset{G}{\longleftrightarrow} J$, t is always placed at some vertex of W. In other words, t can only be slid along edges of $G[W]$. In case $W = \{u\}$, t is said to be (G, I)-*rigid*. The token t is (G, I)-*movable* if it is not (G, I)-rigid.

Let H be an induced subgraph of G. H is called (G, I)-*confined* if $I \cap H$ is a maximum independent set of H and all tokens in $I \cap H$ are $(G, I, V(H))$-confined. In particular, if H is a clique of G, we say that it is a (G, I)-*confined clique*. Note that if H is a clique then $|I \cap H| \leq 1$. We denote by $\mathsf{K}(G, I)$ the set of all (G, I)-confined cliques of G. For a vertex $v \in V(H)$, we define G_H^v to be the (connected) component of G_H containing v, where G_H is obtained from G by removing all edges of H.

3   Some Useful Observations

In this section, we present several useful observations. These observations will be implicitly used in many statements of this paper. The next proposition characterizes some properties of a (G, I)-confined induced subgraph.

Proposition 1 ([6, Lemma 1]). *Let I be an independent set of a graph G. Let H be an induced subgraph of G. Then the following conditions are equivalent.*

(i) H is (G, I)-confined.

(ii) For every independent set J satisfying $I \overset{G}{\longleftrightarrow} J$, $J \cap H$ is a maximum independent set of H.

(iii) $I \cap H$ *is a maximum independent set of* H *and for every* J *satisfying* $I \overset{G}{\longleftrightarrow} J$, *any token* t_x *placed at* $x \in J \cap H$ *is* $(G_H^x, J \cap G_H^x)$-*rigid.*

The next proposition says that when G is disconnected, one can deal with each component separately. In other words, when dealing with SLIDING TOKEN, it suffices to consider only connected graphs.

Proposition 2 ([6, Proposition 2]). *Let* I, J *be two given independent set of* G. *Assume that* $G_1, \ldots, G_k$ *are the components of* G. *Then* $I \overset{G}{\longleftrightarrow} J$ *if and only if* $I \cap G_i \overset{G_i}{\longleftrightarrow} J \cap G_i$ *for* $i = 1, 2, \ldots, k$.

In the next proposition, we claim that in certain conditions, a TS-sequence in a subgraph of G can be somehow "extended" to a sequence in G, and vice versa.

Proposition 3 ([6, Proposition 3]). *Let* u *be a vertex of a graph* G. *Let* $\mathcal{S} = \langle I_1, I_2, \ldots, I_\ell \rangle$ *be a TS-sequence in* G *such that for any* $I \in \mathcal{S}$, $u \in I$. *Let* $G'' = G - N_G[u]$. *Then* $I_1 \cap G' \overset{G''}{\longleftrightarrow} I_\ell \cap G'$. *Moreover, for any TS-sequence* $\mathcal{S}' = \langle I_1', \ldots, I_l' \rangle$ *in* G'', $I_1' \cup \{u\} \overset{G}{\longleftrightarrow} I_l' \cup \{u\}$.

In case G is a block graph, we also have:

Proposition 4. *Let* I *be an independent set of a block graph* G. *Let* B *be a block of* G *and suppose that* $I \cap B = \{u\}$. *Let* $\mathcal{S} = \langle I_1, I_2, \ldots, I_\ell \rangle$ *be a TS-sequence in* G *such that for any* $J \in \mathcal{S}$, $u \in J$. *Let* $G' = G - B$. *Then* $I_1 \cap G' \overset{G'}{\longleftrightarrow} I_\ell \cap G'$. *Moreover, for any TS-sequence* $\mathcal{S}' = \langle I_1', \ldots, I_l' \rangle$ *in* G' *such that* $N_G(u) \cap I_i' = \emptyset$, *where* $i \in \{1, 2, \ldots, \ell\}$, $I_1' \cup \{u\} \overset{G}{\longleftrightarrow} I_l' \cup \{u\}$.

Proposition 5. *Let* G *be a block graph and let* I *be an independent set of* G. *Let* $v \in V(G)$ *be such that no token in* $N_G(v) \cap I$ *is* $(G, I, N_G[v])$-*confined. Then there exists an independent set* J *of* G *such that* $I \overset{G}{\longleftrightarrow} J$ *and* $N_G[v] \cap J = \emptyset$.

Proposition 6. *Let* I *be an independent set of a block graph* G. *Let* $w \in V(G)$. *Assume that no block of* G *containing* w *is* (G, I)-*confined. If there exists some vertex* $x \in N_G[w] \cap I$ *such that the token* t_x *placed at* x *is* $(G, I, N_G[w])$-*confined, then* x *is unique. Consequently, there must be some independent set* J *such that* $I \overset{G}{\longleftrightarrow} J$ *and* $N_G[w] \cap J = \{x\}$. *Moreover, let* H *be the graph obtained from* G *by turning* $N_G[w]$ *into a clique, called* B_w. *Then* t_x *is* $(G, J, N_G[w])$-*confined if and only if* B_w *is* (H, J)-*confined.*

4 Confined Cliques in Block Graphs

In this section, we show that one can compute $\mathsf{K}(G, I)$ in polynomial time, where G is a block graph and I is an independent set of G. First, we prove an useful characterization of (G, I)-confined cliques.

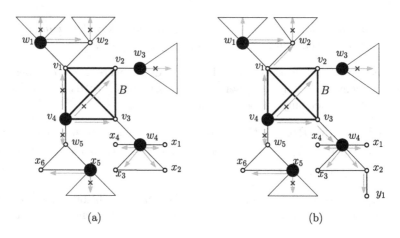

Fig. 2. (a) B is (G, I)-confined and (b) B is not (G, I)-confined.

Lemma 1. *Let I be an independent set of a block graph G. Let B be a block of G with $I \cap B \neq \emptyset$. Let $G' = G - B$. Then B is (G, I)-confined (see Fig. 2(a)) if and only if either $G = B$ or for every cut vertex $v \in V(B)$, one of the following conditions holds.*

(i) *There exists a block $B' \neq B$ of G containing v such that $B' - v$ is $(G', I \cap G')$-confined (for example, the vertices v_1 and v_2 in Fig. 2(a)).*

(ii) *For every block $B' \neq B$ of G containing v, $B' - v$ is not $(G', I \cap G')$-confined; and for every $w \in N_G(v) \setminus V(B)$, either*

    (ii-1) *there exists a block B'' of G' containing w such that B'' is $(G', I \cap G')$-confined (for example, the vertex v_4 in Fig. 2(a)); or*

    (ii-2) *every block B'' of G' containing w is not $(G', I \cap G')$-confined; and there exists $x \in N_{G'}[w] \cap I$ such that the token t_x placed at x is $(G', I \cap G', N_{G'}[w])$-confined (for example, the vertex v_3 in Fig. 2(a)).*

Next, we characterize (G, I)-rigid tokens.

Lemma 2. *Let I be an independent set of a block graph G. Let $u \in I$. The token t placed at u is (G, I)-rigid (see Fig. 3) if and only if for every $v \in N_G(u)$, there exists a vertex $w \in (N_G(v) \setminus \{u\}) \cap I$ such that one of the following conditions holds.*

(i) *The token t_w placed at w is $(G'', I \cap G'')$-rigid, where $G'' = G - N_G[u]$ (for example, the vertex w_1 in Fig. 3(a)).*

(ii) *The token t_w placed at w is not $(G'', I \cap G'')$-rigid; and the block B' of G containing v and w satisfies that $B' - v$ is $(G'', I \cap G'')$-confined (for example, the vertices w_3 and w_4 in Fig. 3(a)).*

The next lemma says that one can compute all (G, I)-confined blocks in polynomial time, where G is a block graph and I is an independent set of G.

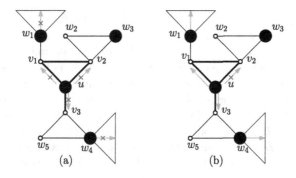

Fig. 3. (a) The token placed at u is (G, I)-rigid and (b) the token placed at u is (G, I)-movable.

Lemma 3. *Let I be an independent set of a block graph G. Let $m = |E(G)|$. Let B be a block of G with $I \cap B \neq \emptyset$. Then, one can check if B is (G, I)-confined in $O(m)$ time. Consequently, one can compute $\mathsf{K}(G, I)$ in $O(m^2)$ time.*

Proof. We describe a recursive function CHECKCONFINED(G, I, H) which returns YES if an input induced subgraph H is (G, I)-confined, where I is an independent set of G and H is either a clique or a vertex. Otherwise, it returns NO and a TS-sequence $\mathcal{S}_H$ in G which slides the token in $I \cap H$ (if exists) to a vertex in $\bigcup_{v \in V(H)} N_G(v) \setminus V(H)$. Clearly, if $I \cap H = \emptyset$ then CHECKCONFINED(G, I, H) returns NO and there is no such $\mathcal{S}_H$ described above. Thus, we now assume that $I \cap H \neq \emptyset$. Note that since H is either a clique or a vertex, $|I \cap H| = 1$. By definition, it is clear that if $G = H$ then CHECKCONFINED(G, I, H) returns YES. Then, we now consider the case when $G \neq H$, i.e., G contains more than one block. Let u be the unique vertex in $I \cap H$, and t_u be the token placed at u. Let $G' = G - H$ and $G'' = G - N_G[u]$. If H is a clique, we will use Lemma 1 to check if H is (G, I)-confined. On the other hand, if H contains only vertex u (i.e., $H = (\{u\}, \emptyset)$), we will use Lemma 2 to check if H is (G, I)-confined (by definition, it is equivalent to checking if t_u is (G, I)-rigid).

If H is a clique, then by Lemma 1, for every cut vertex $v \in V(H)$, we need to check if one of the conditions (i), (ii) of Lemma 1 holds. Note that since v is a cut vertex, there is at least one block $B' \neq H$ of G containing v. To check if Lemma 1(i) holds, we recursively call CHECKCONFINED(G', $I \cap G'$, $B' - v$) for every block $B' \neq H$ of G containing v. If CHECKCONFINED(G', $I \cap G'$, $B' - v$) returns NO for all blocks $B' \neq H$ of G containing v, i.e. Lemma 1(i) does not hold, we can construct a TS-sequence $\mathcal{S}_v$ in G that slides t_u to v as follows. If $u = v$ then nothing needs to be done. Thus, we assume that $u \neq v$, which then implies that $v \notin I$. In order to slide t_u to v, we need to make sure that for every block $B' \neq H$ of G containing v, if $I \cap (B' - v) \neq \emptyset$, the token in $I \cap (B' - v)$ need to be moved to a vertex not in $B' - v$ first. To do this, note that for each such B', the function CHECKCONFINED(G', $I \cap G'$, $B' - v$) also returns a TS-sequence $\mathcal{S}_{B'-v}$ in G' that slides the token in $I \cap (B' - v)$ to a vertex in $\bigcup_{x \in V(B'-v)} N_{G'}(x) \setminus V(B'-v)$. By Proposition 4, such a sequence $\mathcal{S}_{B'-v}$ can indeed be performed in G. Hence, $\mathcal{S}_v$ can be constructed (using the results from CHECKCONFINED$(G', I \cap G', B'-v)$) by first performing all $\mathcal{S}_{B'-v}$, then performing a single step of sliding t_u to v. If Lemma 1(i) does not hold, for every $w \in N_G(v) \setminus V(H)$, we need to check if Lemma 1(ii) holds. We first need to check whether there exists a block B'' of G' containing w such that B'' is $(G', I \cap G')$-confined. This can be done by calling CHECKCONFINED$(G', I \cap G', B'')$ for all blocks B'' of G' containing w such that $I \cap B'' \neq \emptyset$. If the result is NO for every such B'', i.e., Lemma 1(ii-1) does not hold, we still need to check if Lemma 1(ii-2) holds. To do this, we consider the following cases.

- **Case 1:** $|N_{G'}[w] \cap I| = 0$. In this case, Lemma 1(iii) does not hold, which then implies that CHECKCONFINED(G, I, H) returns NO. To see this, we shall construct a TS-sequence $\mathcal{S}_H$ in G that slides t_u to $w \in N_G(v) \setminus V(H)$. Indeed, $\mathcal{S}_H$ can be constructed by simply performing two steps of sliding: t_u to v, and then t_u from v to w (since $|N_{G'}[w] \cap I| = 0$).

- **Case 2:** $|N_{G'}[w] \cap I| = 1$. Let K be the block graph obtained from G' by turning $N_{G'}[w]$ into a clique, called B_w. By Proposition 6, checking if Lemma 1(iii) holds is equivalent to checking if B_w is (K, I)-confined. In case Lemma 1(iii) holds, the construction of $\mathcal{S}_H$ can be done by first sliding the token in $N_{G'}[w] \cap I$ to some vertex not in $N_{G'}[w] \cap I$ (converting a TS-sequence in K to a TS-sequence in G' as in Proposition 6, and extending that TS-sequence to a TS-sequence in G using Proposition 4), and then use the process described in **Case 1** to slide t_u to w.

- **Case 3:** $|N_{G'}[w] \cap I| \geq 2$. We first show how to construct an independent set J such that $I \overset{G}{\leftrightsquigarrow} J$ and $|N_{G'}[w] \cap J| \leq 1$. Note that since $|N_{G'}[w] \cap I| \geq 2$, we have $w \notin I$. The idea of this construction comes from Propositions 5 and 6. Proposition 6 indeed implies that there is at most one token t_x in $N_{G'}[w] \cap I$ that is $(G', I \cap G', N_{G'}[w])$-confined. In other words, all tokens in $N_{G'}[w] \cap I$ except t_x (if exists) can be slid to a vertex not in $N_{G'}[w]$. Now, for each block B'' of G' containing w with $I \cap B'' \neq \emptyset$, from the results of calling CHECKCONFINED$(G', I \cap G'', B'')$, we obtain a TS-sequence $\mathcal{S}_{B''}$ in G' (which can also be extended in G using Proposition 4) that moves the token in $I \cap B''$ to a vertex not in B''. Note that $\mathcal{S}_{B''}$ may or may not contain the step of sliding the token in $I \cap B''$ to w. If for some block B'' of G' containing w with $I \cap B'' \neq \emptyset$, $\mathcal{S}_{B''}$ contains such a step, then clearly it will move all other tokens in $I \cap N_{G'}[w]$ "out of" $N_{G'}[w]$ first, and then moves the token in $I \cap B''$ to w. Stop at this point, we obtain an independent set J such that $I \overset{G}{\leftrightsquigarrow} J$ and $|N_{G'}[w] \cap J| = 1$. The only token in $N_{G'}[w] \cap J$ is now indeed the token placed at w. On the other hand, if for all blocks B'' of G' containing w with $I \cap B'' \neq \emptyset$, $\mathcal{S}_{B''}$ does not contain the step of sliding the token in $I \cap B''$ to w, then we simply perform all such $\mathcal{S}_{B''}$. Since G is a block graph, all such $\mathcal{S}_{B''}$ can indeed be performed independently, i.e., no sequence involves any vertex that is involved by other sequences. At the end of this process, we obtain an independent set J such that $I \overset{G}{\leadsto} J$ and $|N_{G'}[w] \cap J| = 0$. Once we have J, the checking process can indeed be done using either **Case 1** or **Case 2**. Keep in mind that the construction of J uses only the results that can be obtained from the recursive callings of the CHECKCONFINED function.

In the above arguments, we have analyzed the cases that CHECKCONFINED(G, I, H) returns NO using Lemma 1, where H is a clique. In all other cases, CHECKCONFINED(G, I, H) indeed returns YES (by Lemma 1).

If H contains only a single vertex u, then by Lemma 2, we need to check that for every $v \in N_G(u)$, whether there exists a vertex $w \in (N_G(v) \setminus \{u\}) \cap I$ such that one of the conditions (i), (ii) of Lemma 2 holds. Clearly, if $(N_G(v) \setminus \{u\}) \cap I = \emptyset$, one can construct a TS-sequence $\mathcal{S}_H$ that slides t_u to v by performing the single step of sliding t_u to v, and hence CHECKCONFINED(G, I, H) returns NO. Next, we consider the case when $(N_G(v) \setminus \{u\}) \cap I \neq \emptyset$. In this case, for every $w \in (N_G(v) \setminus \{u\}) \cap I$, we recursively call CHECKCONFINEDG'', $I \cap G''$, $\{w\}$ to check if Lemma 2(i) holds. If CHECKCONFINED(G'', $I \cap G''$, $\{w\}$) = NO for all $w \in (N_G(v) \setminus \{u\}) \cap I$, we still need to check if Lemma 2(ii) holds by calling CHECKCONFINED(G'', $I \cap G''$, $B_w - v$) for all $w \in (N_G(v) \setminus \{u\}) \cap I$, where B_w denotes the (unique) block of G containing both v, w. If CHECKCONFINED(G'', $I \cap G''$, $B_w - v$) returns NO for all $w \in (N_G(v) \setminus \{u\}) \cap I$, we can indeed return NO for the function CHECKCONFINED(G, I, H). The TS-sequence $\mathcal{S}_H$ that moves t_u to v in this case can be constructed as follows. For each $w \in (N_G(v) \setminus \{u\}) \cap I$, since CHECKCONFINED($G''$, $I \cap G''$, $B_w - v$) returns NO, there must be a TS-sequence $\mathcal{S}_{B'-v}$ in G'' (which can be extended to G using Proposition 3) that slides the token in $I \cap (B' - v)$ to a vertex in $\bigcup_{z \in V(B'-v)} N_{G'}(B' - v) \setminus V(B' - v)$. $\mathcal{S}_H$ then can be constructed by first performing all such $\mathcal{S}_{B'-v}$, and then performing a single step of sliding t_u to v. In the above arguments, we have analyzed the cases that CHECKCONFINED(G, I, H) returns NO using Lemma 2, where H is a vertex. In all other cases, CHECKCONFINED(G, I, H) indeed returns YES (by Lemma 2).

Next, we analyze the complexity of the described algorithm. First of all, note that all the TS-sequences mentioned in the described algorithm can indeed be construction using the results from the recursive callings of the CHECKCONFINED function. Thus, the running time of our algorithm is indeed proportional to the number of callings of the CHECKCONFINED function. For a vertex $v \in V(G)$, let $f(v)$ be the number of calling CHECKCONFINED *related* to v, in the sense that the function CHECKCONFINED is either called for v or for a block containing v. Thus, the total number of callings CHECKCONFINED is indeed bounded by $\sum_{v \in V(G)} f(v)$. Moreover, from the described algorithm, note that $f(v)$ is at most $O(\deg_G(v))$. Hence, checking if H is (G, I)-confined takes at most $O(\sum_{v \in V(G)} \deg_G(v)) = O(m)$ time, where H is either a clique or a vertex. Consequently, since the number of blocks of G is $O(m)$, computing K(G, I) takes at most $O(m^2)$ time.

5   Sliding Tokens on Block Graphs

Let G be a block graph, and let I, J be two independent sets of G. In this section, we prove the following main result of this paper.

Theorem 1. *Let (G, I, J) be an instance of the* SLIDING TOKEN *problem, where I, J are two independent sets of a block graph G. Then, one can decide if $I \overset{G}{\leftrightsquigarrow} J$ in $O(m^2)$ time, where $m = |E(G)|$.*

To prove Theorem 1, we shall describe a polynomial-time algorithm for deciding if $I \overset{G}{\leftrightsquigarrow} J$, estimate its running time, and then prove its correctness. The following algorithm checks if $I \overset{G}{\leftrightsquigarrow} J$.

- ○ **Step 1:**
  - • **Step 1-1:** If $\mathsf{K}(G, I) \neq \mathsf{K}(G, J)$, return NO.
  - • **Step 1-2:** Otherwise, remove all cliques in $\mathsf{K}(G, I)$ and go to **Step 2**. Let G' be the resulting graph.
- ○ **Step 2:** If $|I \cap F| \neq |J \cap F|$ for some component F of G', return NO. Otherwise, return YES.

We now analyze the time complexity of the algorithm. Let m, n be respectively the number of edges and vertices of G. By Lemma 3, **Step 1-1** takes at most $O(m^2)$ time. **Step 1-2** clearly takes at most $O(n)$ time. Hence, **Step 1** takes at most $O(m^2)$ time. **Step 2** takes at most $O(n)$ time. In total, the algorithm runs in $O(m^2)$ time.

The rest of this section is devoted to showing the correctness of the algorithm. First of all, the following lemma is useful.

Lemma 4. *Let I be an independent set of a block graph G. Let $w \in V(G)$. Assume that every block of G containing w is not (G, I)-confined. Then, there is at most one block B of G containing w such that $B - w$ is $(G', I \cap G')$-confined, where $G' = G - w$.*

The next lemma ensures the correctness of **Step 1-1**.

Lemma 5. *Let (G, I, J) be an instance of the* SLIDING TOKEN *problem, where I, J are two independent sets of a block graph G. Then, it is a* NO-*instance if $\mathsf{K}(G, I) \neq \mathsf{K}(G, J)$.*

In the next lemma, we claim that **Step 1-2** is correct.

Lemma 6. *Let (G, I, J) be an instance of the* SLIDING TOKEN *problem, where I, J are two independent sets of a block graph G satisfying that $\mathsf{K}(G, I) = \mathsf{K}(G, J)$. Let G' be the graph obtained from G by removing all cliques in $\mathsf{K}(G, I) = \mathsf{K}(G, J)$. Then, $I \overset{G}{\leftrightsquigarrow} J$ if and only if $I \cap G' \overset{G'}{\leftrightsquigarrow} J \cap G'$. Furthermore, $\mathsf{K}(G', I \cap G') = \mathsf{K}(G', J \cap G') = \emptyset$.*

Proof. Let $\mathcal{S} = \langle I = I_1, I_2, \ldots, I_\ell = J \rangle$ be a TS-sequence in G that reconfigures I to J. We claim that there exists a TS-sequence $\mathcal{S}'$ in G' that reconfigures $I \cap G'$ to $J \cap G'$. Note that for any independent set I of G, $I \cap G'$ forms an independent set of G'. Moreover, for $i = 1, 2, \ldots, \ell - 1$, let uv be an edge of G such that $u \in I_i \setminus I_{i+1}$ and $v \in I_{i+1} \setminus I_i$, then clearly u and v must be either both in G' or both in some block $B \in \mathsf{K}(G, I)$. Hence, the sequence $\mathcal{S}' = \langle I_1 \cap G', \ldots, I_\ell \cap G' \rangle$ reconfigures $I_1 \cap G' = I \cap G'$ to $I_\ell \cap G' = J \cap G'$.

Let $\mathcal{S}' = \langle I \cap G' = I_1', I_2', \ldots, I_l' = J \cap G' \rangle$ be a TS-sequence in G' that reconfigures $I \cap G'$ to $J \cap G'$. We claim that there exists a TS-sequence $\mathcal{S}$ in G that reconfigures $I = (I \cap G') \cup \bigcup_{B \in \mathsf{K}(G,I)} (I \cap B)$ to $J = (J \cap G') \cup \bigcup_{B \in \mathsf{K}(G,I)} (J \cap B)$. Note that for an independent set I' of G' and a block $B \in \mathsf{K}(G, I)$, it is not necessary that $I' \cup (I'' \cap B)$ forms an independent set of G, where I'' is an independent set of G such that $I \overset{G}{\rightsquigarrow} I''$. For a component F of G', one can construct a TS-sequence $\mathcal{S}'_F = \langle I_1' \cap F, \ldots, I_l' \cap F \rangle$ in F. We now describe how to construct $\mathcal{S}$. Let $A = \bigcup_{B \in \mathsf{K}(G,I)} \bigcup_{v \in I \cap B} (N_G(v) \cap V(F))$. For a component F of G', we consider the following cases.

- ○ $\mathcal{S}'_F$ **does not involve any vertex in** A. In this case, note that for every independent set I_F of F and a block $B \in \mathsf{K}(G, I)$, the set $I_F \cup (J \cap B)$ forms an independent set of G, where J is any independent set of G satisfying $I \overset{G}{\rightsquigarrow} J$. Thus, such a sequence $\mathcal{S}'_F$ above indeed can be "extended" to a TS-sequence in G.

- ○ $\mathcal{S}'_F$ **involves vertices in** A. Note that for a block $B \in \mathsf{K}(G, I)$, since G is a block graph, there is at most one vertex $v \in V(B)$ satisfying that $N_G(v) \cap V(F) \neq \emptyset$. Moreover, if there exists two vertices $u_1, u_2 \in V(F)$ such that $N_G(u_i) \cap V(B) \neq \emptyset$ $(i = 1, 2)$ then they must be adjacent to the same vertex in B. Let v be the unique vertex in $I \cap B$ and assume that $N_G(v) \cap V(F) \neq \emptyset$. Then, the token t_v placed at v must not be (G, I)-rigid. To see this, note that, if the token t placed at $u \in I$ is (G, I)-rigid, then by definition of confined cliques, any block of G containing u must be in $\mathsf{K}(G, I)$. Hence, for a block $B \in \mathsf{K}(G, I)$ and $v \in I \cap B$ with $N_G(v) \cap V(F) \neq \emptyset$, there exists a TS-sequence $\mathcal{S}'(B, v)$ in G that moves the token t_v placed at v to some other vertex in B. Since G is a block graph, if there are two of such block B, say B_1 and B_2, with $v_1 \in I \cap B_1$ and $v_2 \in I \cap B_2$, then clearly $\mathcal{S}'(B_1, v_1)$ does not involve any token which is involved by $\mathcal{S}'(B_2, v_2)$ (and vice versa).

Now, we construct a TS-sequence $\mathcal{S}$ in G that reconfigures I to J as follows. First, we perform all TS-sequence $\mathcal{S}'_F$ that does not involve any vertex in A. Next, for a component F with the corresponding sequence $\mathcal{S}''_F$ involving let $B \in \mathsf{K}(G, I)$ such that there exists a (unique) vertex $v \in I \cap B$ satisfying that $N_G(v) \cap V(F) \subseteq A$. For such component F and such block B, we first perform $\mathcal{S}'(B, v)$, then perform $\mathcal{S}'_F$, and then perform $\mathcal{S}'(B, v)$ in reverse order. Note that if after performing $\mathcal{S}'(B, v)$, the token t_v (originally placed at v) is placed at some vertex $w \in J$, then in the step of reversing $\mathcal{S}'(B, v)$, we do not reverse the step of sliding t_v to w. At this moment, we have reconfigured $I \cap G'$ to $J \cap G'$ in G.

It remains to reconfigure $I \cap B$ to $J \cap B$ in G for each block $B \in \mathsf{K}(G, I)$, which can be done using the observation that for any vertex $v \in J \cap B$, $N_G(v) \cap J \neq \emptyset$.

Finally, we claim that $\mathsf{K}(G', I \cap G') = \emptyset$. Similar arguments can also be applied for showing $\mathsf{K}(G', J \cap G') = \emptyset$. Assume for the contradiction that there exists some block $B' \in \mathsf{K}(G', I \cap G')$. Let v be the unique vertex in $I \cap B'$, and let B be the block of G containing B'. We consider the following cases.

- $B = B'$.

  Note that since B' is a block of both G and G', it follows that B' is not (G, I)-confined. In other words, there exists a TS-sequence $\mathcal{S}$ in G that slides the token t_v placed at $v \in I \cap B'$ to some vertex not in B'. Moreover, as before, we have proved that such a TS-sequence can indeed be "restricted" to G' based on the observation that for any independent set I of G, $I \cap G'$ forms an independent set of G' and any sliding step is performed either along edges of G' or along edges of some (G, I)-confined block. Therefore, B' is not $(G', I \cap G')$-confined, a contradiction.

- $|V(B) \setminus V(B')| = 1$.

  Let w be the unique vertex in $V(B) \setminus V(B')$. Note that since w is a vertex of some (G, I)-confined block $C \neq B$, the token t_v placed at v cannot be slid to w in G. Since B is not (G, I)-confined, as before, there exists a TS-sequence $\mathcal{S}$ in G that slides the token t_v placed at $v \in I \cap B'$ to some vertex not in B'. Moreover, $\mathcal{S}$ does not move t_v to w, which means that it moves t_v to some vertex of G' that is not in B'. Thus, $\mathcal{S}$ can indeed be "restricted" to G', which means that B' is indeed not $(G', I \cap G')$-confined, a contradiction.

Before proving the correctness of **Step 2**, we need some extra definitions. Let B be a block of a block graph G. A block $B' \neq B$ of G is called a *neighbor* of B if $V(B) \cap V(B') \neq \emptyset$. B is called *safe* if it has at most one cut vertex and at most one neighbor having more than one cut vertex. A vertex $v \in V(G)$ is called *safe* if it is a non-cut vertex of a safe block of G.

The next two lemmas are useful for showing the correctness of **Step 2**.

Lemma 7. *Let I be an independent set of a block graph G such that $\mathsf{K}(G, I) = \emptyset$. Let v be a safe vertex of G. Then, there exists an independent set J of G with $I \overset{G}{\longleftrightarrow} J$ and $v \in J$.*

Lemma 8. *Let I be an independent set of a block graph G such that $\mathsf{K}(G, I) = \emptyset$. Let $v \in I$ be a safe vertex of G and let B_v be the (unique) safe block of G containing v. Let G^* be the subgraph of G obtained by removing B_v. Then, $\mathsf{K}(G^*, I \cap G^*) = \emptyset$.*

The following lemma ensures the correctness of **Step 2**.

Lemma 9. *Let (G, I, J) be an instance of the SLIDING TOKEN problem, where I, J are two independent sets of a block graph G satisfying that $\mathsf{K}(G, I) = \mathsf{K}(G, J) = \emptyset$. Then, $I \overset{G}{\longleftrightarrow} J$ if and only if $|I| = |J|$.*

Proof. The only-if-part is trivial. We shall prove the if-part, i.e., if $|I| = |J|$ then $I \overset{G}{\rightsquigarrow} J$. More precisely, we claim that there exists an independent set I^* such that $I \overset{G}{\rightsquigarrow} I^*$ and $J \overset{G}{\rightsquigarrow} I^*$. Indeed, I^* can be constructed as follows. Initially, $I^* = \emptyset$.

- Pick a safe vertex v of G. (Note that the "tree-like" structure of a block graph ensures that one can always find a safe block, and hence a safe vertex.)
- Slide a token from I and a token from J to v. Then, add v to I^*. This can be done using Lemma 7. Let I' and J' be the resulting independent sets.
- Let G' be the graph obtained by removing B_v – the (unique) block of G containing v.
- Repeat the above steps with the new triple $(G', I' \setminus \{v\}, J' \setminus \{v\})$ instead of (G, I, J). The procedure stops when there is no token to move.

The correctness of this construction is followed from Lemmas 7 and 8.

Acknowledgement. The first author would like to thank Yota Otachi for his useful discussions. This work is partially supported by MEXT/JSPS Kakenhi Grant Numbers 26330009 and 24106004.

References

1. Bonsma, P., Kamiński, M., Wrochna, M.: Reconfiguring independent sets in claw-free graphs. In: Ravi, R., Gørtz, I.L. (eds.) SWAT 2014. LNCS, vol. 8503, pp. 86–97. Springer, Heidelberg (2014). doi:10.1007/978-3-319-08404-6_8
2. Demaine, E.D., Demaine, M.L., Fox-Epstein, E., Hoang, D.A., Ito, T., Ono, H., Otachi, Y., Uehara, R., Yamada, T.: Linear-time algorithm for sliding tokens on trees. Theor. Comput. Sci. **600**, 132–142 (2015)
3. Diestel, R.: Graph Theory. Graduate Texts in Mathematics, vol. 173, 4th edn. Springer, Heidelberg (2010)
4. Fox-Epstein, E., Hoang, D.A., Otachi, Y., Uehara, R.: Sliding token on bipartite permutation graphs. In: Elbassioni, K., Makino, K. (eds.) ISAAC 2015. LNCS, vol. 9472, pp. 237–247. Springer, Heidelberg (2015). doi:10.1007/978-3-662-48971-0_21
5. Hearn, R.A., Demaine, E.D.: PSPACE-completeness of sliding-block puzzles and other problems through the nondeterministic constraint logic model of computation. Theor. Comput. Sci. **343**(1), 72–96 (2005)
6. Hoang, D.A., Uehara, R.: Sliding tokens on a cactus. In: Hong, S.H. (ed.) ISAAC 2016. LIPIcs, vol. 64, pp. 37:1–37:26. Schloss Dagstuhl-Leibniz-Zentrum fuer Informatik, Wadern (2016)
7. Ito, T., Demaine, E.D., Harvey, N.J.A., Papadimitriou, C.H., Sideri, M., Uehara, R., Uno, Y.: On the complexity of reconfiguration problems. Theor. Comput. Sci. **412**(12), 1054–1065 (2011)
8. Kamiński, M., Medvedev, P., Milanič, M.: Complexity of independent set reconfigurability problems. Theor. Comput. Sci. **439**, 9–15 (2012)
9. Mouawad, A.E., Nishimura, N., Raman, V., Wrochna, M.: Reconfiguration over tree decompositions. In: Cygan, M., Heggernes, P. (eds.) IPEC 2014. LNCS, vol. 8894, pp. 246–257. Springer, Heidelberg (2014). doi:10.1007/978-3-319-13524-3_21
10. van den Heuvel, J.: The complexity of change. In: Blackburn, S.R., Gerke, S., Wildon, M. (eds.) Surveys in Combinatorics 2013, pp. 127–160. Cambridge University Press, Cambridge (2013)

Author Index